Handbook of Communications Systems Management

James W. Conard, Editor

Fourth Edition

AUERBACH PUBLICATIONS
Boca Raton New York

Other publications edited by James W. Conard:
Auerbach Information Management Series **Data Communications Management Broadband Communications Systems**

Library of Congress Cataloging-in-Publication Data

Catalog record is available from the Library of Congress.

International Standard Book Number 0-8493-9941-6
Printed in the United States of America 1 2 3 4 5 6 7 8 9 0
Printed on acid-free paper

Contributors

HOWARD C. BERKOWITZ, *Manager, Open Systems Technology, PSC International, McLean VA*

CHARLES BREAKFIELD, *Senior Network Architect, Andersen Consulting, Dallas TX*

GLENN R. BROWN, *Manager, Network Services, Lithonia Lighting, Conyers GA*

ROXANNE BURKEY, *Independent Consultant, Dallas TX*

EDWARD BURSK, *Vice-President of Marketing, FastComm Communications, Sterling VA*

DONALD D. DECAMP, *Vice-President of Franchise and License Operations, Interim Services, Inc., Fort Lauderdale FL*

SUNIL DHAR, *Product Manager, Cisco Systems, Inc., San Jose CA*

THOMAS FLEISHMAN, *Vice-President and Director, Information Technology Services, Kaiser Permanente Medical Care Program, Pasadena CA*

DONALD R. FOWLER, *Year 2000 R&D Manager, IBS Conversions, Largo FL*

LOUIS FRIED, *Vice-President, Information Technology Consulting, SRI International, Menlo Park CA*

FREDERICK GALLEGOS, *MSBA-Information Systems Audit Advisor and Adjunct Professor, Computer Information Systems Department, California State Polytechnic University, Pamona CA*

HARVEY GOLOMB, *President, NETSCAN Technology Group, Fairfax VA*

ANURA GURUGE, *Strategic Independent Analyst, New Ipswich NH*

BILL HEERMAN, *Manager, Creative Services, Attachmate Corp., Atlanta GA*

GILBERT HELD, *Director, 4-Degree Consulting, Macon GA*

SHEILA M. JACOBS, *Assistant Professor of Management Information Systems, Oakland University, Rochester MI*

SAMI JAJEH, *Director, Market Development, ExcelleNet, Inc., Atlanta GA*

KEITH A. JONES, *Certified Quality Analyst and Director of the Institute for Software Quality Automation, Palm Harbor FL*

VINCENT C. JONES, *Consultant, Tenafly NJ*

RANDALL KENNEDY, *President, Competitive Systems Analysis, Inc., Albany CA*

BYUNG G. KIM, *Associate Professor, Computer Science, University of Massachusetts, Lowell MA*

KEITH G. KNIGHTSON, *Associate Director, Telecom Architect Program, Canadian Government Telecommunications Agency, Kanata Ontario*
ANDRES LLANA, JR., *Consultant, Vermont Studies Group, King of Prussia PA*
JACK T. MARCHEWKA, *Assistant Professor, Management Information Systems, Northern Illinois University, Dekalb IL*
SCOT MCLEOD, *Senior Manager, RLN Product Marketing, Attachmate Corp., Atlanta GA*
MARK MONDAY, *Director, RLN Product Management, Attachmate Corp., Cincinnati OH*
NATHAN J. MULLER, *Consultant, The Oxford Group, Huntsville AL*
JOHN P. MURRAY, *Technical Resource Manager, Compuware, Madison WI*
ED NORRIS, *Senior Security Consultant, Digital Equipment Corp., Littleton MA*
JAMES A. PAPOLA, *Director of Facilities Management, Windon Capital Management, Reading PA*
STEVEN R. POWELL, *Professor, Computer Information Systems Department, California State Polytechnic University, Pomona CA*
RICHARD ROSS, *Managing Director, Decision Strategies Group, Greenwich CT*
STEVEN J. ROSS, *Senior Manager, Deloitte & Touche LLP, New York NY*
A. PERRY SCHWARTZ, *Principal, TechStar International Corp., Flowery Branch GA*
JOHN P. SLONE, *Senior Systems Architect, Lockheed Martin Corp., Orlando FL*
KENNETH A. SMITH, *Director of Eastern Region Consulting Operations, Sungard Planning Solutions, Wayne PA*
RALPH R. STAHL, JR., *Chief Global Information Security Officer, Architecture Design and Engineering, Worldwide Information Systems, AT&T, Dayton OH*
WILLIAM STALLINGS, *Independent Consultant, Brewster MA*
DICK THUNEN, *Marketing Representative, Sync Research, Inc., Irvine CA*
STEVEN E. TURNER, *Manager, Technical Staff and Research Department, UDS Motorola, Inc., Huntsville AL*
ROBERT E. UMBAUGH, *Principal Consultant, Carlisle Consulting Group, Carlisle PA*
MARY VAN SELL, *Associate Professor of Management, Oakland University, Rochester MI*
LEO A. WROBEL, *President and CEO, Premiere Network Services Inc., Dallas TX*
WILLIAM A. YARBERRY, JR., *Project Director, Technology Assessment and Communications, Enron Corp., Houston TX*

Contents

INTRODUCTION ... ix

SECTION 1 PLANNING FOR COMMUNICATIONS SYSTEMS ... 1

1-1 Information Technology Trends and Business Issues ... 5
1-2 The Economics of a Strategy for Applying Information Technology ... 19
1-3 Security in Value-Added Networks ... 29
1-4 Improving Communication Between End Users and System Developers ... 43
1-5 Network Disaster Recovery Planning ... 55

SECTION 2 BUSINESS MANAGEMENT ISSUES ... 79

2-1 Business Trends on the Internet ... 83
2-2 Strategies for Developing and Testing Business Continuity Plans ... 93
2-3 Corporate Lessons for the Communications Systems Manager ... 117
2-4 Hiring and Retaining Valued Employees ... 127
2-5 Cost Allocation for Communications Networks ... 145
2-6 Managing Morale When Downsizing ... 155

SECTION 3 NETWORKING TECHNOLOGY ... 167

3-1 Inverse Multiplexing: Technical Foundations and Applications ... 171
3-2 The Benefits of Implementing Frame Relay ... 185
3-3 Global Frame Relay ... 209

3-4 Traffic Control Functions in ATM Networks 221
3-5 A Guide to Conferencing Technologies 231
3-6 Choosing a Remote Access Strategy 247
3-7 Considerations for Implementing Corporate Intranets 265

SECTION 4 INTEROPERABILITY AND STANDARDS ISSUES 277

4-1 Worldwide Activities in Standards Conformance Testing ... 281
4-2 Open Network Addressing 299
4-3 The HSSI Standard for High-Speed Networking 311
4-4 RMON: The SNMP Remote Monitoring Standard 321

SECTION 5 COMMUNICATIONS SERVICES 329

5-1 An Overview of Switched Digital Services 333
5-2 High-Speed Services for LAN Interconnection 343
5-3 A Management Briefing on Virtual Private Networks 359
5-4 A Guide to Network Access and Pricing Options 373
5-5 Auditing Telecommunications Costs for Data and Video Communications 383
5-6 The 1996 US Telecommunications Act and Worldwide Deregulation 399

SECTION 6 THE INTERNET AND INTERNETWORKING 407

6-1 The Essentials of Enterprise Networking 411
6-2 Data Compression in Routed Internetworks 427
6-3 Accessing and Using the Internet 443
6-4 Internet Security Using Firewalls 457
6-5 Managing Coexisting SNA and LAN Internetworks 467
6-6 SNA Over Frame Relay 487
6-7 IPv6: The New Internet Protocol 503

SECTION 7 MOBILE COMMUNICATIONS SYSTEMS 511

7-1 Wireless Communications for Voice and Data 515
7-2 The Technical Challenge of Cellular Data Transmission .. 527
7-3 Developing a Cost-Effective Strategy for Wireless Communications 537
7-4 Mobile User Security 551
7-5 Managing a Telecommuting Program 569

SECTION 8 IMPLEMENTATION AND CASE STUDIES 577

8-1 Acquiring Systems for Multivendor Environments 581

8-2 Evaluating Vendor Support Policies 589
8-3 WAN Network Integration: A Case Study 597
8-4 Frame Relay in an IBM Environment 611
8-5 Network Controls 621
8-6 Remote LAN/WAN Connections: A Case Study 639

SECTION 9 NETWORK OPERATIONS AND MANAGEMENT 661

9-1 Managing Distributed Computing 665
9-2 An Overview of Network Management Systems 681
9-3 Introduction to LAN Efficiency 695
9-4 Protecting Against Hacker Attacks 713
9-5 Documenting a Communications Recovery Plan 725
9-6 Points-of-Failure Planning 733

SECTION 10 DIRECTIONS IN COMMUNICATIONS SYSTEMS 743

10-1 Remote Computing: Technologies and Trends 747
10-2 Information Security and New Technology 757
10-3 Voice Recognition Interfaces for Multimedia Applications 769
10-4 The Superhighway: Information Infrastructure Initiatives 785

LIST OF ACRONYMS USED IN THIS BOOK 797

ABOUT THE EDITOR 807

INDEX 809

Introduction

CERTAIN SYMBOLIC MILESTONES, such as a birthday or the beginning of a new calender year, are good places to pause, take stock of the current situation, and take a deep breath before beginning the next segment of the journey. As that very symbolic milestone called the turn of the millennium approaches, it is appropriate that communications systems managers should also pause to reflect and take a deep breath before plunging into the future.

As the second millennium comes to a close, the communications industry is embarking on a new era. This broadband era, as it is being called, is based on a new model of telecommunications. As managers pause to reflect, they ask questions. What does this new model look like? Who are the players in the new game? What are the rules? On what technologies will the new model be based? What will be the impact on the role of a communications manager? The fourth edition of this handbook examines the new model now emerging as the framework for the next generation of communications systems.

GROWING CAPACITY DEMANDS

The communications industry is approaching the end of the wideband era of communications. This era spanned the past two decades and followed the narrowband era of the 1960s and early 1970s. The hallmarks of the wideband era have been local area networks, open systems, and client/server architectures. During this period, end-user systems have migrated from host-dominated groups of dumb terminals to client/server environments of distributed microcomputers and workstations often connected to local networks. Longer-haul, wide area networks at rates of around two megabits per second are now routine. Transport facilities are largely copper-based digital shared networks.

Although this current model works very well, it cannot satisfy the emerging demand for very high-speed transfer of multimedia information required by applications such as local and metropolitan network interconnect, telecommuting, personal workstations with video and graphic capability, and Internet access.

New application needs are driving the demand for increasing capacity from the desktop up to the local network and on to the wide area infrastructure. Service providers and equipment vendors must develop and deploy a new generation of communications systems. This capacity demand is, however, only one aspect of the new model. While new ways to handle vast amounts of information are being explored, other forces are propelling the industry toward a new, competitive, unregulated economic model. It is the combination of these forces—one technical, the other political and economic—that is shaping the model for the next generation of telecommunications.

Technically, the new model will be characterized by the use of broadband transport systems offering user interface rates routinely up to 600 megabits per second. Longer-haul circuits will operate into the tens of gigabits per second. The transport media will be overwhelmingly fiber structured as an extremely reliable, low error-rate synchronous optical network. Segregated voice, data, and video overlay networks will gradually merge into a single cell-switched core network structure. Surrounding this core will be a periphery supporting voice circuit switching, data frame switching, and other services.

THE ECONOMICS OF NEXT-GENERATION NETWORKS

The economic model will move away from the current duality of tightly regulated, segmented, national markets combined with loosely regulated private operating agencies. It will move toward a wide open, competitive market that is global in scope. The current service structure, where the manager buys different services from different providers, will fade away to be replaced by bundled offerings combining local, long-distance, and mobile services in one package. These bundled services will be tailored to individual customer profiles and will be available from competing providers, albeit initially only in the high-volume business corridors.

There will not be much change in equipment suppliers on either the customer premises or network side of the user interface. There will certainly be mergers and alliances, but the dominant system vendors will be those who can react most nimbly to a dynamic marketplace.

On the service provider side of the equation, those who now play major roles in their sectors of the market will continue to do so. A major change, however, is that all of these providers will be trying to expand into other market segments. Interexchange carriers (IXCs), for example, now providing long-distance, cellular, and Internet access services can be expected to move also into local service and into entertainment services. Local service providers will add long-distance and entertainment services to their current product lines. Cable television providers will try to capture shares of the local, long-distance, and Internet access markets.

Competitive access providers have traditionally been providing bypass services in the local data markets. They can now be expected to expand into both long-haul and mobile segments of the market. The Internet service providers (ISPs) have used packet switching and the Internet protocol suites to capture a large share of national and international data and messaging traffic. They could easily begin to offer flat-rate local services. The communications manager could be in a position of buying all of the necessary transport and switching services from any of a half dozen or more suppliers who are currently doing business in only one segment of the market.

If every service provider offers the entire range of services, it may be difficult to choose among them. When making these choices, consumers should remember that each service provider is evolving from its core business and brings to the new model a set of experiences and viewpoints that can slant its approach to the expanded arena. The traditional long-distance providers, for example, are very experienced in competitive environments, have internationally recognized brand identity, and in many cases, have modern facilities with global reach. On the other hand, they do not, for the most part, have access to the local users without either leasing or building such access.

The former competitive access providers have the advantage of state-of-the-art technology. They have installed switches and transport capability, although the latter tends to be limited geographically. The competitive access providers are very nimble and flexible; however, at the moment they are generally limited to a small customer base and may not have the operational support systems in place to support expanded, bundled, usage-based services.

The existing local service providers (i.e., the incumbent local exchange carriers) have the huge advantage of very deep pockets. Money flows in from local service and access fees paid by long-distance carriers. They also have the advantage of incumbency and a reputation for excellent service levels. Their negatives include a lack of national and international facilities and, at least to some degree, a lack of experience in a highly competitive environment. The cable companies have great experience in delivering entertainment and have access to almost as many residential customers as the phone companies. On the other hand, they have little access to business users and have very limited telecommunications experience.

A likely scenario in this situation is the cross-discipline merger of these providers. The idea would be to combine strengths and overcome weaknesses, which has already begun. MFS WorldComm is a telecommunications company comprised of a former long-distance provider (LDDS), a data network service provider (Wiltel), an Internet service provider (UUNet), and a competitive access provider (MFS Communications). Clearly the intent is to position the company for success in a new environment.

Other mergers and partnering ventures will occur as the providers jockey for position. Experience shows that there will probably be three major players—this has been the case in the aircraft business, in automobiles, in the steel industry, and in long-distance telephone service. There will be one dominant supplier who will make lots of money from a large customer base. The holder of the second-largest market share will also make money. The third-place company will struggle but probably survive. The rest of the players will service niches or make alliances in an attempt to increase market share and aim at one of the top slots.

A SERVICE FOCUS

To prepare for this new model of telecommunications, managers should focus more on applications and implications and less on specific technologies. The communications manager, while wearing the technical expert's hat, may care about whether frame relay or cell relay is being used in the core network, but the communications manager wearing the business manager's hat does not. Although it is helpful to know, when using the Internet, that the TCP/IP protocol is providing connectivity and routing, it is more important, as a communications manager, to be aware of connect time, transfer delays, and the reliability of the connection—attributes that directly impact the user.

The manager must be prepared to cope with a veritable blizzard of hyperbola as equipment and service providers vie for business. It will, at least initially, be very difficult to separate the hyperbola from the reality. Sales pitches will tout the new, bundled, all-services-from-a-single-supplier communications nirvana. The manager's best defense is awareness. Managers should ask questions and decide if they even want to consider a single vendor.

Communications services prices will drop as competition drives them down. The providers, however, will try to increase capability to keep the price up. In other words, this year's cell switch will handle 1.5 million cells per second for $30,000. Next year, the same capability should sell for $20,000, but the vendor will, instead, offer a 2 million megacell switch for the same $30,000. The same will be true of services, whether bundled or not. As technology matures, the prices fall until the item becomes a commodity. It is then necessary for the vendor to upgrade the capability, reinvent the product, or develop an entirely new product to maintain revenue streams.

The systems manager will be in the catbird seat in dealing with vendors and service providers for the new telecommunications era. The caveat is that managers must remain knowledgeable about the industry.

HOW TO USE THIS BOOK

The intent of this fourth edition of the handbook is to aid managers in their quest for knowledge. This edition includes material on all aspects of the communications system and networking disciplines. The handbook explores planning requirements and reviews the business aspects of communications systems. Advances in technology, interoperability, and standards are covered. Readers are updated on new communications services, including the Internet and emerging intranets. This edition also provides some interesting case studies. Day-to-day network management tasks may be made easier when readers review the chapters on network operations. Finally, readers will enjoy a peek into the future with the selections on directions in communications systems.

Each of the contributing authors has worked hard to share the benefit of his or her experience. They are users, consultants, and network managers. These authors, and I, as your editor, sincerely hope that you will enjoy and benefit from this handbook. Keep it handy on your bookshelf and refer to it often.

In addition to the many contributors, I must express my sincere thanks and much appreciation to my editors, Rebecca Mabry and Karen Brogno, and to the entire editorial and production staffs at Auerbach who turn concepts and ideas into the reality that you are now enjoying. And, as always, my gratitude and love to my best friend and wife—thank you, Elizabeth, for being there.

JIM CONARD
Venice, Florida
June 1997

Section 1
Planning for Communications Systems

ANY DATA COMMUNICATIONS MANAGER planning to upgrade, modify, or expand a communications network over the next few years must be cognizant of the Telecommunications Act of 1996. This reform legislation, which took effect on January 1, 1997, contains approximately 800 pages of fine print that has the potential to completely change the dynamics of the telecommunications industry.

The objective of the Act is to open up the telecommunications market to all comers. Any telecommunications service provider can enter any service market. Thus, local telephone carriers, long-distance telephone carriers, cable operators, broadcast companies, and online service providers may offer any, or all, telecommunications services. The Act defines guidelines on how long-distance carriers can access customers via local carriers' subscriber loops and how local carriers can provide long-distance services.

The key elements of the Act that will allow competition are provisions for wholesaling and interconnection. Wholesaling requires that a service provider allows other providers to resell their services on a nondiscriminatory basis. This would allow AT&T, for example, to buy local access capacity from Bell Atlantic and resell it as part of an end-to-end service package. The interconnection provisions require that a service provider may not charge other carriers unreasonable rates for calls terminating on its network.

This virtual deregulation of the industry should lead to a wild scramble as the service providers seek advantage, try to raise their market share, try to defend their customer base, and attempt to maintain reasonable profit margins. The communications system manager should, ideally, see more products and services, and because of the intense competition, lower prices. In the case of prices, managers should see them much more closely related to provider costs than is currently the case. There will also be more flat fee, bundled services and customer-specific packages.

Before customers start celebrating their position of strength, a couple of major caveats should be considered. First, the implementation regulations must be written by the Federal Communications Commission (FCC). This monumental task includes interpreting the Act and trying to predict the intent of Congress. Once the rules are written, they may be challenged in court by one party or another. This has already happened in the case of access-charge discounts that must be offered by the local carriers. There are also issues associated with the role of the individual state Public Utility Commissions that regulate within state boundaries. Customers must remember Edmund Burke's admonition that "You can never plan the future by the past." Just because the intent is to provide an open, dynamic, lower-cost environment does not mean that it will develop quite as intended. Often the law of unintended consequences takes hold and leads to radically different outcomes.

What should the communications system manager do to prepare for this new model? How can planning be accomplished in an uncertain environment? The safest route is to apply a cautious but informed approach. Managers should avoid the "early obsolescence" trap—they should be absolutely sure that circuits or equipment need upgrading or replacement. Management should be amenable to the concept of "open" and "standard" solution packages, but at the same time be practical. "Open" is the antithesis of secure, and that standard often negates advantages of proprietary single vendor solutions. It is helpful to plan for the long term but implement for the short term. This provides maneuvering room as circumstances, requirements, and service offerings change.

This section is a collection of useful chapters on the planning function. Chapter 1-1, "Information Technology Trends and Business Issues," provides an insight into the differing concerns of the suppliers and the users of information technology. Both short term and longer term issues are outlined and discussed.

Chapter 1-2, "The Economics of a Strategy for Applying Information Technology," takes the view that the purpose of applying new technology is not solely to reduce costs but to increase profitability and to provide a platform for growth. This chapter offers a model that will guide readers in targeting activities that can be improved by new technology. It then discusses criteria to ensure that advanced technology projects actually accomplish their objectives.

"Security in Value-Added Networks," Chapter 1-3, discusses security concerns that arise when an organization turns over some or all of its networking responsibility to a third party. The chapter looks at how value-added service providers view their responsibilities, what weak spots exist in network security, and how the manager can plan for an assessment of the provider's security apparatus.

Chapter 1-4, "Improving Communication Between End Users and System Developers," examines issues surrounding gaining user acceptance of new systems and technology. Users are often reluctant, and sometimes openly hostile, when it comes to using new technologies. This problem can be alleviated by improving communication between systems developers and the ultimate users with methods described in this chapter.

Chapter 1-5, "Network Disaster Recovery Planning," reviews a subject that becomes even more vitally important in the new deregulated environment. Users sometimes have a blasé attitude toward the subject of disaster recovery. The discussions in this chapter help communications managers address that attitude and provide an overview of the network protection services that are available.

1-1 Information Technology Trends and Business Issues

Louis Fried

THROUGHOUT THE REMAINDER of this decade, information technology and its supporting communications systems are likely to present more challenges than ever to business. To build a logical view of information technology (IT) issues, it is necessary to divide the broad scope of this problem into two principal segments. The first of these is the suppliers of information technology products and services; the second is the users of those products and services. Suppliers include computer manufacturers, manufacturers of computer-related products, systems integrators, software package vendors, telecommunications carriers, value-added network suppliers, and many others. Users consist of manufacturers, financial services companies, government agencies, transportation companies, and almost every business imaginable. Although a large market of private individuals using microcomputers and networks is clearly influencing both suppliers and user organizations, this chapter does not attempt to address the private user segment.

The issues addressed in this chapter are segregated into those that are of primary interest now and those that will become challenges in the near future. Corporate managers responsible for making IT investment decisions must clearly consider these investments within the context of their company's needs, but those needs are conditioned by general industry trends as well as the unique requirements of their company.

The sometimes huge investments in information technology must also be made with a view toward the life cycle of the investment—that is, not only the payback for the investment but also the profit derived from the continued workability of the system. For this reason, the selection of suppliers may be critical to obtaining the maximum benefit from an IT investment. It is therefore important to understand the issues facing suppliers and to evaluate suppliers according to their demonstrated ability to respond to the challenges raised by those issues.

SUPPLIERS OF IT PRODUCTS AND SERVICES: SHORT-TERM ISSUES

The critical issues for most suppliers are strategic business issues rather than technical issues—that is, the ability of the supplier to advance its competitive position in a highly innovative field. The following sections describe some of the trends that will affect suppliers in the short term.

Consolidation of the Industry Through Mergers, Acquisitions, and Failures Combined with Expansion of the Industry Through New Start-Up Businesses. The number of companies in the IT field has continued to expand, primarily because of start-ups. Start-ups enter the field primarily in two venture areas: new technology-based products for either hardware or software, or new service companies, primarily in software development or systems integration. Start-ups are not burdened with a prior customer base of products that must be continually maintained or for which compatibility must be ensured. In addition, the software and service companies have the advantage of low start-up investment. For the buyer of their products and services, the start-ups may provide advanced technology or innovative applications that can improve the buyer's competitive position. On the other hand, start-ups may not survive to support their products over the useful life of the application. Nevertheless, the competitive advantages to be gained are often worth this risk.

At the other end of the industry are larger firms that are heavily involved in mergers, acquisitions, alliances, and investments. During the past several years, IBM Corp. has poured billions of dollars into investments in software firms that market application packages or utility software. Although other manufacturers have not been able to invest on the same scale or breadth, they are hard at work forming alliances with systems integrators and software developers. In the software area, Computer Associates International, Inc., has grown to a billion-dollar business through the acquisition of its financially weaker competitors. In both the hardware and software areas, financially weaker firms are tempting takeover targets, especially as markets reach sales plateaus and expansion is needed. A takeover target need not always be financially weak, only desired by a stronger company. Mergers and acquisitions often increase the financial strength of the supplier, but the buyer must be aware that specific products may be discontinued or lose support as a result.

Globalization of Competition as Suppliers Strive to Keep Pace with the Businesses of Their Customers. The clients and customers of suppliers are globalizing, and it is clearly necessary for their suppliers to globalize as well if the suppliers expect to retain their preferred positions. This represents a major problem for companies that are primarily local or regional (e.g., Bull HN Information Systems Inc., NEC Technologies, Inc., Toshiba America Information Systems, Inc., and others). To enter new markets, they often resort to acquisition. However, this strategy can be dangerous. A company that does well before acquisition may be a disaster after acquisition and a subsequent attempt to integrate it into the acquiring company. One example of such an acquisition gone awry is the acquisition and subsequent sale of ROLM Corp. by IBM. The major concern of companies that intend to deploy a new technology globally is the ability of the vendors to provide continuing product support wherever their products are installed.

Vertical Consolidation as Suppliers Try to More Comprehensively Meet the Needs of Customers for Overall Solutions to Problems. Examples of this are Electronic Data Systems Corp.'s purchase of a major interest in National Data Systems, IBM's purchase of minority shares in many software houses, and Unisys Corp.'s move into the systems integration business. Every hardware manufacturer recognizes that software and services provide more revenues than hardware. Because of the importance of open systems, hardware is more of a commodity, with attendant lower profit margins.

From the supplier side, vertical expansion is seen as necessary to maintain profit margins; however, vendors also seem to believe that there is a market pull for them to enter the services business. Continued strong competition from existing service companies, despite the favorable marketing position of the manufacturers, seems to bring the market pull consideration into question. The buyer can benefit from this condition in two ways. First, a long-range strategy of moving to open systems rather than proprietary systems can lead to dramatic reductions in hardware costs and software development and maintenance costs.

Second, the buyer may take advantage of one-stop shopping and obtain the application development, integration, and hardware from a single source. On the other hand, the experiences of computer manufacturers in supplying the application needs of users has not been an unqualified success. Application implementation is a different business from computer manufacturing, one with which the manufacturers are not familiar.

Sustaining the Pace of Product Innovation. New products from new vendors can ignore the need for backward compatibility to old products. On the other hand, existing vendors must ensure that new products do not destroy their customer base by making old products obsolete before their useful life is over. A prime example was IBM's introduction of OS/2 for its new line of PCs. The DOS market is so large that IBM, in short term, was unable to drag its old customers away from it. As a result, the clone makers that continued to produce DOS machines profited at IBM's expense.

All established hardware and software manufacturers face the problem of maintaining the pace of innovation (to attract new customers and maintain their image in the market) while not damaging their customer base. This situation has often led to substantial delay in new product introduction. Another factor is financial strength. This primarily affects vendors of packaged software. Many software packages, especially in the mainframe market, are technologically obsolete, because the seller was unable to reinvest the sums necessary to rewrite the package. New suppliers of new products face the problem of educating the potential customers, sustaining revenues during their early years, and avoiding premature market entry. Most of their problems focus on cash flow. Existing, strong suppliers can sustain a long market-entry period when they introduce products with new capabilities that do not affect their existing products.

Understanding Customer Needs. This is one of the weakest areas for hardware manufacturers. All suppliers have access to reports from survey firms that describe markets and attempts to forecast trends. Still, they repeatedly encounter surprises because of the inherent limitations of trend-based forecasting. It is necessary to examine customer needs in depth and to understand how industries apply technology. Many firms attempt to remedy the shortcomings of trend forecasting by using input from their own sales forces.

Unfortunately, internal sources are biased for several reasons. Sales personnel often respond in terms of the deficiency that caused them to lose the last major sale. In addition, they are not objective about the products and services of their own companies. Finally, in many supplier organizations, sales personnel are subject to considerable turnover and rarely get to know a single customer's industry in depth. Even when suppliers' marketing groups have performed in-depth analyses of customer industries and needs, the product planning results are often heavily biased toward existing products, services, and internal organization and methods. It is difficult for them to break out into new thinking without outside help.

In acquiring new products, especially applications, the buyer is well-advised to make sure that the vendor's staff that supports the product has a comprehensive understanding of the buyer's industry and business needs. Most successful procurements are based on the buyer's protecting itself through a careful specification of requirements, evaluation of alternatives, and evaluation of the vendor's ability to provide long-term support for the product.

A Slower Market, Especially in Mainframes, Compounded by the Need to Invest in R&D for New Products. Technological advances during the late 1980s and early 1990s in computers and communications have dramatically affected the mainframe market and may have similar effects on the minicomputer market as applications and the resultant equipment purchases move toward greater distribution of desktop systems and workstations. During the late 1980s, mainframe sales reached a plateau. Minicomputers are currently experiencing a similar plateau. At the same time, sales of microcomputers, workstations, local area networks (LANs), and server devices are increasing. These trends are causing computer vendors to reexamine their product lines and place greater R&D emphasis on the low end of the hardware spectrum.

Similarly, software vendors are finding that the greatest increase in sales is for applications and utilities that reside on low-end equipment. Mainframe applications are reaching a plateau. Another problem faces the established vendors of application packages. Many of these applications were developed during the 1970s and early 1980s and are therefore architecturally obsolete. The cost of redesigning and rebuilding major applications to take advantage of up-to-date hardware and software technology is prohibitive. In fact, many such firms cannot afford the investment. Buyers must examine the architectural basis of products (both hardware and software) to ensure that they are not investing in a technologically obsolescent product that may lose its competitive benefits to the buyer in a short time.

Erosion of the Proprietary Operating System Marketplace and the Need to Develop Open Systems Offerings. The financially weaker computer manufacturers, and most start-up manufacturers, are moving (or have moved) to UNIX as a standard operating system. This move is taking place for several reasons:

- The cost of developing and supporting a proprietary operating system has become prohibitive and, for start-ups, beyond reach.
- There is an increasing market demand for heterogeneous open systems that permit portability of applications among a variety of computers and that support easy connectivity.
- Customers are seeking solutions to their needs that require both hardware and applications software. Most third-party software package vendors need to find the largest potential market base for their products. As a result, they have primarily built applications for such major vendors as IBM or Digital Equipment Corp. One of the few ways that a hardware vendor can attract software vendors to build applications is to provide an applications environment that has a less restricted potential market than its proprietary operating system.

As the demand for open systems increases, even the largest computer vendors will have to respond. This creates two problems for manufacturers: first, how to migrate to open systems without making their existing customer base vulnerable to intrusion by other vendors offering similar products; second, how to cope with the erosion of profit margins as hardware becomes more of a commodity. As previously mentioned, these problems are forcing computer vendors into greater vertical integration of their business. From a buyer perspective, the question appears to be how to take advantage of this trend toward open systems without necessarily discarding existing systems. A well-planned migration strategy is needed for the next decade.

SUPPLIERS OF IT PRODUCTS AND SERVICES: LONG-TERM ISSUES

Managing a Large, Vertically Integrated Business Over Diverse Geographic Regions. As an outgrowth of the trends toward vertical integration, globalization, acquisition, and investment during the early 1990s, the larger surviving IT suppliers may encounter the same problems that many of their customers encountered in the 1980s. Managers need to know how to control a diversified group of businesses spread around the world. The business community can expect to see repeated organizational changes as these companies try to meet this challenge and as the situation is further compounded by the demands and regulation imposed by the governments of the countries or regions in which they operate.

Maintaining the Customer Base Through a Transition from Proprietary Operating Systems to Open Operating Systems. As the trends started in the late 1980s continue, it is expected that this issue will reach its greatest impact on suppliers in the next few years. As computer manufacturers shift away from proprietary systems that locked in the customer to systems that permit customers to integrate hardware from any supplier to move applications among hardware sets, there will be an even stronger impact on packaged software suppliers. Applications designed for proprietary operating systems will need to be entirely reconstructed to take advantage of new technology as well as to operate in an open systems environment. The scale of investment required will create turmoil in the software industry and a new wave of failures, mergers, and acquisitions.

Introducing New Technologies for New Applications or Enhancements of Old Applications While Preserving Their Customer's Investments in Existing Applications. Because it would create havoc among users that have huge investments in applications software, the change from proprietary operating systems cannot be abrupt. Transition strategies that permit continued operation of existing applications while new applications are constructed in open systems environments will be necessary. New paradigms in software development are likely to appear in time to meet the conversion needs of users.

However, there will also be a demand for the introduction of new technology into business use. Further distribution of applications, organizational restructuring to take advantage of groupware, use of imaging technology, extended use of artificial intelligence, integration of varied systems, and rapid adoption of hand-held computer devices will all serve to pull vendors into new product and service areas as they strive to retain their customer base.

Transforming the Skills of Their Marketing and Technical Support Staffs from Selling and Supporting Hardware to Selling and Supporting Integrated Systems Solutions (Mostly Software). Obviously, as vendors look to vertical integration of software and service sales to maintain their profit margins, they will need to modify the skills of their sales and marketing forces. Such previously ignored areas as in-depth knowledge of customer industries and project management for large implementation projects will become critical to the expansion (or even survival) of suppliers. Suppliers may find themselves recruiting professional personnel from their customers or developing major in-house training programs to meet these needs. They may also find themselves more frequently forming alliances with consulting firms that already have those skills.

Expansion into New Business Areas to Address Vertical Markets. This represents the continuation of a trend begun in 1989 among vendors to offer consulting services, systems integration, facilities management, value-added networks, or public data bases. It is possible that vertical information will ease computer and communications vendors into such areas as supplying information to their customers as well as hardware, software, systems integration, and facilities management. Buyers of such services should remain alert to changes in the services offered by their existing or potential suppliers.

USERS OF IT PRODUCTS AND SERVICES: SHORT-TERM ISSUES

Users of information technology in business and industry must contend with many of the same broad issues as suppliers of IT products and services. However, within those broad issues lies a range of challenges that affect the manner in which information technology is applied to enhancing the firms' competitive postures. The 1990s is the decade in which it has become impossible to operate a business or government agency of any size without depending on information systems for the operation and management of the organization. The generalized issues described in the following sections have been identified through numerous consulting assignments with large, often global corporations and through published surveys of the information systems concerns of senior executives.

Controlling IT Costs While Reshaping Business Processes Through the Use of Information Technology. The current trend is restructuring, reengineering, or operations redesign of the corporation through use of information technology. However, the capability of information technology to reshape the corporation has existed and has been applied for at least 20 years. Large financial institutions and airlines could not operate on their current scale without the intensive use of information systems, but the combination of communications and computers that exists today has created new opportunities in organization for corporations.

For example, in the past, large global corporations had to balance their organizational form against their need to respond quickly to market conditions throughout the world. With today's technology, the corporation is free to choose among centralized or decentralized forms for any function to suit any management style. Rather than being constrained by traditional organizational forms, information technology makes it possible for an organization to address such real issues as time to market, customer relationships, cost trade-offs, or competitive positioning, to name a few.

Within this context of restructuring the organization and its functions, business conditions make it imperative that the costs of information technology be controlled to maintain the profitability of the firm. Executives responsible for information systems in most companies have always had difficulty in justifying the apparently high costs of information technology as compared with the costs of other functions in the firm. On the other hand, investments in computer-integrated manufacturing (CIM) or process control, for example, have been viewed as a capital investment in production capability. To relieve this cost-control pressure, many IT executives have implemented cost chargeback procedures in which the user is budgeted and charged for the development and use of information systems. Intellectually, this should work; however, when the total IT budget is presented each year for approval, the number still appears larger than most senior managers wish.

It is time for IT management to adopt the accounting conventions that drive much of the rest of the firm. Not only should IT costs be charged back to the users (i.e., the application owners) but IT budgets should also be related to business costs on the basis of direct cost, indirect cost, and general and administrative overhead. In this manner, it becomes clearer to management how IT costs relate to the production and delivery functions of the company.

Whatever conventions are adopted, there will always be countervailing pressures in the firm when it comes to allocation of investment and operating funds. The job of management is to make the investment decisions that will be of most benefit to the firm. Considering the size of IT investments and their usual life spans, it is often necessary for management to take a long-range view of potential benefits to the firm rather than the unfortunately short-range view of responding to quarterly or annual profit performance needs.

Meeting the Need of Executives for Real-Time Management Information from Both Internal and External Sources. Size and global span have changed the firm from one in which the president could walk the floor of the company to gauge its climate to one in which the executive staff is separated by many layers and miles from its operations and field conditions. Nevertheless, the ability of senior management to respond quickly to tactical needs and develop the understanding to conduct strategic planning is vital to its success. Information technology now provides a range of tools that can be applied to improving the information flow to senior management. These include executive information systems, data base access, videoconferencing, and other related technologies.

In addition to internal information, it is now more necessary for the executive to have access to such external information as stock market reports, competitive information, and news. Managing the corporation in real time requires a blend of both internal and external information. To continue managing the widespread business interests of the firm by reading monthly or quarterly reports is like trying to drive a car by looking in the rearview mirror.

Unfortunately, the executive is, more than ever, open to the potential for information overload from these various sources. Current technology requires substantial intervention by staff personnel or the careful design of management information systems to screen and reduce the information finally presented to senior management. Such intervention in the information flow creates a risk that vital decision-making information may not reach the correct level in the organization. During the next few years, artificial intelligence tools are likely to improve the ability of executives to establish their own information-screening criteria (through so-called software agents) and view only that information considered important to the individual. Management must establish mechanisms to ensure that appropriate technology is introduced into the firm on a timely basis and in an effective manner.

Creating Information Systems that Provide the Corporation with a Strategic Advantage and Aligning IT Strategy with Corporate Goals. A constant complaint by executives is that information systems analysts do not understand the business. This is coupled with a similar complaint from systems analysts that executives do not understand information technology or have the patience and commitment to adequately define their requirements. In Japan, many corporations rotate personnel on the management track through successive promotions and tours of duty with the different functional areas of the company (a spiral of training and promotion). One of those tours of duty is usually in the information systems organization. Over the long run, this rotation (from an information systems viewpoint) helps to turn managers into knowledgeable users; however, few persons in the management track achieve any substantive level of technical excellence. In the West, the pervasive distribution of microcomputers to middle management, combined with their need to work with information systems, is gradually breeding a corps of management that has a deeper understanding of the uses of IT.

Despite this progress, the language of a business remains the language of specialized industry terms and conditions, which many information technologists do not understand. User management has become increasingly favorable to bringing in external consultants to assist in bridging the gap between IT knowledge and business knowledge. This has, in part, accounted for the phenomenal growth of the consulting industry.

Alignment of information technology strategy with corporate goals and strategic objectives faces still another problem: in many cases, the goals and objectives of corporations are not clearly set forth and documented. They are understood (or misunderstood) by management. It is often necessary to first obtain a clear statement of goals and objectives before any reliable planning can be accomplished for the use of information technology.

Rationalizing Information Systems as a Result of Mergers, Acquisitions, or Divestitures and Instituting Cross-Functional Systems in the Corporation. As competition increases, some firms divest themselves of holdings or divisions and return to their concentration on a core business. Other firms find a need to capture broader markets through vertical integration. Others acquire companies that have some major focus on their core business and spin off or sell those divisions or product lines in which they have no interest.

During the 1980s, IS organizations made significant efforts to minimize costs and improve effectiveness through development of integrated information systems or the use of common systems for serving common functions throughout the company. When presented with the need to isolate a segment of their systems to accommodate a divestiture or incorporate the systems of an acquisition into existing systems, response has been tragically slow. Most applications have not been designed with such potential conditions as design criteria. This situation has added further complexity to the already complex processes of acquisition or divestiture.

Cross-functional systems may exacerbate this problem; yet, if properly designed, they provide significant advantages. For example, over many years the CIM and process control systems of companies have been separately developed and managed from the information systems. It has become increasingly clear to management that these formerly disparate developments have neglected to satisfy the needs of management for real-time operating information. There is a demonstrable need for the integration of manufacturing information with management information. In banking, there is a pressing need to integrate marketing and sales information systems with such operational systems as demand-deposit accounting. In fact, cross-functional systems design presents the potential for aiding in the restructuring of the firm. Modern data flow analysis techniques can be successfully applied to the design of cross-functional systems.

Managers in companies that are strategically inclined to modify their portfolios of strategic business units must be especially wary when integrated corporatewide systems are proposed. It will be key to their success to be able to easily divest the company of business units that no longer meet portfolio needs and sell them as intact businesses, not as divisions that have little or no independent information systems to support them. Similarly, an integrated systems solution must not be so closed that it presents major problems to the smooth embedding of a new acquisition into the company.

Improving the Effectiveness of Systems Development Activities. Computer-aided software engineering (CASE) tools have proved to be most effective in the initial stages of application development—that is, in requirements definition and application specification as contrasted to coding and testing. Many adopters of CASE tools have been disappointed in their implementation because they do not generally address the maintenance of old applications well and they seem to add to the cost of development of new applications.

Although the first of these conditions generally holds true, the second is deceptive. The true cost of application development must include maintenance and enhancement of the application over its lifetime. In fact, history has shown that maintenance and enhancement activities represent more than 70% of the lifetime cost of an application. CASE tools that, combined with appropriate methodology, provide more precise definition of the application from inception may dramatically reduce the lifetime cost of the application.

Other advances in query languages and fourth-generation languages (4GLs) have made it easier to respond quickly to user requests for new capabilities and often for entire applications. However, despite these advances in tools, systems developers are often perceived as being unresponsive in terms of quality, timeliness, and meeting business requirements. Tools and methodology are of some help, but the basic problems are frequently to be found in organization, management, and resource availability. For the foreseeable future, there will continue to be a shortage of capable software application developers. Management will be increasingly forced to turn to packaged application solutions or external systems development resources to meet their needs.

Maintaining the Security and Integrity of Corporate Information. In the US, as court decisions have increasingly made corporate board members responsible for the prudent management of their company's assets, the issue of protecting the company's information assets and ability to do business has become critically important. Significant losses of information or information processing capabilities can irreparably damage the competitive position of a firm or its financial condition. Guidelines for protecting these assets have been developed, audit procedures have become more important, and security functions have been established in many organizations to ensure the security and integrity of information. Despite these moves, many companies remain deficient in protecting these critical assets.

Managers must ensure that their internal audit capabilities are sufficiently staffed and skilled to evaluate the security of the company's information assets, information processing, and information transmission to ensure against loss. If this capability is deemed insufficient, management should rely on external consulting support to establish appropriate safeguards.

USERS OF IT PRODUCTS AND SERVICES: LONG-TERM ISSUES

Creating Transparent Communications and Information Processing Utilities for End Users. The increasing ability of users to develop their own applications, increasing cooperative use of computing among users with common projects, expanding use of telecommuting (through electronic mail, videoconferencing, and other means), and need to link heterogeneous equipment into common accessed media and information will force IS functions in companies to take another look at how they provide services to users. Users will demand that their applications operate responsively without regard to the physical platform on which those applications reside, or to the locations of those devices. Corporate information resources will become utilities in the same way that telephone service is. These utilities will be expected to work reliably, respond immediately, provide sufficient capacity, and protect the information of the user.

Again, users of information technology should establish regular programs to monitor technology and introduce appropriate technology to the firm. Senior management must reexamine the relationships between the IS organization and its customers to anticipate the changes that are coming.

Creating Applications that Will Enhance the Competitive Stance of Corporations. Business and industry will continue to grow more dependent on information technology to deliver goods and services to their customers. Trends indicate that customers are demanding not only more responsiveness and quality but also increasingly customized and personalized goods and services. Information technology will be critical in meeting these demands and maintaining the competitive position of the business. Customers will also be expecting greater ease of use and flexibility in products, so explosive growth can be expected in the use of small computers embedded in products and in handheld devices.

Creating Information Bases and Integrating Tools that Permit Users to Meet Their Own Needs for Information Access and Manipulation. In conjunction with the trends identified in regard to transparent information utilities, client/server has soared in popularity. Client/server assumes that a repository (i.e., the server) exists for large volumes of data and can provide extra processing capacity if required. Concurrently, smaller amounts of data and significant processing capacity exist on the client machine on the desktop of the end user. Clients and servers are linked by LANs, which are linked to similar client/server groups, to large computers as required and to external data services. In fact, in many instances, the mainframe acts as a server to provide a computing utility. Management must be alert to opportunities for using information technology to improve the performance of cooperative work groups by making them independent of the separations caused by time and distance.

Introducing Tools that Enhance International Communication and Reduce the Need to Travel. Examples include videoconferencing, multimedia workstations, and automated voice language translation. Again, following trends initiated in the early 1990s, it is anticipated that extensive use of communications will become less costly and more effective than the heavy travel schedules that were previously imposed on executives and professionals in many industries. In addition, as fiber-optic channels and satellite links proliferate, it is expected that the cost of communications will continue to drop.

Obtaining Skilled Software Applications Developers and System Integrators. The shortage of systems developers will persist throughout the 1990s, forcing users to continue to turn to outside resources to supplement their major applications development and implementation efforts. Planning and design for overall system architectures, long-range application planning, judicious introduction of new technology, and ranking application development and enhancement activities will become of critical importance to the firm.

Managing Contractors that Supply Software and Systems Integration Services. As a corollary to the shortage of applications developers, the use of external resources requires a different form of management than internal applications development. Greater stress will be placed on contract management and on the specification of work to be performed to successfully purchase services and control the results.

SUMMARY

Each year, various computer industry publications and consulting organizations conduct surveys of the issues and concerns of the industry and of IS executives in user organizations. The positions of these issues vary each year, but the issues identified in this article's short-term view seem to have remained within the top 10 for several years. The long-term issues are, of course, harder to predict. Long-term issues may be affected by world economic conditions, major changes in technology, and a host of other factors. Such issues must be addressed on the basis of assumptions about trends and technological advances.

There are clearly intelligent steps that may be taken by senior management to minimize the risks associated with their investments in information technology and its continuing operational costs. Some of these have been set forth in this chapter. The remainder of the decade will see accelerating change and challenge for both suppliers and users of information technology.

1-2
The Economics of a Strategy for Applying Information Technology

A. Perry Schwartz

IN COMPANY AFTER COMPANY, advanced information and communications technology are hot topics; in many of these same companies, however, there is a tremendous gap between talk and action. That gap exists as a direct result of the failure of too many technology systems projects to have immediate, tangible payoffs. In fact, many projects that were sold to management as a way to reduce costs have actually increased costs; by now, every company has experienced its share of these projects. On the other hand, projects that have produced tangible increases in revenue of profitability are few and far between. Expectations of increased revenue or profitability are often so low that many firms fail to keep even the most basic data for determining whether such increases have occurred.

Consequently, the typical executive's eye has been preoccupied with the cost side of the equation, ignoring the benefit side. This means that the executive is essentially ignoring the fact that the fundamental economic purpose of introducing new technology is not really to reduce costs but to increase a firm's growth and profitability potential. Cost savings are only one part of this equation, and arguably, now even the most important part.

ENHANCING GROWTH AND PROFITS

Some firms have already discovered the connection between technology and profitability. For example, one manufacturing firm, Computer Associates, Inc., conducted several studies of the possibility of using computers to store and disseminate information it was already gathering in the normal course of business. After a large initial investment in new technology, this company now derives a significant share of its revenue from selling this information to business and individual subscribers. Costs rose, but revenues increased much more.

A second manufacturing firm tracked five years' worth of financial experience after a large investment had been made in a new information system. In this case, the additional fixed cost for the new system was substantial: nearly $1 million for improving the support of 30 engineers. As a result of the new technology, however, these engineers were able to speed the pace of key development projects. After five years, the additional net revenue attributable to this department's work exceeded $30 million.

During the past several years, an old information system in a division of a large financial services corporation tracked the implementation of substantial upgrades and new features, costing upward of $2 million. This division had an historical growth rate of 2.5% per month. In the past three years of the study, following the installation of the new system modules, this growth rate rose to more that 4% per month. At the same time, the number of staff members has remained the same. The result has been a substantial increase in staff efficiency and division profitability. These companies' philosophies are typical of firms that have experienced great success in exploiting advanced technologies: They give cost savings low priority while seeking to maximize growth and profit.

PACED AND PACING ACTIVITIES

The key to making this philosophy work is to place the initial emphasis on understanding the activities of the business rather than on understanding the technology. As most companies are beginning to realize, this is because technology itself does not have a payoff; payoff can be realized only to the extent that the firm's activities are amenable to enhancement by application of the technology.

On the basis of these experiences, the author has helped to develop, and subsequently applied, a model that explains the choices these firms made and can guide the successful targeting of activities for improvement by advanced technology. For this mode, it is useful to distinguish between two types of activities by introducing two new terms: "paced" and "pacing." These terms were chosen deliberately to avoid connotations that might result from the use of more conventional terms. In almost all profit-making businesses, certain activities provide support for the firm's operation but do not drive revenue. These activities are paced activities. Alternatively, activities for which increased or improved performance does drive revenue are pacing activities.

Some examples help to distinguish between paced and pacing activities. In a bank, prospecting for new corporate accounts is a pacing activity; credit file

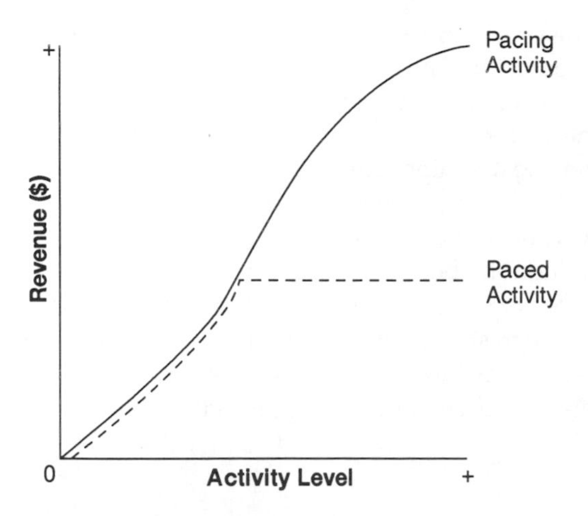

Note:
This model of paced and pacing activities is based on the standard economics of the firm under consideration and observations from several studies of divisions of major companies. This graph and the others in this chapter are derived directly from a study of a major manufacturing firm; the graphs have been smoothed, however, to preserve confidentiality while illustrating the effect under discussion. Data from several follow-up studies have confirmed the curves in the graph.

Exhibit 1-2-1. Relationship Between Revenue and Activity Level for a Paced and a Pacing Activity

maintenance is a paced activity. More prospecting will generally increase revenues; more credit file maintenance, above the level required for the current amount of business, will not contribute to an increase in revenues. For firms that depend on product innovation to attract customers, research and development is pacing and production is paced. In most firms, the sales activity is pacing and accounting, personnel, maintenance, legal, and staff functions are generally paced.

By definition, there is an asymmetry between paced and pacing activities as illustrated in Exhibit 1-2-1. For a pacing activity, an increase in activity enhances revenues while a decrease in activity reduces revenues. Restrictions or limits to pacing activities constrain growth. A decrease in paced activity may also reduce revenue, but this is a result of the paced activity acting as a constraint on a pacing activity. For example, if a banker's computer cannot handle additional accounts, there is no point in making an extra effort to attract new accounts.

On the other hand, an increase in the level of a paced activity, absent an increase in the pacing activity supported by the paced activity, does not increase revenue or constrain growth. If the sample bank's computer system capacity is increased but no new accounts are opened, the change in computer capacity has only increased costs and reduced profits.

Although the difference between pacing and paced activities seems obvious, all but the most senior managers have a tendency to overlook pacing activities and to focus on paced activities when they are asked to consider the prospects of advanced communications or information technology projects. This is understandable, because most of these managers are more concerned with cost control, as measured by their departmental budgets, than growth, and they have a corresponding narrow view of the firm. On the other hand, higher-level executives have a more comprehensive view of the firm and can distinguish between paced and pacing activities.

SELECTING TARGET PACING ACTIVITIES

Investments in advanced information systems are usually made for the purpose of enhancing economic outcomes. If pacing activities were targeted, increased revenues would be likely. Alternatively, if paced activities were targeted, decreased costs would usually be the only reasonable goal. Ordinarily, the incremental increase in revenues from leveraging pacing activities far exceeds the potential cost savings from improvements in paced activities. Therefore, pacing activities represent the greater opportunity for advanced information systems, and as the first step in developing a technology strategy, these activities should be identified.

Once identified, four criteria can help select the pacing activities that are likely to be successful advanced information technology and communications projects:

- *Will increased emphasis on the activity actually result in more business?* This confirms whether the activity is really one that is pacing. If the sales staff believe that the market is saturated, for example, new product development rather than sales may really be the critical pacing activity.
- *Will the additional capacity for quality or quantity of work be directed toward the pacing activity?* For example, if freed time will be soaked up by administrative work rather than by increased sales efforts, the investment in advanced technology is not likely to pay off.
- *Will the additional business resulting from enhancing the pacing activity actually yield more profit?* This may not be true in cases in which a costly production and manufacturing infrastructure must be created to support a large increase in sales or in which prices must be lowered to induce additional purchases.
- *Does the pacing activity contribute to a line of business that is a high priority?*

The firm should have a fundamental interest in enlarging the aspect of their business that will be targeted by the advanced technology.

In general, the more positive the answers to these questions, the greater the likelihood for significant payoff.

Analyzing a Target Activity

After the key pacing activities have been identified, it is time to determine whether opportunities, problems, or bottlenecks exist that can be resolved by advanced information technology. Those activities that cannot be improved through the use of advanced technology may still be identified as targets for improvement but are not relevant to the technology strategy of the firm. Alternately, those pacing activities amenable to enhancement by technological means should be subjected to thorough systems analysis.

In addition to the usual systems analysis concerns, there are three major considerations in the analysis of a target activity:

- *The technical feasibility of any proposed solution.* This assessment must be made in light of the firm's current level of information or communications technology. Solutions requiring the firm to make technological leaps will have high failure rates and should be considered to be research and development efforts.
- *The level of support for the introduction of new technology by the managers of the activity.* If the managers are not enthusiastic, successful implementation will be jeopardized.
- *The level of acceptance of the proposed solution by those who will be asked to use it.* Strong resistance by the target users will increase the risk of failure.

The result of these steps will be a list of one or more pacing activities that represent the firm's high-priority advanced technology investment opportunities.

PEOPLE OR COMPUTERS?

Executives may wonder whether they can get the same increase in output just by adding more staff. The question reflects a course of action every prudent executive should consider as an alternative to a technological proposal. After all, if increased activity is the goal, it can certainly be generated by additional staff. There is a three-part answer to the issue of people versus computers:

1. One of the major benefits of advanced technology is the capability to

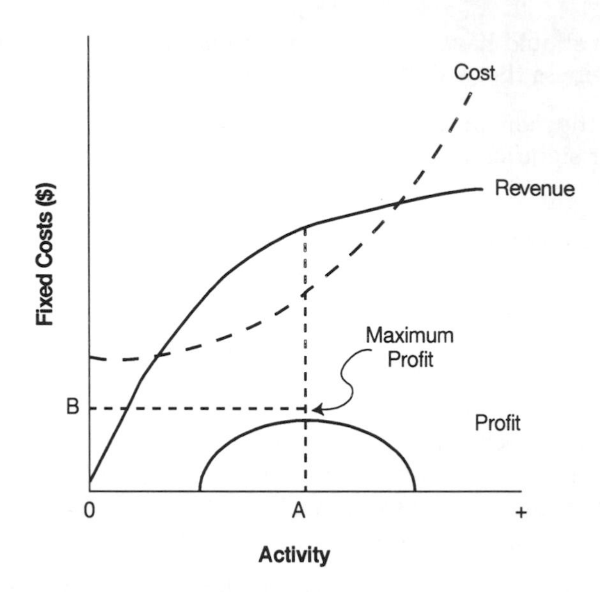

Exhibit 1-2-2. Profit Maximization Based on Activity Level

---

perform tasks more quickly and thoroughly. An increase in staff size often degrades responsiveness because of the increased need for communication and management control within the organization. The hidden costs of organizational control can seriously drain profits.

2. Nevertheless, if an increase in activity through the addition of staff would increase revenues at a much greater rate than it increased costs, yielding increasing incremental profits, a staff increase would likely be desirable.
3. On the other hand, if the resources committed to the pacing activity have been set to maximize profit, expanding the resource commitment along conventional lines by adding staff would only decrease profits. Because advanced information and communications technology can change the underlying cost structure of the business, use of technology to increase effective activity without a corresponding increase in staff may provide the basis for increasing profits.

Exhibit 1-2-2 helps explain these last two points. In this exhibit, the x-axis is the level of activity and the y-axis is dollars. The revenue curve is typical for a pacing activity. The cost curve, showing fixed and variable costs, reflects the general concave upward pattern that the standard economic theory of the firm projects for costs. The profit curve is simply the difference between the revenue curve and the cost curve.

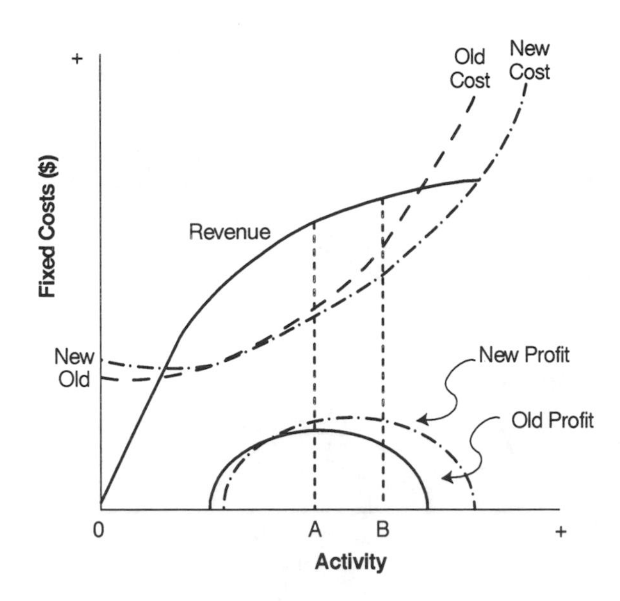

Exhibit 1-2-3. Effect of Change in Cost Structure on Profit

The maximum profit occurs at point A on the x-axis in Exhibit 1-2-2. If the level of activity were less than A, it might be appropriate to add staff, thereby increasing profits. This is the circumstance described in item 2 in the preceding list. However, if the level of activity were greater than A, adding staff is a poor strategy. This is the circumstance described in item 3.

ACCELERATION OF THE FIRM

Effective executives ask tough questions and demand sound answers. For advocates of new information or communications technology projects, one of the tougher questions is: How is the new technology going to increase profits? The answer if rooted in the basic economics of the firm. In fact, the economics are much the same as those used as a basis for the introduction of most industrial automation in factories.

The basic economics are illustrated in Exhibit 1-2-3. The figure in this exhibit superimposes two new curves on those in Exhibit 1-2-2 (i.e., new cost and new profit) to illustrate one possible effect of the introduction of advanced technology. In this case, the technology increases fixed costs while it decreases the variable costs of performing an activity. Thus, the new cost curve starts above the old cost curve but has a lower slope. The break-even point occurs where the two cost curves intersect. From this point on, the

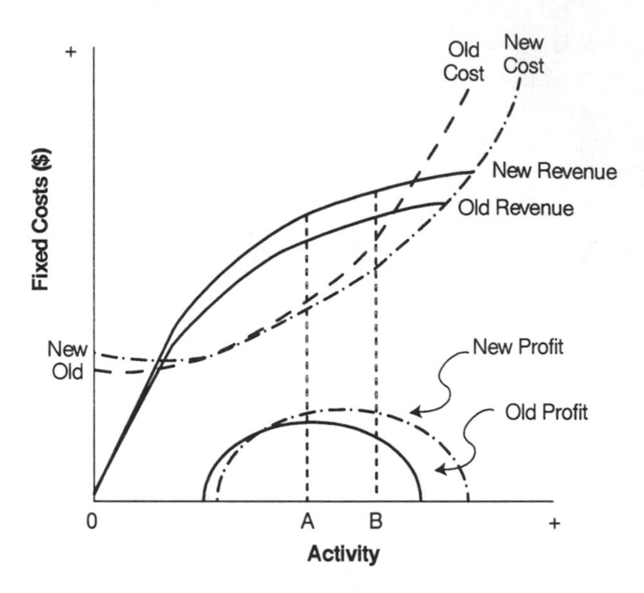

Exhibit 1-2-4. Acceleration of the Firm

advanced technology will produce profits that are uniformly higher than those that would have been achieved without it and the new, higher maximum profit occurs at activity level B rather than level A.

In addition to this long-run cost savings, the advanced technology holds out the promise of increased revenues. For example, technology might be used to create new features or services. The firm may anticipate additional revenue per unit of activity to the extent that the basis of competition is altered.

Exhibit 1-2-4 illustrates the combined impact of both decreased costs and increased revenues. In this case, the effect of introducing advanced technology is reflected by two structural curves. The first is the new cost curve. This reflects the increase in fixed costs and the decrease in variable costs associated with the new technology and is the same curve as shown in Exhibit 1-2-3. The second curve reflects new revenue that would result from new or improved products and services. The new profit is then the difference between the new revenue and the new cost; there is an associated new profit-maximizing activity level B, reflecting profit levels substantially higher than those achieved at the previous maximizing level A. The overall result is a change in the cost structure of performing the activity.

The alteration of the cost structure illustrated in Exhibit 1-2-4 is called acceleration. The acceleration effect is the most desirable outcome, occurring only when the firm takes advantage of both the cost-reducing and the profit-enhancing potential of a new technology.

The typical executive is not likely to have reviewed the benefits of advanced information and communications technology proposals or project outcomes in terms of acceleration. The reason for this is not the absence of acceleration but the inadequacy of systems for measuring the economic contribution of information and communications technology projects. Yet a program of honest and complete tracking of the impact of dollars spent on information and communications technology in terms of cost savings and revenue increases would certainly yield invaluable data and would be the basis for more sound technology investment decisions.

SUMMARY

In the face of an explosion of advanced information and communications technologies, many firms are approaching new systems opportunities without a guiding set of principles for leveraging the results. Traditional strategies focus on costs and savings related primarily to paced activities. However, the potential for payoff would be far greater if the firm could establish a strategy for applying the advanced technology to the improvement of pacing activities.

The choice of pacing activities as targets for advanced technologies is only the start. Over the long term, a successful strategy for the introduction of technology must be driven by fundamental economic considerations. Obviously, no firm can continue for long if it sacrifices profits to buy advanced systems. On the other hand, the inclination to apply advanced technology will be strengthened greatly as initiatives clearly and demonstrably result in economic acceleration of the firm. The challenge is to ensure that projects are chosen to maximize this possibility and that project tracking systems are adequate to measure the economic outcomes.

1-3 Security in Value-Added Networks

Steven J. Ross

COMPANIES OPERATING COMPUTERS FOR OTHERS are known as outsourcers or service bureaus. Private companies, the common carriers, have operated communications systems on behalf of others for a century or more; the telephone companies are the best example. Certain vendors of communications services combine the message transmission of the common carriers with the specializing processing of the service bureau. These are known as value-added networks (VANs).

CONTROL-RELATED VAN SERVICES

VANs offer a number of control-related services, from packet switching to diverse routing, that enhances communications security.

Packet Switching. By breaking long messages into shorter segments, or packets, and sending them over diverse routes within a network, greater advantage can be taken of network availability and diversity. Of course, the network must have the ability to reassemble the packets at the intended destination to reconstitute the original message.

User Identification. Access to and use of VANs should be limited to authorized users. Many VANs offer the capability to restrict the ability to send or receive messages to specific individuals, who must be uniquely identified to the network.

Message Authentication. In such applications as money transfer or electronic data interchange, it is essential that messages be correctly transmitted and sent to the intended recipient. This may be ensured by authentication—

the process of passing the message, as data, through an algorithm to produce a result. The result is appended to the message, and the authenticating process is repeated by the recipient. If the two results match, the message is intact and the sender is correctly identified.

Encryption. Encryption is much like authentication, except that the result of the algorithm replaces the message text instead of being appended to it. This process provides confidentiality of the message as well as integrity. Encryption may be thought of as value-added authentication.

Store-and-Forward Mechanisms. Because communications networks are built of physical components—cables, fibers, satellites, switches—they are subject to failure. Any messages in transit during the failure would be lost. To ensure delivery, many networks store messages in a static device, such as a computer, and then forward them to the recipient. Thus, if the network is interrupted, the message can be redelivered.

Diverse Routing. As further protection against loss of messages, networks are built with multiple, diverse paths for the messages to follow. If there should be a physical breakdown anywhere in the network, messages can follow alternative paths and the network can arrange for bypassing and repairing the network breach.

Application Processing. Such applications as EDI and electronic mail are carried out as adjuncts to network communications. The VANs often provide these services to their customers.

THE VAN PERSPECTIVE ON SECURITY

A VAN provides many services that are directly involved in the day-to-day processes of the companies that acquire its services. To that extent, the security of the VAN is a part of the overall security of the customer organization. And yet the security of the VAN is, for all practical purposes, beyond the customers' control. The security afforded to the customer is no greater than that which the vendor builds for its own purposes or that the customer is willing to pay for expressly.

From the perspective of the VAN vendors, there are three ways to view their responsibility for providing security:

- *Operators of a business.* VANS have a considerable investment in their networks and information processing services. As would any prudent businessperson, mangers of VANs must take precautions to protect their assets and the viability of their businesses.

- *Holders and processors of customer information.* As the holders of others' information, VANs must be able to protect that information while in their possession. This is not only ethics, it is hard-nosed business sense. Managers of all VANs realize that if the confidentiality and integrity of messages transmitted over their facilities are not assured, they will soon enough have no messages to protect at all.
- *Providers of services for a fee.* VANs provide added services if a customer is willing to pay for it. If many customers want the same service, it will be developed and offered to all on an optional basis. Security is a form of value-added functionality that VANs can, but do not always, provide.

SECURITY OF VAN BUSINESS OPERATIONS

The first requirement of VANs, as profit-making entities, is to protect their principal assets; the network, the software used to provide value-added services, and their own data. As such, they have a vested interest in ensuring the availability, integrity, and privacy of their information resources.

Availability. Vendors generally achieve a high level of availability. They do so by means of quality control, line diversity and component redundancy, and backup transmission capabilities.

They manage tightly controlled processes to ensure that the operations of the network are not disrupted, either while the network is being maintained or as a result of the changes themselves. This requires testing, checking, and controlled introduction of changes so that the effect of any error can be instantly identified, rectified, or withdrawn. The ongoing capability of the network and its services depends on management's ability to control quality and enforce standards of service.

Redundant Components, Alternative Routing, and Backup Transmissions. Recognizing that some percentage of error will survive even the most rigorous quality control, network vendors must plan in advance for failure. They do so by providing alternative routing for the transmission and receipt of messages in the event one pathway becomes blocked, as well as maintaining redundant components, such as switches or frames, that can be used if the primary component fails or is taken out of service.

In addition to providing alternate pathways for messages to take, a prudent network operator also provides alternative means of transmitting the messages. Many alternatives exist today, such as terrestrial cable, microwave, satellite, and cellular radio. Although terrestrial lines are the least expensive, they are also the most vulnerable to weather, earthquakes, and manufactured disturbances. Microwave transmissions are subject to breakup over water, and satellites occasionally fail. No one means of transmission is invulnerable; that is why network vendors must provide backup transmission capabilities.

Protection of Proprietary Software. For competitive purposes, the vendors protect their proprietary software from both external intrusion and internal misuse, primarily by means of access controls. The vendors must establish an internal security administrative function to control who has access to specific resources. It is necessary to keep careful control over systems programmers to prevent misuse, innocent or otherwise, over the system software. This is accomplished by limiting the types of privilege functions they have, monitoring all activity, and the separation of duties.

Physical Security. Just as the VAN software must be protected, so too must the physical components of the network. Customers should expect vendors to provide physical protection of their major nodes. Although terrestrial lines are usually routed through a shielded conduit, it is not possible to prevent all cable cuts in a far-flung network. This is the reason that the vendors use store-and-forward techniques, and why they absorb the cost for backup transmission facilities.

Disaster Recovery Plans. In the same way, they use business continuity and disaster recovery methods to protect their computer installations. Shared direct-access storage devices (DASD) to allow multiple computers to access the same data will keep the value-added services available when a single processor fails. Use of multiple processors provide redundancy of processing capability. It is in the nature of intelligent networks that processing capability is not only multiplied but distributed to diverse locations.

The VAN should have a disaster recovery plan to ensure recovery if an entire facility is damaged or destroyed. Alternate processing sites, regular backups, and off-site storage should be supported. Each location should provide fire and water protection to prevent physical destruction.

Auditability and Use Monitoring. One component of security is auditability, the means of determining how and by whom the network was used. All VANs (in fact, all commercial networks) have this capability. Its prime purpose is for billing. In the case of a VAN, the vendor must know which network facilities and value-added services were used and for how long.

There is software in each node for accounting and billing purposes. Although customers do not usually have access to this software, the vendor will usually cooperate with customers who have a need to review the audit trail after a suspected security breach.

SECURITY OF CUSTOMER INFORMATION

Although VAN vendors must protect their businesses, they also recognize that there are distinct requirements for security in an enterprise that takes control of others' resources in the regular course of business. These include ensuring that customers do not receive messages intended for another customer, that they cannot access other customers' data, that their data is protected from the vendor's employees, and that the overall network service is not degraded by the actions of a customer.

Ensuring the Customers Do Not Receive One Another's Messages. VAN vendors ensure that messages are routed correctly by building in redundant destination controls. At each stage in a message's progress from sender to recipient, the network management system constantly checks that the message is going where it is supposed to go. This is particularly the case for packet switching, in which the message must be reassembled after being routed over multiple independent paths.

Terminal Authentication. In some cases, vendors use terminal authentication to ensure that messages are sent to the correct location. This hardware-oriented feature ensures that the receiving device is the intended one. This level of security is not often requested and, where available, is usually provided for an extra charge.

Legal Consequences. Depending on the circumstances and consequences of misrouting, network vendors may have a legal or contractual liability to their customers. In financial activities, it is of little use for the sending party to blame a service vendor such as a VAN. The recipient is entitled to compensation, and the originator and the VAN must then settle the damages. Even if the VAN might elude reimbursement, failure to provide the essential service the customer paid for—transmitting messages from one point to another—is a distinct source of customer discontent. In a competitive marketplace, maintaining customer satisfaction is a key to profitability.

Ensuring that Customers Cannot Access One Another's Stored Data. A considerable amount of information belonging to or about a customer is stored on the VAN's computers. Some of this is for billing purposes, some for message routing; these sorts of data might be of interest to a competitor. But most critical in a VAN environment is the information that is being held for later retransmission (e.g., data in E-mail and EDI mailboxes). Attempts to tamper with or disclose such information must be detected and stopped.

Access Control Software. To accomplish this, the VAN vendors use various access control software products, both at the system level (i.e., software such as RACF and CA-ACF2) and within their data bases. Customers must take the responsibility for administering access to their own data. If the vendor performs this service, it is only as an agent for a customer, who retains the responsibility for how its employees use the data. Restricting access to customer data requires the vendor to establish and maintain a naming convention that associates customers with their files. In this way, the access control mechanism can limit the use of those files to individuals within the customer organization.

Protecting Customers from VAN Employees. Because the data transmitted over a VAN may be of significant value, either for the financial assets represented or for the date itself, customers need assurance that their data is safe even from their vendors. Ultimately, there are some employees who can access any resource on the network or the computer system. A VAN would be unmaintainable if that were not so. In an emergency, the management of a VAN allows these employees access even to customer data to continue providing service.

It is at the vendor's discretion to determine what constitutes an emergency. The vendor should notify customers if the confidentiality or integrity of their data is to be breached, even for valid purposes.

VAN vendors protect their customers from their employees by encouraging customers to take as much responsibility as possible for their own security. For example, where possible, customer personnel, not vendor staff, should administer access to the VAN.

It is possible to buy insurance against misuse of the VAN by vendor staff, but this is a control of last resort. By the time insurance is invoked, the damage is already done and the damage may have more than financial consequences.

Protecting the Network from Customers. The vendors should ensure that no one customer can affect the service level of another. Each customer must be prevented from degrading or preventing use of the network and the value-added services.

In providing computer services, VANs must have strict controls to ensure process-to-process isolation. To each person sitting before a terminal attached to the network, it should seem that he or she is the only user present. Some of this is achieved through access control software, but more by the multiuser capabilities of the operating system on each platform. The primary role of access control software in this case is to prevent customers from reaching system software and files. In some cases, specific hardware or lines are dedicated to specific customers.

Although help desks are usually thought of as providing service for confused users, they also have a powerful protective role in securing the VAN. Help desks are often the first to realize that a potential security breach has occurred. Because they monitor the network and the computers in real time, they are in position to detect when system or network resources are being tampered with or misused.

SECURITY OF FEE-BASED SERVICES

The successful introduction of new services often depends on demonstrating their security. This has been the case with E-mail and especially with EDI. Expanded network services are made more marketable by making them more secure. If for no other reason, VAN vendors want the good public relations that come with a claim of advanced security. But if there is no clamor for message authentication or encryption or virus protection, these features will only be available as options, not as standard parts of the service.

Encryption and Message Authentication. Where secrecy is the primary consideration of protecting the information on the network, encryption is the only choice. All other means of protection limit access to the information; only encryption protects the information itself, by rendering it unreadable by the unauthorized. In some cases, the confidentiality of the information has little value but there is a great risk if a message cannot be read. For example, if the data in an electronic funds transfer cannot be read, the money cannot be moved. In that case it is better to authenticate the message than to encrypt it.

To the extent that encryption and message authentication are used, the customer must be relied on to make them work. The customer must control the encrypted line, must manage the keys, bear the risk of equipment failure making messages unreadable, and accept the limitations on the ability to communicate with those who do not have such protection. Still, these techniques are the best available to protect the data in transmission. Most VAN vendors provide encryption or authentication equipment for a fee. The service may also include key management techniques. But it is up to the customer to ensure that they are in place and are used properly.

Password Protection. Passwords are the most commonly used form of security in communications systems. Yet many VAN vendors have found that their customers would like to eliminate passwords. Passwords mean keystrokes: memorization and inconvenience; and administration. Thus the VAN vendors must strike a balance between what they know their customers need and what they want.

Many VANs do not enforce any standards of length or structure for passwords, nor do they require their regular change. Some VANs may even fail to prohibit printing passwords in the clear. If customer management insists on protection for the passwords, these features can be made available, but often only at an additional charge.

Security Reports. VAN vendors collect a substantial amount of information about network and service use. In the course of gathering that information, they also get indications of attempted misuse of the VAN. If those attempts harm the vendors' interest, they must take action themselves. However, there are sometimes attempted breaches within a customer's own domain. For example, it is not the vendors' responsibility to detect misuse of an EDI mailbox by a customer's own users or systems.

A vendor would willingly pass along selections from its internal security reports if a customer could make use of them to manage its own security. This is easier to do when access to customer information is controlled by a commercial access control package. But in many cases, the information a customer would need to manage its own internal security must be provided through custom coding and reports. Creating these reports may require special fees to be charged.

Segregation of Test and Production Environments. Keeping developers out of production processing has long been recognized as an important means of ensuring security. This may be especially difficult for such VAN applications as EDI, in which customer development staff are constantly devising new automated interfaces to the VAN's services.

Some VANs allow (and many encourage) separate test and production facilities on the network and on the computers. This is like having two separate contractual arrangements with the VAN vendor. It is then the customer's responsibility to keep developers out of the production environment, through policy, password control, and enforcement. Although having two separate sets of services does not increase a customer's use of the VAN, some vendors do charge additional fees for this segregation, for their administrative costs.

VAN FEATURES THAT CREATE EXPOSURE

As conscientious as VAN vendors may be, they are also under pressure from customers to provide features that are detrimental to security. Security practitioners should be aware of such features and discourage their use.

Auto Log-on Terminals. Many customers view passwords as an unnecessary nuisance. To eliminate them, it is possible to have an intelligent terminal or microcomputer sign itself on when it connects to the network. Thus, anyone with access to the device has access to the VAN and its services. Individual accountability is lost and so, potentially, is the confidentiality of the data displayed on the terminal.

Directed Sign-ons. Many VANs offer the facility of a directed signon, which takes the uninitiated user through all the steps needed to establish connection with the VAN. As a result, hackers, as well as new users, find it much easier to learn how to access resources, even those for which they have no authority.

Optional Access Control Software. Many VANs have either acquired or developed access control software to manage user identification, rule-based access to resources, audit trails, and protection of system resources. However, for those customers who find passwords too burdensome, the vendor may make the access control software optional. In other words, where security is available, the vendor may be pressured to turn it off.

Forever-Fixed Passwords. VAN vendors make it possible for a customer to bypass the change process and never have to change the initial password given to the customer's staff. Over time, the confidentiality of these passwords degrades, but there is little evidence of concern on the customer's part until security is breached.

ASSESSING VAN SECURITY

VAN customers should access the security provided by their vendors. Such a review may be difficult to perform and may be best done by an independent third party. The assessment should address the following categories of control objectives:

- *Authorization.* The managerial practices that ensure that all access to the network, by both customers and VAN employees, has been approved.
- *Access control to the network.* Limitations on the ability of unauthorized users to access the network.
- *Access control to processes.* Restrictions on the value-added services a customer may use and that VAN employees may affect.
- *Access control to functions within those processes.* Restrictions on particular functions within the services.

- *Auditability.* The ability to monitor and reconstruct activity on the network.
- *Recoverability.* The ability of the VAN to provide uninterrupted service or, if interruptions occur, to recover rapidly.
- *Systems development and maintenance.* The processes by which the network and its software are implemented, maintained, and upgraded.
- *Operational controls.* The procedures by which the network is operated and controlled.

Authorization. The vendor should establish procedures for adding and maintaining users on the network. Such procedures control who is allowed access to the network and its underlying computer systems and the kind of accesses assigned to different users. These procedures must be performed by a vendor when adding a new customer. Thereafter, the customer should perform administrative procedures for its own users.

A help desk or other interface should be able to ensure that the network is used as intended. The help desk is often the first to recognize that an operations problem is actually caused by a security weakness or even a breach. These customer interface personnel are the first line of defense for security.

There should be internal processes to control access or use of network services by vendor personnel. Formal authorization procedures provide assurance that access is granted only after review of appropriate supporting documentation. This is the basis for prevention of unnecessary and unauthorized access being assigned to internal personnel. The most significant restrictions should be placed on network engineers and systems programmers.

A vendor's organizational functions should be well documented to ensure appropriate segregation of duties. Documentation of functions instructs users of the systems as to their proper activities, so that no one can completely dominate any process. The pattern of control and the functions allowed to personnel should be embodied in policy and standards with regard to security and control. Documentation of these procedures ensures that standards are applied thoroughly and consistently.

Access Control to the Network. There should be access control over operating systems on host and node processors where the value-added services are applied. These controls should address the implementation, testing, maintenance, and administration of those operating systems. This provides a reliable method of mediating access to the system and, by extension, to the network so that unauthorized users are not allowed on it.

Network access controls should be used to control communication processing and delivery of messages. These controls include terminal authentication and personal authentication by means of smart cards and challenge-response systems. Monitoring and surveillance capabilities should be used to detect attempts to gain unauthorized access to the system. Attempted breaches of the network's security should be detected as soon as possible so that appropriate action can be taken.

Access Control to Processes. Access to network software should be limited to authorized individuals. This ensures that only people who are familiar with the network and are trusted individuals are able to modify its software. Even trustworthy individuals can make mistakes, and in a commercial network environment error can be very harmful.

Access Control to Functions Within Processes. Network features should restrict users' ability to read, write, delete, and create messages and transactions. For example, a user who does not need to make changes to a static billing file but who does need to view it from time to time can be given read privileges but not write privileges. Data and transactions entered onto the network should be checked for authorization, accuracy, and completeness.

Auditability. Sufficient audit trails should be produced to enable users and the network service to reconstruct activities on the network and to recognize attempts at unauthorized activities in a timely manner. Internal audit, quality assurance, or a similar internal function should advise the management of the VAN of the current status and effectiveness of security controls.

Recoverability. The VAN vendor should strive for the highest possible level of continuous service. The physical resources of the network should be adequately safeguarded. The VAN should be provided with adequate physical security over computers and network hardware, stored data, and documentation to prevent unauthorized destruction, modification, disclosure, or use.

Plans for restoring full service in a natural or manmade disaster should be developed and tested. The disaster recovery plan should be both feasible and testable, and those tests should occur frequently.

Systems Development and Maintenance. VAN vendors should employ a formal systems development methodology for the development and maintenance of network software. This should include procedures to ensure that changes to network software are authorized and restricted to approved changes.

Moreover, these procedures should be defined to help ensure that all changes to software are requested by authorized individuals and are restricted to the changes that are requested. Processes should be included in the systems development life cycle for maintaining as well as developing applications. This ensures that security and control are addressed whenever changes are made to the network software.

Independent functions should be assigned responsibility for ensuring the integrity and security of network software. Such a separation of duties in the development and deployment process is critical to a trustworthy VAN operation.

Network software should be well documented and the documentation itself protected. In particular, the security and control features of the software should be documented. Documentation provides a base of knowledge that can be referenced in the future when changes need to be made. It is also needed when knowledge of the security controls that exist within the network is required.

Operational Controls. There should be processes to secure data at the points of capture and during transmission to and from a VAN's computers. This helps to ensure that data is not intercepted by unauthorized users while it is being entered or routed to its destination. Network controls should be implemented so that messages are received as sent, are complete, and are correctly routed.

OBTAINING THIRD-PARTY ASSESSMENTS

Because value-added networks are, by definition, the property of a vendor, it is particularly difficult for a user to determine just how much (or how well) security is provided. The VAN vendor is unlikely to allow users to conduct a firsthand examination of premises or review its procedures for acquiring and monitoring lines. The vendor might open its doors for a single customer, but is unlikely to make it a general policy to allow access to all customers.

For practical purposes, the only viable option available to most VAN users is reliance on the opinion of a third party—not the third party that operated the VAN but rather an independent assessor who has the access that the user would want if allowed. In all likelihood, this third party will be an independent accounting firm. Many communications specialists could perform such an assessment, but only accounting firms have generally accepted standards by which their opinions may be judged.

Many VAN vendors make service auditor reports available to their customers. These reports, developed by accounting firms, are obtained by the VAN vendors on behalf of their customers. There is a degree of self-interest in this. By publishing the results of an audit, vendors can keep customers' internal and external auditors from requesting access to their facilities to satisfy themselves concerning security and control in the VAN. The customers' auditors can turn to one source for satisfaction, and the vendor is not besieged by numerous requests for the same information.

From the point of view of the customer, there are certain shortcomings to a service auditor report. The vendor selects the criteria or control objectives under which the review is to be conducted. The review is in effect only at a point in time or over a period of time that has necessarily passed by the time the customer reads the report. The auditing standards are intended to apply to service bureaus, and in a sense that is what a VAN is, a communications rather than a processing service. But the requirements of the standards are focused not on the exigencies of network operation but rather on computer operations. Most critically, the intended audience for a service auditor report is the external auditors of the VAN's customers. As a result, the basis of the review (depending in large measure on both the vendor and the auditor) is likely to be accounting controls rather than communications controls.

Interpreting the Service Auditor Report: The Customer's Viewpoint

The service auditor report should be treated with healthy skepticism. It is not enough that controls meet an auditor's objectives, they must meet the customers' as well. Even though all security objectives have been met in the aggregate, it is still possible that the cumulative effect from noncritical weaknesses might constitute a serious security problem. Unfortunately, a service auditor report is simply a report, not a mechanism to effect changes if a customer is not satisfied with the network service's security and control.

Despite these shortcomings, it is recommended that customers request these reports from their VAN vendors. A service auditor report can be a valuable tool if properly used and interpreted. Following are the key elements of the report communications systems managers should be concerned with.

The Opinion Letter. The opinion letter is intended for accountants and need not concern most customers. The letter also establishes the dates for which the opinion is in effect.

Network Description. This section is not written by the auditor but by the VAN vendor. It is the set of assertions that the auditor subsequently attests to. This is a concise technical overview of hardware, software, protocols, interfaces, and other information about the VAN that enables a customer to determine that appropriate use is being made of the available network service features.

Description of Policies and Procedures. This is the core of the report, the detailed description of the security controls. It tells what the service is doing to provide internal controls. From the point of view of customers, these security features are an extension of their operations and thus of their security system. Just as the determination of the adequacy of security is keyed to the achievement of stated control objectives, so too must customers consider their own objectives and whether these are being met.

This section of the report also gives some indication of the availability of external security and control features customers may take advantage of. Most critically, this section identifies weaknesses that might prevent meeting a control objective.

Client Considerations. In all cases, a vendor's security must fit within a customer's system of controls. A service auditor provides a guide to network features that are available to a customer and that should be considered as part of an organization's system of control. The report also points out security features that are made optional by the vendor but should be taken advantage of. This section effectively places the burden on customers to incorporate available security and control measures and to carry out the appropriate tests of security.

SUMMARY

Value-added networks are becoming an increasingly important component of business operations, providing both information processing and communications services. The customer for these services places itself at risk if it simply assumes that the network vendor's controls are adequate. This chapter provides a checklist of controls that should be implemented by the vendor and a method for assessing vendor compliance.

1-4
Improving Communication Between End Users and System Developers

Jack T. Marchewka

THE TRADITIONAL APPROACH to information systems development (ISD) assumes that the process is both rational and systematic. Developers are expected to analyze a set of well-defined organizational problems and then develop and implement an information system (IS). This, however, is not always the case.

The full extent of IS problems and failures may not be known, as most organizations are less than willing to report these problems for competitive reasons. However, a report by the Index Group indicates that 25 out of 30 strategic system implementations studied were deemed failures, with only five systems meeting their intended objectives. Moreover, it has been suggested that at least half of all IS projects do not meet their original objectives.

A lack of cooperation and ineffective communication between end users and system developers are underlying reasons for these IS problems and failures. Typically, the user—an expert in some area of the organization—is inexperienced in ISD, while the developer, who is generally a skilled technician, is unacquainted with the rules and policies of the business. In addition, these individuals have different backgrounds, attitudes, perceptions, values, and knowledge bases. These differences may be so fundamental that each party perceives the other as speaking a foreign language. Consequently, users and developers experience a communication gap, which is a major reason why information requirements are not properly defined and implemented in the information system.

Furthermore, differences in goals contribute to a breakdown of cooperation between the two groups. For example, the user is more interested in how information technology can solve a particular business problem, whereas the system developer is more interested in the technical elegance of the application system.

On the other hand, users attempt to increase application system functionality by asking for changes to the system or for additional features that were not defined in the original requirements specifications. However, the developer may be under pressure to limit such functionality to minimize development costs or to ensure that the project remains on schedule.

Subsequently, users and developers perceive each other as being uncooperative, and ISD becomes an "us versus them" situation. This leads to communication problems that inhibit the user from learning about the potential uses and benefits of the technology from the developer. The developer, on the other hand, may be limited in learning about the user's functional and task requirements. As a result, a system is built that does not fit the user's needs, which, in turn, increases the potential for problems or failure. Participation in the ISD process requires a major investment of the users' time that diverts them from their normal organizational activities and responsibilities. An ineffective use of this time is a waste of an organizational resource that increases the cost of application system.

The next section examines the conventional wisdom of user involvement. It appears that empirical evidence to support the traditional notion that user involvement leads to IS success is not clear cut. Subsequently, this section suggests that it is not a question of whether to involve the user but rather a question of how or why the user should be involved in the ISD process. In the next section, a framework for improving cooperation, communication, and mutual understanding is described.

USER INVOLVEMENT AND COMMON WISDOM

The idea that user involvement is critical to the successful development of an information system is almost an axiom in practice; however, some attempts to validate this idea scientifically have reported findings to the contrary. Given the potential for communication problems and differences in goals between users and developers, it is not surprising, for example, that a survey of senior systems analysts reported that they did not perceive user involvement as being critical to information systems development.

Moreover, a few studies report very limited effects of user involvement on system success and suggest that the usual relationship between system developers and users could be described as one in which the IS professionals are in charge and users play a more passive role.

There are several reasons why involving the user does not necessarily guarantee the success of an information system. These include:

- If users are given a chance to participate in information systems development, they sometimes try to change the original design in ways that favor their political interests over the political interests of other managers, users, or system developers. Thus, the potential for conflict and communication problems increases.

- Users feel that their involvement lacks any true potential to affect the development of the information system. Consequently, individuals resist change because they feel that they are excluded from the decision-making process.

Despite equivocal results, it is difficult to conceive how an organization could develop a successful information system without any user involvement. It therefore may not be a question whether to involve the user, but how or why the user should be involved. More specifically, there are three basic reasons for involving users in the design process:

1. To provide a means to get them to "buy in" and subsequently reduce resistance to change.
2. To develop more realistic expectations concerning the information technology's capabilities and limitations.
3. To incorporate user knowledge and expertise into the system. Users most likely know their jobs better than anyone else and therefore provide the obvious expertise or knowledge needed for improved system quality.

While the more traditional ISD methods view users as passive sources of information, the user should be viewed as a central actor who participates actively and effectively in system development. Users must learn how the technology can be used to support them, whereas the system developer must learn about the business processes in order to develop a system that meets user needs. To learn from each other, users and developers must communicate effectively and develop a mutual understanding. This leads to improved definition of system requirements and increased acceptance, as the user and developer co-determine the use and impact of the technology.

COOPERATION, COMMUNICATION, AND MUTUAL UNDERSTANDING

Effective communication can improve the ISD process and is an important element in the development of mutual understanding between users and system developers. Mutual understanding provides a sense of purpose to the ISD process. This requires that users and developers perceive themselves as working toward the same goal and able to understand the intentions and actions of the other.

To improve communication and mutual understanding requires increased cooperation between users and developers. As a result, many of the inherent differences between these individuals are mitigated and the communications gap bridged; however, the balance of influence and their goals affects how they communicate and cooperate when developing information systems.

The Balance of Influence

In systems development, an individual possesses a certain degree of influence over others by having a particular knowledge or expertise. This knowledge or expertise provides the potential to influence those who have lesser knowledge. For example, a system developer uses his or her technical knowledge to influence the design of the information system. If the user has little or no knowledge of the technology, the system developer, by possessing the technical knowledge needed to build the application system, has a high degree of influence over the user. On the other hand, the user has a high degree of influence over the system developer if the user possesses knowledge of the domain needed to build the application system. By carefully employing their knowledge or expertise, the user and the developer cultivate a dependency relationship. Subsequently, the balance of influence between the user and developer determines how these individuals communicate with each other and how each individual tries to influence the other.

Reconciling the Goals Between the User and Developer

Even though users and developers may work for the same organization, they do not always share the same goals. More specifically, the nature of the development process creates situations in which the user and developer have different goals and objectives. For example, the developer may be more interested in making sure that the IS project is completed on time and within budget. Very often the developer has several projects to complete, and cost/schedule overruns on one project may divert precious, finite resources from other projects.

Users, on the other hand, are more interested in the functionality of the system. After all, they must live with it. A competitive situation arises if increasing the system's functionality forces the system to go over schedule or over budget or if staying on schedule or within budget limits the system's functionality.

In 1949, Morton Deutsch presented a theory of cooperation that suggests cooperation arises when individuals have goals linked in a way that everyone sinks or swims together. On the other hand, a competitive situation arises when one individual swims while the other sinks.

This idea has been applied to the area of information systems development to provide insight as to how goals might affect the relationship between users and developers.

Cooperation arises when individuals perceive the attainment of their goals as being positively related (i.e., reaching one's goals assists other people in attaining their goals). Cooperation, however, does not necessarily mean that individuals share the same goals, only that each individual will (or will not) attain their goals together. Here the individuals either sink or swim together.

The opposite holds true for competition. In competition, individuals perceive their goals as being negatively related (i.e., attainment of one's goals inhibits other people from reaching their goals). In this case, some must sink if another swims.

Cooperation can lead to greater productivity by allowing for more substitutability (i.e, permitting someone's actions to be substituted for one's own), thus allowing for more division of labor, specialization of roles, and efficient use of personnel and resources. Cooperative participants use their individual talents and skills collectively when solving a problem becomes a collaborative effort. Conflicts can be positive when disagreements are limited to a specific scope, and influence tends to be more persuasive in nature.

Cooperation also facilitates more trust and open communication. In addition, individuals are more easily influenced in a cooperative situation than in a competitive one. Communication difficulties are reduced when persuasion rather than coercion is used to settle differences of viewpoints. Honest and open communication of important information exemplifies a cooperative situation. Competition, on the other hand, is characterized by a lack of communication or misleading communication.

The competitive process also encourages one party to enhance its power while attempting to reduce the legitimacy of the other party's interests. Conflict is negative when discussions include a general scope of issues that tend to increase each party's motivation and emotional involvement in the situation. Defeat for either party may be less desirable or more humiliating than both parties losing. In addition, influence tends to be more coercive in nature. Competitive individuals tend to be more suspicious, hostile, and ready to exploit or reject the other party's needs and requests. The cooperative process supports trust, congenial relations, and willingness to help the other party's needs and requests. In general, the cooperative process encourages a convergence of similar values and beliefs. The competitive process has just the opposite effect.

A CLASSIFICATION OF USER AND DEVELOPER RELATIONS

Exhibit 1–4–1 provides a classification scheme for viewing potential user/developer relationships based on the interdependency of goals and their

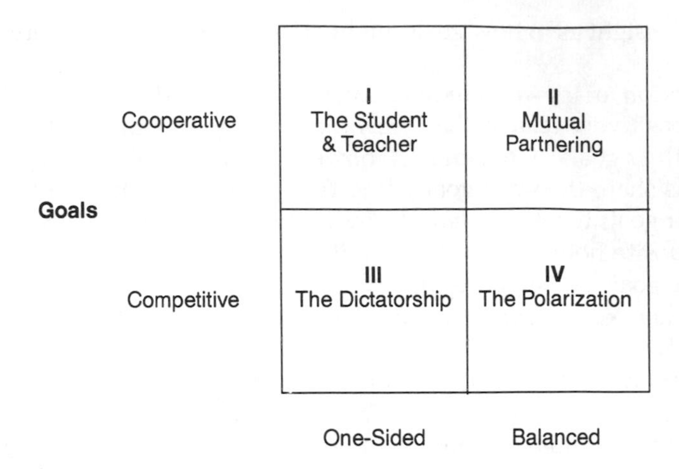

Exhibit 1-4-1. Classification of User/Developer Relationships

balance of influence. Classification of relationships clarifies the social process of user involvement (i.e., how the user currently is involved or how the user should be involved) in ISD.

Quadrant I: The Student and Teacher

In this quadrant, the balance of influence is one-sided; however, the goals between the user and the developer are positively related. Subsequently, this relationship resembles a teacher/student relationship for two reasons.

First, because the balance of power is one-sided, the more experienced or knowledgeable individual most likely leads the ISD process. Because they both perceive their goals as being positively related, the less-experienced individual most likely follows the advice of the more influential individual.

The second reason has to do with a one-way model of learning. If the more influential individual leads the ISD process, he or she has more to offer in terms of being able to share his or her knowledge or expertise than the less-experienced individual. As a result, learning generally takes place in one direction as in a typical teacher/student relationship.

An example of this type of relationship is an experienced developer teamed with a novice user. The users' limited knowledge or experience may make it difficult to specify their requirements. The developer may then attempt to control or lead the ISD process, in which users contribute to the best of their knowledge or expertise.

Since these individuals perceive their goals as being positively related, the potential for resistance may be low. The user may view the development process as an opportunity to learn about the technology from the developer and may be easily influenced by the developer. An information system may be developed and viewed as successful; however, as the user becomes more experienced and familiar with the system, he or she may begin to request changes. Subsequently, these requests may result in higher maintenance costs later on.

Quadrant II: Mutual Partnering

In this quadrant the user and developer share the same degree of influence and have positively related goals. Here users play a more active role in the ISD process than a novice, as their knowledge and level of expertise is greater. Because the developer also is experienced and knowledgeable in ISD, the potential for a two-way model of learning exists.

Users, for example, learn how the technology supports their needs, whereas the developer learns about the business processes. Because the goals of these individuals are positively related, the potential for resistance is low. Subsequently, a two-way model of learning suggests a higher degree of mutual learning and understanding where a system is being built successfully with lower maintenance costs later on.

Quadrant III: The Dictatorship

In the third quadrant the individuals exhibit more of a dictatorial relationship, in which the individual with the greater potential to influence leads the ISD process. Resistance is high, because the goals of these individuals are negatively related. If the developer has the greater potential to influence the user, for example, he or she may view the user as a passive source of information. Users may perceive themselves as lacking any real chance to participate and then subsequently offer a high degree of resistance. The developer may build a system that fits the developer's perception of user needs or wants, to attain his or her goals. As a result, a system is developed that does not meet the initial requirements of the user and ultimately exhibits high maintenance costs. The system might be characterized as a technical success but an organizational failure.

On the other hand, if the user has the greater potential to influence the developer, the user tries to increase, for example, the functionality of the system to attain his or her goals. As a result, the developer offers passive resistance when asked to comply with the user's requests to minimize what the developer perceives as a losing situation. Conflicts may be settled through coercion with limited learning occurring between these individuals.

Quadrant IV: The Polarization

The fourth quadrant suggests a situation in which both the user and developer have an equal balance of influence but negatively related goals. Mutual learning is limited or exists only to the degree needed by an individual to attain his or her goals. Settlement of conflicts are achieved through political means, and a high degree of resistance results if one side perceives themselves as being on the losing side. Conflicts increase each individual's motivation and emotional involvement in the situation, making defeat less desirable or more humiliating than both parties losing. These individuals may become more suspicious, hostile, and ready to exploit or reject the other party's needs or requests. Subsequently, this type of relationship may potentially be the most destructive and could lead to the abandonment of the IT project if neither side is willing to capitulate.

STRUCTURING THE USER-DEVELOPER RELATIONSHIP

The framework presented in the previous section may be used to assess and structure the relationship between users and developers. A three-step process is now presented to assess, structure, and then monitor the social relationship between these individuals.

Assessment

Assessment of the user/developer relationship is useful for choosing project participants as well as for gaining insight during the project if problems begin to arise. Using the framework presented in the previous section, a manager may begin by determining the potential balances of influence. Examples of such factors that affect the balance of influence for both developers and users include:

- The level of technical knowledge.
- The level of domain knowledge (i.e., knowledge of the business processes or functions that are the core of the IS project).
- The years of experience in the company or industry.
- The prior involvement in system development projects.
- The rank or position within the organization.
- The reputation (i.e, how the individual's level of competency is perceived by others in the organization).

Other factors relevant to the project or organization can and should be used to assess the balance of influence among individuals. Subsequently, the manager should begin to get a clearer picture as to whether the balance of influence will be one-sided or balanced.

The next step is to assess individuals' goals. An easy way to make this assessment is to ask each individual to list or identify factors that are important to him or her. Examples include:

- What do you have to gain should the project succeed?
- What do you have to lose should the project fail?
- How would you determine whether the project is a success or failure?

After having both users and developers list what is important to them, a manager can compare these items to determine whether the individuals have potentially conflicting goals. End users and developers need not have exactly the same items of interest; these items need only not cause a win/lose situation. Asking the individuals to list how they would determine whether the project is a success or failure may uncover other potentially conflicting goals. For example, a manager may discover that users value functionality over cost, whereas the developer is concerned with ensuring that the project is developed by a specific date.

Structuring

A manager has several alternatives that can alter the balances of influence. The first alternative—choosing project participants—has the greatest impact. In an ideal situation, a manager can choose from a pool of personnel that includes both users and developers with varying degrees of skill, expertise, and knowledge. Unfortunately, if the pool is small, the number of possible combinations is reduced. As a result, training may prove to be a valuable tool for users becoming more knowledgeable about technology and developers becoming more knowledgeable about the business processes and functions.

As suggested in the framework presented, the goals of these individuals may be the most important factor in improving the social process of system development. For example, involving novice or less experienced individuals on a project is desirable when the project participants perceive their goals as positively related; however, serious communication problems arise if these same participants have negatively related goals.

To increase cooperation, then, the goals of the development team should be structured so that the individuals' goals are positively related. This may be accomplished in a number of ways.

Make Project Team Members Equally and Jointly Accountable. This may be in terms of a bonus or merit system under which each of the project team members is equally and jointly accountable for the success or failure or the information system. In other words, both the users and developers should both be concerned with such issues as the functionality of the system and with cost/schedule overruns.

Goals Should Be Made Explicit. Each individual involved in the development of the information system should have a clear, consistent perception that his or her goals are related in such a way that all sink or swim together. It is important not only that users and developers be held accountable using the same reward or merit system but also that they are aware that each is held accountable in the same way.

Management's Actions Must Reinforce Project Team Goals. It is important that management not allow project team members the opportunity to point fingers or assign blame. Subsequently, the actions of management must be consistent with the goals of the project team members. This is the most difficult challenge of all, because a change in values, attitudes, and possibly culture is required. For example, if goals are to be positively related, there can be no "us versus them" ideology. Instead, both users and developers should see themselves as part of the same team.

Monitoring

The goals and perceptions of the individuals may change over the course of the project. Just as the allocation of time and resources must be monitored during a project, the social process between the project participants should be monitored as well. Monitoring should be continual during the project to identify any problems or negative conflict before they adversely affect the project. Similar to assessment, a manager may want to look for warning signs. Examples of warning signs include:

Finger Pointing When Problems Arise. Project team members should fix the problem, not the blame.

Negative Conflict. Individuals focus on petty issues that do not move them closer to their goals but only serve one party at the expense of the other. However, users and developers should agree to disagree. Conflict can be positive, especially when developing innovative approaches or refining new ideas.

Lack of Participation or Interest. All members of the project should be involved actively. However, even when all members perceive themselves as having a cooperative relationship, some individuals may become less involved. This may occur when the balance of influence is one-sided. Too often systems developers take control of the IS project and attempt to act in the best interest of the users. Although the developers may mean well, they may attempt to develop a system that the user never asked for and does not want.

Assessment, structuring, and monitoring should be a cycle. If specific problems are identified, a manager should assess the balance of influence and goals of the project team members. Changes can be made to alter or fine-tune the balances of influence or goals among the members. By managing this social process between users and developers, a manager increases the likelihood that systems that meet the objectives originally envisioned are developed on time and within budget.

SUMMARY

This chapter suggests that even though users and developers may work for the same organization, they do not always share the same goals. Subsequently, problems in ISD arise, especially when the user is more interested in functionality and the developer is more interested in maintaining cost/time schedules.

Cooperation facilitates improved communication and leads to greater productivity, because individuals perceive their goals as being positively related. In addition, the goals of the individuals provide some insight as to how each party uses its influence in the development of an information system. This idea was presented through a classification of user and developer relationships that considered their interdependency of goals and their balance of influence. Using this framework, a manager can assess, structure, and monitor the social relationship between users and developers. Managing this social relationship may result in more systems being developed within budget and on schedule and that meet the needs of the user.

1-5

Network Disaster Recovery Planning

Nathan J. Muller

BECAUSE THERE ARE MANY MORE LINKS than host computers, there are more opportunities for failure on the network than in the hosts themselves. Consequently, a disaster recovery plan that takes into account such backup methods as the use of hot sites or cold sites without giving due consideration to link-restoral methods ignores a significant area of potential problems.

Fortunately, corporations can use several methods to protect their data networks against downtime and data loss. These methods differ mostly in cost and efficiency.

NETWORK RELIABILITY

A reliable network continues operations despite the failure of a critical element. The critical elements are different for each network topology.

Star Topology

With respect to link failures, the star topology is highly reliable. Although the loss of a link prevents communications between the hub and the affected node, all other nodes continue to operate as before unless the hub suffers a malfunction.

The hub is the weak link in the star topology; the reliability of the network depends on the reliability of the central hub. To ensure a high degree of reliability, the hub has redundant subsystems at critical points: the control logic, backplane, and power supply. The hub's management system can enhance the fault tolerance of these redundant subsystems by monitoring their operation and reporting anomalies. Monitoring the power supply, for example, may include hotspot detection and fan operation to identify trouble before it disrupts hub operation. Upon the failure of the main power supply, the redundant unit switches over automatically or manually under the network manager's control without disrupting the network.

The flexibility of the hub architecture lends itself to variable degrees of fault tolerance, depending on the criticality of the applications. For example, workstations running noncritical applications may share a link to the same local area network (LAN) module at the hub. Although this configuration might seem economical, it is disadvantageous in that a failure in the LAN module puts all the workstations on that link out of commission.

A slightly higher degree of fault tolerance may be achieved by distributing the workstations among two LAN modules and links. That way, the failure of one module would affect only half the number of workstations. A one-to-one correspondence of workstations to modules offers an even greater level of fault tolerance, because the failure of one module affects only the workstation connected to it; however, this configuration is also a more expensive solution.

A critical application may demand the highest level of fault tolerance. This can be achieved by connecting the workstation to two LAN modules at the hub with separate links. The ultimate in fault tolerance can be achieved by connecting one of those links to a different hub. In this arrangement, a transceiver is used to split the links from the application's host computer, enabling each link to connect with a different module in the hub or to a different hub. All of these levels of fault tolerance are summarized in Exhibit 1–5–1.

Ring Topology

In its pure form, the ring topology offers poor reliability to both node and link failures. The ring uses link segments to connect adjacent nodes. Each node is actively involved in the transmissions of other nodes through token passing. The token is received by each node and passed on to the adjacent node. The loss of a link not only results in the loss of a node but brings down the entire network as well. Improvement of the reliability of the ring topology requires adding redundant links between nodes as well as bypass circuitry. Adding such components, however, makes the ring topology less cost-effective.

Bus Topology

The bus topology also provides poor reliability. If the link fails, that entire segment of the network is rendered useless. If a node fails, on the other hand, the rest of the network continues to operate. A redundant link for each segment increases the reliability of the bus topology but at extra cost.

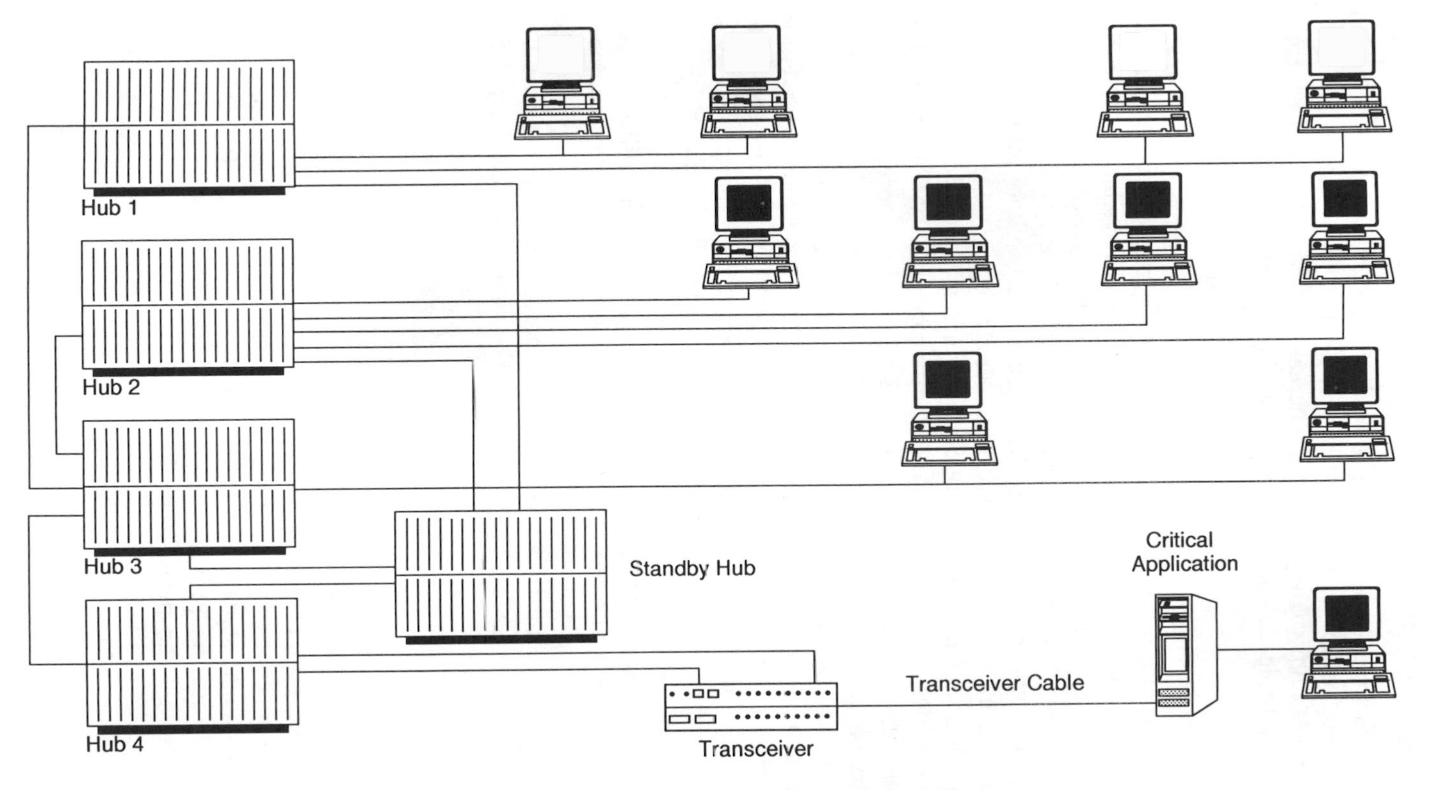

Exhibit 1-5-1. Fault Tolerance of the Hub Architecture

NETWORK AVAILABILITY

Availability is a measure of performance dealing with the LAN's ability to support all users who wish to access it. A network that is highly available provides services immediately to users, whereas a network that suffers from low availability typically forces users to wait for access.

Component Availability

Availability on the bus topology depends on the load, the access control protocol used, and length of the bus. With a light load, availability is virtually ensured for any user who wishes to access the network. As the load increases, however, so does the chance of collisions. When a collision occurs, the transmitting nodes back off and try again after a short interval. The chance of collisions also increases with bus length.

With its multiple paths, a mesh topology, which is a variation of the bus topology, provides the highest degree of interconnectivity, which implies that the network is always available to users who require access.

A network based on a star topology can only support what the central hub can handle. In any case, the hub's LAN module can handle only one request at a time, which can shut out many users under heavy load conditions. Hubs equipped with multiple processors and LAN modules can alleviate this situation somewhat, but even with multiple processors, there is not usually a one-to-one correspondence between users and processors. Such a system would be cost-prohibitive.

The ring topology does not provide the same degree of availability as does a mesh topology but still represents an improvement over the star topology. The ring has a lower measure of availability than the mesh topology because each node on the ring must wait for the token before transmitting data. As the number of nodes on the ring increases, the time interval allotted for transmission decreases.

METHODS OF PROTECTION

In today's distributed computing environments, with so much information traversing public and private networks, network managers must be acquainted with the available protection methods to ensure uninterrupted data flow and guard against data loss. On a wide area network (WAN), the choices include carrier-provided redundancy and protection services, customer-controlled reconfiguration, bandwidth on demand using ISDN, and dial backup. On the LAN, the choices include various recovery and reconfiguration procedures, the use of fault-tolerant servers and wiring hubs, and the implementation of redundant arrays of inexpensive disks. All these methods are discussed in detail in the following sections.

Tariffed Redundancy and Protection

Among the traditional methods for protecting WAN facilities are the tariffed redundancy and protection services offered by such interexchange carriers as AT&T, MCI Communications Corp., and US Sprint Communications Co.

A reliable method for minimizing downtime on the WAN is to have redundant lines ready and waiting. When a link goes down, the standby facility can be activated until the source of the failure is determined and appropriate action taken to restore service. Having duplicate facilities is a prohibitively expensive option for most businesses because monthly charges accrue whether or not the facilities are used.

To minimize the effects of failed facilities on the same route, AT&T, for example, offers two special routing methods in conjunction with its digital and analog service offerings: diversity and avoidance.

Diversity. Diversity is available for ACCUNET T1.5, ACCUNET Spectrum of Digital Services (ASDS), 56k-bps Dataphone Digital Service (DDS), and voicegrade private lines. With diversity routing, designated groups of interoffice channels (i.e., AT&T's portion of the circuit) are furnished over physically separate routes. Each route entails installation and routing charges. A custom option for diversity furnishes the interoffice channels partially or entirely over physically separated routes when separate facilities are available. In this case, AT&T applies a special routing charge to each channel.

Avoidance. The avoidance option allows the customer to have a channel avoid a specified geographical area. The customer minimizes potential impairments, such as delay, that might be exacerbated by long, circuitous routes. It also enables the customer to avoid potential points of congestion in high-use corridors, which can block traffic. This option also gives customers the means to avoid high-risk environments that can be prone to damage from floods, earthquakes, and hurricanes.

Further Protective Capabilities

Although special routing can minimize the damage resulting from failed facilities by allowing some channels to remain available to handle priority traffic, special routing makes no provision for restoring failed facilities. AT&T has attempted to address this issue with its automatic protection capability and network protection capability.

Automatic Protection Capability. Automatic protection capability is an office function that protects against failure for a local channel or other access for the ACCUNET T1.5 and ACCUNET T45 services. Protection of interoffice channels is provided on a one-to-one basis through the use of a switching arrangement that automatically switches to the spare channel when the working channel fails. To implement this capability, a separate local access channel must be ordered to serve as the spare, and compatible automatic switching equipment must be provided by the customer at its premises.

Network Protection Capability. Whereas AT&T's automatic protection capability guards against the failure of a local access channel, its network protection capability is designed to guard against the failure of an interoffice channel. Protection is furnished through the use of a switching arrangement that automatically switches the customer's channel to a separately routed fiber-optic channel on failure of the primary channel.

For both ACCUNET T1.5 and ACCUNET T45, an installation charge is incurred for the network protection capability. For the amount of protection promised, however, it may not be worth the cost, because most, if not all, of AT&T's interoffice channels are automatically protected, whether or not they use the network protection capability. When AT&T circuits go down, traffic is automatically switched to alternative routes.

Dial Backup

Over the years, dial backup units have come into widespread use for rerouting modem and digital data set transmissions around failed facilities. Dial backup units are certainly more economical than leasing redundant facilities or opting for reserved service or satellite-sharing arrangements.

This method entails installing a standalone device or an optional modem card that allows data communication to be temporarily transferred to the public switched network. When the primary line fails, operation over the dial backup network can be manually or automatically initiated. At the remote site, the calls are answered automatically by the dial backup unit. When the handshake and security sequence are completed and the dial backup connection is established, the flow of data resumes. On recovery of the failed line, dial backup is terminated in one of two ways: a central site attendant manually releases the backup switch on the dial backup unit, or, when in the automatic mode, the dial backup unit reestablishes the leased line connection and disconnects the dial network call upon detection of acceptable signal quality.

Rerouting on T1 Lines

To support applications requiring a full T1, dial backup over the public switched network is available with AT&T's ACCUNET T1.5 reserved service. With this service, a dedicated T1 facility is brought online after the customer requests it with a phone call. An hour or more may elapse before the reserved line is finally cut over by AT&T. This option may be acceptable under certain conditions, however, as a possible alternative to the loss of network availability for an indeterminate period. This is an effective alternative for the customer who has resubscribed to the ACCUNET T1.5 reserved service. If the customer has not resubscribed, this service is not a suitable alternative for routing traffic around failed facilities, if only because the local access facilities must already be in place at each end of the circuit.

Customer-Controlled Reconfiguration

Management capabilities, such as customer-controlled reconfiguration (CCR) available using AT&T's Digital Access and Crossconnect System (DACS), can be a means to route around failed facilities. Briefly, the DACS is a routing device; it is not a switch that can be used for setting up calls (i.e., a PBX switch) or for performing alternate routing (i.e., a multiplexer switch). The DACS was originally designed to automate the process of circuit provisioning. With customer-controlled reconfiguration, circuit provisioning is under user control from an on-premises management terminal.

With customer-controlled reconfiguration, however, a failed facility may take a half hour or more to recover, depending on the complexity of the reconfiguration. This relatively long period is necessary because the carrier needs time to establish the paths specified by the subscriber through use of a dial-up connection.

A recovery time of 30 minutes may seem tolerable for voice traffic, in which the public switched network itself is a backup vehicle, but data subscribers may need to implement alternate routing more quickly. Therefore, AT&T's DACS and customer-controlled reconfiguration service, and the similar offerings of other carriers, are typically used to remedy a long-term failure rather than to rapidly restore service on failed lines.

ISDN Facilities

T1 multiplexers offer many more functions than does DACS with customer-controlled reconfiguration. In fact, the instantaneous restoral of high-capacity facilities on today's global networks calls for a T1 networking multiplexer with an advanced transport management system.

An ISDN-equipped T1 networking multiplexer offers yet another efficient and economical means to back up T1 and fractional T1 facilities. With ISDN, the typical time required for call setup is from 3 to 10 seconds. An appropriately equipped T1 multiplexer permits traffic to be rerouted from a failing T1 line to an ISDN facility in a matter of seconds rather than hours or days, as is required by other recovery methods.

Using ISDN facilities is more economical than other methods, including ACCUNET T1.5 reserved service. With the reserved service, the user pays a flat fee for a dedicated interoffice circuit over a certain length of time, including the time when the circuit remains unused. With ISDN, the user pays for the primary rate local access channels and pays for the interoffice channels only when used because these charges are time and distance dependent—just like ordinary phone calls.

With AT&T's high-capacity H0 (384K bps) and H11 (1.544M bps) ISDN channels, users can avail themselves of the ISDN for backing up fractional or full T1 lines rather than pay for idle lines that may only be used occasionally during recovery. This is accomplished through a T1 multiplexer's capability to implement intelligent automatic rerouting, which ensures the connectivity of critical applications in an information-intensive business environment.

When confronted with congestion or impending circuit failure, the intelligent automatic rerouting system calculates rerouting on the basis of each likely failure. During a failure, the system automatically recalculates optimal routing, based on current network conditions. After restoration, the system again automatically calculates the most effective rerouting, should a second failure occur on the network. In this way, the system is always ready to handle the next emergency.

Because applications require different grades of service to continue operating efficiently during line failures, circuits must be routed to the best path for each application, not just switched to available bandwidth. This ensures that the network continues to support all applications with the best response times.

To avoid service denial during rerouting, voice transmissions can be automatically compressed to use less bandwidth. This can free up enough bandwidth to support all applications, both voice and data.

DDS Dial Backup

Despite carrier claims of 99.5% availability on digital data services (DDS), this seemingly impressive figure still leaves room for 44 hours of annual downtime. This amount of downtime can be very costly, especially to financial institutions, whose daily operations depend heavily on the proper operation of their networks. A large financial services firm, for example, can lose as much as $200 million if its network becomes inoperative for only an hour.

An organization that cannot afford the 44 hours of annual downtime might consider a digital data set with the ability to "heal" interruptions in transmission. Should the primary facility fail, communication can be quickly reestablished over the public switched network by the data set's built-in modem and integral single-call dial back unit.

Sensing loss of energy on the line, the dial-backup unit automatically dials the remote unit, which sets up a connection through the public switched network. Data is then rerouted from the leased facility to the dial-up circuit. If the normal DDS operating rate is 19.2K bps, dial restoral entails a fallback to 9.6K bps. For all other DDS rates—2.4K, 4.8K, and 9.6K bps—the transmission speed remains the same in the dial-backup mode. Downspeeding is not necessary.

While in the dial backup mode, the unit continues to monitor the failed facility for the return of energy, which indicates an active line. Sensing that service has been restored, the unit reestablishes communication over it. The dial-up connection is then dropped.

RECOVERY OPTIONS FOR LANs

The LAN is a data-intensive environment requiring special precautions to safeguard one of the organization's most valuable assets—information.

The procedural aspect of minimizing data loss entails the implementation of manual or automated methods for backing up all data on the LAN to avoid the tedious and costly process of recreating vast amounts of information. The equipment aspect of minimizing data loss entails the use of redundant circuitry as well as components and subsystems that are activated automatically upon the failure of various LAN devices to prevent data loss and maintain network availability.

Recovery and Reconfiguration

In addition to the ability to respond to errors in transmissions by detection and correction, other important aspects of LAN operation are recovery and reconfiguration. Recovery deals with bringing the LAN back to a stable condition after an error, and reconfiguration is the mechanism by which the network is restored to its previous condition after a failure.

LAN reconfigurations involve mechanisms to restore service upon loss of a link or network interface unit. To recover or reconfigure the network after failures or faults requires that the network possess mechanisms to detect that an error or fault has occurred and to determine how to minimize the effect on the system's performance. Generally, these mechanisms provide:

- Performance monitoring.
- Fault location.
- Network management.
- System availability management.
- Configuration management.

These mechanisms work in concert to detect and isolate errors, determine errors' effects on the system, and remedy these errors to bring the network to a stable state with minimal impact on network availability.

Reconfiguration. Reconfiguration is an error-management scheme used to bypass major failures of network components. This process entails detection that an error condition has occurred that cannot be corrected by the usual means. Once it is determined that an error has occurred, its impact on the network is assessed so an appropriate reconfiguration can be formulated and implemented. In this way, normal operations can continue under a new configuration.

Error Detection. Error detection is augmented by logging systems that keep track of failures over a period of time. This information is examined to determine whether trends may adversely affect network performance. This information, for example, might reveal that a particular component is continually causing errors to be inserted onto the network, or the monitoring system might detect that a component on the network has failed.

Configuration Assessment Component. This component uses information about the current system configuration, including connectivity, component placement, paths, flows, and maps information onto the failed component. This information is analyzed to indicate how that particular failure is affecting the system and to isolate the cause of the failure. Once this assessment has been performed, a solution can be worked out and implemented.

The solution may consist of reconfiguring most of the operational processes to avoid the source of the error. The solution determination component examines the configuration and the affected hardware or software components, determines how to move resources around to bring the network back to an operational state or indicates what must be eliminated because of the failure, and identifies network components that must be serviced.

Function Criticality. The determination of the most effective course of action is based on the criticality of keeping certain functions of the network operating and maintaining the resources available to do this. In some environments, nothing can be done to restore service because of device limitations (e.g., lack of redundant subsystems) or the lack of spare bandwidth. In such cases, about all that can be done is to indicate to the servicing agent what must be corrected and keep users informed of the situation.

Once an alternate configuration has been determined, the reconfiguration system implements it. In most cases, this means rerouting transmissions, moving and restarting processes from failed devices, and reinitializing software that has failed because of some intermittent error condition. In some cases, however, nothing may need to be done except notify affected users that the failure is not severe enough to warrant system reconfiguration.

For WANs, connections among LANs may be accomplished over leased lines with a variety of devices, typically bridges and routers. An advantage of using routers for this purpose is that they permit the building of large mesh networks. With mesh networks, the routers can steer traffic around points of congestion or failure and balance the traffic load across the remaining links.

Restoral Capabilities of LAN Servers

Sharing resources distributed over the LAN can better protect users against the loss of information and unnecessary downtime than a network with all of its resources centralized at a single location. The vehicle for resource sharing is the server, which constitutes the heart of the LAN. The server gives the LAN its features, including those for security and data protection, as well as those for network management and resource accounting.

Types of Servers. The server determines the friendliness of the user interface and governs the number of users that share the network at one time. It resides in one or more networking cards that are typically added to microcomputers or workstations and may vary in processing power and memory capacity. However, servers are programs that provide services more than they are specific pieces of hardware. In addition, various types of servers are designed to share limited LAN resources—for example, laser printers, hard disks, and the RAM mass memory. More impressive than the actual shared hardware are the functions provided by servers. Aside from file servers and communications servers, there are image and fax servers, electronic mail servers, printer servers, SQL servers, and a variety of other specialized servers, including those for videoconferencing over the LAN.

The addition of multiple special-purpose servers provides the capability, connectivity, and processing power not provided by the network operating system and file server alone. A single multiprocessor server combined with a network operating system designed to exploit its capabilities, such as UNIX, provides enough throughput to support 5 to 10 times the number of users and applications as a microcomputer that is used as a server. New bus and cache designs make it possible for the server to make full use of several processors at once, without the usual performance bottlenecks that slow application speed.

Server Characteristics. Distributing resources in this way minimizes the disruption to productivity that would result if all the resources were centralized and a failure were to occur. Moreover, the use of such specialized devices as servers permits the integration of diagnostic and maintenance capabilities not found in general-purpose microcomputers. Among these capabilities are error detection and correction, soft-controlled error detection and correction, and automatic shutdown in case of catastrophic error. Some servers include integral management functions (e.g., remote console management). The multiprocessing capabilities of specialized servers provide the power necessary to support the system overhead that all these sophisticated capabilities require.

Aside from physical faults on the network, there are various causes for lost or erroneous data. A software failure on the host, for example, can cause write errors to the user or server disk. Application software errors may generate inaccurate values, or faults, on the disk itself. Power surges can corrupt data and application programs, and power outages can shut down sessions, wiping out data that has not yet been written to disk. Viruses and worms that are brought into the LAN from external bulletin boards, shareware, and careless user uploads are another concern. User mistakes can also introduce errors into data or eliminate text. Entire adherence to security procedures are usually sufficient to minimize most of these problems, but they do not eliminate the need for backup and archival storage.

Backup Procedures. Many organizations follow traditional file backup procedures that can be implemented across the LAN. Some of these procedures include performing file backups at night—full backups if possible, incremental backups otherwise. Archival backups of all disk drives are typically done at least monthly; multiple daily saves of critical data bases may be warranted in some cases. The more data users already have stored on their hard disks, the longer it takes to save. For this reason, LAN managers encourage users to offload unneeded files and consolidate file fragments with utility software to conserve disk space, as well as to improve overall system performance during backups. Some LAN managers have installed automatic archiving facilities that move files from users' hard disks to a backup data base if they have not been opened in the past 90 days.

Retrieving files from archival storage is typically not an easy matter; users forget file names, the date the file was backed up, or in which directory the file was originally stored. In the future, users can expect to see intelligent file backup servers that permit files to be identified by textual content. Graphics files, too, are retrieved without having the name, backup date, or location of the file. In this case, the intelligent file backup system compares the files with bit patterns from a sample graphic with the bit patterns of archived files to locate the right file for retrieval.

As the amount of stored information increases, there is the need for LAN backup systems that address such strategic concerns as tape administration, disaster recovery, and the automatic movement of files up and down a hierarchy of network storage devices. Such capabilities are currently available and are referred to as system storage management or hierarchical storage management.

Levels of Fault Tolerance

Protecting data at the server has become a critical concern for most network managers; after all, a failure at the server can result in lost or destroyed data. Considering that some servers are capable of holding vast quantities of data in the gigabyte range, loss or damage can have disastrous consequences for an information-intensive organization.

Depending on the level of fault tolerance desired and the price the organization is willing to pay, the server may be configured in several ways: unmirrored, mirrored, or duplexed.

Unmirrored Servers. An unmirrored server configuration entails the use of one disk drive and one disk channel, which includes the controller, a power supply, and interface cabling, as shown in Exhibit 1–5–2. This is the basic configuration of most servers. The advantage is chiefly one of cost: the user pays only for one disk and disk channel. The disadvantage of this configuration is that a failure in either the drive or anywhere on the disk channel could cause temporary or permanent loss of the stored data.

Mirrored Servers. The mirrored server configuration entails the use of two hard disks of similar size. There is also a single disk channel over which the two disks can be mirrored together, as shown in Exhibit 1–5–3. In this configuration, all data written to one disk is then automatically copied onto the other disk. If one of the disks fails, the other takes over, thus protecting the data and ensuring all users have access to the data. The server's operation system issues an alarm notifying the network manager that one of the mirrored disks is in need of replacement.

The disadvantage of this configuration is that both disks use the same channel and controller. If a failure occurs on the channel or controller, both disks become inoperative. Because the same disk channel and controller are shared, the writes to the disks must be performed sequentially—that is, after the write is made to one disk, a write is made to the other disk. This can degrade overall server performance under heavy loads.

Disk Duplexing. In disk duplexing, multiple disk drives are installed with separate disk channels for each set of drives, as shown in Exhibit 1–5–4. If a malfunction occurs anywhere along a disk channel, normal operation continues on the remaining channel and drives. Because each disk uses a separate

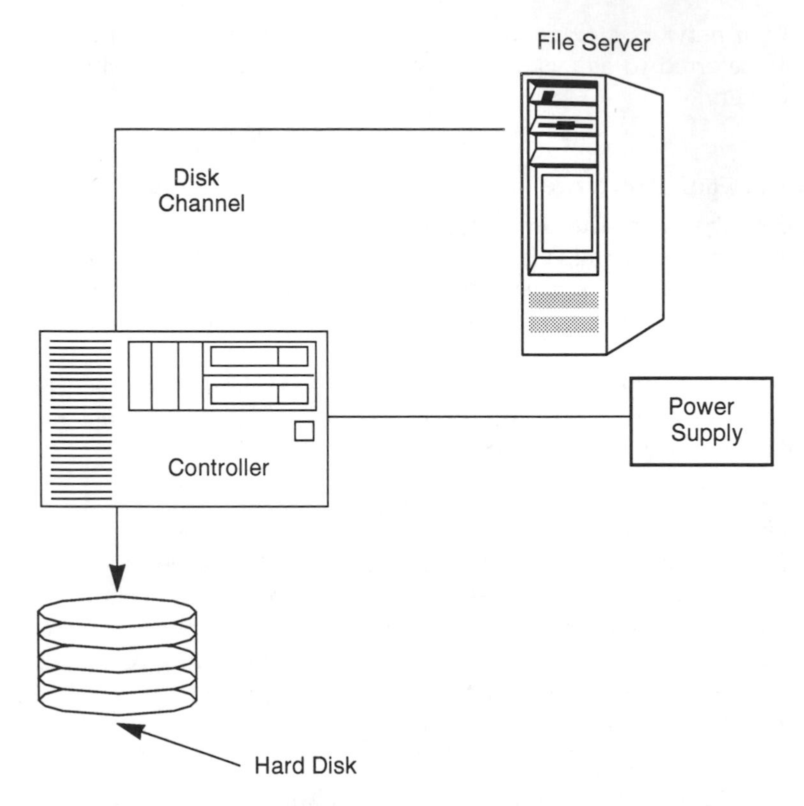

Exhibit 1-5-2. Unmirrored Disk Drive Configuration

disk channel, write operations are performed simultaneously, offering a performance advantage over servers using disk mirroring.

Disk duplexing also offers a performance advantage in read operations. Read requests are given to both drives. The drive that is closest to the information responds and answers the request. The second request given to the other drive is canceled. In addition, the duplexed disks share multiple read requests for concurrent access.

The disadvantage of disk duplexing is the extra cost for multiple disk drives, also required for disk mirroring, as well as for the additional disk channels and controller hardware. However, the added cost for these components must be weighed against the replacement cost of lost information plus costs that accrue from the interruption of critical operations and lost business opportunities. Faced with these consequences, an organization might discover that the investment of a few hundred or even a few thousand dollars to safeguard valuable data is negligible.

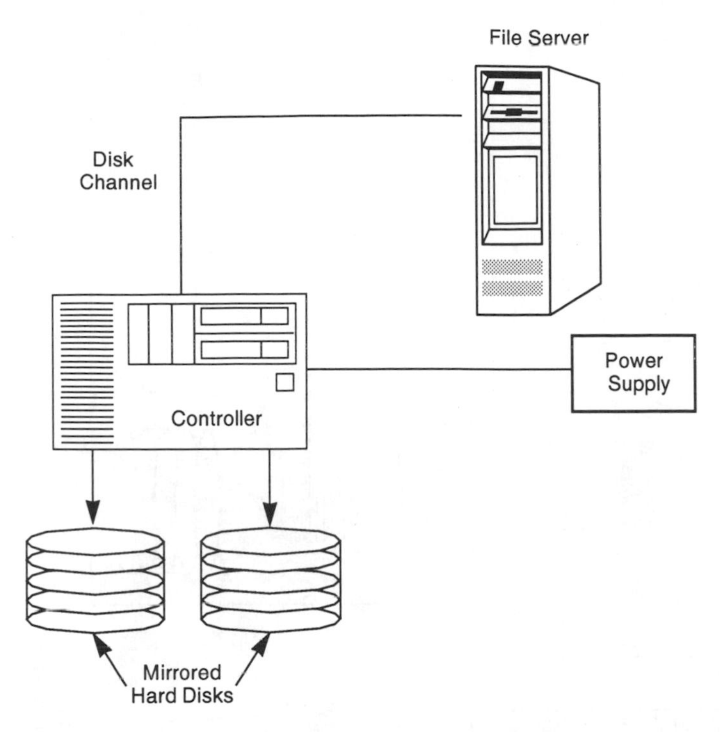

Exhibit 1-5-3. Configuration for Disk Mirroring

REDUNDANT ARRAYS OF INEXPENSIVE DISKS

One method of data protection is growing in popularity: redundant arrays of inexpensive disks (RAID). Instead of risking all of its data on one high-capacity disk, the organization distributes the data across multiple smaller disks, offering protection from a crash that could wipe out all data on a single, shared disk. Exhibit 1–5–5 illustrates redundant arrays of inexpensive disks. Other benefits of RAID include:

- Increased storage capacity per logical disk volume.
- High data transfer or input/output rates that improve information throughput.
- Lower cost per megabyte of storage.
- Improved use of data center floor space.

RAID products can be grouped into the categories described in the following sections.

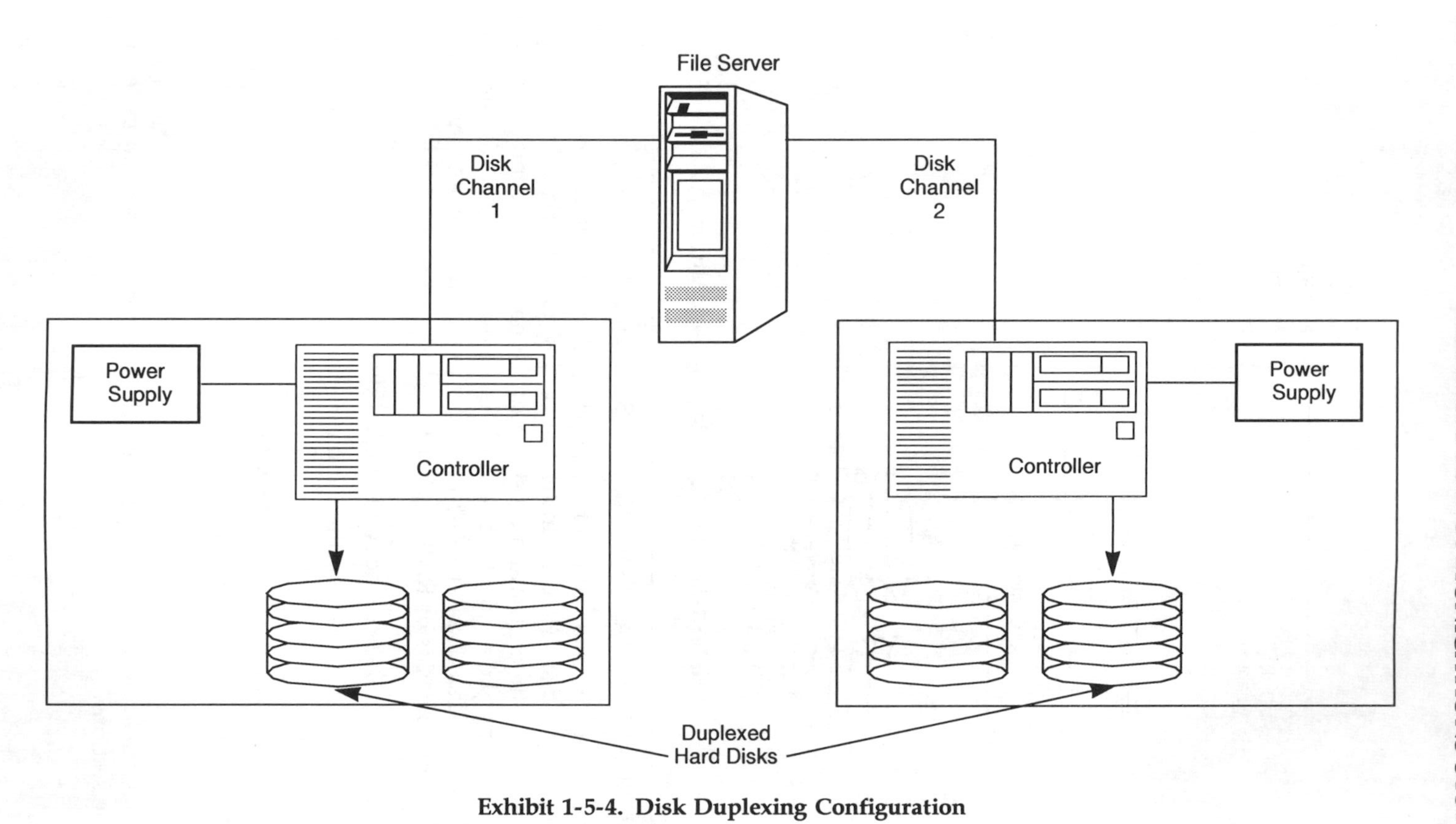

Exhibit 1-5-4. Disk Duplexing Configuration

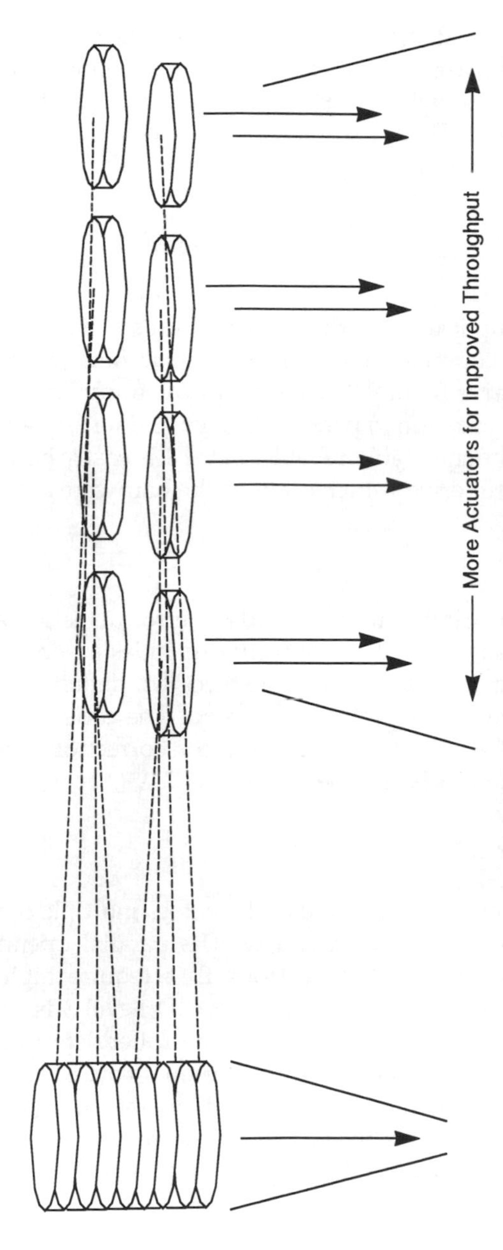

Exhibit 1-5-5. Redundant Arrays of Inexpensive Disks

RAID Level 0

Technically, these products are not RAID products at all, because they do not offer parity or error-correction data to provide redundancy in the event of system failure. Although data striping is performed, it is accomplished without fault tolerance. Data is simply striped block-by-block across all the drives in the array. There is no way to reconstruct data if one of the drives fails.

RAID Level 1

These products duplicate data that is stored on separate disk drives. Also called mirroring, this approach ensures that critical files are available in case of individual disk drive failures. Each disk in the array has a corresponding mirror disk, and the pairs run in parallel. Blocks of data are sent to both disks simultaneously. Although highly reliable, Level 1 is costly because each drive requires its own mirror drive, which doubles the hardware cost of the system.

RAID Level 2

These products distribute the code used for error detection and correction across additional disk drives. The controller includes an error-correction algorithm, which enables the array to reconstruct lost data if a single disk fails. As a result, no expensive mirroring is required. The code, however, requires that multiple disks be set aside to do the error-correction function. Data is sent to the array one disk at a time.

RAID Level 3

These products store user data in parallel across multiple disks. The entire array functions as one large, logical drive. Its parallel operation is ideally suited to supporting imaging applications that require high data-transfer rates when reading and writing large files. RAID Level 3 is configured with one parity (i.e., error-correction) drive. The controller determines which disk has failed by using additional check information recorded at the end of each sector. However, because the drives do not operate independently, every time an image file must be retrieved, all the drives in the array are used to fulfill that request. Other users are put into a queue.

RAID Level 4

These products store and retrieve data using independent writes and reads to several drives. Error-correction data is stored on a dedicated parity drive. In RAID Level 4, data striping is accomplished in sectors rather than bytes or blocks. Sector-striping offers parallel operation in that reads can be performed simultaneously on independent drives, which allows multiple users to retrieve image files at the same time. Although multiple reads are possible, multiple writes are not because the parity drive must be read and written to for each write operation.

RAID Level 5

These products interleave user data and parity data, which are then distributed across several disks. Because data and parity codes are striped across all the drives, there is no need for a dedicated parity drive. This configuration is suited for applications that require a high number of input/output operations per second, such as transaction processing tasks that involve writing and reading large numbers of small data blocks at random disk locations. Multiple writes to each disk group are possible because write operations do not have to access a single common parity drive.

RAID Level 6

These products improve reliability by implementing drive mirroring at the block level so data is mirrored on two drives instead of one. Up to two drives in the five-drive disk array can fail without loss of data. If a drive in the array fails with RAID Level 5, for instance, data must be rebuilt from the parity information spanned across the drives. With RAID Level 6, however, the data is simply read from the mirrored copy of the blocks found on the various striped drives. No rebuilding is required. Although this results in a slight performance advantage, it requires at least 50% more disk capacity to implement.

Vendors continually tout the effectiveness of various RAID solutions. In truth, the choice among RAID solutions involves tradeoffs between cost, performance, and reliability. Rarely can all of these requirements be satisfied simultaneously, especially when trying to address high-availability, large-scale storage needs.

OTHER CONSIDERATIONS

As more businesses interconnect their computers at remote locations and run critical applications over WANs, they are discovering that financial and operational losses can mount quickly in the event of internetwork downtime. Businesses of all types and sizes are recognizing that disaster recovery plans are essential, regardless of the particular computing environment. The disaster recovery plan should be a formal document that has been signed off by senior management, IS management, and all department heads. The following items should be addressed in any disaster recovery plan.

Uninterruptible Power Supplies (UPSs)

UPSs are designed to provide temporary power so attached computer systems and servers can be shut down properly to prevent data loss. UPSs are especially important in WANs. Because of the distance among links, sometimes reaching thousands of miles, WANs are more susceptible to power problems than LAN segments. Therefore, using battery backups to protect against fluctuations and outages should always be the first line of defense.

Although most central sites have UPSs, many remote sites typically do not, usually as a cost-savings measure. However, battery backup can be very inexpensive, costing only a few hundred dollars, which is cheap compared to the cost of indeterminate network downtime. Moreover, some UPSs have simple network management protocol (SNMP) capabilities, which lets network managers monitor battery backup from the central management console. For instance, every UPS can be instructed using SNMP to test itself once a week and report back if the test fails. The network manager can even be notified if the temperature levels in wiring closets rise above established thresholds.

Generators

To keep computers operating during a prolonged loss of power, a generator is required. A generator is capable of supplying much more power for longer periods of time. Using a fuel source, such as oil, a generator can supply power indefinitely to keep data centers cool and computers running. Because generators can cost tens of thousands of dollars, many companies unwisely decide to skip this important component of the disaster recovery plan.

Unless an organization has experienced a lengthy outage that has disrupted daily business operations, this level of protection is often hard to justify. However, many office buildings already have generators to power lighting and elevators during electrical outages. For a fee, tenants can patch into the generator to keep data centers and networks operating.

Off-Site Storage

Mission-critical data should be backed up daily or weekly and stored off site. There are numerous services that provide off-site storage, often in combination with hierarchical storage management techniques. In the IBM environment, for example, this might entail storing frequently used data on a direct access storage device (DASD) for immediate usage, whereas data used only occasionally might go to optical drives, and data that has not been used in several months would be archived to a tape library.

Carriers, computer vendors, and third-party firms offer vault storage for secure, off-site data storage of critical applications. Small companies need not employ such elaborate methods. They can back up their own data and have it delivered by overnight courier for storage at another company location or bring it to a bank safety deposit box. The typical bank vault can survive even a direct hit by a tornado.

In addition to backing up critical data, it is advisable to register all applications software with the manufacturer and keep the original program disks in a safe place at a different location. This minimizes the possibility of both copies being destroyed in the same catastrophe. Software licenses, manuals, and supplementary documentation should also be protected.

Surge Suppressors

In storm-prone areas like the Southeast, frequent electrical storms can put sudden bursts of electricity, called spikes or surges, on telephone lines. These bursts can destroy router links and cause adapters and modems to fail. To protect equipment attached to telephone lines, surge-suppression devices can be installed between the telephone line and the communications device. Surge suppressors condition the power lines to ensure a constant voltage level. Many modems and other network devices have surge suppressors built in. The disaster recovery plan should specify the use of surge suppressors whenever possible, and equipment should be checked periodically to ensure proper operation.

Spare Parts Pooling

Most companies can afford to stockpile spare cables and cards but not spare multiplexer and router components that are typically too expensive to inventory. Pooling these items with another area business that uses the same equipment can be an economical form of protection should disaster strike. Such businesses can be identified through user group and association meetings. The equipment vendor is another source for this information.

After each party becomes familiar with the disaster recovery needs of the other, an agreement can be drawn up to pledge mutual assistance. Each party stocks half the necessary spare parts. The pool is drawn from as needed and restocked after the faulty parts come back from the vendors' repair facilities.

Switched Digital Services

Carriers offer an economical form of disaster protection with their networks of digital switches. When one link goes down, voice and data calls are automatically rerouted or switched to other links on the carrier's network. Examples of switched digital services are switched 56K bps, ISDN, and frame relay. Many routers now offer interfaces for switched digital services, allowing data to take any available path on the network. The same level of protection is available on private networks, but it requires spare lines, which is often a very expensive solution.

Multiple WAN Ports

A well-planned internetworking system avoids single points of failure. This entails equipping nodes with redundant subsystems, such as power, control logic, network cards, and WAN ports. Routers, for instance, need multiple WAN ports so if a primary line goes down, the router can automatically use the backup line on another port.

Even branch sites with remote-access routers should have multiple WAN ports. If the first line goes down, the remote router is programmed to autodial into a second line on another port, which remains inactive until needed. The second port can dial up a switched 56K line that is paid for on a usage basis.

Links to Remote Sites

WANs of three or more sites often link the remote locations to the primary site but not to each other. Though this strategy saves linkage costs, it risks leaving remote workers stranded should the main office's services go down. To keep branches up and running, inexpensive links should be established among them. The links' bandwidth should be adequate to keep critical systems communicating.

As long as backup circuits are available, routers can run a routing protocol that understands link states and can reroute around points of network failure. The routing information protocol (RIP), often used in smaller WANs, does not support link states, but the open shortest path first (OSPF) protocol does. OSPF runs on TPC/IP networks and is the protocol of choice for larger internetworks.

Periodic Testing

It is advisable to test the disaster recovery plan periodically to check assumptions and to find out whether the plan really works. After giving users advance notice, the network manager can come in after business hours, unplug one of the communication links, and see what happens. If something unexpected occurs, it is necessary to fine-tune the disaster recovery plan and test again.

With certain types of network equipment, such as multiplexers and switches, several disaster-simulation scenarios can be programmed in advance and stored for emergency implementation. With the integral network modeling capability of some T1 multiplexers, network planners can simulate various disaster scenarios on an aggregate or node level anywhere in the network. This offline simulation allows planners to test and monitor changing conditions and determine their precise impact on network operations.

Any outage should be treated as an unannounced test of the disaster recovery plan. Network managers should determine if the response was adequate and if the response can be improved.

Worst-Case Scenarios

Planning how to provide communications connectivity based on the assumption that the entire network is inoperable is a sound business practice, especially for organizations in areas of the country that can suffer widespread damage from hurricanes, tornados, floods, and earthquakes. Whole nodes may have to be replaced to get the network back into proper operation, requiring advance arrangements with suppliers so the necessary equipment can be obtained on very short notice instead of when it comes off a production run.

Many vendors of switches, hubs, routers, and multiplexers offer network recovery services at a reasonable cost and guarantee equipment delivery within 24 hours of the request. Even carriers offer disaster recovery services. AT&T Global Information Solutions, for example, offers crisis management services designed to allow customers to occupy a regional AT&T crisis center within two to four hours after a disaster. The customer gets fully restored voice and data communications and computer-ready floor space.

Some companies specialize in disaster recovery services. SunGard Recovery Services Inc., for instance, offers customers a hotsite backup center for mainframe computers. The company also addresses the needs of PC-based environments and offers PC software that helps identify which services are critical to operations. Its overnight assistance program rushes equipment—PCs, servers, and modems, as well as bridges, multiplexers, and routers—to any customer site. As part of its mobile recovery program, SunGard maintains a fleet of trailers that are ready for dispatch to customer sites for use as temporary work space.

Training

Often overlooked in the disaster recovery plan are provisions for training, essential because in an emergency reactions to impending disaster must be automatic. End users must save data and shut down applications at the first sign of trouble. Network managers must be able to assess quickly the criticality of numerous alarms and respond appropriately. Help desk operators must be able to determine the nature and magnitude of the problem to address end user concerns. LAN administrators must be able to determine the impact of the problem on local networks.

Insurance

The disaster recovery plan should provide for the periodic review of the organization's insurance policy. Of particular concern is to what extent information systems, network components, and applications software are covered in the event of a disaster. The policy should be very specific about what the insurance company does and does not cover. Ambiguities must be resolved before a disaster occurs, not after.

Reviews also provide an opportunity to add provisions for system or network upgrades and expansions that occurred in the time since the policy was initially written. Though this may add to the policy's cost, it is still much less compared to the wholesale replacement of equipment that falls outside of the insurance contract.

Risk Assessment

The disaster recovery plan should include a review of the physical layout of data centers, wiring closets, floor and overhead conduits, and individual offices that contain computers and data communications equipment. All equipment and cabling should be kept in a relatively safe place and not exposed to objects that are likely to fall and cause damage should disaster strike. Desks, tables, shelves, and cabinets should be solid enough to safely fasten equipment to them, to survive minor earthquakes. Whenever possible, equipment should not be positioned under water sprinklers. These and other precautions can even lower insurance costs.

SUMMARY

The methods mentioned in this chapter are only a few of the many network restoral and data protection options available from carriers, equipment vendors, and third-party firms. As more businesses become aware of the strategic value of their networks, these capabilities are significantly important to organizations. A variety of link-restoral and data-protection methods can provide effective ways to meet the diverse requirements of today's information-intensive corporations that are both efficient and economical. Choosing wisely from among the available alternatives ensures that companies are not forced into making cost/benefit trade-offs that jeopardize their information networks and, ultimately, their competitive positions.

Section 2
Business Management Issues

THE *HANDBOOK OF COMMUNICATIONS SYSTEMS MANAGEMENT* has always emphasized the importance of business management skills to the communications systems manager. In this era of exploding technology and rapidly changing user demands framed by corporate restructuring, the business aspects of the manager's job can assume an even higher level of importance and visibility. Some observers argue that the task of managing communication systems and networks is 50% technical and 50% business.

This increase in the business side of communications management is due simply to the fact that a network is a business. Corporate communication facilities, whether referred to as a network, an intranet, or an internetwork, whether confined to local environments or encompassing global elements, is a service business with suppliers and customers. As an agency of the enterprise it serves, it may not be required to earn a profit, but it certainly is not expected to be a massive drain on revenues. As with any business, the manager's task is to perceive user needs, develop a product to meet that need, "sell" the product to the customers (i.e., users), and then maintain and support that product.

The manager is also often expected to grow and expand the business. It is in this role that the manager must recognize the strategic role of the communications system. Many managers think of communications networks and the associated systems only in the tactical sense. In the tactical role, the network manager focuses almost exclusively on the day-to-day operations of the system. Emphasis is on meeting user demands for service, costs, and performance. There is little time for exploring alternatives. The tactical role is reactive in nature—tasks are tilted in the technical direction.

The business-oriented manger, on the other hand, also has the skills to recognize the potential strategic value of the network infrastructure. In the strategic mode, the manager is constantly thinking of ways to use the network and systems to increase enterprise revenue or to create an advantage over the competition. The corporate network, for example, might be used to add value to an existing product. AT&T did something similar when they added telephone calling card numbers to their universal credit cards, thus expanding the potential for revenue. It may be possible to use the corporate network infrastructure to support or improve customer service by linking customers to help desks associated more directly with their product, even if the customers are in their offices attached to a local network.

Another example might be to use the network to combine communications capability with an existing product to create an entirely new product. There is also the possibility of selling any excess bandwidth on the corporate network. In this case, the corporation becomes a carrier (i.e., a transmission service provider), which creates all sorts of value-added possibilities.

The idea here is not to suggest specifics, but to urge the communications system manager to set aside some quiet time to think about innovative ways to use the communications facilities to the greater benefit of the organization. It might also be advantageous to gather members of the staff for a brainstorming session on the subject.

Section 2 of the handbook further explores of some interesting communication business-related issues. The first of these issues deals with that now ubiquitous concatenation of networks called the Internet. Chapter 2-1, "Business Trends on the Internet," discusses the use of the Internet as a strategic business tool. It provides examples of the three major classes of businesses linked to Internet services. These are the Internet service providers (ISPs), the software developers best known for products that facilitate locating information, and those that provide and publish the information that may be found on the Internet. Important issues of retailing, marketing, and security are also examined.

Business continuity is the result of efficient disaster recovery planning. "Strategies for Developing and Testing Business Continuity Plans," Chapter 2-2, is an excellent in-depth analysis of disaster recovery. Beginning with a model for an effective program, it details the methodology of developing and implementing a plan. Many useful checklists are also included.

Communications managers, recognizing the need to address strategic business issues, are likely to find that they are assuming additional duties and responsibilities. Identifying and assessing the risks associated with this expanded role is the subject of Chapter 2-3, "Corporate Lessons for the Communications Systems Manager." Issues associated with the changing managerial role are identified, and guidelines for understanding and coping with these issues are delineated.

Employees are an organization's most valuable resource, yet it is amazing how casually managers sometimes approach the hiring process. Studies indicate that managers most often select a new employee based on personal chemistry. Hiring a new employee is an expensive process, but making a mistake and selecting the wrong candidate is at least twice as expensive. "Hiring and Retaining Valued Employees," Chapter 2-4, tries to overcome the tendency toward an unstructured hiring process by providing an organized approach to defining the job and the work environment. This is followed by detailed techniques and checklists on organizing and conducting the interview. Because it is often better and usually cheaper to retain a good employee than to replace one, this chapter also offers guidance on this aspect of the issue.

Another aspect of the business side of communications systems management is the need to allocate network costs to the users with some form of chargeback system. Chapter 2-5, "Cost Allocation for Communications Networks," looks at this question. The communications manager must first identify all chargeable network resources. In addition to the obvious, such as leased-line costs and modems, these resources include cabling, servers, storage, and Internet access. Once the resources have been identified, their fixed and variable cost components can be determined as the basis for establishing a chargeback formula.

In this turbulent period of rapidly changing technology, corporate restructuring, outsourcing, and downsizing, it is not surprising that maintaining a high level of professional employee morale can be a daunting task. "Managing Morale When Downsizing," Chapter 2-6, recommends techniques by which management and employees can cooperate in building the open, candid, communications environment that is so essential to an effective organization.

2-1
Business Trends on the Internet

Harvey Golomb

THE INTERNET, THAT VAST COMPUTER NETWORK connecting thousands of computers and millions of users, is rapidly becoming a fertile incubator for new business ventures. Many new businesses are capitalizing on the growth of the Internet itself, and many more businesses use the Internet to improve existing operations or to create new ventures. This chapter describes how the Internet is being used by businesses today and suggests areas where opportunities for future business development lie.

THE INTERNET: AN OVERVIEW

Simply stated, the Internet is a computer network connecting other computer networks as well as individual computers, be they PCs, Macintoshs, workstations, or mainframes. Any computer connected to the Internet can communicate with any other computer on the network, nationwide or globally, independent of platform. This communication can be as simple as electronic mail (E-mail) or as rich as the interactive, multimedia World Wide Web (WWW).

The power of the Internet lies in its being an open network; any computer network willing to conform to the Transmission Control Protocol/Internet Protocol, or TCP/IP, communications convention used by the Internet can connect to it. Because such connections are widely and inexpensively available, there is a virtual stampede among businesses and other organizations to connect to the Internet.

THE INTERNET AS A STRATEGIC BUSINESS TOOL

The Internet's quick transformation to a business tool is reminiscent of the fax machine's leap to prominence in the office in the mid-1980s. Two Internet services—E-mail and the World Wide Web—are already near necessities to effective business operations. Few large companies operate without internal E-mail, and most find it necessary to be connected to the Internet for global E-mail. Likewise, many large companies have WWW sites for marketing and technical support. Small companies and even individuals are adding E-mail and the WWW to their arsenal of business tools so that they can communicate on an equal footing with larger competitors. Companies that do not learn how to use the Internet effectively will be at a major competitive disadvantage in the very near future.

INTERNET BUSINESS USE TODAY

Businesses use the Internet for various forms of electronic communication, marketing, and publishing. In addition to E-mail, communications services include systems for transferring files between computers. The most popular of these is the file transfer protocol, or FTP. A related communications service is the USENET newsgroup system—a huge collection of computerized bulletin boards. Thousands of newsgroups cover every conceivable topic from computers to agriculture to fly fishing.

The Internet is widely used for marketing. The World Wide Web supports such applications as displaying electronic brochures and online product catalogs. Web brochures can include text, full-color photographs and drawings, and audio and video. The WWW also supports electronic publishing, which is now being tested and used by such companies as Time, Inc., The New York Times, and ESPN Television, to name only a few.

Several thousand companies have already established Web sites on the Internet to post current information about themselves and their products. They have addresses like www.ibm.com and www.ford.com (these are actually the addresses for IBM Corp. and Ford Motor Co., respectively). Internet users entering these addresses can learn about companies, place orders for products, request printed information, or even receive such products as software and movie clips immediately through the network.

Many observers believe that by the end of this century, nearly all computers will be connected to the Internet and it will be imperative for most professionals to have E-mail addresses and to be accessible over the Internet. An Internet E-mail address on a business card will be as common as fax numbers are today. Similarly, over the next few years, most companies, from government agencies to nonprofit institutions, are expected to establish Web sites.

INTERNET BUSINESS MODELS

Companies are expanding their use of the Internet as a communications tool to transact business through the interactive, multimedia World Wide

Web. Thus far, three types of businesses have developed around the Internet: Internet access providers, Internet software developers, and Internet publishers. Internet access providers are the largest of the three classes today, but publishing holds the greatest promise for long-term growth and profits.

Internet Access Providers

Internet access providers (IAPs) were the first big businesses to develop around the Internet. As the companies through which businesses and individuals connect to the Internet, the IAPs perform functions similar to those of local phone companies when connecting phone users to the long-distance carriers. IAPs provide the connection from an individual's or business's front door to the worldwide Internet.

The IAP connects its clients to the Internet through a high-speed connection. The TCP/IP packet protocol used by the Internet allows data from many users to be mixed on one high-speed line. Communications processors throughout the Internet route each user's packet and deliver it to the proper destination. IAPs link their clients to the so-called Internet backbone, which is a high-speed network operated by five large communications companies.

IAPs provide two basic types of connections: dial-up access and dedicated access. Dial-up access is typically used by individuals or occasional business users. Subscribers install Internet access software on their computers and dial into their IAP. Typically, they enter a password and are then connected to the Internet. Dial-up connections cost about $25 per month.

A dedicated Internet connection supports many users simultaneously. Typically, a company wanting access for its employees links its local area network or mainframe computer to the IAP using a dedicated phone line. Such a connection costs between $500 and $2,000 per month, depending on many factors, including line speed.

The Internet access business is like the telephone business. In each business, companies must install relatively expensive equipment to accept, process, and route calls or data, but once networks are in place, the cost of processing calls is very low. As volume grows in this high capital cost, low operating cost business, profits increase dramatically. Thus the Internet access business is attractive for any company that can achieve volume and remain competitive.

As the number of companies and individuals rushing to obtain access to the Internet has grown, so too have the revenues of Internet access providers. Many of them have been doubling their revenues and customer bases annually for the past several years. Although such growth rates cannot continue indefinitely, these companies may well build substantial businesses before the rates level off. To support these high growth rates, nearly all the IAPs have been investing heavily; as a result, none has, as yet, reported substantial profits.

There are three types of IAPs: local providers offering access within a small geographic area, national providers offering access in many areas of the country (and in some cases, offering national 800-dial-in service), and commercial online service providers such as America Online, Prodigy, and CompuServe, which now offer their customers Internet access.

Local Internet Access Providers. Local providers exist in all major metropolitan areas and in many smaller towns. They vary in size from very small companies operating out of the metaphorical garage to large publishing companies that have added Internet access to their local newspapers as an additional service to their communities. In the Washington DC area, for example, Capital Area Internet Service is a small, early-stage company providing local access. InfiNet, a part of the much larger Landmark Publishing concern, provides local Internet access in many of the cities in which Landmark publishes newspapers. In smaller towns, local access providers offer users the only option for obtaining Internet access without paying high long-distance telephone charges.

Entrepreneurial opportunities exist to provide local access where it is unavailable, unreliable, or otherwise of poor quality. Once users find a quality access provider, they are not likely to switch to another provider. Business opportunities also lie in merging local providers into larger networks of regional or even national providers.

National Internet Access Providers. Several well-financed and well-managed companies provide national access. Three young companies—UUNET, Performance Systems International (PSI), and NETCOM On-Line Communications—are reported to be eyeing international expansion. Having gone public only in 1994 or 1995, each of the three companies has the substantial financial resources to fund rapid expansion. All three are targeting service to 100 to 200 US cities.

MCI Communications, the long-distance telephone giant, has entered the access business in a substantial way. In addition to providing Internet access services, MCI recently joined with Rupert Murdoch's News Corp. to jointly produce content for the Internet as well as to provide other, international Internet services.

All of the national and international IAPs anticipate achieving substantial economies of scale as their large investments in computers, communications equipment, software, and service organizations are spread over exploding customer bases. If the Internet access business follows the example of the long-distance telephone business, the market can probably support two or three national IAPs. Becoming a national IAP now requires substantial capital investments and the marketing savvy and clout to compete on a national scale with experienced and well-financed competitors. In the long-term, the national IAP business may well become a commodity business with one or two marginally profitable competitors. Entering this segment of the Internet business is best left to the well-heeled and bold.

Online Service Providers. The three major online service providers—America Online (AOL), CompuServe, and Prodigy—support closed networks accessible only to subscribers. Thus these networks differ from the Internet, which is open to anyone with Internet access. They also differ from the national IAPs like UUNET and PSI because they offer their subscribers content as well as access. Although the online services and the IAPs compete for the same subscriber, the competitive advantage of the online services ultimately lies in the quality and value of the content and services they have developed for their proprietary networks. Unlike the IAPs, online service providers have been profitable, although margins have been low.

The big three of the online business are bracing for a potentially bigger fourth contender—Microsoft Corp. With the introduction of Windows 95, Microsoft began offering the Microsoft Network (MSN), which competes directly with the current online services. The outcome of the new MSN was anything but obvious as of this writing. Starting a new online service today is not an attractive proposition . The market is already crowded with experienced, aggressive, well-financed players. Microsoft's entrance into the market is expected to erode the online service providers' already low profit levels.

Internet Software Developers

The Internet runs on more than the wires and modems of the IAPs. It takes software at both ends of the wires and all along the way.

Browser and Server Software. The best-known Internet software is the browser that supports each computer attached to the WWW. The browser allows the user to specify the address of the Web site to be displayed and then displays that information on the screen. In addition to browsers, the Internet requires server software to store, process, and transfer data such as World Wide Web pages, files, and E-mail. All Internet applications require browsers and servers.

The first popular Web browser, Mosaic, was developed by the government-funded National Center for Supercomputing Applications (NCSA) and given away to Internet users. NCSA also developed a server to support Mosaic. As Mosaic grew in popularity, NCSA decided to license it to ensure widespread use with adequate commercial support.

The licensing of Mosaic generated a flurry of activity; numerous companies now operate under the NCSA license (through NCSA's master licensee, Spyglass, Inc.) and market their own enhanced versions of Mosaic. In addition, several companies have developed Web browsers and Web servers independent of the NCSA license. Currently the major players in this market are Netscape Communications, Spyglass, Spry (now a division of CompuServe), Quarterdeck, and America Online and Prodigy, both of whom offer their clients proprietary Web browsers. IBM and Microsoft offer proprietary Web browsers bundled with versions of OS/2 and Windows, respectively. Also in this market are vendors of servers only, including OReilly & Associates, Open Market, Inc., and several data base publishers with enhanced versions of their own data bases that include Web servers.

Netscape has taken a commanding early lead in popularity among users in the market for desktop Web browsers. It also leads the field in commercial-grade server software, particularly where security is an issue. Spry, Spyglass, and Quarterdeck are also establishing reasonable Web browser market shares. All of the competitors are well financed and the markets are expanding rapidly and will probably support several competitors. Issues facing participants in this market include Microsoft's entry with a browser bundled with Windows 95, Netscape's large market share (70%), and the fact that there are six other competitors, several of whom are fast-moving, entrepreneurial software companies. Most of the survivors in this market already deliver product; the unknown is which ones will survive the competition.

Publishing on the Internet

The Internet has tremendous potential as both a low-cost publishing medium and a low-cost distribution system, but these same features are a double-edged sword. Publishers find the market an attractive one to enter, but they are challenged to create defensible businesses with reasonable barriers to competitive entry. Thus, the would-be Internet publisher must establish a competitive advantage such as proprietary information, brand awareness (which is difficult for information products), or some other market edge.

Internet publishers are developing three business models:

- The subscription revenue model.
- The advertising model.
- The unit-pricing model (where users pay for an article or book).

The advertising model is used when the information itself has relatively low value (such as news or sports information) but the Web site attracts a large number of viewers. The unit-pricing model is often used by traditional data base information vendors when the information itself is perceived to be valuable and users are willing to pay for specific facts or reports. Several publishers entered the market without charging any fees but are expected to begin charging after an appropriate product shakedown period.

Several traditional information publishers now deliver information over the Internet. International Data Corp. (IDC) offers numerous online reports and E-flashes for immediate notification. Dialog (owned by Knight-Ridder) delivers information by Internet E-mail but as of this writing did not have interactive Internet access.

One of the most aggressive publishers on the Internet is Individual, Inc., which offers NewsPage, a current awareness service drawing from hundreds of news sources. The user selects topics from a list and then receives the relevant stories, either in summary format or as full text. The service is offered at fixed monthly prices that vary based on the number of sources selected. In addition, some full-text information is billed at a premium. NewsPage covers a broad range of topics and has received a great deal of coverage. Its subscriptions are growing rapidly.

NewsPage also sells advertising on its Web site. Thus users retrieving information can select Ad buttons that give them a full advertisement from the vendor. An interesting wrinkle here is that subscribers are offered a discount if they agree to allow their names to be given to advertisers. Thus far, NewsPage seems to be a good model for low-cost information services on the Internet.

Another type of information being marketed over the Internet is information about the Internet. Several directories list Internet sites, and a number of search tools allow users to describe what they are looking for and receive a list of sites matching that description. Several of these sites are very popular, receiving numerous "hits" or "visits" daily. The best known of them—Yahoo—is one of the busiest sites on the entire Internet. Yahoo plans to continue to offer its search service free of charge, but the company will sell advertising on the site to generate revenue.

Giving news or other information away (or at low cost) and selling advertising is the traditional business model of the newspaper. Hundreds of newspapers are experimenting with information delivery over the Internet using a similar universal model: access to the news is either free or nearly free, and the papers sell advertising. Some of the early players here are the *San Jose Mercury News*, the *Raleigh News and Observer*, the *San Francisco Chronicle* and *San Francisco Examiner*, and a recent entry from the *Wall Street Journal*. Newspapers have a natural constituency as well as a product (i.e., news) ideally suited to the fast pace of the Internet. They also have existing relationships with advertisers, so they may be able to readily generate the needed advertising revenues.

MARKETING PRODUCTS ON THE INTERNET

The beauty of marketing information over the Internet is that information products can be delivered electronically for instant gratification. Physical products can be marketed over the Internet too, but ultimately, they must be delivered by less exotic means. Products of all sorts are sold. From the convenience of a computer, consumers can order flowers, hiking boots, pizza for delivery (if they live close to an Internet-aware pizza parlor), pen sets, and toolkits (the kind with hammers and screwdrivers, not software for sorting and summing). Large companies marketing over the Internet include 1-800 Flowers, Lands End, and Hammacher Schlemmer, among many others.

Some retailers are setting up their own Web sites and attempting to market products themselves. Most, however, put their electronic storefronts (i.e., their electronic catalogs) into electronic malls, just as they put their stores in malls. MCI has a popular mall as does the Home Shopping Network with its Internet Shopping Network.

Unlike the obvious consumer benefits offered by real malls (such as the convenience of having stores close together), electronic malls have yet to prove their value. Because consumers are never more than a few mouse clicks from any site on the Internet, consolidating stores in an electronic mall may not offer many advantages.

Are the Retailers Too Early?

Although there has been much discussion in the popular press about marketing products on the Internet, there are not many success stories. The business model of the printed catalog that offers consumers the benefits of shopping at home has not worked well on the Internet—yet. The Internet is still too slow and most computer screens too small to allow for the kind of convenient browsing consumers do in paper catalogs. Because prices on the Internet are not generally lower than store or print catalog prices, there is no compelling reason to shop on the Internet.

The problem of slow browsing as well as of price may be eliminated over time. Technology will certainly improve, facilitating better graphics and faster response times. As the volume of transactions on the Internet increases, many products may become cheaper than the same items in stores or print catalogs. Merchants do not have to pay for the cost of mailing the catalogs or for expensive retail rent space.

INTERNET SECURITY

Internet users are appropriately concerned about the privacy and security of their Internet transactions. People ordering products, services, and information do not want the nature of their transactions compromised, nor do they want sensitive information, such as bank account and credit card numbers, to be divulged.

The current generation of Internet hardware and software provides good transaction security, and the next generation will be even better. Today's secure browsers and servers allow a merchant to request a credit card number (or other sensitive information) and to have that information encrypted before transmission. Thus the data sent over the phone lines (i.e., over the Internet) is scrambled and cannot be translated into a valid credit card number even if it is intercepted. Once received by the merchant, the number is decoded and submitted to the credit card clearing system. Over the next few months, today's systems, although not entirely foolproof, should become almost universal throughout the Internet and largely eliminate this form of data vulnerability.

In addition to transaction security, Internet users are concerned about the security of their data once it is received, processed, and stored by the merchant. Newly developed software systems will provide a much higher degree of security in this area. These new systems integrate advanced data base software with Web servers so that traditional, well-developed mainframe computer data security procedures are maintained. Information stored in these systems will be as secure as the transaction data currently stored by major retailers and nearly as secure as bank records.

SUMMARY

The Internet nurtures some of the biggest business opportunities of the 1990s. There are 20 million Internet users today, a figure expected to reach 100 million by 1999. Supplying these users with access holds tremendous promise for rapid growth. Likewise, Internet users will require software for everything from browsing the Internet, to ordering products, to sending and receiving electronic mail. Software developers could hardly ask for a better opportunity to develop products for such a rapidly growing, easily entered new market.

Beyond Internet service and software, publishers are scrambling to take advantage of the new world of low-cost publishing and virtually free information distribution represented by the Internet. Nearly all major publishers are readying products for the Internet. Small publishers are quickly following suit, and new Internet publishing ventures are being announced nearly every day. Finally, traditional retailers are establishing beachheads on the Internet and marketing everything from shoes to flowers. The major unknown in all these ventures is where exactly the profits lie.

2-2
Strategies for Developing and Testing Business Continuity Plans

Kenneth A. Smith

COMPREHENSIVE BUSINESS RESUMPTION PLANNING IS GROWING beyond the walls of the data center. Business's growing dependence on multiple computing platforms and increasingly sophisticated communications networks is making companies more vulnerable to disaster. Actual disasters are highlighting the importance and vulnerability of distributed computing systems, even while mainframe recovery is becoming more reliable. Business continuity must address these evolving integrated and networked business environments.

Public awareness of the need for comprehensive business recovery is on the rise as well, probably largely because of the increase in major regional catastrophes during the past few years. The public sector is also getting involved, with increasing federal, state, and local government participation in comprehensive disaster planning.

Companies are reacting to this need by developing enterprisewide disaster recovery plans but are discovering a host of problems associated with comprehensive business recovery planning. These plans require far more participation on the part of management and support organizations than mainframe recovery planning. The scope is no longer limited to recovery; plans must integrate with existing disaster prevention and mitigation programs. These companywide resumption plans are costly to develop and maintain and are frequently prolonged, problematic, and unsuccessful.

Fortunately, there have been successes from which security specialists can learn. This chapter presents some of the lessons learned, including some of the tools, techniques, and strategies that have proved effective.

COMPREHENSIVE BUSINESS RECOVERY STRATEGIES

Successful recovery from disaster often depends on workable and timely alternative operating recovery strategies. A well-known set of recovery strategies, including some simple and relatively inexpensive commercial products and services, has made recovery planning feasible for large-scale mainframes and midrange computers. New solutions are becoming increasingly available for small computer configurations and business resumption (e.g., work group) recovery. These evolving solutions are based on experiences gained from a multitude of single-site and regional disasters.

Formal recovery planning and testing is an essential part of the total recovery solution. Companies that develop, maintain, and regularly test their business resumption plans recover far better and faster than those that are not prepared.

Exhibit 2-2-1 illustrates the scope needed for a companywide business continuity program. Business resumption should be thought of as an ongoing program, much like existing avoidance programs (e.g., security) and loss-mitigation programs (e.g., insurance).

Resumption planning is moving into departmental work groups. Executive management and the support organizations are also taking part in business resumption through integrated crisis incident management planning.

Numerous planning tools and strategies are available to help companies develop plans. The tools include commercially available software, consulting services, and several public-domain information sources. These tools and services are invaluable in reducing the effort, elapsed time, and cost involved in developing business resumption plans.

Before discussing how to develop a plan, it would be useful to review the available strategies for both computer and work group recovery. The following sections examine these strategies.

Mainframe Systems

The choice of which mainframe recovery strategy to follow is based primarily on business recovery timing requirements, cost, and reliability. Hot sites are fast becoming the recovery strategy of choice for companies requiring rapid recovery. The commercial hot site is second only to an internally owned redundant site in terms of reliability and timing, and usually at approximately 10% to 30% of the annual cost. In some areas (e.g., disaster support and communications infrastructure), hot sites can actually provide a more reliable strategy than an internally owned redundant facility.

The potential strategies for mainframe and midrange computer recovery are listed in Exhibit 2-2-2. Most organizations use multiple strategies, depending on the severity of the disaster and expected outage duration. During

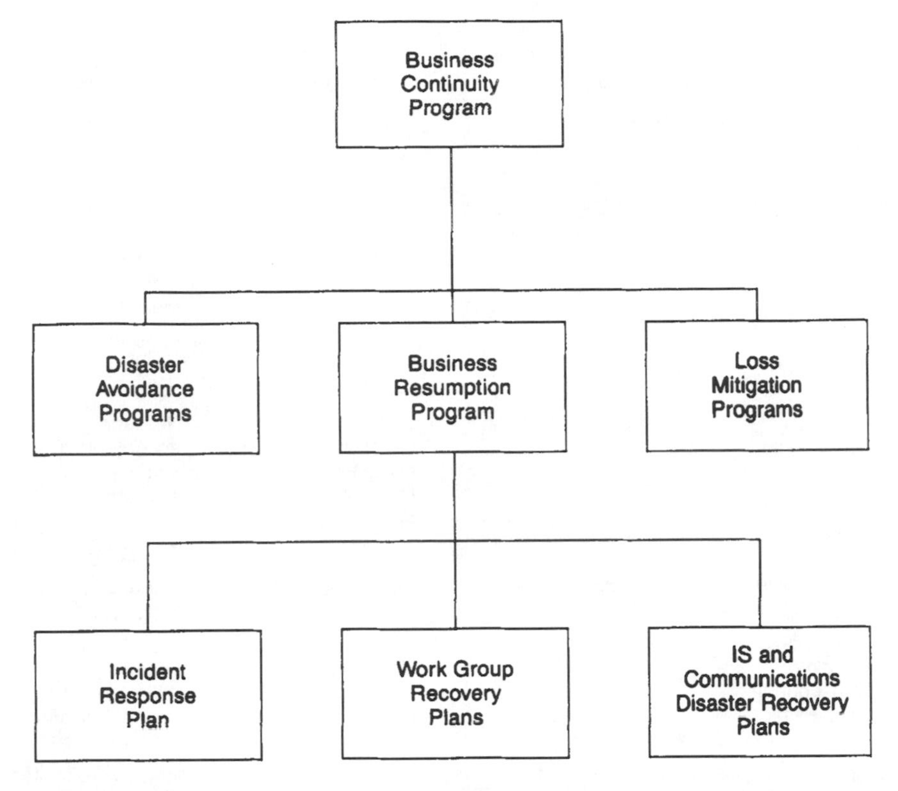

Exhibit 2-2-1. Scope of the Companywide Business Continuity Program

planning, different strategies may be identified for different applications, depending on the potential business consequences.

The recovery time frames in Exhibit 2-2-2 do not imply either minimum or maximum times; these figures represent actual company experiences following disaster. For example, in two recorded examples of total loss of a data center, the rebuilding time was 6 months for one company and 12 months for the other. Most recoveries using commercial hot sites have been accomplished within 8 to 24 hours.

When costs for mainframe recovery strategies are analyzed, it is important to realistically estimate personnel and equipment costs. In addition, the strategies should be planned out over a minimum three- to five-year period to ensure that the cost of maintaining and upgrading equipment is considered. Equipment resources for mainframe strategies should be defined at a fairly detailed level, because the cost of the incidental infrastructure and network can significantly affect actual costs.

Strategy	Recovery Time Frame	Advantages	Disadvantages
Repair or Rebuild at Time of Disaster	6-12 months	• Least cost	• Time to recover, reliability, and testability
Cold Site (private or commercial)	1-6 weeks	• Cost-effective	• Testability
		• Time to recover	• Detailed plans are difficult to maintain • Long-term maintenance costs
Reciprocal Agreement	1-3 days	• Useful for specialized equipment in low-volume applications	• Not legally acceptable in some environments • Testability
Service Bureau	1-3 days	• For contingency planning (e.g., backup microfilm)	• Not available in large CPU environments
Shippable or Transportable Equipment	1-3 days	• Useful for midrange computing	• Logistical difficulties in regional disaster recovery
Commercial Hot Site	Less than 1 day	• Testability • Availability of skilled personnel	• Regional disaster risk
Redundant Facility	Less than 1 day	• Greatest reliability	• Most expensive • Long-term commitment and integrity

Exhibit 2-2-2. Mainframe and Midrange Recovery Strategies

Evaluating potential hot-site or cold-site recovery vendors is less clearcut. Quality of service is most important but difficult to evaluate. Decisions are often made on the basis of technical and pricing criteria.

The checklist shown in Exhibit 2-2-3 covers several criteria that may be used to evaluate disaster recovery vendors. The checklist may also be used in deciding whether to adopt an internal or an external solution.

Questions	Vendor 1	Vendor 2	Vendor 3	Vendor 4
Section 1: Business Issues How many years has the vendor supplied disaster recovery services to the commercial marketplace?				
What percentage of the vendor's business is in disaster recovery?	%	%	%	%
Can the vendor's contract be assigned or transferred without the subscriber's consent?	Yes No	Yes No	Yes No	Yes No
Does the vendor provide an audited, ongoing technology exchange program?	Yes No	Yes No	Yes No	Yes No
Does the vendor provide an annual survey of subscriber satisfaction and critical components?	Yes No	Yes No	Yes No	Yes No
Are nonmainframe recovery solutions available at the primary hot-site facility?	Yes No	Yes No	Yes No	Yes No
Does the vendor provide an account management approach or philosophy?	Yes No	Yes No	Yes No	Yes No
Is there easy access to the vendor's facility? Is there sufficient, secure parking?	Yes No	Yes No	Yes No	Yes No
Can the subscriber notify the vendor of a potential disaster without incurring hot-site declaration fees?	Yes No	Yes No	Yes No	Yes No
Does the vendor have a proven track record of having restored every declared disaster within 24 hours, including full network restoration and application processing?	Yes No	Yes No	Yes No	Yes No
Has the vendor ever lost a customer subsequent to the customer declaring a disaster? If yes, attach a list of those customers.	Yes No	Yes No	Yes No	Yes No
Does the vendor provide hot-site-related services, including remote bulk printing, mailing, and inserting services?	Yes No	Yes No	Yes No	Yes No
Does the vendor have a proven, proactive crisis management approach in place?	Yes No	Yes No	Yes No	Yes No
Section 2: Contractual Commitment Does the vendor contractually accept full liability for all direct damages caused by its negligent or intentional acts, without monetary limit?	Yes No	Yes No	Yes No	Yes No

Exhibit 2-2-3. Alternative-Site Vendor Comparison Checklist

Questions	Vendor 1	Vendor 2	Vendor 3	Vendor 4
What is the limit on the liability for direct damages?				
What is the statute of limitations on claims made by the subscriber?				
Does the vendor contractually guarantee comprehensive technical support to the subscriber during recovery operations?	Yes No	Yes No	Yes No	Yes No
Will the vendor pretest the subscriber's operating system, network control programs, and communications circuits outside the subscriber's test shifts?	Yes No	Yes No	Yes No	Yes No
Section 3: Audit Does the vendor contractually give the subscriber the right to audit the recovery center?	Yes No	Yes No	Yes No	Yes No
Will the vendor contractually guarantee to have its disaster recovery facilities and contracts regularly reviewed by an independent, third-party auditor to verify compliance with all contractual commitments made? If yes, have the vendor provide a copy of the current audit. If no, why not?	Yes No	Yes No	Yes No	Yes No
Will the vendor allow any regulatory authority having jurisdiction over the subscriber to inspect the recovery center? If no, why not?	Yes No	Yes No	Yes No	Yes No
Section 4: Backup Network Are the subscriber's circuits connected on a full-time basis to the front-end processors?	Yes No	Yes No	Yes No	Yes No
Are the subscriber's circuits immediately switchable (using a matrix switch) to: • All other front ends within the primary recovery center?	Yes No	Yes No	Yes No	Yes No
• All other front ends within the secondary recovery centers?	Yes No	Yes No	Yes No	Yes No
• The primary cold site?	Yes No	Yes No	Yes No	Yes No
• All other vendor-provided cold sites?	Yes No	Yes No	Yes No	Yes No
Can the subscriber perform comprehensive network testing at the secondary recovery centers?	Yes No	Yes No	Yes No	Yes No

Exhibit 2-2-3. *(continued)*

Questions	Vendor 1	Vendor 2	Vendor 3	Vendor 4
Can the subscriber test remotely from any subscriber-designated remote location using a channel-attached or equivalent system console?	Yes No	Yes No	Yes No	Yes No
Can the subscriber perform IPIs from a remote system console?	Yes No	Yes No	Yes No	Yes No
What are the vendor's fees for backup network consulting?				
How many central offices support the vendor's recovery center?				
How many AT&T points of presence are directly accessible to and from the vendor's hot site?				
Which common carriers are currently providing service to the vendor's hot site?				
Is diverse local loop routing available at the hot site?	Yes No	Yes No	Yes No	Yes No
Are multiple vendors providing local loop routing to and from the hot site?	Yes No	Yes No	Yes No	Yes No
Does the vendor have diverse access capacity for: • T3 access?	Yes No	Yes No	Yes No	Yes No
• Accunet Reserve T1 access?	Yes No	Yes No	Yes No	Yes No
• Accunet Reserve Switched 56K-bps access?	Yes No	Yes No	Yes No	Yes No
• 9.6K-bps access?	Yes No	Yes No	Yes No	Yes No
• Live dial tones available for disaster recovery?	Yes No	Yes No	Yes No	Yes No
Section 5: Access Rights What are the subscriber's rights of access to the hot-site facility (e.g., guaranteed access or standby access)?				
If standby access (first-come, first-served), can the subscriber be denied access in the event of a multiple disaster?	Yes No	Yes No	Yes No	Yes No
Does the subscriber need to declare a disaster to be guaranteed access to the recovery center?	Yes No	Yes No	Yes No	Yes No

Exhibit 2-2-3. (*continued*)

Questions	Vendor 1 ___	Vendor 2 ___	Vendor 3 ___	Vendor 4 ___
Has the vendor ever allowed a company that is not a hot-site customer to recover from a disaster?	Yes No	Yes No	Yes No	Yes No
Will the vendor allow a company that is not a hot-site customer to recover at the time of a disaster? If so, how does the vendor guarantee recovery for hot-site customers?	Yes No	Yes No	Yes No	Yes No
Does the vendor contractually limit the number of subscribers per hot site?	Yes No	Yes No	Yes No	Yes No
Can the subscriber verify the limitation of subscribers per hot site?	Yes No	Yes No	Yes No	Yes No
Are any of the vendor's hot sites located in an area susceptible to floods, hurricanes, or earthquakes?	Yes No	Yes No	Yes No	Yes No
Will the vendor contractually limit the number of subscribers per building? If so, how many?	Yes No	Yes No	Yes No	Yes No
Does the vendor contractually guarantee access to the hot site immediately upon disaster declaration?	Yes No	Yes No	Yes No	Yes No
Does the vendor provide a cold site at the primary recovery center?	Yes No	Yes No	Yes No	Yes No
If so, how many square feet does it have?	___ sq. ft.	___ sq. ft.	___ sq. ft.	___ sq. ft.
Is the vendor's cold site capable of immediately supporting the installation of multiple CPU configurations?	Yes No	Yes No	Yes No	Yes No
If so, how many and what type?				
Does the vendor provide any subscriber with greater rights than any other subscriber?	Yes No	Yes No	Yes No	Yes No
Can the vendor provide a list of all customers that have preemptive hot-site access rights? If yes, attach a list of them.	Yes No	Yes No	Yes No	Yes No

Exhibit 2-2-3. *(continued)*

Questions	Vendor 1	Vendor 2	Vendor 3	Vendor 4
Section 6: Technical Issues Does the vendor use physical or logical partitioning of hot-site CPUs to support customer processing?				
If physical partitioning is used to partition the vendor's hot-site CPUs, what is the maximum number of subscribers per CPU partition?				
If logical partitioning is used to partition the vendor's hot-site CPUs: • What is the maximum number of partitions per CPU?				
• What is the contractual limit of customers per partition?				
• Does the vendor provide additional resources to accommodate hardware partitioning (e.g., IBM's PR/SM)?	Yes No	Yes No	Yes No	Yes No
• How does the vendor resolve problems associated with testing multiple subscribers simultaneously on the same mainframe?				
What is the exact physical or logical hardware configuration to be provided to the subscriber: • How many megabytes of central storage are dedicated to the customer?				
• How many megabytes of expanded storage are dedicated to the customer?				
• How many channels are dedicated to the customer?				
Does the vendor supply a comprehensive electronic vaulting program?	Yes No	Yes No	Yes No	Yes No

Exhibit 2-2-3. *(continued)*

Midrange Systems

Effective business resumption planning requires that midrange systems be evaluated with the same thoroughness used with mainframes. The criticality of the midrange applications is frequently underestimated. For example, an analysis of one financial institution found that all securities investment records have been moved to a midrange system previously limited to word processing. Because it was viewed as office support, this system's data was not protected off site. Its loss would have meant serious, if not irreparable, damage to this company and its investors.

Midrange systems share the same list of potential recovery strategies as mainframes. In addition, shippable and transportable recovery alternatives may be feasible. Shippable strategies are available for recovery of several specific hardware environments, including DEC/VAX, IBM AS/400, and UNISYS.

Cold-site and repair or replacement recovery time frames can be much shorter for midrange systems (e.g., days instead of weeks), because many systems do not require extensive facility conditioning. However, care should be taken to ensure that this is true for specific systems. Some systems are documented as not needing significant conditioning but do not perform well in nonconditioned environments.

Special considerations should be given to turnkey systems. Turnkey software vendors often do not package disaster recovery backup and off-site rotation with their systems. On the other hand, other turnkey vendors provide disaster recovery strategies as an auxiliary or additional cost service.

Companies using midrange systems frequently have mixed hardware and network platforms requiring a variety of recovery strategies and vendors. When recovery strategies are being evaluated, some cost savings can be realized if all of the midrange systems are considered at the same time. Another planning consideration unique to the midrange environment is the limited availability of in-house technical expertise. Recovery at the time of the disaster often requires people with extensive skills in networking, environmental conditioning, and systems support. A single hardware vendor may not be able to supply these skills in the mixed platform environments.

In an evaluation of the recovery timing strategy, special attention should be given to recovery timing issues on midrange systems. Some platforms are notoriously slow in restoring data.

Work Group Systems

The computer recovery strategies can be borrowed and adapted to work group recovery planning. However, the optimum choices frequently differ because of the different technical and logistical issues involved. As a result,

Strategy	Recovery Time Frame	Advantages	Disadvantages	Comments
Repair or Rebuild at Time of Disaster	1-3 days	• Ongoing cost for office space and equipment	• Availability risk	Rent space and buy replacement equipment
			• Limited availability of special equipment and space	
Shippable or Transportable Equipment	1-3 days	• Ease of use	• Ongoing cost	Use commercial products and services
		• Reliability		
Hot Site or Cold Site	Immediate	• Testability	• Availability in regional disaster	Use of commercial backup office space
Reciprocal Agreement	1-3 days	• Useful for specialized equipment in low-volume applications	• Limited application capacity	Arrange office space (internal)and specialized facilities (external)
Service Bureau	1-3 days	• Useful for daily contingency planning	• Not availablefor large CPU environments	Use commercial services (e.g., print shops and microfilm companies)
Redundant Facility	Immediate	• Greatest reliability	• High cost	Internal use only
			• Long-term commitment and integrity	

Exhibit 2-2-4. Work Group Recovery Strategies

numerous commercially available products and services are becoming available for work group recovery.

The goal of work group recovery planning is to establish essential day-to-day business functions before the consequential effects occur. To accomplish this, most organizations find it necessary to relocate their employees to an alternative location or to relocate the work itself. Exhibit 2-2-4 lists the most common work group strategies.

In addition to these alternative operating strategies, work group planning has some unique and difficult computer recovery challenges. Businesses' dependence on desktop computing is growing far faster and with less control than did their dependence on mainframe and midrange systems. Disaster experiences are showing that many businesses are absolutely dependent on these systems and that the degree of disaster awareness and preparation is seriously and dangerously lacking.

Desktop Computers and Local Area Networks

Currently, the most common method of information protection is to back up data at a file server level and accept the business risk of the loss of microcomputer workstation data. In actual disasters, many companies have been found to be inadequately protecting their microcomputer-based information. The ultimate solution for desktop and local area network information recovery rests in two major areas: standards and standards enforcement.

Planning for LAN recovery is made more difficult by the absence of standardized backup devices. Unlike mainframes and minicomputers, no backup media (e.g., no equivalent to standard mainframe nine-track tape) has been accepted industrywide. Backup device technology changes frequently and is not always downward compatible. Some companies have found it difficult to acquire older compatible technology at the time of a disaster. Redundant equipment and meticulous testing may be the only feasible solution.

The two most common hardware recovery alternatives for individual workstations are replacement at the time of the disaster and having shippable microcomputers. Ordering, packaging, assembling, and installing workstations is a long and labor-intensive process during recovery. Use of commercially shippable microcomputers is becoming more common because of the prepackaging of these systems. In addition, some disaster recovery vendors are providing LAN capability as part of the shippable offering.

Unfortunately, solutions for file server configurations are less clear-cut. Because these customized machine configurations are frequently not stocked in quantity by local computer suppliers, replacement can be quite difficult. Internal reciprocal and redundant options are being used for the file servers. One network software company and some recovery vendors are also making file servers available as a shippable alternative. This reduces the redundant hardware requirements to a backup device and software.

Technological obsolescence must be considered in any long-term LAN recovery strategy. Equipment purchased and stored off-site (e.g., redundant strategy) rapidly becomes obsolete. Reciprocal agreements require that hardware remain compatible over time, which is often difficult.

An even more difficult planning consideration is network wiring. Companies are wiring their buildings with special network facilities (e.g., Token

Ring and Ethernet), making relocation to dissimilar facilities difficult. Companies with multiple facilities (e.g., campus environments) can sometimes use reciprocal arrangements if capacities are sufficient. In the absence of these facilities or in a regional disaster, shippable microcomputers that include preinstalled network capabilities are the safest alternative.

Lack of industry-standard communications hardware is a problem in local and wide area network recovery, making rapid replacement at the time of the disaster risky. Several shippable products (e.g., shippable bridges and gateways) are commercially available to assist local and WAN recovery. When these tools are unavailable, stockpiling of redundant equipment is usually the only recourse.

Wide Area Networks

Disaster recovery planning for WANs is still in its infancy. Even though few companies are addressing recovery of WANs, these networks are often installed to support vital business missions. For example, they are being installed to support such mission-critical functions as LAN-resident business applications, electronic data interchange (EDI), and gateways to mainframes.

Recovery of a WAN is primarily a network planning issue. WANs are typically connected using communications lines with massive bandwidth capabilities (e.g., 56K-bps or more). Typically, the same type of network solutions for large mainframe-based networks are available for WAN connections. Unfortunately, that massive bandwidth can also equate to large network expense.

Networking

That the communications infrastructure (both voice and data) is essential to daily business operations is well understood and accepted. Business impact studies have shown that networks must be restored in most locations at near-full production capacities, usually in a very short time. Some companies have found that the need for voice and data communications is actually higher than normal during a disaster.

Network recovery strategy decisions are driven by business timing requirements, choice of alternative processing decisions for computer recovery, work group recovery, and cost. The technical strategies and the menu of products and services are far too complicated to discuss here; however, the network strategy planning criteria are quite simple to describe.

Simply stated, network recovery strategies should address all technology and facilities required connectivity. This includes person-to-person, person-to-computer, and computer-to-computer connections. All network components should be addressed and appropriate strategies decided on. For most components, the same recovery strategies previously described for computer and work group recovery can be applied.

The following sections discuss some of the special requirements of work group facilities and communications equipment.

Work Group Facility. Loss of a work group facility requires replacing all equivalent network components. These include telephones, terminals, control units, modems, LAN wiring, and the private branch exchange (PBX). These may be obtained at time of disaster using replacement or shippable strategies. They may already be in place in an existing redundant, reciprocal, or commercial hot-site or cold-site facility. The same set of planning issues and network disaster recovery strategies can be employed.

Access to Communications. A disaster may affect the communications infrastructure outside the work group facility (e.g., loss of phone lines or a central office). In this case, an entirely different set of strategies comes into play.

Two possible recovery strategies can be used: relocating to an alternative facility in which the infrastructure is in place or reconnecting to the surviving infrastructure through alternative facilities. Because of timing, these alternative communications facilities are usually redundant and can be quite expensive.

Electronic Vaulting

Electronic vaulting has gained wide attention during the past couple of years as an emerging disaster recovery strategy. Electronic vaulting allows critical information to be stored off site through means of a network transfer rather than traditional backup and off-site rotation. Electronic vaulting brings two major benefits: decreased loss of data and shortened recovery windows.

Electronic vaulting is currently available. Commercial disaster recovery vendors provide both remote transaction journaling and data base shadowing services. Several companies with multiple computer sites are using electronic vaulting on a variety of computer platforms. Electronic archiving is becoming fairly common in the microcomputer arena. Although use has been limited because of significant hardware and communications costs, these costs are expected to decline, making electronic vaulting more attractive in the future.

Until the costs become more reasonable and standardized technology is in place, however, electronic vaulting will be limited to selected applications needing its unique benefits. The business impact analysis process helps determine when this strategy is justified.

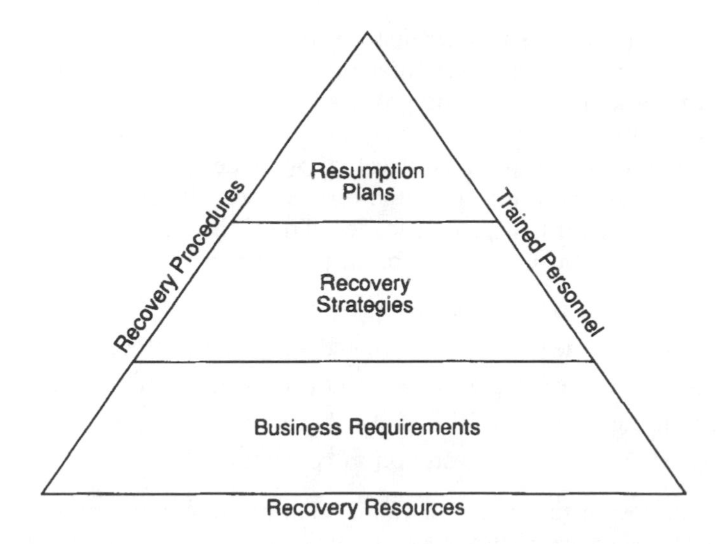

Exhibit 2-2-5. Recovery Planning Phases and Elements

A DEVELOPMENT APPROACH

Presented in Exhibit 2-2-5 is a graphical representation of a simple but effective three-phase approach for developing a business resumption plan. The foundation phase of the development methodology is identification of disaster recovery business requirements. Once these requirements are fully understood, appropriate recovery strategy planning can be conducted. Finally, detailed resumption plans, or action plans, may be developed and documented. All three of these recovery phases involve the surrounding elements of personnel, recovery, resources, and planned recovery action.

Project Planning and Management

Before the first planning phase can be initiated, some thought should be given to project planning and management. Two of the first crucial activities within project planning are clearly defining the scope of the project and enlisting management support.

In larger companies (e.g., those with 500 to 1,000 or more employees), the sheer magnitude of the task may justify staging work group recovery planning. Usually computer disaster recovery planning should be done before or at the same time as work group recovery planning. The business requirements phase helps identify the areas that need to be planned first as determined by the consequences of losing those organizations.

Important business decisions must be made during the planning process regarding preparedness and eventual recovery issues. Active management support throughout the planning process is essential if the planning project is to be successful.

The success of the project is affected by selection of the development project team and distribution of planning responsibilities. Care should be taken to select a qualified project leader. The skills needed for an effective project manager or business resumption planner include:

- Extensive large-project management skills.
- A thorough understanding of records and data protection concepts.
- A basic understanding of network concepts and recovery capabilities.
- Outstanding communication skills, both written and verbal.
- Knowledge of business resumption concepts.

Delegating responsibility for planning to individual work groups usually is not practical. Because of the learning curve and documentation skills, it is more cost-effective to lend a specific work group these skills in the form of a qualified project manager. In addition, many recovery planning decisions involve centralized strategy planning (e.g., determining how the voice network will be recovered or how office equipment and departmental computers will be replaced), which would be better managed by a centralized project team.

On the other hand, some planning activities are more effectively conducted at the individual group level. For example, inventorying of equipment resources, identifying minimum acceptable configurations, and identifying specific implementation requirements are all handled best by departmental management. Responsibility for maintaining the recovery capability must remain the individual work group's responsibility because it must use the plan when a disaster occurs.

The following sections discuss the three phases of developing the business resumption plan.

Defining the Business Requirements

Once the project plan has been developed, actual business resumption planning can begin. The first and most essential step in this process is to gain an understanding of the business recovery requirements—that is, determining which functions must be recovered and when.

The business impact is often intuitively obvious without a formal analysis. For example, loss of a reservation system is a catastrophic event to an airline company. Other businesses and functions may need to conduct a formal business impact analysis to quantify the potential impact and timing requirements. In either case, it is necessary to understand the consequences and timing of the negative business impact of a disaster to each work group. Most of the negative effects on the business develop in spurts over time. For example, an outage of a few hours may be acceptable, but cessation of business operations for a few days may be intolerable.

The business impact may be quantified in tangible values, such as revenue (e.g., underwriting support). The impact may also be intangible, affecting such areas as company reputation, client satisfaction, and employee morale.

Gathering and analyzing business impact information and the current level of preparedness is best done by the project management team working with work group management. Care should be exercised to identify the appropriate level of management; the size of the organization often dictates the level of analysis. Questionnaires should be used sparingly; besides the obvious difficulty in getting responses, questionnaire responses almost always miss unique characteristics and effects on some work groups.

Selecting Appropriate Recovery Strategies

Once the business recovery timing requirements are understood, the choice of an appropriate recovery strategy becomes a business risk issue rather than an emotional issue. Recovery strategies vary by cost and the speed at which business operations can be resumed. Recovery strategies providing the very fast recovery (e.g., redundant facilities) are usually expensive. At the other end of the scale, strategies for replacement at the time of the disaster may lower ongoing costs but can take considerably longer. The choice of recovery strategies must weigh the potential impact from loss of the business functions, relative timing of the impact, and the cost of protection compared to the business risk.

A company can immediately eliminate those strategies that do not meet the business needs. The remaining strategies can then be analyzed on the basis of their relative merits and costs, resulting in an informed business decision.

It is important to note that companies always make a decision about recovery strategies. Companies without a formal recovery plan have implicitly chosen to use either a repair and rebuild strategy, depending on how successful they are at identifying and storing essential data off-site.

It is important to determine all strategies, not just the alternative computer processing strategy. For example, recovery of the voice network, the level of detailed planning, and the degree of training are all strategy issues that must be considered. As these recovery strategy decisions are weighed and decisions made (either explicitly or implicitly), these decisions should be documented.

Developing Detailed Resumption Plans

The final phase of business resumption planning is the detailed resumption planning. Planning must address all three recovery elements: personnel, recovery resources, and recovery actions. A common mistake is to focus on personnel and recovery resource planning and to overlook planning and documenting the actual business activities needed to resume operations.

Planning is best done in a series of working sessions conducted jointly by the project team and work group management. Testing is an integral part of the development process. Each of the following steps represents a work group planning session:

1. Formulate the strategy.
2. Analyze the implementation.
3. Validate the implementation.
4. Approve the recovery plans.

The project team may assist the work groups in carrying out these steps. In addition, the project team is responsible for much of the preparatory work for these steps, because these team members have more knowledge of disaster recovery and more experience in writing plans. The work groups bring special knowledge of their particular areas to the planning process. The following sections discuss these steps in more detail.

Formulating the Strategy. The work group must review the business requirements and high-level recovery strategies and then formulate implementation plans and strategies. As a result of this planning session, recovery management and logistical issues will be defined and documented.

Analyzing the Implementation. A work-flow analysis is a useful tool with which to conduct this planning. By reviewing how work is processed on a daily basis, the work group can identify which recovery actions would be required to recreate this environment in an alternative operating location.

Detailed planning should identify those individuals responsible for managing the recovery process. In addition, any key technical resources (internal or external) necessary for effecting recovery should be identified and documented within the recovery plan.

Logistical arrangements and administrative activities must be clearly documented in a plan. A frequent complaint of companies recovering from a disaster is that the logistical and administrative activities, particularly in regard to personnel, are inadequately planned.

Testing should be considered in planning the recovery activities. Resources should be documented in such a way that their presence may be validated during exercises. Information resources (e.g., vendor contact and emergency telephone numbers) should be usable during exercises.

Validating the Implementation. Once plans have been defined and documented, they should be validated through testing. This can be done in a manual exercise of the plan, comparing the plan's recovery actions against a hypothetical disaster scenario. Following this validation, interactive implementation planning sessions may be required.

Approving the Recovery Plans. In each step, recovery strategies, actions, and resources are documented. As a final step, the plans should be formally reviewed, accepted, and turned over from the project team to the respective work groups.

TESTING THE PLAN

There is no surer way to turn a disaster recovery manual into waste paper than to fail to frequently and periodically test the plan. Testing is crucial in the development and ongoing maintenance of business resumption plans.

There are five important reasons why business resumption plans should be tested periodically. These reasons, which apply equally to traditional mainframe and midrange planning and work group recovery planning, include:

- Testing proves whether the recovery plan works and whether it can meet the business recovery requirements.
- Testing the plan identifies the weak links in the plan, allowing them to be corrected before the plan is actually needed.
- Periodic testing is necessary to comply with legal and regulatory requirements for many organizations. Although this is especially relevant in the banking industry and some insurance companies, it is fast becoming a de facto regulatory requirement for all industries.
- Testing is a prudent banking practice. The testing program protects the initial development investment, reduces ongoing maintenance costs, and protects the company by ensuring that the plan will work when a disaster occurs.

Testing is a universal term used in the disaster recovery industry. Unfortunately, testing has a negative pass/fail connotation carried over from school days. The term "testing" would be better replaced by such terms as "exercising" or "rehearsing." Finding problems during a test should not be viewed as failure when it is really the basic reason for conducting the exercise. Attention should be focused on the positive, not the punitive. For the testing program to be successful, this attitude should be carefully communicated to employees.

Testing Approaches

An effective testing program requires different types of tests to cost-effectively examine all components of the plan. For the purposes of this discussion, testing can be categorized into four types:

- *Desk checking or auditing.* The availability of required recovery resources can be validated through an audit or desk check approach. This type of test should be used periodically to verify stored, off-site resources and availability of planned time-of-disaster acquisitions. Unfortunately, desk checking or auditing is limited to validating the existence of resources and may not adequately identify whether other resources are required.
- *Simulations by walkthroughs.* Personnel training and resource validation can be performed by bringing recovery participants together and conducting a simulated exercise or plan. A hypothetical scenario is presented, and the participants jointly review the recovery procedures but do not actually invoke recovery plans. This type of exercise is easy to conduct, inexpensive, and effective in verifying that the correct resources are identified. Most important, this testing approach helps train recovery personnel and validate the recovery actions through peer appraisal.
- *Real-time testing.* Real-time testing is frequently done on the mainframe or hot-site backup plan and is gaining popularity in work group recovery planning. Real-time testing provides the greatest degree of assurance but is the most time-consuming and expensive approach. If only limited real-time testing is planned, priority should be given to high-risk areas.
- *Mock surprise testing.* Surprise tests are a variation of the other three approaches but with the added dimension of being unanticipated. This type of test is frequently discussed but rarely used. The inconvenience for personnel and the negative feelings it generates tend to outweigh its advantages. The benefits derived from a mock surprise disaster can be achieved by careful attention and implementation of controls to avoid the possibility of cheating during planned exercises.

These testing approaches can be combined into an unlimited variety of tests. For example, walkthroughs can be extended by actually performing some recovery activities (e.g., notifying vendors). Training facility equipment can be used to test replace-at-time-of-disaster strategies, alleviating the need to actually purchase replacement computers, desks, tables, and chairs.

TEST PLAN / PLAN RECOVERY COMPONENTS	Type of Test	Frequency of Testing	Comments
Crisis Management Plans			
Data Center • Phase 1			
• Phase 2			
• Phase 3			
Work Group 1			
Work Group 2			
⋮ **Work Group *n***			

Exhibit 2-2-6. Test Planning Matrix

A Matrix Approach to Testing

An orderly and organized approach to testing is necessary to ensure that all recovery strategies and components are being adequately validated. A matrix approach may be used to ensure that all plan components were adequately considered, as determined by their level of importance (e.g., business impact) and risk (e.g., reliability and complexity of recovery strategy). The matrix presented in Exhibit 2-2-6 illustrates this concept.

In this approach, one or more tests are identified for each component of the plan. The organization can develop a long-range (e.g., two-year) test program during which each element and component of the plan is verified or validated. The test program can then be reviewed and revised periodically (e.g., annually) on the basis of testing results, identified exposures, and training requirements.

Work groups testing approaches and frequency depend on the complexity of the organization. For ease of testing and awareness purposes, some plans may be separated and tested by phase (e.g., test alert notification and alternative-site restoration). In general, technical resources (e.g., computer systems or network recovery) require frequent real-time testing. Work group computing needs occasional real-time testing to ensure that the recovery strategies work when they are needed.

Some departments (e.g., outside sales and long-term strategic planning) have fairly low risk in recovery, allowing less rigorous testing to be done. Process-oriented departments (e.g., order processing, credit and collections, and plant scheduling) have greater risk and recovery complexities, justifying more frequent personnel training and recovery testing.

Conducting the Test: Guidelines and Techniques

There are several important ground rules that must be followed in developing a testing program.

Limit Test Preparation. Test preparation should be limited to developing a test disaster scenario, scheduling test dates and times, and defining any exceptions to the plan. Exceptions should be limited to defining the test scope (e.g., testing only the notification component) or resource acquisition (e.g., substituting a training center for rented office space). Actual testing should follow the documented recovery procedures.

Avoid Cheating. An independent observer should be identified for each recovery test. Controls should be put in place to ensure that only resources identified in the recovery plans are used for the recovery effort. Exceptions to the recovery plan should be noted for the subsequent follow-up. The object of limiting cheating is not to be punitive but to ensure that all activities and essential resources have been identified and will be available at time of disaster.

Document and Communicate Test Results. The Results of the recovery test should always be documented, including follow-up activities, when possible. Corrective actions should be identified, responsibilities defined, and dates set.

Test results should be communicated to the participants and management in a positive manner. Successes should be clearly documented and recognition given to those contributing to the success. Likewise, identified problems should be stated in a positive manner with emphasis on the corrective actions.

Test Information Reconstruction. Difficulties in data restoration and recreation are usually discovered only through real-time testing. The off-site storage facility should be periodically audited to ensure that backups are present and safely stored. The ability to restore should be tested using the actual off-site backup media. When information recreation depends on other facilities (e.g., paper in branch offices or microfilm at the off-site vault), the ability to access this information should be verified. Sufficient volume should be tested to ensure that recovery actions are effective at production volumes.

SUMMARY

To develop an effective and comprehensive business resumption plan, companies should take advantage of the lessons learned from other companies and should approach disaster planning from a business perspective as well as a technical perspective. The course of action depends on the status of the organization's current disaster recovery plans and business priorities.

The following steps summarize how a company should expand its current plans into a comprehensive business resumption plan:

1. Conduct realistic and critical evaluation of the current recovery program. This evaluation should clearly define the status of the entire business resumption plan scope, including incident crisis management, computer disaster recovery, and work group business resumption.
2. Develop a project plan to expand the current program, using the development guideline presented in this chapter. This involves:
   - Analyzing the business recovery requirements.
   - Adopting appropriate recovery strategies to meet the recovery requirements.
   - Developing detailed recovery action plans necessary to implement the proposed recovery strategies.
3. Develop an ongoing testing, training, and maintenance program for the business resumption plan.

2-3
Corporate Lessons for the Communications Systems Manager

Robert E. Umbaugh

THE DRIVE TO INTEGRATE new and existing technologies, a need to become even more competitive, and a more complex organizational environment is making the job of communications systems managers even tougher than it has been traditionally. There are many forces at work in the communications field, and it is management's job to respond to these changes in a positive way that brings value to the organization.

Although the trade press is filled with articles that articulate the problems that the communications manager faces, there are few articles that offer practical advice on what to do about them. In order to survive and prosper, managers should broaden their outlook to include the organizational and personnel changes thrust on all levels of management: operations, systems development, end-user computing support, programming, data management, communications management, and general management. To achieve this broadened outlook, managers will have to learn a few lessons about their changing role in the corporate environment.

THE CHANGING ROLE OF COMMUNICATIONS

The expanding role of communications management has several components. First is the role at the corporate level; many communications executives who once ran supportive, overhead functions are now running functions that contribute more directly to the organization's corporate goals. Sometimes this means acting as product line managers: information is offered as a product of the organization. Having product line responsibility is substantially different from managing a staff group, if for no other reason than the increased attention that responsibility receives from senior management. It also differs in the way performance is measured. Product lines are almost always measured on a profit-and-loss basis—a more direct and immediate measure than most managers are accustomed to.

Another difference is found in the allocation of resources. The equipment and personnel needed to prepare products (in this case, communications services) for sale are usually allocated with a high priority—often with urgency—as opposed to the lower priority of products or systems for inhouse use. If a communications group sells services as a profit-and-loss center directly to paying customers and at the same time continues to provide staff support in the form of systems development and operations for inhouse use, conflict is almost certain to arise, and the communications systems manager will be responsible for eliminating such problems.

Responsibility for voice communication, office systems, data communications planning, LANs, multimedia integration, remote computing, broadband, Internet connections—the list goes on and on and adds to the traditional duties of a communications systems manager. Assuming responsibility for other groups means managing organizational integration, and starting up support groups that do not now exist can place inordinate stress on an already busy manager. None of these tasks are easy and they demand the attention of the senior team within the communications department.

The communications systems manager is even more likely to report at a very high level in the organization. Organizations in which the communications systems department continues to report at the IS, controller, or financial levels can be considered old-fashioned by modern standards. More organizations are placing the communications systems manager at the corporate-officer (e.g., vice-president or higher) level; many are even forming separate subsidiaries for providing communications services. Moving toward one of these two structures is clearly the trend.

Risks and Costs of the Changing Role

All of this additional responsibility and increased visibility is not without both its costs and risks. The risks are often partially offset by higher pay, more prestige, and other incentives. But savvy managers should not lose sight of the need to be aware of the changed environment in which they work.

Politics. Although there is a greater degree of politics at the higher levels of many bureaucracies, politics is not always a dirty word. Politics is nothing more than goal-oriented behavior that has human interchange and social reaction as its primary elements. Politics takes on its darker side when it is used to further personal goals to the detriment of the goals of the organization or its individual members. The best advice for the newly elevated manager is to be highly observant and to avoid, at all costs, giving the impression of manipulating others for personal gain.

Responsibility Versus Risk. Having responsibility for a greater number of corporate functions brings with it both greater potential for success and greater risk of failure. Risk management is a well-structured process used to identify areas of possible or probable weakness and to help managers take steps to mitigate those weaknesses. Managers are tempted to try to eliminate risk altogether, which is a futile effort. Risk comes with the job, and the emphasis should be on understanding risk and managing it. Most successful managers take risks, but they do so on the basis of conscious calculations. These managers usually perform the following actions in their risk-management efforts:

- *Assessing risk.* Managers can estimate the degree of risk with a simple classification scheme of high, medium, or low. A more sophisticated risk classification scheme would include severity, probability, timing and the most likely consequences.
- *Identifying the consequences of an action.* Can immediate and decisive action mitigate the risk? Reduce the severity or impact of the risk? Defer the risk to another time frame? Replace the risk with one of lesser impact or result in some other, less-negative consequence?
- *Identifying the risk-versus-regard ratio.* Making management decisions involves making assumptions. In many cases, assumptions must be made to form a risk-versus-reward ratio. Precision is not the goal; practicality is. The primary question to be answered is: Do the potential rewards outweigh the risk involved?
- *Mitigating negative effect.* Managers should identify ways to reduce the negative effect of complete or partial failure. In finance, this is called hedging, which is a respectable financial ploy used to reduce risk to predictable and manageable terms. In a communications system, redeploying resources can be a hedge as can realigning political alliances. The action taken to hedge against potential failure depends on the environment and the degree of risk, but planning for the potential for partial or full failure is a legitimate management activity.
- *Raising the odds of success.* After studying the risk and the need to mitigate against failure, communications systems managers should use resource application to help raise the odds of success. The cost of the resources must be weighed against the cost of potential failure and some careful judgement is needed. An example of applying resources to reduce risk of failure is the use of consultants to offer needed but rare skills to solve a unique problem.

- *Measuring results.* Too often, managers fail to learn from their past experiences. Communications systems managers should think back to the results of a risky situation and critically assess how well they did in taking the preceding six actions. Were one or more steps skipped? Was the risk adequately understood and did the risk-versus-reward ratio prove true?
- *Taking action.* Passivity is often more destructive than taking action. If a project is headed toward failure, managers should have the determination to terminate the project if necessary. Many managers have wasted valuable resources trying to rescue projects that they know or suspected were headed for failure. Taking action will prevent this wasted effort.

Integration. The more functions managers are responsible for, the more need they have to integrate those activities skillfully. Unfortunately, this is not easily accomplished. In the attempt to integrate technologies, some areas do not fit together naturally, and they do not function well when the fit is forced. This is also true of organizations with long histories and many layers of management.

One example of this occurs when end-user departments choose LANs that are not compatible with established communications architecture and the task of building bridges to the corporate network falls to the communications manager. Still other instances of the need to integrate will come up when departments build their own Internet applications and then need to integrate them with existing corporate systems.

Integrating a new organization or function is not a matter of adding it on as an appendage. The same considerations given to the merger of two corporations should also be given to the merger of two internal departments. These considerations include thinking about different cultures, taking steps to eliminate or reduce redundancy, looking for economies of scale, and streamlining operations.

Time Demands. Perhaps the greatest cost of moving up the management hierarchy is the increased demand on the manager's time—everyone wants just a minute, which, when translated, means at least 20 minutes. In addition to the need to spend time with all the functions and projects for which communications systems managers have direct responsibility, two other specific activities also place increased demands on their time: corporate activities and social responsibility.

Corporate-level committees, ad hoc problem-solving task forces, staff meetings, board meetings, and other sometimes ritual activities will consume huge amounts of time if executives allow this to happen. Some activities cannot be avoided, but managers should not make the mistake of confusing membership on every possible corporate committee and task force with prestige or success. Success comes from results, and managers who devote too much time and effort to many lower-priority tasks often do not produce results.

Corporate officers are usually expected to devote time to social activities that reflect well on their organizations. These activities include serving on boards for hospitals, universities, and other organizations as well as fund raising and running for local office. All of these activities are worthwhile, and although communications systems managers may feel they have a responsibility to their community, they must recognize when they are being offered more opportunities to serve than they can accept.

There are many time management classes currently available to managers for the purpose of teaching them to make better use of their time. These programs include such techniques as :

- Segmenting time into 15-minute slices and scheduling activities to fit into these small time slices.
- Grouping similar activities together. For example, returning all telephone messages at the same time of day, usually the end of the day.
- Taking the first 15 minutes of the day to make a plan of things to do rather than just letting the day happen.
- Requesting an agenda for every meeting to decide whether attendance is required and to better prepare in case the decision is made to attend.
- Learning to say no when that is an acceptable response to an invitation to attend meetings, seminars, and other gatherings that require inordinate travel time.

These and other time-management techniques can make the entire staff more effective. Managers should consider attending these seminars and then passing some of the techniques on to their staff members.

Need to Delegate. Because of additional demands on their time, managers must delegate to subordinates many of the activities they previously performed themselves. Delegation is not without risk. Effective delegation carries with it authority to act, and when managers delegate authority, they must be sure that the recipient of the authority is capable and that effective control and feedback mechanisms exist so that nothing goes wrong. A manager who delegates an activity still retains residual responsibility for the actions of those to whom the activity is delegated.

Other Corporate Issues

The forces that are shaping the job of the chief information officer could also have an impact on newly elevated executives in other areas within the organization, though the nature of communications systems is rarely replicated in other areas. The organizational impact of technology implementation is an added responsibility, however. Not the least of these technologies is end-user computing and the organizational implications of its expanding role. Rarely has such a revolution taken place in the corporate world, and the ultimate fruition of end-user computing has not yet been seen. Most informed observers believe that moving full computing power directly into the hands of users at all levels of the organization will change the way business is transacted in the US.

LESSONS FOR THE COMMUNICATIONS SYSTEMS MANAGER

Several lessons can be drawn from the issues discussed. These lessons include planning, training lower-level managers, broadening horizons, using formal processes, organizational impact, moving from technician to integrator, and assessing and managing risk.

Planning. Books, seminars, and the trade press all extol the importance of careful planning, and yet many organizations do little, if any, truly effective planning. One thing is clear, however: all successful communications systems managers understand and use planning. Consultants point out that they rarely find a communications department in significant trouble if it has a well-conceived plan and capable management.

No substitute exists for carefully thinking through the organization's mission and objectives and setting down specific plans to achieve those objectives. The most important advice that the manager or aspiring manager can follow is to learn how to plan and to do it well. Execution is much easier when the manager has a well-constructed plan to follow.

Training Lower-Level Managers. A traditional method for training lower-level managers is the apprenticeship method. Trainees work under one experienced manager until they are ready to assume more responsibility. More likely, however, an opening occurs because of turnover, and then everyone in the management chain moves up one step, ready or not. The shortcomings of this systems are obvious. More formal methods for preparing future communications systems managers must be developed because the demands on these managers will be even greater than those on current managers.

One area not covered in most training programs for technical managers is basic business functions, including a better grounding in planning, budgeting, decision making, change management, finance, economics, organizational theory, behavioral science, work planning, ethics, and the effects of governmental regulations. All of these subjects can be studied in either a formal or informal program designed to develop the skills of technical managers so that they can manage technology. Just as managers must learn to do a better job of planning, they must train their staff to do a better job as well. As managers delegate more responsibility to others, it becomes even more important to ensure that the members of their staff are well trained and ready to move into positions of responsibility.

Broadening Horizons. Many of today's communications systems managers have come up through the data or voice communications ranks, and this experience has served them well. As they take on greater responsibilities, however, they must broaden their vision of the organization for which they work. This is particularly true if they want the technology they manage to make a direct contribution to the strategic direction of that organization.

In marketing, the term "work the territory" means to understand the marketplace, it needs, its opportunities, and its quirks. Managers must do the same thing—learn more about the business and strengthen the bonds to those who manage other parts of it. Understanding the management network helps ensure a smoother transition into it.

Using Formal Processes Within Communications Systems Management. Communications systems departments have been quite successful in using technology to make other departments more efficient and more effective. The time has come to apply that same technology to the work of the communications department. Development centers with programmer's or designer's workbenches properly supported with hardware and staff can substantially increase productivity. Systems to help work load scheduling, project management, change control, and configuration management are available and should be used. Managers must strive to develop processes that help move systems analysis and system design activities toward repeatable success.

Structured techniques, prototypes, and standard systems architecture are all practical and proven devices that yield better-quality products and greater productivity. The integration of computer technology for use by the department itself must be a major objective for managers.

Another source of help in the search for better control and efficiency is the internal auditor. Managers at all levels should view the auditor as a management tool to be included in all activities.

Organizational Impact. Most action taken by the communications department affects some other part of the organization, and, in today's environment, it is common for that action to affect entities outside that organization. Managers should be aware of the impact their department's actions can have on the rest of the organization and, if possible, should make sure the impact is positive.

Current technology allows organizations to be linked electronically, and most are taking advantage of this technology to become more closely aligned with their customers and suppliers. These linkages make it imperative that managers take a global view of their organizational responsibilities and that they be especially sensitive to the immediacy with which their organization can either enhance or ruin customer and supplier relations.

Moving from Technician to Integrator. Although the manager's need to move away from being a technician might be obvious, the need to move toward being an integrator might not. On a macroscopic level, an efficient manager is a master integrator—one who can integrate technology, management information, strategy, organizational change, and people.

Assessing and Managing Risk. Risk is inherent in all human activity. Managers should not avoid it, because only by taking some risk can they move forward. Taking risk recklessly is foolish, but managing risk is a sound practice. The business of managers is rife with risk, and the better they learn to manage it, the more successful they will be.

UNDERSTANDING THE CHALLENGES

Managers and their staff live in a world of change. Understanding the forces causing that change will make the transition to the inevitable restructuring of the organization at little easier. Some of the trends affecting management are changing organizational structure, cost effectiveness, creative use of technology, and implementation of strategic business systems. Understanding and studying these trends as well as the lessons described previously should position managers for their future challenges.

Changing Organizational Structure. As end-user computing takes on greater importance, and as technology moves into more and more parts of the organization, there will be continual shifts in the balance of power. Although centralized IS will control less of the computing resources, it will continue to be responsible for the backbone systems that, to a large extent, drive the organization. The communications systems manager, on the other hand, will continue to be responsible for the network that links the organization's components. The result will be a blurring of responsibilities between communications systems and information systems. The challenge is to predetermine who is responsible for what in a formal document that spells out roles, responsibilities, and authority. These changes should not be left to chance.

Cost Effectiveness. There is considerable pressure on all parts of the organization to keep costs down and to become more productive. In many cases, there have been across-the-board budget cuts. In the communications systems department, this is difficult to manage because there is usually a high ratio of fixed costs to variable costs. The risk management techniques discussed previously can be helpful in addressing this issue.

Creative Use of Technology. Although end-user computing has opened many new doors for the application of technology, it remains true that most new technology enters an organization through IS or communications. This is especially true of heavy-duty technology—that which requires a higher level of technical sophistication to assess and implement. The responsibility for adopting new technology will continue to be a centralized one in most organizations and in some, in which that responsibility has evolved to users and mistakes have been made, it is being returned to the centralized staff.

Implementing Strategic Business Systems. The push to use information processing and communications technology for strategic advantage (and that term has many different meanings depending on application, business line, and economic environment) is catching some communications systems managers unprepared. Rapid response with little room for error is called for and some managers are not comfortable operating in that mode. The need to support the primary business product is critical. Managers are learning that they will have to be tied more closely to the primary business product if they are to continue to be key players on the senior management team.

SUMMARY

There can be no question that the job of communications management will become more complex and demanding in the coming years. The impact of the year 2000 will be felt by communication management as will the continuing growth of the Internet. The risks associated with the Internet (and these are not limited to questions about security and reliability) are not well understood by the user community. The skills mentioned in this chapter will be needed by every communications manager to survive and prosper in this rapidly changing and complex environment.

The role of a high-level manager or corporate officer with primary responsibility for communications technology is undergoing change. Members of the staff, at any level, who wish to improve their management skills should be alert to the changing environment in which they work.

The intensity of the challenges facing managers and their staffs will not diminish; in fact, they will in all likelihood become even more intense as information becomes an increasingly vital part of the corporate arsenal. Managers should review the key points in this chapter, especially the section on risk management, and formulate objectives that focus on improvement in these areas.

2-4
Hiring and Retaining Valued Employees

Donald D. DeCamp

COMPANIES DEVELOP STANDARD procedures to conduct business; they set and define goals, develop strategies for such activities as research and development and marketing, allocate resources to implement these strategies, and set up checkpoints to meet schedules. When hiring personnel, however, most companies do not use goals, strategies, and checkpoints, and intuition takes over. The results of such hiring practices are high turnover, a bottom line weakened by the costs to replace personnel, and job dissatisfaction among employees.

In most cases a person is hired on the basis of personal chemistry, and the decision to hire is usually made less than two minutes into the interview. About 85% of a typical job interview is devoted to determining technical competence, but the reasons personnel leave an employer are usually not related to technology but to incompatibility with the organization.

The goal of an interview is to determine how a candidate will perform in a position. Human resource authorities list five primary ways to determine future performance, and these techniques are all based on finding out the candidate's past accomplishments:

- Asking the candidate the right questions.
- Asking others (e.g., coworkers or references).
- Administering technical and psychological tests.
- Observing the candidate.
- Guessing (i.e., relying on intuition).

Guessing is the method on which most managers rely when interviewing job candidates. Less than 50% of employers ever check references.

Hiring a new employee is hard work, and because of the personalities involved in the hiring process, hiring entails an amount of guesswork. There are, however, effective procedures for increasing the chances of hiring the right person the first time, reducing employee turnover, and successfully competing for the shrinking pool of skilled professionals.

Managers are becoming more directly involved in the hiring process as it becomes essential to find and retain a core group of key employees. They are looking for the control and accountability that characterize other disciplines. Hiring technical personnel that fit an organization's culture and systems environment is key to retaining valued personnel. This chapter describes in detail how a manager can conduct interviews to find the right employee for a position. The article also lists some steps that can be taken to ensure that the organization retains its current personnel.

DEFINING THE JOB

The interviewing and hiring process should begin by defining a zero-base job description that carefully defines the tasks to be performed now, six months from now, and a year into the future. Often, the manager must construct a job profile but does not know exactly what the position entails; and this poses some difficulties.

The job description should detail the actual day-to-day tasks performed, the technical qualifications required to perform them well, and the personal characteristics necessary for accomplishing the work while being able to work with the supervisor, coworkers, and clients. The description should also include any drawbacks associated with the job, such as long or irregular hours, extensive travel, or a hectic pace. The manager must also know possible changes to the job to determine the personal characteristics of the ideal candidate. For example, some people are content with doing the same thing all the time; others are rapidly bored by such routine and subsequently leave for new challenges.

Constructing such a job description is difficult and time-consuming, but it is the only way to find the right person to fill an open position. The best course of action for constructing this description is obtaining firsthand information about what the job requires. The manager should talk to the employee who is leaving the position or to employees who perform tasks similar to the ones to be performed by the new employee. Managers who are hiring need a detailed description of what tasks had to be done during a workday, what level of interaction with others was essential, and the types of decisions that had to be made regularly. Perhaps the position is open because its previous holder did not fulfill all responsibilities. On the other hand, an employee leaving the organization might be exemplary and a similar person is wanted.

For example, an organization is hiring an applications programmer. At one firm, such a person may work on coding 90% of the time and on analysis the other 10%. At another company, the applications programmer's job could be divided evenly between coding and analysis. In writing the job description, the systems development manager should know the current mix of responsibilities and what they will be six months from now. There is no such thing as a standard job description, unvarying from company to company, and this especially applies to jobs in communications networking.

The Importance of Stories. Also needed are descriptions of how the departing employee or other, valued employees in similar positions dealt with extraordinary situations. How periodic crises are handled affects the business and also reveals a great deal about the personal characteristics needed to perform effectively under stress. This type of information is conveyed through anecdotes and short stories, which can reveal skill and character or the lack thereof. The candidate's personal stories are also the key to hiring the right people. How to use them is described in a subsequent section of this chapter.

Using Exit Interviews. Another vital source of information is the exit interview, which is sorely neglected in most organizations. Interviewing the person who performed the job is probably the best way to find out what the job entails. That person may emphasize the drawbacks instead of the advantages, but such information is also useful. The departing employee may have insightful ideas about how the job could be improved. Except for those terminated for incompetence, departing employees can reveal in the exit interview character traits that are not compatible with a specific job assignment.

Creating a New Position

It might seem easier to write a job description for a new position, but this is not usually the case. Tasks that are actually performed in a new position are often not those in the job description. This is usually done to the new employee's character and the circumstances at the time. In addition, supervisors may exaggerate the qualifications needed or describe them in general terms. The manager who is hiring should establish a set of minimum criteria and then evaluate the management style of the department to obtain an idea of the required personality traits.

When hiring for a new position, managers should talk with other employers who have filled a similar position to determine which type of candidate has worked out best for them. Professional recruiters specializing in IS personnel can also provide valuable information. Often, an employer develops a mental image of the ideal candidate, which can hinder the selection of equally suitable people with different backgrounds.

High employee turnover is likely if a manager relies too much on paper qualifications. Effective hiring practices weigh these qualifications at about 30%, usually during the screening process rather than the interview. Past performance and behavior on the job are more than twice as important in evaluating a candidate.

Because no one can predict exactly what a new job will entail, hiring for it can be a process of trial and error. Although a well-researched job description improves the chances of a fit the first time, many companies are exploring other alternatives. One is to train existing employees, who are motivated and already familiar with the corporate culture. Another is to use professional temporaries. By actually performing the duties associated with the new position, a temporary employee can help to define them more accurately. Those who do not work out can be replaced without the problems of a layoff; those who do work out can be offered permanent employment.

DEFINING THE WORK ENVIRONMENT

The reason that 85% of job changes do not work out is incompatibility. The employee can do the job but does not get along with the boss, with fellow employees, or with the corporate culture. The management style of the person to whom a new employee reports is also critical in determining compatibility and can be quantified.

Defining the Corporate Culture

Companies generally fall into three broad categories:

- Charismatic.
- Democratic.
- Systematic.

Examples of charismatic firms are Apple Computer, Inc., shortly after its founding, and the New York Yankees under the direction of George Steinbrenner. Jobs in such a company are characterized by lots of responsibility but little authority. A democratic company, such as Hewlett-Packard Co., manages by consensus using quality circles, management teams, and decisions by vote of task committees. The systematic organization, such as IBM Corp. or the US federal government, is highly structured and relies on set procedures, channels of communication, and chains of command.

Even these highly general characteristics can result in ineffective hiring if they are not factored into the interview. The interview environment is not the working environment, and both parties are wearing masks to some extent. For example, a newly hired systems development manager quit after just six weeks on the job, claiming that the chief information officer acted like a tyrant. The systems development manager, a take-charge type, could not work for a CIO who would not relinquish decision-making authority. During the interview, however, the CIO had talked about independence and the need to delegate responsibility.

Companies, divisions, and departments may also be classified as either theory X or theory Y. Theory X managers believe that people are basically lazy and require a kick to get them started. A Theory Y manager is interested primarily in motivating employees through carrot-and-stick techniques. Theory Y managers believe that people want to be successful and do a good job. This type of manager tries to create an environment that enables or inspires employees to succeed.

All organizations are a blend of cultures and philosophies, but over the years one or the other usually prevails. Corporate cultures can also change. For example, formerly paternalistic Eastman Kodak Co. has become a lean-and-mean organization. The hiring authority that reduces turnover identifies the corporate culture and management style correctly and matches them to candidate styles and behavior. As in evaluating candidates, evaluating a corporation's culture is performed by examining the corporation's past behavior. For example, how layoffs, strategic planning, employee suggestions, and monitoring of employees are handled can all be explored to determine corporate culture and management style.

DETERMINING MANAGEMENT STYLES

The final task before the interview is determining the individual management style of the hiring authority and the person to whom the successful candidate will report. Today, because of the downsizing of personnel departments, the two may be the same person. The management style of the hiring authority biases the evaluation of the candidate, and that of the immediate boss determines the boss's relationship with the new employee.

As stated previously in the chapter, approximately 70% of hiring decisions are based on personal chemistry, a like or dislike reaction that takes place just a few minutes after the initial encounter. Often, interviewers who will not be the new employee's supervisor have had to admit that the candidate they dislike the most is the most suitable candidate for the job. A worst-case scenario, for example, might be an entrepreneurial, go-getter department head interviewing a methodical detail-oriented computer programmer.

What Won't Work. Such calculated techniques as the high-stress interview do not work. They erect a barrier between the personalities of both the interviewer and the interviewee. Those who perform best in the interview may perform the worst on the job, because they have become adept at using calculated interviewing techniques to their advantage. The most appropriate candidate could have problems with interviews because of a distaste for such game playing. Those hiring often request straight-arrow, no-nonsense candidates, but these candidates are usually rejected because they are considered too low key.

Determining Compatibility. Human resources departments are aware of these types of problems and consequently conduct compatibility interviews or use extensive psychological testing. To successfully conduct compatibility interviews, it must be determined with whom the new candidate must be compatible. Likewise, it must be determined which profile is required. Candidates who cannot pass a test to reflect their conceptions of what the company is seeking obviously lack business experience.

The grid in Exhibit 2-4-1 can be used to quantify management styles effectively. Managers can use the grid to measure concern for task orientation against concern for people and relationships. The degree of each concern is rated on a scale from 1 to 10. At the extremes of the grid are the impersonal and abdicating management style, the democratic style, the autocratic style, and the coaching style. In the center of the grid is the compromising and political style. Any managerial style is characterized by two numbers on the grid.

It would not be worthwhile to develop a computer program for making compatibility judgments, because measures of corporate culture, management philosophy, and management style are not absolute. No two people rate the same work environment equally. These measures serve as a checklist to ensure that the most significant aspects of the workplace have been considered. These measurements should be taken at least to prompt thinking about whether valued employees leave or stay on the job.

STRUCTURING THE INTERVIEW

The anecdotes collected as part of the job description (called critical incidents by human resources professionals) can be separated into four or five types of behavior required by the open position. Examples include initiative, communicative, analytical, and efficient types. For convenient rating of candidates, these types can be arranged in order of priority. Exhibit 2-4-2 presents a list of questions devised to rate entry-level candidates. Each staff position, however, should have its own interview questions pertinent to that job.

After the interviewing homework has been done to determine exactly which behavior patterns are necessary or desirable, the questions should practically write themselves. The objective is to reveal highly specific ex-

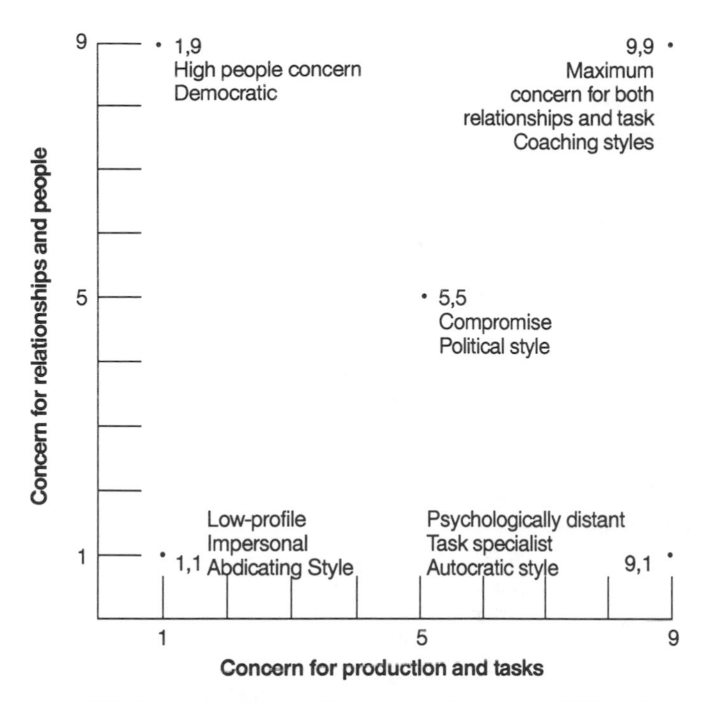

SOURCE: R.R. Blake and J.S. Mouton, *Corporate Excellence through Grid Organizational Development* (Houston: Gulf Publishing).

Exhibit 2-4-1. Grid for Rating Managerial Style

---

amples of past actions and accomplishments. Even if the questions appear obvious, they should be written in sequence so that each candidate can be rated on the same basis. An interviewer should neither improvise questions during interview nor rely only on memory to recall candidate responses.

Behavior-specific questions should be followed on the interview sheet by a brief synopsis of the working environment. This synopsis should also be easy to write with the information from the critical incident interviews, which reveal as much about the supervisor's management style as they do about employee behavior.

Interviewers must also be aware of their management styles and make allowances for this source of bias in rating the candidate. Because self-knowledge is a rare commodity, it is advisable to have a few trusted co-workers fill out copies of the management style grid so that the interviewer can obtain a more independent viewpoint. Making allowances for individual bias is important for conducting successful interviews because personal characteristics are usually amplified in interviews and serious misinterpretations can arise.

Job Skills	Questions
Management Skills	
Ability to plan tasks	Tell me about a big project you had to plan for school work. What steps were involved? What was the outcome?
Ability to set priorities	Describe a situation when you had several things to do in a limited time, such as study for exams. What led up to the situation? How did you handle it? What was the outcome?
Ability to delegate	Tell me about a time when you were in charge of something and had to let others help you. What were the circumstances? How did you assign work? What happened?
Interpersonal skills	
Client relations	Tell me about a time when you've had to deal with the public. Who was involved? What did you do? How did they respond?
Being a team player	Tell me about a time when you helped resolve a group problem. What caused the problem? What did you do? How was it resolved?
Ability to deal with people at all levels	Tell me about a time when you had to work closely with someone in a position above (or below) you. Who was the person? What did you have to do? What was the outcome?
Technical skills	
Problem solving	Tell me about the most difficult work or school problem you ever faced. What steps did you take to tackle it? What were the results?
Knowing limitations	Tell me about a time when you had to turn to someone else for assistance. What was the situation? Whom did you ask for help? What was the outcome?

Exhibit 2-4-2. List of Job Skills and Interview Questions

Job Skills	**Questions**
Growing with the job	
Taking initiative	Tell me about a time when you had to take charge and get a job done or resolve a difficult situation. What did you do? What happened?
Ability to learn on the job	Tell me about a time when you had to learn something new in a short time. What was the situation? What did you have to learn? What was the result?
Communication skills	
Ability to communicate	Tell me about a time when someone misunderstood something you said or wrote. How did you make yourself clear? What was the outcome?
Ability to listen	People sometimes listen but don't hear. Tell me about a time when you misunderstood a teacher or superior. Why did you misunderstand? How did you resolve the misunderstanding?
Commitment	
Work commitment	Tell me about a time when you had to finish a job even though everyone else had given up. How did you manage to finish/ What was the result?
Service commitment	Tell me about work you've done in the community or in a school organization. What did you do? What was the outcome?

SOURCE: *Journal of Accountancy.* September 1991, p.96.

Exhibit 2-4-2. ***(continued)***

Evaluating Communication Style

The last item in the interview sheet should be the grid shown in Exhibit 2-4-2, which evaluates the communication styles of the candidate. A dot at any point on the grid indicates the degree of responsiveness versus assertiveness.

High responsiveness combined with low assertiveness indicates an amiable style. High responsiveness and assertiveness reflects an expressive style. Low responsiveness and high assertiveness characterize a directing style and a low score for both characteristics indicates a detail-oriented analytical style. This method of evaluation correlates with the managerial style grid shown in Exhibit 2-4-1; people with responses in the same quadrant usually are compatible in a supervisor-employee relationship.

The communication-style characterization of personality works well with the behavior-description or zero-base interview. The candidates' positions on the grid can usually be evaluated by the way in which they answer questions about specific incidents as well as by the content of the answers.

Interview Techniques

With an accurate description of the position and working environment, the manager is not only in a better position to find the right employee but also protected from possible charges of discrimination. This is another reason for using a list of specific behaviors rather than a mental picture of the ideal candidate. All of the hard work that created a critical incident interview is for naught, however, if the interviewer asks illegal questions about race, religion, national origin, age, marital status, handicaps, criminal record, or credit rating.

Screening and Preparation. Candidates to be interviewed should be screened for minimum technical qualifications. Although the screening qualifications should not be overstated, they should narrow a field to at most 10 to 12 candidates to save time and avoid interview fatigue. A professional recruiter can be used to limit the field, because it is to the recruiter's benefit to recommend only qualified candidates. A recruiter may also perform some preliminary management-style matching based on the insight of a third-party observer.

Because the interview is conducted with only qualified prospective employees, at least an uninterrupted hour should be allowed for each interview, which should be held in a private, comfortable setting. Candidates should be relaxed enough to minimize role playing, but the interview should be highly structured and similar so that each candidate can be compared. A few questions about the resume, which should be read or reread shortly before the interview, helps to begin the interview.

Questions to Ask: Focus on Performance. Most professionals are familiar with standard interview techniques and have prepared for such questions as: "What do you consider your greatest strengths and weaknesses?" However, they are usually surprised by specific questions about their performance on previous jobs. Ample time should be allowed for the candidate to answer, but the interviewer should not settle for vague answers. Eventually, the candidate should provide specific answers easily, because some people like to talk about themselves. Answers should be sought for each critical-incident question on the job description form. If prompting is necessary, the interviewer might use an actual incident from the critical-incident research to show the type of information needed.

It is important to complete the behavior-specific form for each candidate and to take notes of any other information of interest. After a few interviews, profiles start to blur, and the likelihood of each successive candidate's being hired increases. Initial ratings should not be too high, because that leaves little room for rating a superior candidate.

If not enough information can be found in the answers to the critical-incident questions to make a judgement about communication style, a few standard approaches can help provide the missing information. The candidate's description of favorite and least-favorite managers, likes and dislikes about the current or previous position, and the preferred environment usually provides the necessary information.

Questions that can be answered in generalities should serve only to confirm an opinion based on the answers to previous specific questions. Those new to interviewing sometimes like to pose trick questions, which are included even in reputable books on effective hiring. The temptation to play detective or psychologist should be avoided in all circumstances. The object of the interview is to obtain accurate and specific information about the candidate's past performance, which is used as an indicator of future behavior and a very general impression of the candidate's personality. An attempt to solicit any other information is counterproductive.

Concluding the Interview. The final part of the interview is devoted to answering the candidate's questions about the company and the job. Interviewing is a two-way street, and the employer should promote the company as a good place to work. Because of the competition for skilled employees, this part of the interview is just as important as evaluating a candidate. Although the company as a whole may be painted in glowing colors, a particular position's advantages should never be exaggerated. If a job is portrayed inaccurately, the most effective hiring methods cannot prevent high turnover. If the position has too many drawbacks, maybe it should be changed so that it is easier to fill. The information compiled about the changes likely to occur in the next year in the job can be important in attracting the right person for the position.

The interview should be closed on a positive note but without any commitment. Sometimes, an employer comes across a candidate who seems so right that a job offer is made immediately, because the employer is afraid that the candidate might accept a different job offer. Such impulsive hiring practices should be resisted. A few days consideration probably makes no difference in the candidate's decision, and the next candidate to be interviewed may prove to be even more right.

CONFIRMING IMPRESSIONS

After each candidate has been rated, the results of the interview should be checked for at least the top two or three contenders. One method is psychological testing, which can be useful in confirming impressions of character and communication style or in determining if the candidate profile matches the job description as closely as the interviewer believes. If the candidates are aware of the desired profile, however, they can produce the necessary test results.

Two psychological tests useful in hiring IS personnel are the Comprehensive Personality Profile from Wonderlic Personality Test, Inc., and the Myers-Briggs Type Indicator Report, both of which correlate well with the grid of personality traits given in Exhibits 2-4-1 and 2-4-3.

A survey of one leading employment agency that places IS personnel indicates little use of technical testing except with entry-level candidates. One such technical test is the Wolf Test for logical thinking; others are devised in-house (e.g., a candidate is asked to program in a computer language devised expressly for the test).

For higher-level positions, however, even those companies that once used technical tests have discontinued the practice. They have found that competent IS employees did not necessarily score well on tests and that test performance generally did not indicate future performance. The proliferation of systems, hardware, and software has also made it difficult to test for specific skills. Finally, those candidates with experience in technologically advanced systems may not only refuse to be tested but take offense by the request to be tested.

Importance of Reference Checking

A more effective method is to check references. First, the necessary release must be obtained from the candidate. References are too often neglected today because legal restrictions discourage former employers from giving anything but the most superficial response. Three ways to overcome this obstacle are:

- *References from former clients or customers.* These sources can provide more information than a former employer.
- *Proactive references.* These references call a prospective employer at the candidate's request.
- *Critical-incident references.* The candidate provides the name of someone

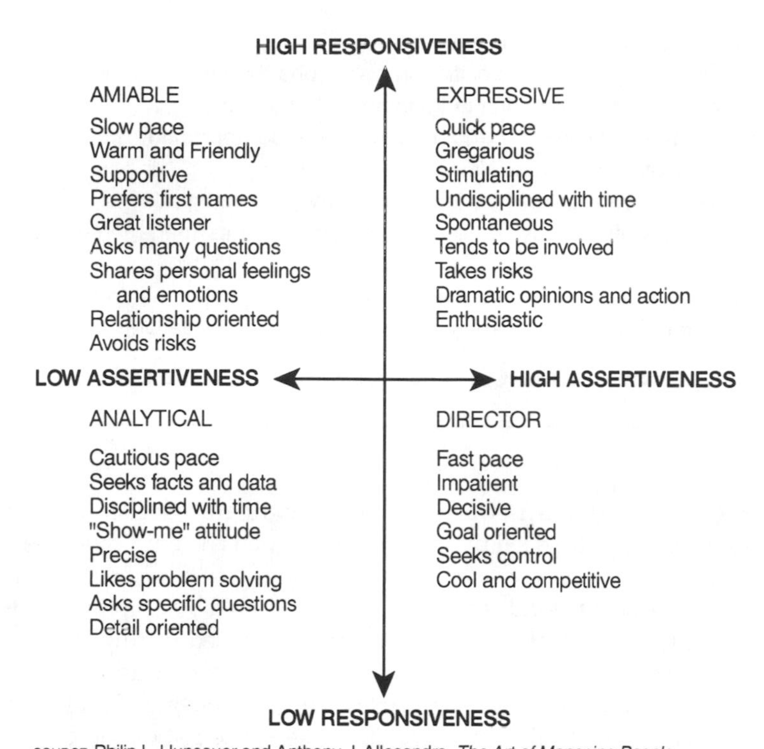

SOURCE: Philip L. Hunsauer and Anthony J. Allesandra, *The Art of Managing People* (New York: Simon & Schuster, Inc.)

Exhibit 2-4-3. Types of Communication Style

who can provide firsthand knowledge of what happened on the candidate's former or current job.

The critical-incident reference is a witness rather than a character reference, and a single such reference can provide enough information to confirm the candidate's credibility.

The candidate to whom employment is to be offered should meet informally with supervisors and coworkers. Unsuccessful candidates should be sent letters of notification, not only because such letters are a matter of courtesy but also because qualified candidates may be considered for other positions in the future.

RETAINING KEY EMPLOYEES

Effective hiring procedures are 90% of the battle of reducing employee turnover, because they place the most suitable candidates in the positions

Will I like the nature of the work I will be doing?
Will I be functional in the new position in a reasonable amount of time?
Is the reputation of the company in keeping with my goals?
Will the chemistry between me and my coworkers be appropriate?
Is the compensation package fair?
Will the opportunity for growth be in line with my personal goals?
Is the company's philosophy of "doing business" and its management style compatible with mine?
Will working there affect my personal lifestyle or family relationship negatively?
Will working there affect my mate's career or education negatively?

Exhibit 2-4-4. Questions for Evaluating a Job Offer

where they can work most effectively. The candidate's decision to accept an offer, however, is influenced by factors over which the employer has little or no control.

Exhibit 2-4-4 lists several common questions asked by employees in evaluating a job offer. Most of the questions are addressed during the job interview or the final negotiation of a wage and benefits package.

The first two questions are significant for two-income families. If both wage earners in the family are pursuing careers, the relocation of one wage earner can present unsurmountable problems for the family. To help ease these problems, an employer should present the organization's location as attractively as possible, be knowledgeable about job opportunities for the other wage earner, and offer assistance to this wage earner in finding a new position nearby. When a candidate travels to another city for a job interview, the employer should recommend that all family members travel along, because they have a strong influence on whether or not a job offer is accepted. The willingness of employees to relocate, even within the same corporation, has greatly decreased during the past 20 years because of the rise in the number of two-income families.

Why Employees Keep a Job

The work environment, which is carefully defined before hiring, often changes after the candidate has been hired. To help retain valued or newly hired employees, managers should know the following key reasons employees stay with an organization:

- Challenge.
- Location.
- Advancement.

- Money.
- Pride in the job and the company.
- Equality (i.e., equal treatment for equal competence).
- Recognition.
- Security and skills enhancement.

These reasons are easily remembered by the acronym CLAMPERS. Skills enhancement is especially important for IS personnel, because technology changes so rapidly. Often, IS professionals change jobs to gain simply technological experience and not higher salaries, greater job security, or improved working conditions. Some technologically advanced companies have little difficulty in attracting highly qualified candidates but seldom retain them for more than a year.

There are two major phases in most IS careers: the skills development phase, which lasts 10 to 15 years, and the return-on-investment phase, in which the employee uses the skills developed earlier to enter management positions or change career tracks. In the development phase, access to new technology, training, and company-supported higher education are the most important incentives an employer can offer to retain key people.

In the return-on-investment phase, equality and recognition become more important tools for retention. Business has traditionally rewarded good performance by advancement to the ranks of management. Management, however, is an entirely different skill from those that produce superior technical performance, and many of those so advanced are unhappy in management positions.

Rewarding Technical Experts. A recent development is the two-track system. In this system, those who are interested in management and develop the necessary skills are promoted to supervisory positions, and those who prefer technical work are moved laterally and receive equivalent improvements in responsibility and compensation.

Because of the prestige given to management, it is difficult to provide equal recognition to those who move laterally. This is an aspect of employee retention that should be watched very carefully, because IS technology is important to today's corporation. If equal prestige cannot be given, incentives in such other areas as compensation and challenge may be improved. Many companies seem to be moving toward a less hierarchical structure, but there is still a tendency to rate a position's importance to the firm by the number of employees that report to it.

Reviewing Incentives. Periodic surveys should be conducted to determine how employees rank the company's incentives. The most reliable surveys are conducted by third parties and have anonymous answers. Under any other circumstances, employees tend to give the answers they believe are expected of them, as they often do in interviews. For example, independent surveys have shown that security is highly valued by most employees, but it is seldom ranked high on a signed questionnaire. If such surveys are conducted, they must lead to positive results, of which all employees are aware, or they become regarded as routine paperwork.

Using Exit Interviews Effectively

One of the best surveys of a company's effectiveness in retaining key people is the exit interview. This is an opportunity to hear the reasons for which employees resign. A few employees may leave because of an overwhelmingly attractive opportunity elsewhere; fewer still are terminated because of incompetence. The majority look elsewhere because of some condition that management can usually correct. Exit interviews are too often performed perfunctorily or not at all.

The employee's immediate supervisor may be the reason for leaving. Therefore, the exit interview should be conducted by an independent third party, such as an executive in a different department in the company. The interview should discuss the issues covered in CLAMPERS and target one or two of the most important factors.

The objective of exit interviews is to correct any deficiencies and match more closely employees and working environment. The interview is not a way to retain employees who have decided to leave, no matter how valuable they may be. Counteroffers are counterproductive. The majority of employees who accept counteroffers remain with the company less than six months and are not very productive during that time. They may also affect the morale of fellow employees or that of their replacement.

The Importance of Challenge

The one factor most affecting job satisfaction is probably challenge. Assignments should be made slightly more difficult that the previous one but well within the employee's capabilities. Often, employers err by hiring overly qualified candidates. These candidates can quickly learn a new job but soon lack a sense of accomplishment and become bored by routine. Those who must strive to meet a goal usually feel that they are making progress, which is probably the greatest source of job satisfaction. No employee, of course, should be given projects that require more skills than they possess.

SUMMARY

Because of structural changes in most organizations, hiring technical personnel is becoming more and more the responsibility of IS managers. Interviewing candidates with equivalent technical qualifications is a highly complex process that is more an art than a science. As with any art, proficiency comes only with much practice. Certain techniques should, however, allow any manager to accurately match jobs and employees and to reduce employee turnover. These techniques include:

- Developing an accurate, firsthand job description.
- Screening candidates for technical qualifications before interviewing.
- Evaluating corporate culture and management philosophy of immediate supervisors.
- Structuring an interview so that a candidate reveals specific past actions, which can be used to indicate future performance.
- Evaluating personality traits to determine compatibility with the working environment.
- Confirming interview results through testing and through reference checking.

Although the hiring techniques described in this chapter should aid greatly in reducing employee turnover, managers can take additional steps. Retaining key employees can be accomplished by periodically evaluating the following motivating factors, which are described by the acronym CLAMPERS:

- Challenge.
- Location.
- Advancement.
- Money.
- Pride in the job and the company.
- Equality (i.e., equal treatment for equal competence).
- Recognition.
- Security and skills enhancement.

The most important factor in retaining technical personnel is the opportunity for skills enhancement. Exit interviews, too often neglected, should not be used to retain employees who have decided to leave. They are, however, valuable tools for locating reasons for unacceptable turnover rates and correcting these reasons.

2-5 Cost Allocation for Communications Networks

Keith A. Jones

DATA CENTERS, WHEN UNDER EFFICIENT MANAGEMENT, provide services to a business enterprise that are recovered by a cost allocation system that distributes these costs to customer business units. Operational expenses can be allocated across data communications network infrastructures along the lines of chargeback systems established for the host processing environments.

Most communications managers are already very much aware of all the host system resources that must be charged back to their users. This includes the most common direct system use charge categories (e.g., host CPU, system printer, DASD storage, and tape drive units) as well as indirect resource costs that must be billed (e.g., space, utilities, and technical support services).

Many communications managers may not be as well aware of the resources that must be charged to recover costs across host-linked enterprisewide information networks—or the impact that such networks can have on their host resource costs. Use of these networks can result in new categories of host data center costs that must be recovered through more complex pricing methods. Although managers can readily accommodate any of the more conventional unexpected changes in data center operating expense (e.g., a rise in utility or paper costs) by using an industry-standard strategy that adjusts availability of host access to match available capacity, they are often uncomfortable at the prospect of chargeout across their host network links. This is primarily because they are not as aware of their options for recovering unexpected changes in costs of supporting an enterprisewide data network.

CHARGEABLE RESOURCES

At first, it is easier for a communications manager to focus on effects to support chargeback of direct network costs on services involving equipment with very short paybacks and high plug-compatibility. The ideal initial resources are modems, terminals, printers, and if already in place, a central help desk telephone line to contact vendors.

The ideal arrangement is to leverage the volume-purchasing power of the central-site data center to obtain the lowest possible direct cost for the business client users. In addition to increasing economies of scale in purchase of hardware, software, and supplies, the data center can offer a central focal point for vendor contacts, negotiations, and coordination of maintenance services—at lower pass-through costs than can be obtained by independent arrangements.

Managers should begin viewing their function as more than just a service bureau for data processing services. The network-linked host data center provides expanded opportunities to offer an enterprisewide range of business information management support services (e.g., computer systems hardware and software purchasing, inventory control, financial analysis, and physical plant engineering).

Especially if existing departments within the organization already offer such basic services, the data center does not have to provide these services directly. Instead, management should increasingly position and present its overall operation as a ready and effective conduit to help enable each business enterprise client to tap into—and leverage—existing packets of expertise within the organization. Communications managers are in a unique position to identify common areas of operational business needs and present both internal and external sources as options to help pool costs.

Regardless of the source of the resources that make up the information enterprise network and host system costs that are to be allocated across the client service base, the primary consideration of the data center manager must be that of defining the categories of use that are to be charged. The chargeable resource categories must be easily understood by the customers who are to be billed and must be measurable by use statistics that can be readily accumulated for the review of customers in their billing statement. This is true regardless of whether the costs are to be allocated according to direct line item unit measure charges or indirect fee assignments.

As with the host chargeback system, it is first of all necessary to be able to identify—and track—all possible network services and resources. Once there is a reasonable assurance that a system of resource accounting procedures has been established that does not allow any network cost or any contingent host system cost to escape unaccounted for, data center managers can consolidate individual categories of network services and resource costs into network cost pools suitable for use in pricing out network chargeback billings.

The definition of these cost pools must be sufficiently detailed to support future strategy for directing patterns of desirable network use in a manner that is predictable and compatible to host cost-efficiency and cost-containment procedures. In other words, categories of resource that are identified to network customers for cost allocation purposes must serve to increase customer awareness of the types of expenses involved in managing the network, regardless of how much is to be billed to recover costs of each resource type.

NETWORK RESOURCE CATEGORIES

The chargeable resources for enterprisewide network cost allocation can absorb some of the data center resources (e.g., printing, data base use, data media, and data center operational services) as well as the resources required for the network. The most important network resource categories are discussed in the following sections.

Cable. The costs of cabling to connect enterprisewide network devices can easily be the largest and most complicated category of all network management expenses. Among the available options are:

- *Coaxial cable.* An example is the cable used in IBM 3270 terminals. It is expensive but extremely reliable and compatible with most of the existing network methods. It is often kept in stock by data centers.
- *Unshielded twisted-pair cable.* Also known as telephone wire, this is inexpensive but subject to interference; it often causes host-linkage error.
- *Shielded twisted-pair cable.* Although similar to the unshielded telephone wire, this is wrapped like coaxial cable. It is moderately expensive and moderately dependable.
- *Fiber-optic cable.* This has the widest possible range of applications, and it is the most reliable cable option.
- *Wireless networks.* These are increasingly feasible options, especially when rapid installation is desired and the physical layout of the business enterprise facility is suitable.

Linkages. In network terms, these are interconnectors, or plug-compatible interfaces, between networks and host lines. Linkage components include:

- *Plugs.* These include cable connectors and data switches, splitters and sharing devices, wire ribbons, and all other interface linkage hardware. This is typically the most overhead-intensive cost category in network management because it requires inventory control and planning to have every possible plug-type on hand.
- *Modems.* Modems and all other dial-up line management devices as well

as communications line multiplexing hardware are most critical to line-speed costs.

- *Wire boxes.* These involve all the business enterprise physical plant or facility-determined access hardware and management.

Workstations. This category includes microcomputer configurations that are connected at each local network node or station. The components are:

- *Terminals.* These can include intelligent workstations through gateway servers.
- *Storage.* Storage elements can include the fixed hard disk (if available at the workstation node), diskettes, and any dedicated tape or CD-ROM storage devices.
- *Personal printer.* This is included if a dedicated printer is attached.

Servers. This category always comprises an intelligent workstation with the minimum CPU and storage required by the particular network control methods used. It includes:

- *File-server hardware.* This includes the file-server workstation and all network boards that it controls, including LAN boards and host gateway link boards as well as fax boards and other specialized hardware.
- *File-server software.* This includes the network file management control software and shared applications.

Storage. High-volume mass storage is frequently needed for backup and archival purposes, which may involve technical support and specialized hardware or software well beyond the usual range of support for LAN administrators. This may include optical memory, imaging, CD-ROM jukeboxes, or other advanced technology with many gigabytes of storage capacity. The need for this technology is increasingly likely if the client network involves graphics, engineering, or any other application requiring intensive backup cycles and complicated version control. The use of high-volume mass storage is usually controlled from a dedicated server that automatically manages files and application software. The client-server may be a minicomputer that functions as a front-end processor to manage enterprise network message requests for real-time access to host programming functions or a back-end processor to manage online access to a host data base.

Communications. This is by far the most cost-intensive network resource. Telecommunications can make or break the network chargeback system, and unfortunately it is an area in which a data center may have limited options.

Much can be learned about network billing from a local telecommunications company, often at great expense.

LAN Administration Support. This can include the salaries of the local LAN administrator and the LAN help desk personnel, the costs of training, and the costs of host-based support of LAN inventory, purchasing, billing or any other service provided by the data center. It can also involve both short-term and long-term leasing of network hardware or software, diagnostics, or user training. This category can be large or small, depending on the client. In some cases, clients may prefer to provide some of the LAN technical support; in others, a data center may provide it all.

Internet Gateway Access

Whether the data center network is SNA or TCP/IP dictates whether it is an option to establish the enterprise's own Internet gateway, which can cost as much as 10 intelligent workstations. That gateway, however, can be cost justified if the enterprise organization does any kind of business on the Internet.

Typically, business on the Internet takes one or more of three forms: E-mail, user news groups, or a home page on the World Wide Web (WWW, or just the Web). A dedicated gateway requires substantial UNIX-related technical expertise to establish and maintain. A more viable option is a dedicated line to an online service account, which normally costs no more than a single workstation, and internal enterprise customers can still be supported in the three major functional business support areas.

Both a dedicated gateway and a dedicated line can usually be financed using long-term methods. If Internet access is a legitimate enterprise business need, it can be more than easily justified. If not, until the demand (and real need) for Internet access can be precisely predicted, it is often feasible to simply pass through direct charge billing of departmental accounts on online services with Internet access, such as Compuserve, Prodigy, or America Online.

EXPENSE CATEGORIES

The next step in developing an enterprisewide network cost-allocation system is to assign all expenses into a manageable number of nonoverlapping categories. Most managers can develop a matrix of network resource categories that has a direct correspondence to their host data center resource assignment matrix. If there is no existing host data center expense matrix to use as a model for defining the network cost allocation expense matrix, the

Resources / Expenses	Stations	Servers	Linkages	Storage	Telecom
Fixed Costs Equipment Facilities Insurance Interest Maintenance Salary					
Variable Costs Consultants Data Base Support Help Desk LAN Support Paper Supplies Telecom Support Vendors					
Surcharges Diagnostics Disaster Recovery Documentation Planning Prevention Tuning Training					
TOTAL COSTS					

Exhibit 2-5-1. Sample Enterprisewide Network Expense Matrix

communications manager can begin by grouping the basic categories of network expenses. Exhibit 2–5–1 shows a basic network expense assignment matrix.

There are two important considerations. First, each of the categories in the expense matrix must correspond to budget line items that will be forecasted and tracked by a business enterprise financial controller or accountants assigned to audit network costs. Second, expense line items must be defined so they are clearly separated from all other items and there is limited opportunity to assign expenses to the wrong category or, even worse, duplicate direct and indirect costs.

Although the manager should consult with the financial controllers who audit the network before preparing the network cost assignment matrix, the data center manager must usually define how the resource allocation procedures is to be administered, which includes assumptions about the methods for measurement and assignment of enterprisewide network resource use and billing obligations. The proposed network resource cost allocation matrix should also be reviewed with the prospective network customers. In addition, the anticipated categories of host-lined data network expenses should be reviewed by the business enterprise organizational management before any pricing strategy or rate structure is determined for the enterprisewide network cost chargeout.

RATE DETERMINATION

To expand enterprisewide business information services and allocate host system costs across interconnected distributed networks, managers must change their fundamental view from a centralized focus to a decentralized one. This change in focus must include a reorientation to accommodate market-driven as well as demand-driven operations planning.

Communications managers are increasingly in a position of competition with alternative outside vendor sources for every product and service that has traditionally been a vital part of their exclusive business organizational domain. It is therefore necessary for data centers to begin defining their pricing structure on not only what chargeable resources there are but how each of the resources is charged to help achieve strategic market advantage over the competition.

Data centers can no longer simply apply a straight distribution formula to recover costs but must also factor in potential future costs associated with the risk that new technology will be available at lower cost and that business enterprise users may bypass the data center. Unless the manager is also an expert in both financial management and risk management, this usually means an increasing emphasis on leasing or subcontracting most new services and resources at short-term premium rates, until a sufficient economy of scale can be achieved by bundling enterprisewide data network resource demands to reduce risks as well as costs.

In some cases, host-linked network domains may be large enough to achieve the break-even economies of scale quickly, especially if sufficiently large proprietary data stores are involved. In most cases, however, the process will be more like a traditional tactical penetration to achieve a dominant share of mature, nongrowth market segments, also known as buyer's markets. The management of the central site host data center must go to the business enterprise customer rather than the other way around.

If a data center cannot package centralized services as superior to the competition, not all costs may be recovered. Furthermore, the ultimate determination of what constitutes a chargeable resource is not simply a data center expense that must be recovered but is now also an information network service or resource that a business enterprise customer must want to buy.

The manner of rate determination is directly determined by the network resource allocation strategy and chargeback system cost-control objectives. The most basic consideration determining the rate structure is whether data center management has decided to fully recover network costs on the basis of actual use or to distribute indirect costs of a network resource pool evenly among all of its users. This consideration applies to each network resource category as well as to each category of network customer.

As a basic rule of thumb, the decision of whether to recover costs on the basis of actual use or to distribute costs evenly among business enterprise users is largely determined by the extent to which a resource is equally available for shared, concurrent use or is reserved exclusively for use by an individual network user. If the resource is shared, the rate structure is based on forecasted patterns of use, with sufficient margin to absorb potential error in the forecasted demand as well as the potential probability of drop-off in demand for network services. On the other hand, rate structures can be based on underwriting network capital equipment or financing service-level agreements, which provide greater security for full recovery of all costs from an individual business enterprise customer on the basis of annualized use charges, with provisions for additional setup fees and disconnection fees to recover unforeseen and marginal costs.

As a matter of practical reality, the decision on which of the two methods of rate determination to use must also take into account the availability of dependable enterprisewide network use measurements to support direct line charges. It is also critical to first determine whether the costs that must be recovered will vary depending on the level of use (e.g., the pass-through costs of data transmission over a public telecommunications carrier line) or will be fairly well fixed regardless of whether they are fully used at all times when they are available (e.g., the on-call network technical support cost of labor). It is also important to attempt to define all assumptions on which each resource cost recovery strategy is based; this identifies all the conditions that would necessitate a change in the rate structure and exactly how each cost will be recovered in the event of major changes in enterprise information network use or underlying rate strategy assumptions.

The degree to which the enterprisewide network chargeback pricing can be easily understood by the customer largely determines how effectively all costs are recovered. It is also critical that business enterprise clients be aware of the goals of each rate decision. Depending on how responsibly they use the network, each individual enterprise network customer can help or hurt overall efficiency of the LAN as well as all interconnected LANs and the host.

SUMMARY

Opportunities exist to broaden the data center's customer base through identifying and establishing chargeback methods for host-linked enterprisewide information networks. These new categories, however, require more complicated pricing methods. Increasing economies of scale make network chargeout a valuable strategy for the communications manager and staff. Managers should:

1. Initially focus on services based on equipment with short paybacks and high compatibility, such as modems, terminals, printers, and a help desk phone line to vendors.
2. Begin viewing their function as more than service bureau for data processing. Instead, the data center manager should position the data center as a resource for where expertise can be found in the organization.
3. Identify and track all network services and resources, including cable, linkages, workstations, servers, storage, communications, and LAN administration support.
4. Establish, if possible, the organization's own Internet gateway.
5. Develop a matrix for defining network cost allocation expenses, to assign all expenses into nonoverlapping categories.

2-6
Managing Morale When Downsizing

Donald R. Fowler

A STRAW POLL OF COMPUTER PROFESSIONALS on what factors affected their morale, either positively or negative, highlighted four issues:

- Career concerns.
- Lack of trust in management.
- Lack of direction.
- Change resistance.

Almost no respondent reported positive reinforcement taking place within their business organization.

Among this small but somewhat representative group of computer professionals (mostly mainframe technical professionals), the most pressing morale issue is their careers outlook. Employees are concerned for both long-term future careers in information systems and in maintaining their current job positions. The next most prevalent issue is lack of trust in senior management. This concern surfaces as a lack of confidence in management to lead, properly communicate, and properly empower personnel. The third issue—the opinion that the business or information systems organization was adrift—leaves employees with fears that no one is in charge. The last issue, resistance to change within an organization or business, usually snowballs into a situation in which individuals increase their resistance and become obstacles to successful change. Other responses as to what affects job morale included personal conflicts, monetary issues, and job evaluation disagreements.

CAREER CONCERNS

The IS professionals' concerns stem from several causes. Respondents believe that their companies are no longer investing in education. Instead of building skills in the current staff in such new areas as client/server computing, LANs, object-oriented technology, or data base management systems, businesses are buying new skills through contractors or currently skilled new employees.

Eroding Skills. Failure to educate employees leaves mainframe professionals with quickly eroding skills that will become less marketable. It is also causing an immediate decrease in the value of these employees within an existing business structure.

The respondents said that they felt unable to perform some new job requirements because of their lack of professional skills. These job requirements include such items as team management, empowering employees to make decisions, and closer contact with user groups. Key concerns included little or no investment in developing employee skills for interpersonal communications, negotiation, writing and public speaking, and decision making. IS professionals feel trapped because they are required to carry out these new responsibilities with no idea how to do them, which makes them feel that senior management is setting them up for failure.

Disappearing Voice. The respondents also felt that the company decision makers were no longer relying on technical professionals to implement technology business solutions. Instead, decision makers were following a solution "du jour," based on what products were deemed hottest in the trade publications.

DEFINING AND MEASURING EMPLOYEE VALUE

All of these factors leave IS professionals with low self-esteem and even lower company value. Even those professionals who are considered important company resources are concerned over their value to the business. A point of contention is, however, what is the definition of value?

Defining Value

Employees should understand exactly how their skills support the business. The business should have a clear definition of value that each employee clearly understands. Management and employees should discuss how current skills and job duties provide value and what future skills are required to increase value. Of equal importance, management should be able to state what value it brings to the business. Value can frequently be equated to the basic competence required by a business.

As information systems change, the basic competency of today is antiquated tomorrow. An example might be a mainframe organization that has developed COBOL applications for the past 20 years. As time went on, this organization learned more about systems analysis and design, structured code, reusable code, and defect elimination before production. The cost of coding declined and the quality of the code improved. There is a current basic competency made up of many different skills (e.g., in-depth knowledge of COBOL, analysis, and testing).

However, if this mainframe COBOL shop looks at the application of technology required by the business in the next two or three years, the need for COBOL skills fades and possibly object-oriented languages and prototyping design approaches are needed. A new set of skills is needed to provide new basic competence.

Measuring Value

Exhibit 2–6–1 shows how employee and organizational value can be defined and identified in terms of basic competence. This method is a combination of function points, critical success factors, competency, skills, and valuation approaches. Purists find fault with this method because it lacks detail for each of the previously mentioned approaches. However, the basic competence approach has been successfully applied and serves the purpose in sufficient depth.

The upper blocks of the figure in Exhibit 2–6–1 show activities to determine the current basic competence and value. Employee value should be stated in realistic and reasonable terms that are consistent across the business.

The lower blocks of the exhibit show activities for figuring out future basic competence and how to acquire it. Each basic competency is broken down to specific technical and professional skills. Performing this breakdown is laborious. No single source provides all of the possible permutations. A fairly quick and successful approach is given in the following paragraphs.

First, identify the competency and value in a short sentence or paragraph. An example is "We achieve a consistent 99.999% availability and subsecond response time for mainframe (host) customer support applications, thus enhancing our external business customers' satisfaction through same day order entry and confirmation, order query and fulfillment information." This statement means that the IS organization has proven and successful systems management processes in availability, performance, and capacity management. The statement also provides a general value measurement, which is customer satisfaction.

To find out which skills are required for performance management, for example, managers should list the skills needed to do this function. Technical skills for performance management include in-depth knowledge of operating systems, network protocols, transaction processors, data and storage management, performance tools, and performance methodologies.

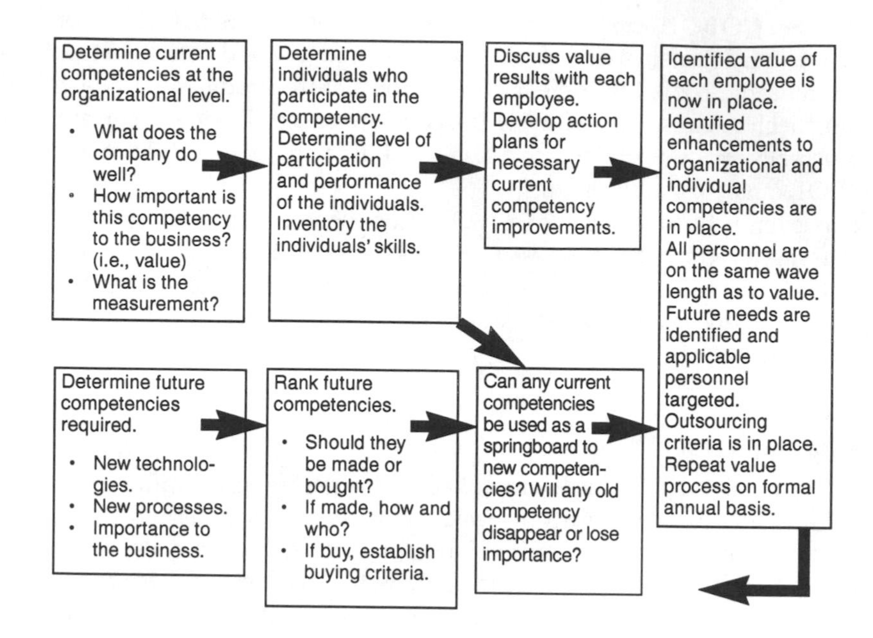

Exhibit 2-6-1. Straw Poll Results of IS Professionals' Concerns

Professional skills include written and oral communications, negotiation, and time management.

Once the skills have been identified, an inventory is taken of employee skills. Important to list are the strengths and weaknesses within a skills set. Training, mentoring, or formal education should be planned to address weak areas. Employees should use strong areas to mentor others. The findings should be discussed and the plan presented to the employees in one-on-one sessions. At the end of a session, an employee should have an agreement with management on how to attain needed improvements. Employees should also have an understanding of their:

- Current and future areas of basic competence.
- Skills that are valued.
- Skills that need to be improved.

Benefits of Knowing Value

Using this or a similar approach helps address many career concerns. It shows employees their value in terms of the skills they possess that are important to the business. It provides a map for increasing their value and helps to show the company's direction. This activity by itself does not answer all of an employee's concerns about the lack of direction.

The first iteration of this value method is usually lengthy and at times frustrating, but the immediate effects on employees' self-esteem are worth the effort. Subsequent iterations become much easier.

SKILLS PLANNING

The process of identifying basic competence can be used as a springboard for skills planing. Skills fall into two main classifications:

- Technical skills, which are hardware, software, and protocol oriented.
- Professional skills, which involve communication, project management, personal interaction, and decision making.

Improving Technical Skills

Technical skills are cultivated through formal classes, self-study, hands-on experience, and certification by specific vendors. Mainframe support personnel can further be split into operations and technical support.

Operations personnel should focus on such new skills as LAN administration, workstation support, and automation tools. Technical support personnel should focus on PC operating systems, network operating systems, and high-level development languages. Many vendors offer certification testing and training. Many mainframe professionals are entering these certification programs on their own and obtaining a solid working knowledge of desktop environments. Their value to the business increases dramatically when they become conversant in PCs.

Management must consider the appropriate company support for these individuals. Managers should decide whether the company will provide education funding for these certifications. The certifications should enable employees to reach the competencies required within the business.

Most training and consulting companies offer formal training courses at client sites. Recommended are training programs structured to immerse prior mainframe personnel into the PC, LAN, and client/server environment. Courses should provide hands-on experience or use case studies. Both types of activities aid in reinforcing the formal lecture materials.

Internet Resources. Many management teams overlook the Internet as an inexpensive source of subject matter expertise for employees. The volume of information available on such services as CompuServe and Internet nodes and bulletin boards is astounding. Authorization of subscription and connect

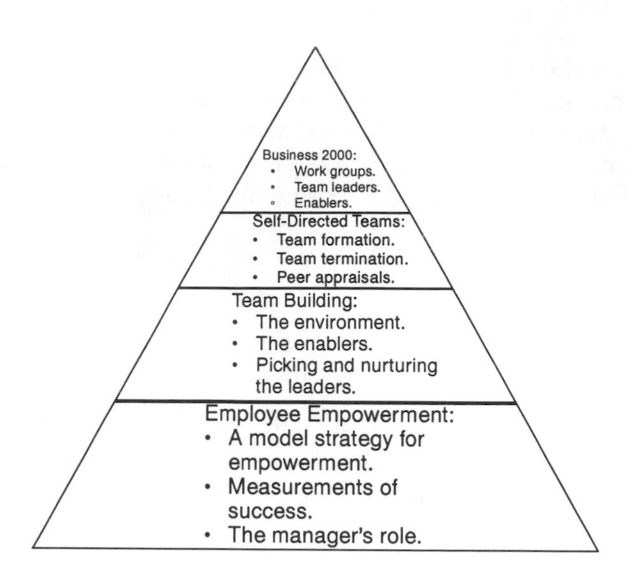

Exhibit 2-6-2. Components of a Successful Downsizing Effect

charges for all staff pursuing new skills should be part of a skills-development plan. Management should place a cap on connect-time charges, and any access charges should belong to the individual. It is helpful for subscribers to operate in the workstation environment they are attempting to learn. Also, employees should install any software themselves, customize it to their own requirements, and read the support documentation. Knowing where to find the answer is a key objective of any training program.

LACK OF TRUST IN MANAGEMENT

Trust is neither inherited nor given; it is earned. In today's downsized business world, many management teams have tried to apply traditional management techniques. With downsizing comes such new necessities as empowered employees and work teams. Each of these concepts can fail miserably if certain old-style management techniques continue to be used. To have a functional and productive downsized organization, both empowerment and teams are critical.

If empowerment or work teams fail, morale spirals downward and management trust diminishes. As shown in Exhibit 2–6–2, employee empowerment should be considered the foundation for organizational success. There is a clear cut and proven path for success called *planned implementation.*

Build the strategy and communicate It
- A clear definition of empowerment
- Policy on how management responds to a poor decision by made empowered employee's

Phases of implementation
Pilot management team.
- Employees selected by established criteria.
- Measurements of success.
- Phased rollout.

Morale issues
- What am I being asked to do?
- What are the risks?
- How do I relate to others?
- When will this affect me?
- Do I perceive management buying in to this?
- How do I know if I am successful?

Exhibit 2-6-3. An Empowerment Strategy

Employee Empowerment

Failure starts with a management team that does not properly enable staff, plan the empowerment effort, and define measurable objectives. Empowerment often fails because:

- Everyone has a different definition of empowerment.
- Success requires skills many employees have never acquired.
- Management's actions often counteract empowerment.

The implementation strategy in Exhibit 2–6–3 shows how an organization can avoid these mistakes. First, an organization defines empowerment. One IS organization defined it as: the offer and acceptance of authority and responsibility showing a trust and confidence in the ability of individuals to act on their own initiative in support of customer service. The key words in this definition are offer and acceptance. Not all employees should be offered an empowered status nor will all employees accept such a status. Many employees are task-oriented and no amount of enablement training can make them change. Empowerment of the masses leads to chaos.

Enablement Training. Some employees thought to be capable of being empowered refused to accept the status. This refusal can usually be overcome by providing enablement training (e.g., negotiation and communications skills courses). A large computer company attempted to empower a group of highly regarded technical experts at one of their main sites. Management became involved in putting out fires in the company's various development labs caused by these experts. The cause for these problems turned out to be the experts' lack of ability to negotiate. They would stake a position and refuse to move from it. By providing these experts effective training in how to negotiate and build consensus, the first died out. If these enabling skills had been taught before the empowerment, many burned bridges could have been saved.

Permission to Make Mistakes. Some employees refuse empowerment because they do not know what will happen if they make a mistake. The result of erroneous decisions must be defined and communicated to all employees. Management should make it clear that everyone makes mistakes and can learn from them. A pattern of making the same mistake should be recognized and addressed. A possible course of action is having management act as the coach. An employee should know why a decision was faulty and how and why management would have made a different decision.

Senior managers must buy into this stated approach to assisting empowered employees in improving decision making skills. An empowerment plan implodes if an employee makes a decision and is attacked by upper management. All levels of senior management must be willing to accept the risk of a bad decision and assist in coaching and mentoring for improving people's decision making skills. Obviously, management may have to correct a bad decision, but it should be done constructively.

Senior management must say how it will handle bad decisions. This helps remove the fear of the unknown and shows employees that careful thought went into the empowerment plan.

Avoiding Chaos. Empowerment of the masses leads to chaos. The initial implementation should include a pilot program where certain select employees are empowered. A set of established criteria determines which people to empower. Such criteria include:

- Professional skills such as negotiation and communication.
- Technical skills.
- Personnel skills such as leadership.

Once these employees are functioning well within the empowerment role, they should help select the next employee empowerment group.

Measures of Success. The final activity that should be considered before any rollout is how to measure success. Without a stated approach to measuring success, many employees felt that management is setting them up for failure. Management must create and communicate measurements of success, which can include:

- Thank you letters from clients or customers.
- A peer-voted award.
- Baseline and delta surveys of customers for such specific metrics as:
  - Organizational response time reductions.
  - Customer satisfaction.
- Baseline and delta surveys of such employee perceptions as:
  - Job satisfaction.
  - Productivity.
  - Communication.

Staged Implementation. A phased roll out is necessary to avoid major problems. One functional area at a time should be carried out. This allows both the employees and management to become comfortable with the empowerment approach. Some managers will not relinquish control and will block empowerment. By phasing the roll out, these managers can be detected more easily and dealt with as needed. Employees will be watching the management team. Peer managers and upper level managers should be prepared to have open and frank discussions with managers that are resistant to empowerment and suggest areas of immediate improvement.

WORK TEAMS

Once empowerment has been achieved to a sufficient level, the implementation of work teams can be seriously considered. Work teams are important because of the emphasis on reducing costs and improving service. An organization must enter the world of teams for any chance of success. There are two forms of teams:

- *Project teams.* These accomplish a specific project or mission and are of finite duration. They are also known as *departmental work groups.*
- *Self-directed teams.* These address ongoing business processes or functions. They are formed across departments and may involve personnel from all aspects of a business.

The breadth and depth of self-directed team concepts could compose an entire chapter. The focus of this chapter is on project teams because this is the first step in developing work teams.

A Team Is Composed of Individuals

A team consists of individuals motivated to seek a product to address the team's mission. All team members must provide a conducive atmosphere for accomplishing the team's mission. Each member must be able and willing to accept specific roles, duties, and responsibilities within the team.

The glue that holds a team together is the team leader. This individual must possess skills in consensus building and personnel management. Some organizations require their team leaders to have proven strong project management backgrounds. Most successful project managers have such skills.

The Team Manager's Role

The manager should be viewed as the arbiter, mentor, and coach for the team. Occasionally, disputes arise within a team that cannot be settled, and the manager should be the arbiter, listen to the dispute, make a decision, and inform all the team members why and how the decision was arrived at.

Mentoring is an extremely important function for a team manager. Many teams become frustrated when they cannot weave their way through the corporate maze. The manager should point team members in the right direction, and possibly first contact the other corporate entity for the team. The manager should ensure that the other party realizes the team represents the manager's functional area and is regarded as speaking for that functional area.

The initial contact is important. People form their perceptions of others in the first few minutes of contact. If the manager knows that a specific person in another department has particular traits that might affect the initial contact, the team should be briefed on these traits and how to ensure the best face forward on the initial contact. The team's mission should be reviewed frequently with the team members to ensure they stay bounded by what has to be done. Teams will frequently try to go beyond their bounds.

The manager should also ensure that the team methodology is being followed. A standard methodology ensures teams are formed, given a mission, and dissolved at the right time and in appropriate ways.

Team Methodology

The methodology (see Exhibit 2–6–4) provides three key activities needed to improve the chances of team success. The first key activity is to decide how teams are formed. For example, one organization wrote a document that defined how to create mission statements, choose necessary team members, and write team rules of order. What was lacking, however, were any controls on when and why teams should be formed, and the number of teams mushroomed. In a department of 38 people, there were, at one point, more than 40 teams. Nobody was getting anything done because all they did was hold team meetings. Each team's mission appeared important, but no general control existed. Team A needed deliverables from Team B; Team C required information from Teams A and B; Team B could not get anything done

Determine the Team Charter:
- Mission.
- Members.
- Tasks.
- Completion.

Determine How Teams are Formed:
- Avoid uncontrolled growth of teams.
- Start small with a few teams.
- Establish scope and authority.

Determine How Teams are Chosen:
- Team member criteria:
  - Subject matter expertise.
  - Knowledge of company history.
  - Growth experience.
- Inter and intra department.

Determine How Teams are Dissolved:
- Team completion criteria.
- Documentation.
- Thank you.

Exhibit 2-6-4. A Strategy for Teams

because Team D was not scheduled to complete its necessary activities for three months. Another team was formed to look into why the teams were gridlocked.

Controlling the Number of Teams. To avoid this situation, an organization must start small and with few teams. The teams being put in place must be essential to organizational success. If there is a list of success items, they should be ranked and the first team should be formed to address the important item.

Importance of Mission Statements. A team must be given a concise mission statement that establishes its objective. This mission statement should bound the scope of the project and establish the authority level. This mission statement should be reviewed by the team frequently to avoid creeping outside the original project scope.

Composing a Team. The next piece of the team methodology is a consistent approach to team content (i.e., the players). This involves answering such questions as:

- What types of subject matter experts are needed? What level of subject matter expert is needed?

- What is the availability of such experts?
- Will excessive time be needed from other functional areas?
- If so, should these people be considered members of the team?
- Can team members with less experience be used and given an opportunity for growth?
- Does the expertise level of people available to fill the team cause unavoidable delays in the target completion date?
- How long does the team have to identify the tasks that must be accomplished?
- Are all tasks assigned to team members?
- Are they correctly assigned?
- How will tasks be tracked?

Disbanding the Team. The team must also know when the team mission is deemed complete, as well as:

- The final deliverables that mark the end of the team's mission.
- The form of the final team documentation.
- How the team will be recognized or thanked for the completed mission.

SUMMARY

This chapter presented suggestions that an organization can use to stand a better chance of success in keeping morale up and accomplishing the needs of the business. Planned implementation and well-thought-out measurements remove much of the fear of the unknown for the employees. Keeping employees in the dark or giving them minimal information negatively affects morale. Managers must be forthright, open, and candid about the needs of the business, employee careers, and implementing empowerment and work teams.

Section 3
Networking Technology

SEARCHING FOR JUST THE RIGHT TECHNOLOGY to apply to a communications system can be a frustrating experience. This is no less true in the new high-speed integrated networking model than it was in the slower, less open, past. If anything, the rush of new technologies and the proliferation of niche product vendors seems to compound the confusion.

The communications industry too often seems to put the technology cart before the application horse. Communication system managers complain that vendors and service providers offer solutions for which there are no problems and seem to ignore real user problems for which there are no commercially available solutions. It is rare that a product or service actually solves an immediate customer problem. When this does happen, the product or service is immediately and overwhelmingly successful. Examples of all three of these cases abound.

Early ISDN offerings may be the best example of a solution in search of a problem. When developed in the 1980s, ISDN was touted as a way to offer subscribers in small office environments as well as in organizations served by PBX or Centrex platforms a way to simultaneously combine data transfer and telephony on a single local loop access line. Telephone service providers kept trying a technical sell of this wonderful new technology. The problem was that no one needed the service. It did not solve a specific user problem. Sales of the equipment and the service languished. The ISDN acronym was said to stand for "Innovations Subscribers Don't Need."

More recently, however, ISDN has seen a dramatic turnaround in its fortunes because of the emergence of customer needs. Telecommuter access to corporate networks and Internet access are examples of newly emerging requirements that can be helped by ISDN.

At the other extreme are those user needs for which no practical technology solution exists. The need to handle high-speed data and graphics transfer in a mobile environment is an example. The immense capability of today's notebook computers and the increased mobility of the working professional is accompanied by a demand for high-speed access to corporate intranets. This access must be wireless for at least some portion of the connection path. Yet, about the best service currently available for wide area wireless data provides access at perhaps 19.2K-bps. There are, of course, technical, spectrum, and regulatory issues involved. The question is, is the same effort being devoted to seeking wireless solutions that was devoted to, for example, ISDN?

A shining example of what can happen when a technology solution matches a customer requirement is frame relay, which was originally developed as an additional packet mode of operation for ISDN networks. Frame relay provided a very efficient way to use the fixed-rate 64K-bps ISDN channels for data and a way to multiplex different connections on a single data link. Frame relay provides an X.25 packet service, but at a lower architectural level, taking advantage of improved, low error rate transmission facilities.

About the time that frame relay was being standardized, a need arose to interconnect geographically distributed corporate LANs by a more cost-effective means than leased T1 lines. Frame relay was an ideal solution to this requirement. The service began to be offered independently of its roots in ISDN and has continued to grow beyond the most optimistic forecasts because the technology meets a real customer need. The lesson here is that communications system managers need to try to convince their equipment and service providers to switch from an "If we build it they will come" philosophy to a customer-driven "ask and you will receive philosophy."

This useful section has gathered a set of chapters explaining several different aspects of networking technologies. The section begins with Chapter 3-1, "Inverse Multiplexing: Technical Foundations and Applications," which is an exploration of techniques enabling a number of lower-capacity communications channels to be combined into what appears to be a single high-capacity channel. This chapter describes the basic principles of this valuable technology and illustrates the framing and synchronization mechanisms. Chapter 3-1 also examines cost issues for various applications.

The next chapter, "The Benefits of Implementing Frame Relay," Chapter 3-2, examines four scenarios in which frame relay service offers significant advantages. These applications range from Internet access to the sharing of wide area trunks. Very useful pointers are provided to help users decide if frame relay is the right solution. Chapter 3-2 also includes guidelines on designing and managing the frame relay network.

Chapter 3-3, "Global Frame Relay," extends the scope of the frame relay discussion into the international arena. This "how to" chapter includes hints on planning and establishing an international network. Chapter 3-3 also discusses dealing with ISDN and ATM support, how to handle voice data, and how to maintain quality frame relay service.

Asynchronous transfer mode (ATM) networks use fixed-length cells to carry information that may be data, voice, video, or graphics. Designed for high-speed transport, these networks are sensitive to traffic control, traffic policing, and end-user behavior. Chapter 3-4, "Traffic Control Functions in ATM Networks," provides an excellent perspective of mechanisms that may be used to manage the flow of information through an ATM local, backbone, or wide area network. Readers will finish this chapter with a firm understanding of constant, variable, available, and unspecified bit rates and their ramifications.

Despite the reluctance of many managers to participate actively in videoconferences and audioconferences, the technology is growing as the work force becomes more separated from the core corporate structure. Telecommuting, satellite offices, global operations, and smaller work units all contribute to a need for videoconferencing. A key element to success will be the quality of the service as perceived by the users. "A Guide to Conferencing Technologies," Chapter 3-5, examines the business benefits of teleconferencing, reviews the various technologies and standards, and offers useful tips for ensuring success.

The very timely "Choosing a Remote Access Strategy," Chapter 3-6, explains various possibilities for remote access including dialup, ISDN, X.25 packet service, and wireless data solutions. The chapter also explores hardware and software requirements, reviews cost factors, and introduces security and management issues.

Perhaps the fastest growing corporate networking technology at the moment is the intranet. Chapter 3-7, "Considerations for Implementing Corporate Intranets," provides readers with an introduction to this important subject along with guidelines on accomplishing a successful transition to this new style of networking.

3-1
Inverse Multiplexing: Technical Foundations and Applications

Steven E. Turner

JUST AS THE COMPUTER INDUSTRY HAS MOVED AWAY from mainframes toward distributed processing, the communications industry is distributing intelligence and processing power throughout the network. Increased processing capacity in customer premises equipment and higher intelligence in the network have combined so that devices at remote network nodes can control how they use the network—not just access the network and rely on centralized network processors to control and manage network traffic.

Because computing and processing power at the customer premises continues to grow, the applications requirements and connectivity demands at remote network nodes are likewise growing. This means moving higher-bandwidth signals across the network and making more critical demands on data transfer. But accessing the network with wider-bandwidth signals costs money, and cost considerations require network managers to minimize network expenditures. The solution is to put as much intelligence as possible at the customer premises (minimizing the use of centralized network capabilities) and access network bandwidth on an as-needed basis. The technology that accomplishes this bandwidth-on-demand requirement today is inverse multiplexing.

INVERSE MULTIPLEXING FUNDAMENTALS

In a typical multiplexer, multiple independent channels are combined into one large channel for efficient transmission. Often, these independent channels are from different sources or represent individual data paths, but they can be intelligently combined into a single data pipe for transport to another location.

Current applications and networks have just the opposite requirement. Customers have wide-bandwidth signals (e.g., bulk data files, video images, interactive communications signals, high-speed local area network data signals) to send. This can be accomplished with dedicated, wideband services (e.g., T1 and T3), but leasing these circuits is expensive, and to justify their cost the circuits must be used to their full capacity a large amount of the time. An alternative is to synchronize a number of smaller-bandwidth digital telephone circuits and use those channels to carry pieces of the wider-bandwidth application signal across the network. Within North America, the most common service used for this is Switched 56 (as the name implies, a 56K-bps switched service). Other services that can be used for this purpose include Switched 64, Switched 384 (also known as ISDN HO), Switched 1536 (also known as ISDN H11), and Switched Nx64 (also known as ISDN Multirate). The process is the inverse of multiplexing, because it involves breaking a wider signal into a number of smaller, independent channels for transmission.

The beauty of this approach is that the underlying channels are part of the existing switched digital telephone network, so they are available on a dial-up basis. The user's networking requirements are evaluated application by application and an inverse multiplexer on the desktop is used to determine how many individual channels are needed to transmit that particular signal. The inverse multiplexer then sets up a call to the chosen destination, dials up the number of switched digital channels required, and sends the wide-bandwidth signal over those circuits. This process is illustrated in Exhibit 3-1-1.

To ensure that the data carried on the individual circuits arrives at the opposite end in the proper order for recombination, careful channel framing and synchronization of the independent circuits are required, because the characteristics and transit delay associated with each channel are different. An inverse multiplexer at the destination end receives the independent circuits, resynchronizes them, and integrates them together to reconstruct the original wide-bandwidth signal.

The result is that the wide-bandwidth signal has been effectively moved from one destination to another and the customer has used the minimum bandwidth needed to move it, and for the minimum time necessary. Within the boundaries of the inverse multiplexer's capabilities, the user has the flexibility to use as much or as little bandwidth as needed, and it can be dialed up between any two points where an inverse multiplexer is present. As with any other toll telephone call, the user has paid only for the time actually used. Because inverse multiplexing provides this capability using the existing

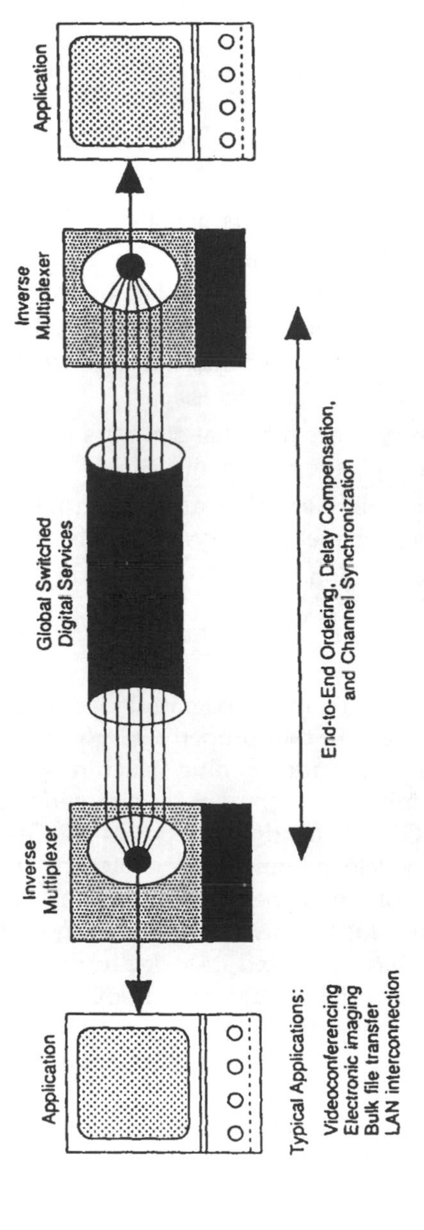

Exhibit 3-1-1. Inverse Multiplexing

dial-up digital network, it is no wonder that inverse multiplexing technology is gaining rapid acceptance in today's network environment.

Basic Requirements

For compatibility and to provide a basic set of features, inverse multiplexers have several operating characteristics, including:

- Automatic synchronization and alignment of individual data channels.
- End-to-end negotiation of such items as operating mode, channel data rate, and phone numbers of circuits used in the call.
- Monitoring of data transfer throughout the call to determine if and when a circuit fails (loses synchronization due to channel errors or is disconnected while a call is in progress).
- A call failure recovery procedure that supports call disconnection, rate reduction, and bandwidth replacement.
- Support of a dynamically variable transmission rate during a data call (adding and deleting channels to vary the bandwidth used).
- A remote loopback function.

Operational Modes

Several years ago, manufacturers recognized that if they are to design inverse multiplexing equipment that properly interoperates, some standardization of the control signaling and framing structure was required. A standardization effort was initiated in the US by a manufacturer's consortium called the Bandwidth-ON-Demand INteroperability Group (BONDING). The BONDING group developed much of the baseline standard for inverse multiplexing, and the work was then moved into a recognized standards body, the Telecommunications Industries Association (TIA) TR-41.4 Committee. That committee has submitted its work to the ITU-TSS and to the International Organization for Standardization (ISO), for the purpose of establishing international standards for inverse multiplexing.

Today, nearly all inverse multiplexers implement at least a subset of the BONDING protocol, which specifies the five operational modes shown in Exhibit 3–1–2. Two modes (transparent mode 1 and mode 1) are mandatory, whereas modes 0, 2, and 3 are operational. Transparent mode is the simplest; it provides direct cut-through of a call between the two end-points, without any channel delay equalization or error recovery. It serves as a simple data pipe between two destinations. The other mandatory mode (mode 1) allocates the full bandwidth to the call; initial setup is furnished, but no in-band call monitoring is provided, and channel resynchronization must be provided manually. These mandatory modes afford a minimal level of service.

Transparent Mode (mandatory)

Provides direct "cut-through" of incoming and outgoing channels to allocate the full bandwidth to the application. No delay equalization or parameter negotiation is provided (transmission pipe only).

Mode 0 (optional)

Provides initial call setup, but leaves channel delay equalization and synchronization functions to the endpoint devices, such as codecs, that support CCITT (now ITU-TSS) recommendations for channel aggregation.

Mode 1 (mandatory)

Provides the full bandwidth for the application. No in-band monitoring is provided; call resynchronization must be done manually.

Mode 2 (optional)

Provides 98.4% of the bandwidth for the application; the remaining portion is used for in-band monitoring of the channels being used for the call. Calls are automatically resynchronized if errors occur during transmission.

Mode 3 (optional)

Provides the full bandwidth for the application, but an additional channel must be used to provide in-band monitoring.

Exhibit 3-1-2. Operational Modes in Inverse Multiplexers

The three optional modes provide additional features. Mode 0 arranges the initial call setup, but leaves channel delay equalization and synchronization to the destination equipment. This mode is used when intelligent devices at the endpoints are capable of providing these features themselves. Modes 2 and 3 provide the highest level of service, with automatic channel delay equalization and resynchronization as well as call monitoring; mode 2 does this in-band, and mode 3 uses an additional channel to provide these features.

Frame Structure and Channel Assignment

To organize and synchronize the individual circuits, the inverse multiplexer divides each of these channels into a sequence of frames. The framing structure is detailed in Exhibit 3-1-3. Each frame consists of 256 octets (groups of 8 data bits), which are numbered for bookkeeping purposes

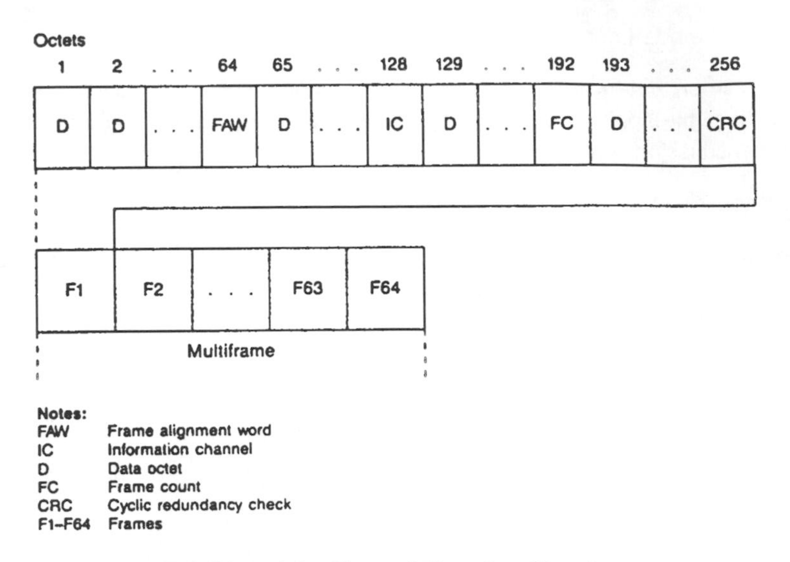

Exhibit 3-1-3. Channel Framing Structure

throughout the call. To establish and maintain the framing structure, four octets (octets 64, 128, 192, and 256) are reserved for overhead; the remaining 252 octets carry user data. As shown in Exhibit 3-1-4, 64 consecutive frames are combined to form a multiframe, which has a signaling duration of 2.048 seconds (256 X 8 X 64 bits per multiframe/64,000 b/s=2.048 seconds).

The frame alignment word (overhead octet 64) is a fixed pattern having zeroes in bit locations 1, 2, 3, and 6 and ones in bit locations 4, 5, 7, and 8. Using this constant pattern in octet 64 of every frame allows the inverse multiplexer receiver to lock on to frame octet 64 and align incoming frames and multiframes. The frame count octet (overhead octet 192) is used to measure the relative delay variance between the individual channels of the N X 56 or 64K bps call. By tracking the delay experienced in receiving the frame count octet from one frame to the next, the inverse multiplexer assesses the delay (and how it is changing as time goes on) associated with each frame of every incoming channel. A six-bit, modular 64 counter in the inverse multiplexer's receiver is incremented each time the frame count is detected, and the counter rolls over to zero and starts again at the end of each multiframe.

The cyclic redundancy check (CRC) octet (overhead octet 256) is used, in the traditional manner, for error checking. The CRC octet indicates whether frame alignment has been checked and verified. It also carries a four-bit CRC sequence based on the actual date and overhead bits carried in the 255 octets that precede the CRC octet in each frame. The transmitting inverse multiplexer performs a cyclic redundancy check on the data in each frame before

Octet	b1	b2	b3	b4	b5	b6	b7	b8
1	0	1	1	1	1	1	1	1
2	1	Channel Identifier						1
3	1	Group Identifier						1
4	1	Operating Mode			Reserved Bits			1
5	1	Rate Multiplier						1
6	1	Subrate Multiplier			Bearer Channel Rate	Reserved	Manufacturer ID Flag	1
7	1	Remote Indicator	Remote Loopback Rqst	Remote Loopback Ind	Revision Level			1
8	1	Call Subaddress ($0 \leq n \leq 63$)						1
9	1	Transfer Flag						1
10	1	1	1	Phone Number Dial Digit 1				1
11	1	1	1	Phone Number Dial Digit 2				1
12	1	1	1	Phone Number Dial Digit 3				1
13	1	1	1	Phone Number Dial Digit 4				1
14	1	1	1	Phone Number Dial Digit 5				1
15	1	1	1	Phone Number Dial Digit 6				1
16	1	1	1	Phone Number Dial Digit 7				1

Exhibit 3-1-4. Information Channel Frame

---

it is transmitted and places the appropriate four-bit sequence in the last octet of that frame. At the destination endpoint, the inverse multiplexer receiver performs the same CRC operation on each frame as was performed at the transmitting end. If the four-bit CRC sequences match, it is likely that no errors have occurred in that frame during transmission. If they do not match, errors are likely, and a message is sent back to the transmitting end requesting that the errored frame be resent. As an error-checking mechanism, this four-bit CRC sequence is somewhat simplistic, but is usually adequate to ensure data reliability on a frame-by-frame basis. Additional error-checking procedures found in most endpoint application equipment, coupled with the high-performance digital lines used by inverse multiplexing equipment work well in concert with the CRC mechanism to provide data integrity.

The information channel (IC) octet (overhead octet 128) is actually part of a 16-octet frame used to communicate the bulk of the necessary control information between the two endpoints. This information is transmitted initially during call setup, and is updated and adjusted as needs dictate throughout the call. By using octet 128 of each frame to provide the control information, in-band parameter exchange is possible, and external control channels are not necessary. However, in transparent mode and other operational modes where in-band parameter exchange is not provided, the information channel is replaced by user data after call setup is accomplished.

The information channel frame (made up of octet 128 from 16 consecutive frames of each multiframe) is depicted in Exhibit 3-1-4. The information carried in the information channel includes the channel identifier, which identifies each individual telephone circuit used in the call, and the group identifier, which maintains a grouping number assigned to various groups of circuits for bookkeeping purposes. The operating mode (transparent, 0, 1, 2, or 3) being used by the call is also carried in the information channel. The rate multiplier and subrate multiplier are used to uniquely define the data rate the application equipment uses at the sending endpoint. The rate multiplier and subrate multiplier are also used in conjunction with the channel identifier (all set to 0 by either the transmitting or receiving inverse multiplexer) to indicate that all user data has been sent and call disconnection procedures can commence.

The bearer channel rate octet indicates the actual data rate being used between the two inverse multiplexers. The manufacturer ID flag is used to identify the vendor of the inverse multiplexing equipment at each end; individual manufacturers are assigned numerical identification codes that are inserted in this octet. The various remote indicators and request flags are used for end-to-end testing of the link and equipment by the inverse multiplexers, and the revision level octet indicates the version of the inverse multiplexing standard implemented in each device.

The last half of the information channel (octets 8 to 16) is used to track the channels being used in the call (call subaddress) as well as the actual telephone numbers of the 56K- or 64K-bps dial-up circuits being used during the call. The transfer flag in octet 9 is used to identify specific telephone circuits as they are added or deleted during the transmission; this enables the inverse multiplexers to dynamically increase or decrease the number of channels used as the call progresses, according to variations in the data rate needed by the sending application. This points out a unique and most useful feature of the inverse multiplexer—it not only dials up the number of switched telephone circuits it needs to transfer the data for each individual application, but also senses changes in that need during the call and adds or deletes circuits as necessary to optimize the transmission rate.

Delay Equalization

With many different 56K- or 64K-bps dial-up channels being used to make up a call, there is often a significant variation in the delay each circuit encounters between the two endpoints. Each of these dial-up circuits takes a different routing, and a call group may even contain some circuits that are terrestrial circuits, whereas others involve microwave or satellite links. Because the application signal must be reconstructed from all of this data sequentially at the receiving end, delay adjustment and equalization are critical.

This is performed through the use of independent channel buffering, as indicated in Exhibit 3-1-5. Once all channels are synchronized at the receiving end, the delay in each channel is measured by time-making the frame count octet (octet 128) in each channel frame, as discussed earlier in relation to Exhibit 3-1-3. The appropriate delay is then imposed on the incoming data from each channel by buffering the channel's data to match the longest delay experienced on any of the incoming channels. In other words, all channels are delayed to match the worst-case delay, and the channels are then realigned and synchronized so that the composite output channel bits are properly sequenced. In the example shown in Exhibit 3-1-5, incoming channel B has the longest transmission delay, so the data from all other channels is delayed and buffered as needed to match the delay in channel B. The time-aligned channels are then combined in the right order to reproduce the wide-bandwidth application signal for the data terminal equipment at the receiving end.

COST COMPARISONS

An obvious issue that must be addressed is how expensive the transmission costs are for inverse multiplexing as compared with other wide-bandwidth services. The solution is found in Exhibit 3-1-6. Not surprisingly, because inverse multiplexing involves switched digital circuits that are dialed up as needed and paid for only as used, the transmission costs for inverse multiplexing depend on how many channels are needed and how long the circuits remain established. Exhibit 3-1-6 illustrates sample costs for cross-country and international circuits, comparing the cost of dialing up six 56K- or 64K-bps switched circuits for inverse multiplexing with the cost of leasing six 64K-bps segments of a fractional T1 circuit and the cost of leasing a full T1 link. The comparison of six inverse multiplexing channels and six channels of a fractional T1 circuit is made because this is the requirement for providing 384K-bps videoconferencing service—one of the most popular real-time, wide-bandwidth applications today.

The cost of the leased circuits is fixed. Clearly, based on the example in Exhibit 3-1-6, leasing a full T1 circuit for six-channel videoconferencing is almost never cost-effective. Leasing six channels of a fractional T1 circuit is cost-effective only if the service is used more than approximately six hours per day domestically or five hours per day internationally. For anyone using the transmission service for (on the average) less than half of each workday, inverse multiplexing is the least expensive option. As a practical matter, this

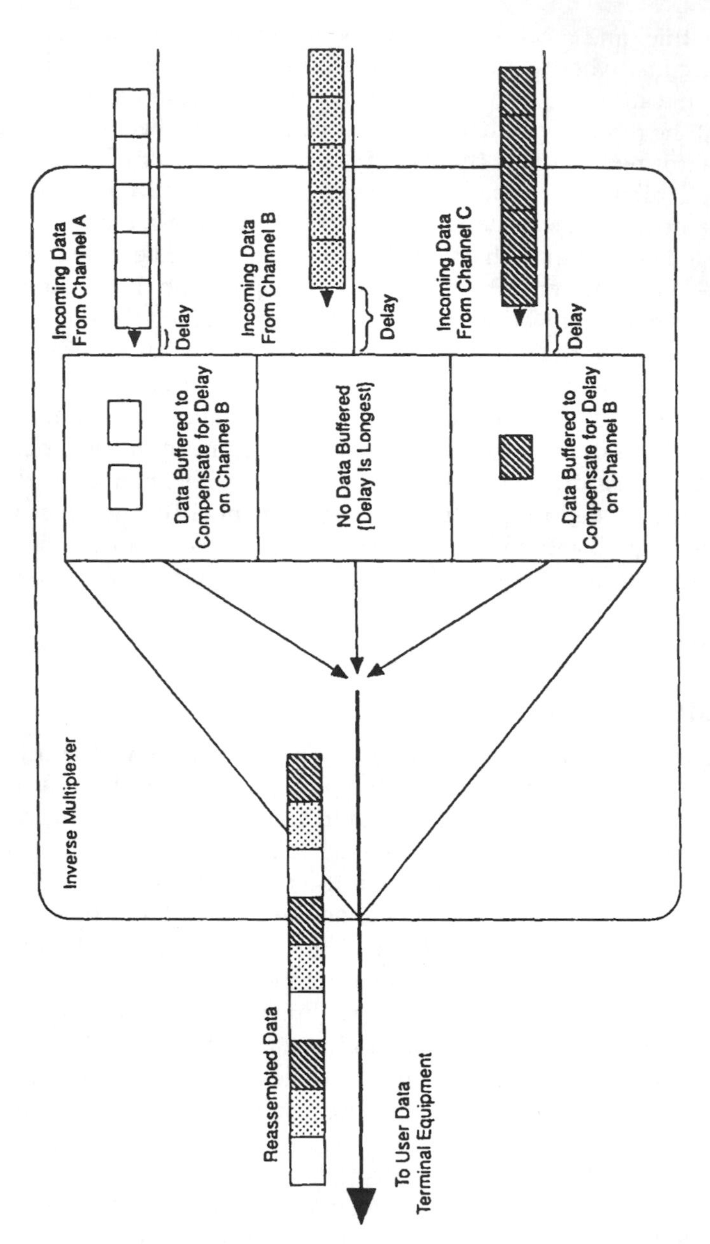

Exhibit 3-1-5. Channel Delay Equalization

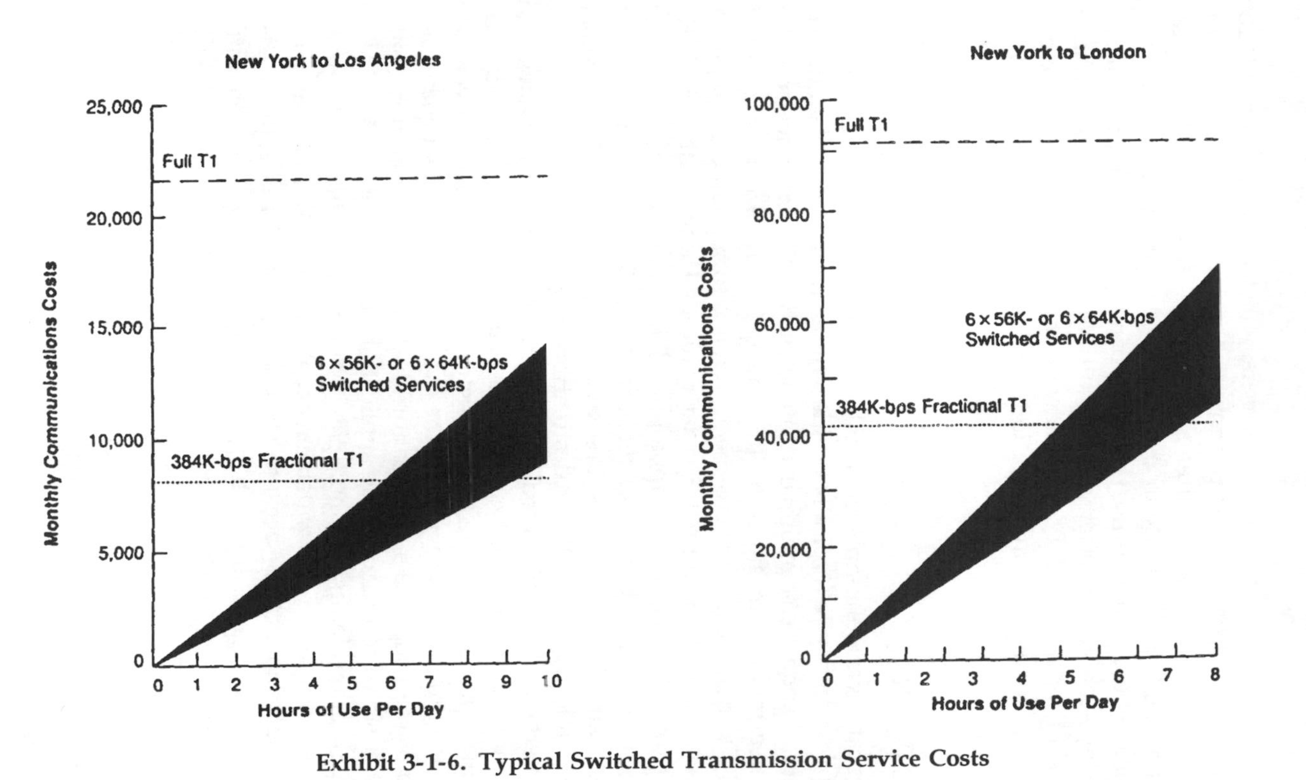

Exhibit 3-1-6. Typical Switched Transmission Service Costs

includes the vast majority of business users, which explains the rapid growth of inverse multiplexer sales. As per-minute pricing of switched digital telephone circuits continues to fall at a faster rate than the decline in leased circuit pricing, this difference in user cost will continue to grow. To cover those cases where customers use their wideband service a large amount of the time, many vendors now offer inverse multiplexers that can connect directly to ISDN, T1, or fractional T1 lines as well; this provides a measure of versatility that few digital communications devices can offer.

APPLICATIONS

A typical inverse multiplexer installation is illustrated in Exhibit 3-1-7. The application device (typically a videoconferencing codec, data device, or LAN router) and inverse multiplexer are located at the customer premises. If the inverse multiplexer supports T1 modes, it may also be connected to external T1 equipment at the customer premises to support other equipment and functions. The inverse multiplexer connects through multiple 56K- or 64K-bps (or other services, if appropriate) telephone circuits to the local exchange carrier's point of presence. From there, it is routed over an interexchange carrier's 56K- or 64K-bps switched lines to the local exchange carrier's point of presence at the destination endpoint.

The interexchange carrier's routing plan may cause variable delays among these circuits, and these delays must be equalized by the receiving end's inverse multiplexer. At the receiving customer's premises, the inverse multiplexer synchronizes and aggregates the incoming independent channels into a replica of the wideband signal that originated at the transmitting end and passes that signal to the receiving applications equipment.

The applications for inverse multiplexing technology continue to grow and expand, as new uses for this wideband capability arise. Videoconferencing is such a popular application because six-channel, 384K-bps signaling is adequate to provide reasonably high-quality video imaging. This application is also driven by the prior existence of international standards for 384K-bps videoconferencing.

Many other applications are coming to the forefront, however, as users recognize the capabilities that inverse multiplexers possess. Pseudo real-time medical and scientific imaging are good examples. Inverse multiplexers are now being placed in hospitals and clinics to provide on-demand X-ray, CAT-scan, and MRI images to and from remote sites. This allows experts located around the globe to view test results almost as soon as the tests are performed, speeding diagnoses and treatment plans, and in many cases, saving lives. In the educational field, interactive video and high-resolution graphics make long-distance, real-time education possible over telephone lines. The

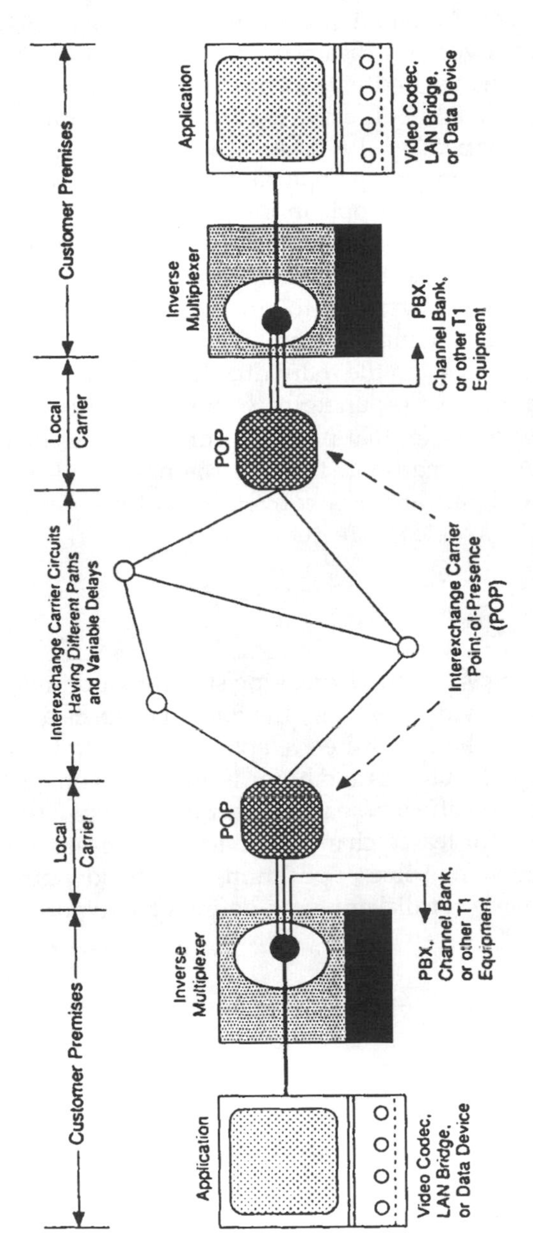

Exhibit 3-1-7. A Typical Inverse Multiplexer Installation

same concepts apply in the entertainment industry, where pictures and images can move freely over long distances in very short time spans.

Business applications beyond videoconferencing include high-speed data communications, efficient bulk file transfer, and business image transfer, and efficient LAN interconnection. LANs are the connection media for most businesses' distributed computing applications, so the network is, more and more, the backbone of business operations.

Businesses, especially large businesses with multiple campuses, are now finding it necessary to connect their LANs together via high-speed, high-capacity links. The goal is typically to find some way to interconnect the LANs in such an efficient manner that they all appear to comprise one cohesive LAN, so that users on the individual LANs communicate across the boundaries as if there were no boundaries. Connecting LANs through the use of inverse multiplexing makes that possible. For businesses that do not need permanent inter-LAN connections, the cost savings over leased inter-LAN connections may be significant. The convenience of dial-up LAN interconnectivity may also be a welcome feature.

SUMMARY

Inverse multiplexing is a ground-breaking step forward in digital communications because it provides a simple, flexible, dial-up means of wideband communications using the established telephone network. Industry research indicates that what both business and home users need is not just bandwidth; users need flexible, cost-efficient, easily accessible bandwidth.

Although the need for leased channel services and leased-line equipment will continue to grow for some applications, the bandwidth explosion is expected in the demand for dial-up, user-defined bandwidth.

3-2 The Benefits of Implementing Frame Relay

Edward Bursk

THERE ARE MANY APPLICATIONS for frame relay in wide area networking. Frame relay connections are in some ways like logical leased lines with special properties—namely, they cost less, support multiple applications, or allow traffic to temporarily burst at high rates. Following are a few discussions of applications that should help network managers understand how frame relay can benefit their networking situations.

THE BEGINNING: LAN INTERNETWORKING

The first application for frame relay to see widespread use was the interconnection of routers. Before the advent of frame relay, router networks typically used leased lines and point-to-point protocol (PPP), or perhaps X.25, to carry traffic between sites. A dedicated leased line was necessary between locations for which a one-hop connection was desired, but as networks grew, more and more ports and leased lines were required in every router in the network, thereby increasing cost. In addition, leased lines often cost more for installation and in per-month charges than frame relay connections do today.

When using frame relay for LAN internetworking, savings can be realized in several areas, including equipment, installation, local loop, and wide area charges. If a smaller router chassis or less expensive port card can be used, then additional savings can be realized.

Newer frame relay access devices (FRADs) from several vendors provide an alternative to routers in some internetworking applications. FRADs offer one or both of the following advantages:

- Equipment cost savings, as compared to access routers.

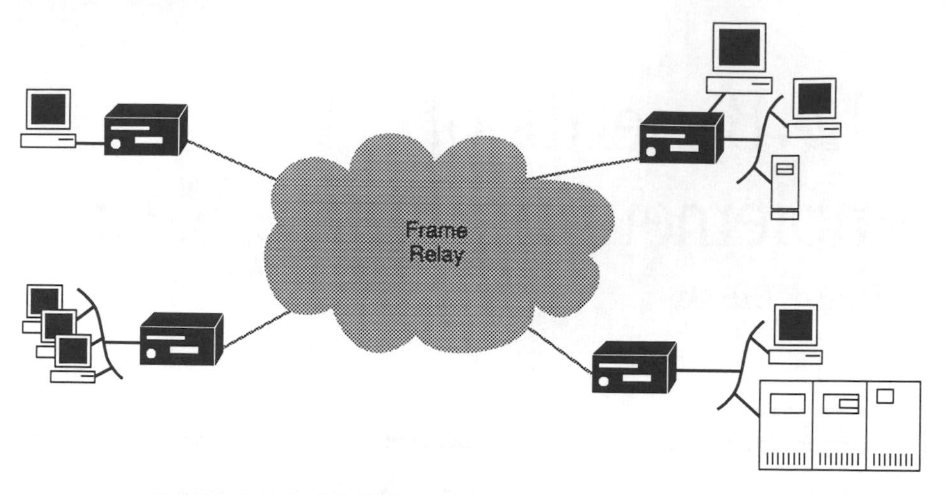

Exhibit 3-2-1. 1 LAN Internetworking over Frame Relay

- The ability to share the frame relay network with non-LAN applications.

Training costs are usually minimal as communications or IT personnel learn to manage frame relay as data link connections (DLCs), in a fashion similar to how they have managed leased lines; virtual circuits replace physical circuits when using packet-switching techniques.

Workstations and workgroups can send mail and access servers over the internetwork riding over frame relay (see Exhibit 3-2-1). As shown in this exhibit, only four physical connections are required to link all sites (using as many DLCs as required for the desired logical topology).

An area to be aware of in this application concerns dynamic-routing updates. All of the various routing update protocols (e.g., RIP, OSPF) generate some traffic over the wide area.

The application in Exhibit 3-2-1 applies to NetWare Internet Protocol Exchange (IPX) as well as Internet Protocol (IP) environments. When using NetWare, some tuning may be required for optimal performance over wide area links, such as frame relay.

SNA/ SDLC COMMUNICATIONS

One application—the replacement of leased lines carrying synchronous data link control (SDLC) traffic by frame relay DLCs—provides Systems Network Architecture (SNA) users with the opportunity to save twice, and improve performance to boot. Users save first by replacing a leased line (typically 19.2K or 56K bps) with a lower-cost frame relay connection and, second, by sharing the frame relay connection with another application (see the section on LAN and Legacy Transport). When the frame relay connection provides more bandwidth to the application than was previously available, performance can be improved as well.

This application has picked up steam more slowly than LAN internetworking over frame relay, probably because of the conservatism of most SNA network managers as compared to LAN designers. However, the maturing of frame relay functionality in IBM front end processors (FEPs) and AS/400s in early to mid-1994 has cemented the practicality of frame relay in the IBM environment.

The task of connecting remote SDLC controllers to FEPs and AS/400 midrange computers can be accomplished in two ways:

- By using a FRAD at both ends of the connection and transporting SDLC over frame relay (usually by "spoofing" the protocol—that is, by locally acknowledging the polls and sending only data across to the other end). Network managers may also use a FRAD or router as the remote end only and communicate directly with the FEP or AS/400 by using LLC2 (logical link connection type 2) encapsulated in frame relay using the RFC 1490 standard. Some FRADs and routers support this direct connection with IBM gear.
- By using a router or FRAD at both ends that supports data link switching, which encapsulates the SDLC/LLC2 traffic into IP.

Both of these methods can use a frame relay network for long-distance communications. Either technique offers the possibility of sharing the frame relay connections with other traffic—either with LAN or other legacy applications in the case of most FRADs, and with other LAN applications only, in the case of most routers—and thus achieving further cost savings.

Token Ring Connectivity. Token Ring has widespread acceptance in the IBM environment, of course. When connecting routers or bridges over frame relay to support source route bridging (SRB), DLCs are used to replace leased lines. Care must be taken in network design to account for the traffic generated with SRB. Sometimes Token Ring is used only at the host end, whereas some or all of the remote ends use SDLC serial connections.

The latter solution is popular when there are many remote sites, but the expense of Token Ring cannot be justified at these sites the way it is at the host, that is, for the savings gained using a single Token Ring versus supporting a large number of serial SDLC ports. This application variant also fits the model previously described when using a FRAD or router at the host site that supports Token Ring connectivity.

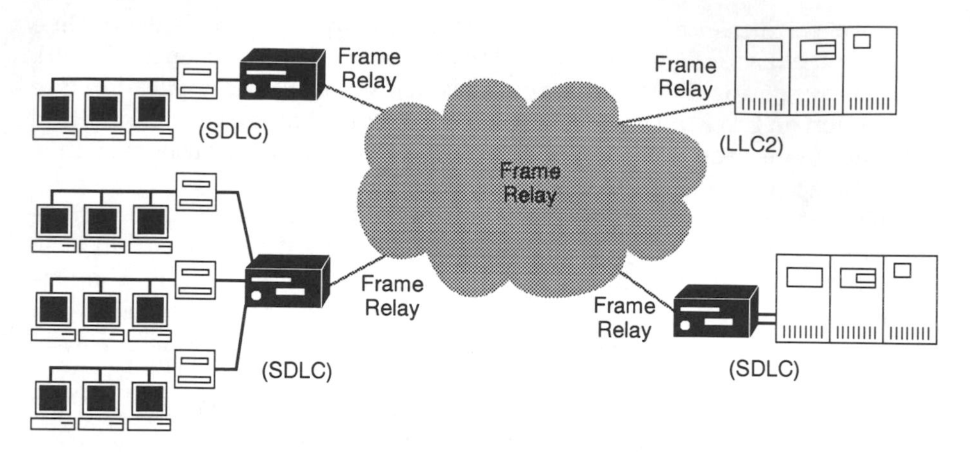

Key:

LLC2 logical link connection type 2

SDLC synchronous data link connection

Note:
SDLC traffic is sent over a frame relay service either as SDLC (using SDLC "spoofing") or using direct frame relay communications with the front end processor (top right). In the latter case, the FRADs on the left convert the SDLC traffic into LLC2. The LLC2 traffic is encapsulated in frame relay using RFC 1490. If Token Ring was used at the remote sites, the traffic would already be LLC2 based.

Exhibit 3-2-2. SNA/ SDLC Communications over Frame Relay

Additional information on SNA over frame relay support should become available as IBM and others introduce products that support IBM's High-Performance Routing (HPR) functionality under the Advanced Peer-to-Peer Networking (APPN) umbrella. Until then, the types of solutions indicated in this application example (and illustrated in Exhibit 3-2-2) can be used to provide practical SNA over frame relay solutions.

INTERNET ACCESS OVER FRAME RELAY

With the ever-increasing popularity of the Internet, the largest public data communications network in the world, the need for cost-effective network access by both businesses and consumers has increased. The issue faced by small and medium-size businesses—and by branch or regional offices of larger businesses—is that the monthly cost of dedicated access may still be prohibitive, yet the performance of more affordable dialup connections is somewhat disappointing.

The situation is exacerbated by the use of powerful information tools such as Mosaic, from the National Center for Supercomputing Applications (NCSA) at the University of Illinois, or its cousin Netscape, from Netscape Communications Corp., both of which are among the popular tools used to access the World Wide Web. With these tools regularly transferring thousands of characters of text and perhaps millions of bytes of graphics to and from users' PCs each and every session, 14.4K-bps or even 28.8K-bps modems quickly become a bottleneck.

Frame relay offers a unique solution to this problem. Dedicated access to the user's Internet service provider is provisioned through a synchronous 56K-bps, Fractional T1, or full T1 link (or from Nx64 to E1), depending on the performance required and budgetary constraints, for example. This connection is usually less expensive than the equivalent leased line service. (If it is not, some users may choose to acquire a leased line to another one of their offices, where the service is more affordable, or to set up a private frame relay network of their own.) A FRAD or router is then used to connect to the Internet service (often associated with a firewall server or bastion host for security). The secret to higher performance than the dialup connections is frame relay's 56K-bps or higher data transfer rate, compared with the underlying 14.4K or 28.8K-bps rates for dialup (56.7, 115.2, and higher asynchronous speeds require compression and are actually peaks as opposed to true long-term transfer rates).

Savings are realized in any or all of the following areas: equipment, installation, local loop, and Internet service charges. If a small FRAD or smaller router can be used, then additional savings are realized. FRADs from several vendors provide an alternative to routers in some applications, making it possible to reap some equipment cost savings and to share the frame relay network with non-LAN applications. For a single workstation, a PC card providing the required serial interface can be used (remembering to include the DSU/CSU, however). Training costs are usually minimal as communications personnel become accustomed to managing DLCs much like they have managed leased lines.

Exhibit 3-2-3 illustrates Internet access over frame relay. Frame relay can also be used for private internetworks and for networking NetWare LANs (IPX).

LAN and Legacy Transport

For some users, it is possible to share the frame relay line for multiple applications. This solution allows the multiple leased lines required for, say, NetWare LAN-to-LAN along with simultaneous Burroughs terminal/host communications to be collapsed into a single frame relay access line. How the applications share the line depends on several factors, including:

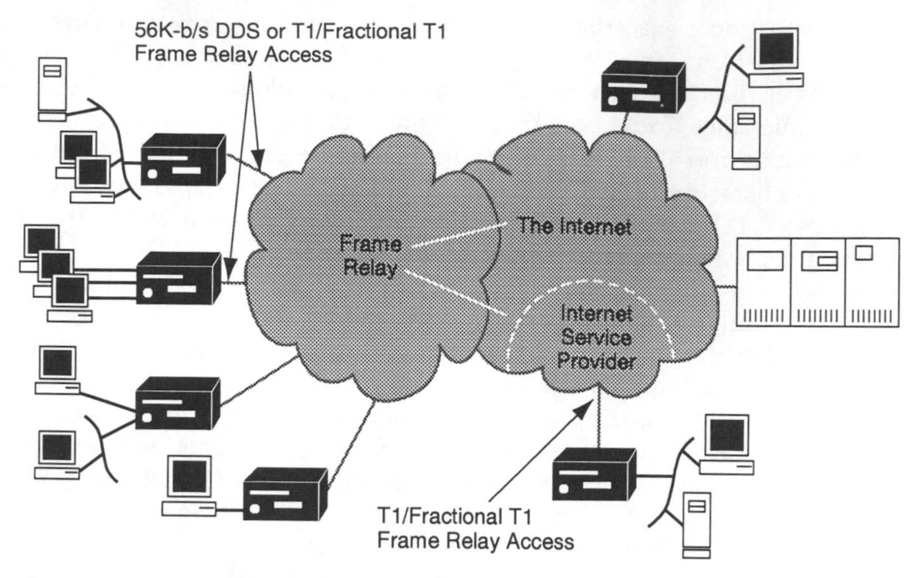

Key:
DDS Dataphone Digital Service

Note:
Numerous sites receive access to the Internet through a single frame relay connection to an Internet service provider (ISP). The ISP then provides connectivity into the Internet (through frame relay or other means).

Exhibit 3-2-3. Internet Access over Frame Relay

- How the network is designed (e.g., the price/performance criteria).
- What frame relay access equipment is selected.
- What the selected frame relay access equipment offers in terms of applications support and for DLC sharing.

The last factor is quite often ill-understood by many network designers who are unfamiliar with either frame relay or X.25. The ability of the frame relay equipment to share a DLC with multiple applications means that, in some cases, a single DLC can serve as a pipe to send more than one application through to a common destination, as is often seen with both hub-and-spoke and hierarchical networks.

Not all frame relay equipment lends itself to supporting the DLC-sharing feature. In the case of SNA/SDLC, routers more commonly support SNA/SDLC transport through data link switching, or DLSw. Some routers offer some X.25 support as well, but it is usually limited to routing IP over X.25. Asynchronous support may be available but limited to Telnet (i.e., asynchronous over IP). That is, routers can share a DLC (acting as a logical leased line) with multiple IP applications. This feature is not surprising, however, because routers were designed for IP applications and use IP as a backbone transport protocol.

Client/ Server Migrations. In the case of a mixed LAN and legacy situation, most FRADs offer a larger degree of legacy application support, adding bisynchronous, Burroughs Poll/Select, Uniscope, asynchronous, and X.25 support to SNA/SDLC, as already discussed. Again, not surprisingly, most FRADs offer less in the way of IP and IPX support than routers; some only support serial line IP (SLIP) and PPP, for example. There are, however, several notable exceptions to this rule.

Some FRADs offer the ability to share a single DLC with heterogeneous applications (e.g., IPX plus Burroughs). These cases arise out of the desire to migrate applications gradually to LAN client/server arrangements, or often simply to take advantage of the lower cost of frame relay versus leased lines (perhaps by merging two leased connections into a single frame relay line) while avoiding complete reengineering of the applications for an IP client/ server arrangement, for example.

Priority Algorithm. No matter how the DLC is being shared, a key factor is the ability of the FRAD or router to keep one of the applications from overwhelming the other. This is accomplished by way of some sort of a fairness or priority algorithm. More and more FRADs and routers offer this capability, which is required when sharing a line for multiple applications over the same DLC or over separate DLCs. When the applications are homed on different destinations (e.g., different hosts or servers in distant facilities), the applications can share the frame relay access line, but not the same DLC. It may also be advantageous from a network management point of view to have, for example, all LAN traffic from a site travel over one DLC and all legacy traffic travel over another (see Exhibit 3-2-4). That way, prioritization or tuning can be done in a way somewhat similar to using separate leased lines. The disadvantage is, of course, the incremental cost of additional DLCs.

SHARING WIDE AREA LINKS: BANDWIDTH ON DEMAND

An overwhelming reason to use frame relay is the ability to send a burst of data at rates above the committed information rate (CIR)—the guaranteed service a customer has contracted for with the frame relay carrier—for those times when the offered traffic from one or a combination of user applications exceeds the CIR.

For example, a customer's original application runs on a 56K-bps leased line. The average traffic is 14.4K bps. However, at certain times, the instan-

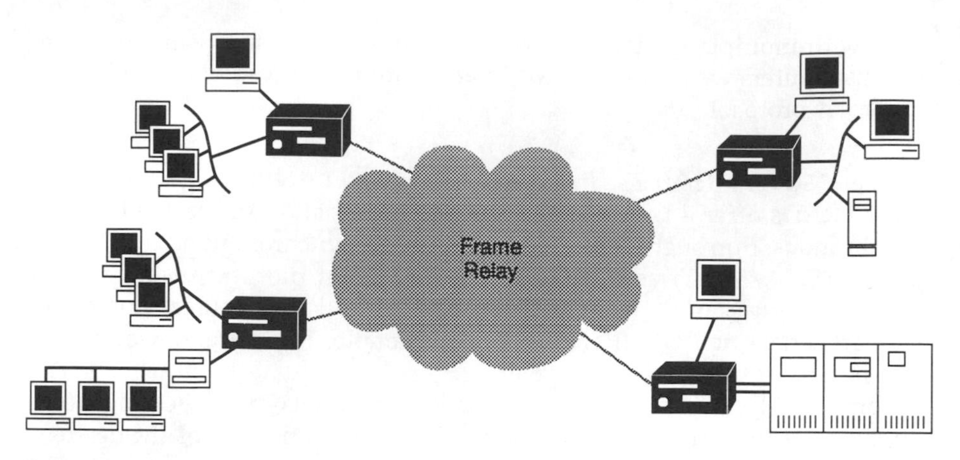

Note:
Depending on the equipment used and the methods employed, traffic can ride over the same DLCs (for cost savings) or use a separate DLC for each application (for ease of management). All traffic is carried over frame relay, allowing only one physical line to be employed at each site. Management traffic, such as SNMP, may also ride over shared or separate DLCs.

Exhibit 3-2-4. LAN and Legacy Traffic Networked over Frame Relay

taneous demand is much higher. The leased line could support 56K bps maximum. A 56K-bps frame relay line can offer approximately the same transmission (actually slightly less, because of a small amount of overhead). But the customer would not necessarily have to pay for a 56K-bps CIR when using frame relay, given that the average traffic is lower. The customer could select a typical 16K-bps CIR offering, for example, to deal with the average traffic rate and allow the traffic to burst up to 56K bps. The downside is that the traffic above the CIR (16K bps in this example) can be discarded by the carrier if the network is in congestion at the time the extra data is sent.

Burst Rates. This arrangement is even more advantageous when the customer considers a T1 frame relay connection versus the original 56K-bps leased line. The burst rate could then approach 1.536M bps, or whatever Fractional T1 rate (e.g., 768K bps) was contracted for with the carrier.

The frame relay carrier can accommodate these bursts up to the line rate, or more typically, up to another limit known as the burst rate, as long as there is excess bandwidth available in the frame relay network at the time it is needed. If the bandwidth is not available, the excess traffic (i.e., the amount over the CIR) is discarded. If there is a burst rate limit set below the line rate, then any traffic above the burst rate is automatically discarded. (The speed the traffic can burst to can never be higher than the line rate, of course, unless some form of data compression is being used.)

There are some pitfalls, of course, which are discussed later. Overall, however, the ability of the frame relay connection to accommodate bursts of traffic becomes even more compelling when users combine multiple applications onto the line. With two applications whose combined average traffic rate is less than or equal to the CIR chosen, the ability to send a burst of data allows more traffic to get through for those instances it is required, while the customer pays only for the average rate. Because most carriers design their frame relay networks to allow for a reasonable level of instantaneous bursting by their customers, the situation should work in the customer's favor.

Congestion Control. A related issue is congestion control. The router or FRAD the network uses should be able to deal with congestion caused by oversubscription at the user side of the device or from the network side, if the carrier's network becomes overloaded either transiently or over the long term.

When network managers measure their traffic throughput, they should be getting at least as much as their CIR. If not, something is wrong, either because of errors or congestion somewhere in the network or in the attached equipment.

Sharing Bandwidth Between Voice and Data. A topic of increasing importance is the sharing of frame relay bandwidth for voice and data applications. The Frame Relay Forum is nearing a formal implementation agreement. The keys to successfully sharing bandwidth among these disparate applications are congestion control, plus a priority or fairness scheme in the voice-capable FRAD, giving it the ability to limit traffic, and often the ability to fragment and reassemble traffic into smaller frames to better interleave the voice and data traffic. This causes less delay between frames carrying voice. Also, an appropriate buffer mechanism is needed to account for the jitter possible in packet-type networks such as frame relay. Sharing of bandwidth, which is accomplished through statistical multiplexing, is illustrated in Exhibit 3-2-5.

POINTERS ON FRAME RELAY NETWORK DESIGN

Although this section is not intended to give blanket guidance on network design, it does offer a few pointers on frame relay network design.

Think of Frame Relay DLCs as Specialized Leased Lines. Cost and performance characteristics of DLCs are predictable (or at least estimable). Network managers should size their traffic and choose their CIR appropriately. Zero (0) CIR service may be priced attractively, but it may not offer the

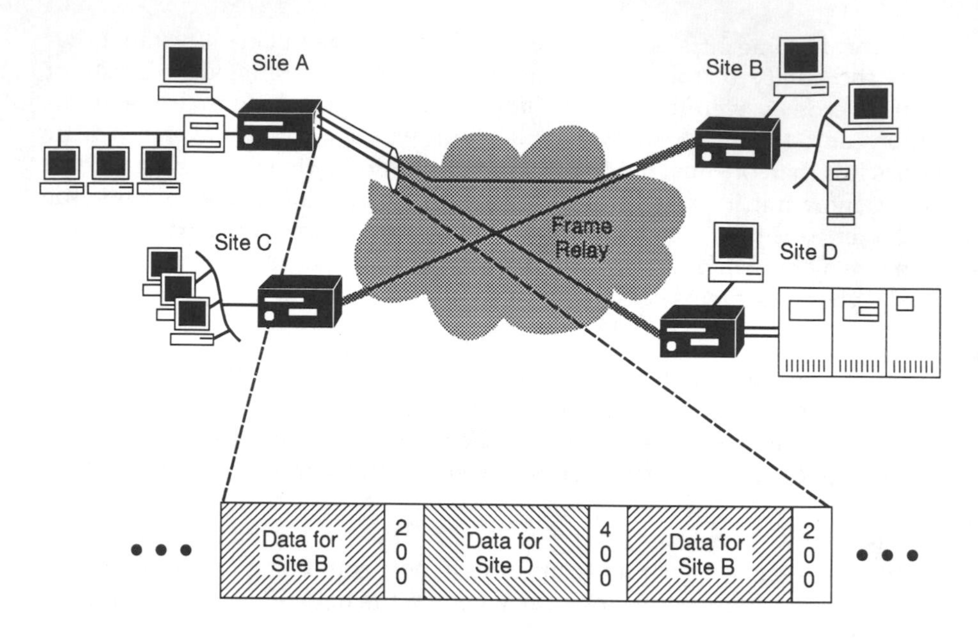

Note:
Sharing of bandwidth is accomplished through statistical multiplexing. All frames associated with a particular DLC carry that number in the header. (Some control information was left out of the exhibit for simplicity.) Data for sites B and D is interleaved on the line. Data going to site B is sent on DLC 200. Data going to site D is sent on DLC 400.

The line should be sized so that the line rate is higher than the aggregate rate required to support all applications, plus peaks. The CIR chosen should reflect the minimum throughput that can be tolerated for the applications sharing the line. Some additional "head room" is usually called for, to allow for peaks and heavy traffic periods. Frame relay services can usually tolerate "bursting" for traffic above the CIR, accommodating the transient need to get higher throughput for peaks.

Exhibit 3-2-5. Sharing Bandwidth over Frame Relay

throughput desired when it is really needed. (As always, when negotiating a contract, network managers should make sure they have service guarantees.)

A Frame Relay Connection Should Have Slightly Higher Latency than the Same-Speed Leased Line. This is especially true over long distances that may route traffic through several switches. Network managers should investigate costs to find out where frame relay's savings provide the opportunity to use a higher-clock speed connection (as opposed to the CIR) in order to reduce latency.

When Comparing a Leased Line Network to Frame Relay, Significant Cost Savings May (or May Not) Be Apparent. Customers typically see a small savings with most frame relay service offerings, but not all. Some frame relay offerings are actually more expensive when compared one-to-one with leased lines, often where the leased line bit rate and the CIR are the same. The largest savings occur when customers build a network of DLCs using fewer physical connections than required by the original leased line topology and size these connections—clock speed and CIR—appropriately.

Factor in Other Charges. When comparing frame relay dedicated access to switched connectivity options such as integrated services digital network (ISDN), switched 56 service ("dial" DDS at 56K bps), and analog modems, bear in mind that these options typically entail a per-minute charge—at least for long-distance traffic, though sometimes for all traffic—in the same ways that X.25 services typically have per-packet charges on the amount of traffic sent. These charges can add up quickly. Network managers should do the math for the connect times that their applications require, at least for an estimate of the average times. Usually, they will find a crossover point in terms of minutes or hours of connect time for their applications that may make dedicated services such as frame relay a better choice. Often this is on the order of only several solid hours of connectivity per day. (In addition, users may need to disconnect from switched services when they are done working or doing work offline. They will then experience a call setup delay when reconnecting. This situation does not occur with dedicated access.)

Frame Relay Services Are Typically Priced as Fixed Costs Based on Local Loop, Number of DLCs, and CIR Chosen. A few services charge in regards to distance, too. Nevertheless, a fixed amount can be budgeted that will not vary for the term of the contract. This pricing can be a major advantage in controlling costs when compared with switched services. (There is something to be said for having predictable communications costs, especially in today's dynamic environment.)

Consider a Hybrid Approach. Several carriers offer (or are planning to offer) ISDN-to-frame relay interworking in their networks, with ISDN used as a switched, cost-effective local loop alternative. Some regional carriers may either drop or significantly reduce per-minute charges (at least for local calls within their service region), making ISDN the best choice for the local loop, as it already is in Germany and Australia, for example. Therefore, some hybrid approaches might be considered, including:

- Dedicated or switched remote connections, chosen depending on the application served, connect-time estimates, and service available (e.g., ISDN).

- Dedicated service at the host/server site, where connect time will be high, easily justifying dedicated service that is sized for predicted average and peak traffic.

Do Not Overlook Standard Facets of Connectivity or Performance. Frame relay will not magically supercharge an application. Businesses should design their distributed applications with networking in mind.

Using higher-speed connections (frame relay or not) or improving the organization's applications are the factors that make the bits travel faster (data compression is another option, of course). What frame relay can do for the corporate network, however, is provide higher-speed connectivity more affordably.

THE ECONOMICS: WHEN DOES FRAME RELAY MAKE SENSE?

The flowchart in Exhibit 3-2-6 is intended to be used as a quick guide for determining when frame relay may be appropriate for a particular networking application. Of course, network managers need to carefully examine and consider several key factors before making a final determination about whether it makes sense to use frame relay. These factors include:

- What is the network topology (e.g., hub and spoke, hierarchical, mesh, or hybrid)?
- What performance criteria does the network use (e.g., transfer times, latency, or response time)?
- What is the networking budget (e.g., for circuits, people, and equipment)?

Cost Components

The following is a summary of cost components in frame relay network design. Readers should use this list as a basis for their own evaluation checklist, adding those items that apply to their unique situation. Each item on the network manager's checklist can be assigned a weight according to its importance or impact for the particular networking scenario. This method can help the networking professional in making a thorough evaluation and decision.

The chief frame relay costs are for communications, personnel, and equipment. Communications costs cover frame relay service, which includes:

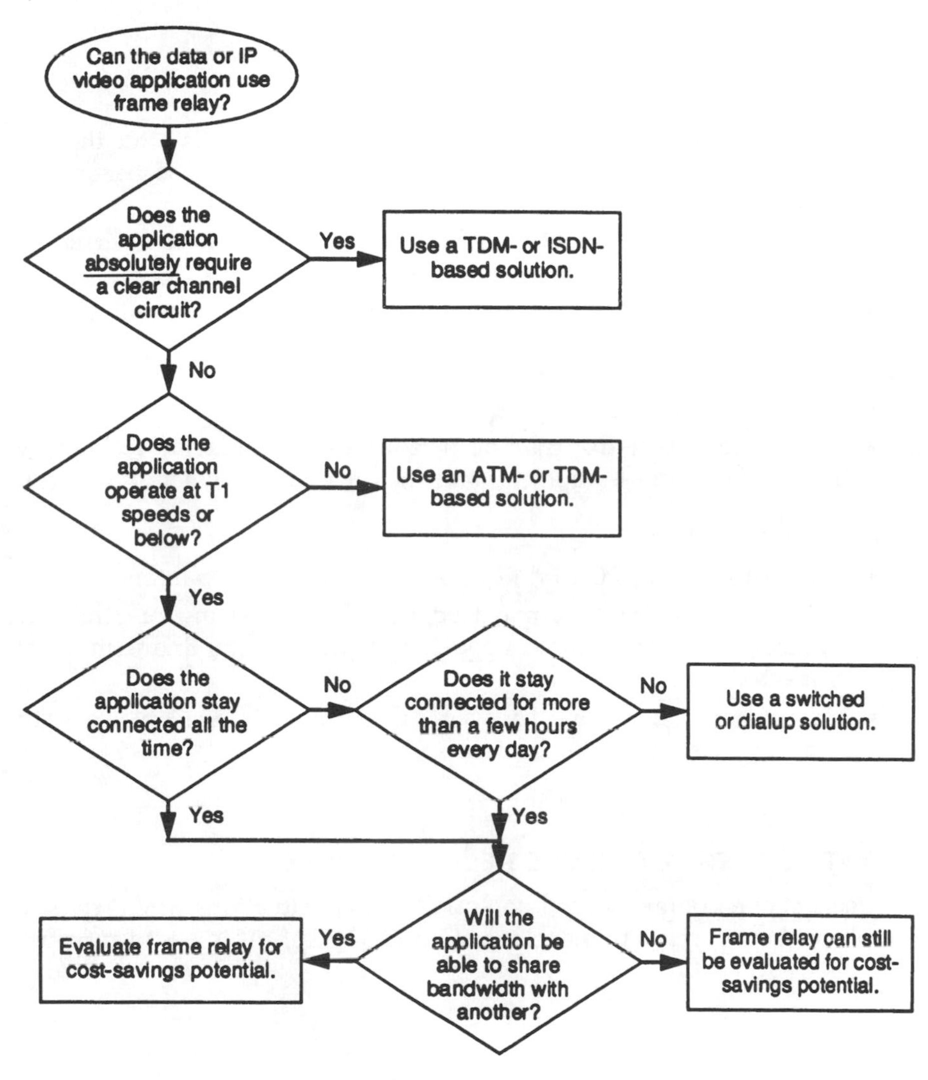

Key:
ATM asynchronous transfer mode
ISDN integrated services digital network
TDM time division multiplexing

Exhibit 3-2-6. A Rule of Thumb for Evaluating Frame Relay

- Access speeds (e.g., 56K bps, T1).
- Burst rate (which may be fixed).
- Distance charges (usually do not apply).
- DLC and CIR (although costs should be less for DLCs after the first charge, charges for succeeding DLCs vary significantly, depending on the carrier).
- Local access (e.g., per month to RBOC or IXC service; these costs may be bundled).
- Installation (e.g., one-time charge for provisioning and installing).

Personnel costs include:

- Salaries.
- Training (e.g., costs for training personnel on frame relay technology, SNMP, the chosen equipment, and the carrier management interface).

Equipment costs include:

- FRAD or router or PC card (e.g., one-time cost plus maintenance).
- DSU/CSU, TA, ISDN terminal adapter (TA), modems, or other line termination and cable (unless equipment with integral line termination is selected).
- Management station and software (unless already available; maintenance costs must also be factored in).

CONTRACTING FOR FRAME RELAY SERVICES

Users may encounter several potential problem situations when trying to implement or operate a frame relay network. Some of the issues to anticipate are discussed next.

The Carrier Does Not Have Any Circuits Available. This situation has occurred from time to time with many carriers over the past few years as demand for frame relay services has increased. Although this problem is usually rectified by the purchase of more frame relay switches by the carrier, delays can be significant—as long as 12 to 16 weeks—while the new switches are installed. Many options are available to customers in this situation, including:

- *Waiting.* Depending on the situation, the time can be filled with other project activities, such as the selection of customer premises equipment.
- *Looking for another carrier.* Sometimes even the initiation of a search can get a customer moved up the waiting list.

- *Choosing another networking option.* If the choice of frame relay is marginal for an application (e.g., remote asynchronous access for PCS and laptops accessing E-mail or files at low speeds or for less than a hour a day), then another technology might be appropriate.
- *Beginning installation of a portion of the network that does not use frame relay.* For example, this might include a defined portion of the network in which leased line service is less expensive than frame relay, or at the central site where LAN connections must be installed.
- *Building a private frame relay network.* After doing the math on the purchased, financed, or leased equipment, network managers can use the results as a guide that might dictate the construction of a hybrid public/private network.

Performance: What to Expect. Network managers should negotiate clear performance guarantees or warranties into the contract with the carrier. Several important items for a contract checklist are outlined in Exhibit 3-2-7. Readers can add their own requirements to customize the checklist for their situation.

In addition, there are certain steps network managers will need to take as they design and implement their overall networking strategy, including the frame relay-related portions. Exhibit 3-2-8 reviews typical steps during network planning and acquisition that should precede any contract.

MANAGING THE PACKET-SWITCHED NETWORK

Management of a frame relay, IP, or any packet-switched network is difficult to imagine for those who have not done it before. It is often helpful to think in terms of managing the virtual circuits (i.e., the DLCs in the case of frame relay) rather than the frames or packets. Each DLC can be thought of as a leased line. The FRAD or a router should offer information on frames transmitted and received, errored frames, and a number of other statistics. Likewise, carriers may offer statistics to their customers (the carrier should have statistics available from the switches used to provision the service—after all, that is how the carrier's equipment polices the CIR and burst rate activity). Anyone who has managed a time division multiplexing (TDM) network that supported multiple connections on a T1 (e.g., DS0s or fractional connections) will have done something conceptually similar. Those with ISDN or X.25 experience will also have worked with similar concepts.

Most frame relay equipment available today supports either local management interface (LMI) or ANSI Annex D, both of which are standards governing the management and status of active and inactive DLCs. Another version of this specification is the ITU-TSS's version of frame relay called

Communications Circuits

☐ Does the carrier offer the access speed desired (e.g., 56K b/s, Fractional T1)?

☐ Do distance charges apply? To answer this question, always draw the geographic network and ask the chosen carriers to price it.

☐ What are the charges for the first DLC (by CIR)? What are the charges for succeeding DLCs?*

☐ What are the options for burst rate, for the access speeds chosen? How fast can the network react to bursts — immediately or over some time? What happens to frames that are sent above the CIR — are they tagged DE** or not?

☐ What guarantees are there for availability? Are these guarantees in writing? How will they be monitored?

☐ What are the problem escalation procedures? Has the carrier explained how its procedures work?

☐ What are the local access charges — are they per month to the regional Bell operating company or IXC service? Are they bundled with the IXC service?

☐ Are there one-time charges for provisioning and installing the communications circuits? Are installation dates and guarantees in writing?

☐ What is the length of the contract (in months or years)? How long are the selections guaranteed at the specified costs?

Personnel

☐ Is training on equipment (including the carrier management interface, if necessary) provided by the carrier?

Equipment

☐ Does the carrier offer frame relay access devices (FRADs) or routers? Does the carrier recommend equipment?

☐ If recommended equipment is used, are installation, warranty, and maintenance explained?

☐ If carrier-recommended equipment is not used, is the equipment chosen "approved" for use on the carrier's frame relay service (it should be)?

☐ What are the problem escalation procedures?

☐ Does the equipment provided/selected support a fairness or priority algorithm? Offer congestion control? Support enough LAN and serial ports, and protocols, required now and in the future?

☐ Is the equipment SNMP manageable?

☐ Does the equipment provide the DSU/CSU, TA, or other line termination? If it does not, is a cable provided?

*Charges should be less for DLCs after the first, but charges for succeeding DLCs will vary significantly among the carriers.

**DE, or discard eligible, refers to frames to be discarded during congestion.

Exhibit 3-2-7. Frame Relay Contract Checklist

- **Develop the business plan.** Justify why the network needs updating, what the plans are, and what the schedule and budget will be.
- **Design the network from the topological, geographical, and applications perspectives.** In detail, describe what is required from all aspects of connectivity and application support. If there are opportunities to combine applications over the wide area or migrate to LANS, this effort should help the planner or its vendors identify them. This detailed information is essential to getting accurate quotes from carriers, vendors, outsourcers, or integrators.

  Details on how to run the applications over frame relay emerge at this stage. Preliminary decisions can be made in regard to routing protocols, encapsulation, and link sizing, for example. Other key issues center on: network management, disaster recovery, and access security (i.e., dial backup and alternate routing, backup computing, password protection and authentication).
- **Develop an RFP for service and equipment.** Assign weights to the categories and judge the readers' and carriers' responses accordingly. Include time and cost constraints as appropriate.
- **Select a carrier and an equipment vendor.** Develop and maintain a relationship with both . . . good communications will prevent headaches during installation, trial and commissioning, and on through ongoing operations.
- **Trial-run the equipment and service with the application (if appropriate).** This requirement should be part of the contract (it should show up in the RFP and associated schedule first). Success criteria help everyone understand if the service and equipment are meeting the user's needs. The scope of the testing should be realistic, however. The larger the network, the more complex they can be, but for a small network, exercise restraint. The resource implications are serious for everyone involved.
- **Implement the plan.** Once the carrier and vendor have been selected and the trial completed, it will be time to implement. Work the plan!

Exhibit 3-2-8. Overview of Network Planning and Acquisition Process

Annex A, which is used internationally. Frame relay service typically requires that the network support at least one of these options, though usually Annex D. When problems occur with DLCs, customers should first look at the status available from the frame relay access equipment before contacting the carrier.

Installation and Configuration

There are several places things could go wrong during installation of frame relay. The most typical problem occurs with the configuration of DLCs in the frame relay access equipment. When the wrong DLCs are configured, the network will not be able to pass data traffic. Because the wrong DLCs may be configured because of human error on the customer side or on the carrier side, customers need always check the access equipment's configuration versus the one the configuration carrier has made in its switch. Because DLC identifiers (DLCIs) have local significance only, it is easy to accidentally configure the local side with DLCIs associated with the other side of the connection, which may or may not be the same as the local one.

If the network is configured correctly according to the information the carrier has provided, it is still possible that the customer received incorrect information from the carrier. Therefore, it is prudent to contact the carrier and have them verify the DLC numbers they have provided. If they are correct, the DLCs will come up after the carrier checks on them (perhaps they were not already activated or the carrier's port required initialization). As a last resort, if the DLC is the first or only one at this connection, the user might try configuring from DLCI 16, because many connections start from this number. It always helps to have a datascope available to check on the circuit or to have sufficient status display ability in the chosen frame relay equipment.

Some frame relay devices have the ability to display the DLCs that are active on the network side, but that have not been configured on the customer side. This ability is a real advantage in debugging this sort of installation problem. Alternatively, some devices automatically configure the active DLCs, so the user does not have to check.

Exhibit 3-2-9 is an installation checklist for frame relay networks. Once again, the checklist considers three cost factors: communications circuits, personnel, and equipment.

Keeping Costs Under Control

The key to maintaining control over costs is proactive management of network resources. Proper network design is a good start. Monitoring of activity and errors will reveal how to maintain and grow the network as required. Establishing a good relationship with the carrier will give the customer a leg up when dealing with service problems, or when it comes time to renegotiate the contract.

Provided that the chosen frame relay carrier is giving proper service based on the contracted CIR and burst rate, the most typical problem stems from the CIR chosen and the actual traffic rate. When underestimated, the amount of traffic actually being offered to the network will be higher than the CIR, and there is a chance that this traffic will be discarded if it comprises a higher average rate as opposed to just an occasional burst. If this is the case, throughput could suffer, even when higher-level protocols such as Transmission Control Protocol (TCP) adapt to the seemingly high error line, due to the discarded traffic. Customers must then revisit their traffic calculations and adjust the CIR to be contracted from the carrier appropriately. If the monthly cost of the increased CIR is not within budget, a customer might consider adding equipment or software that compresses data. This option

Communications Circuits

- ☐ Is frame relay service ordered and are installation date and DLCIs confirmed (in writing)?
- ☐ Is local access ordered and the installation date confirmed? (May be bundled with frame relay service.)
- ☐ Are installation charges known and PO number issued?

Personnel

- ☐ Have installation or operations staff been trained? Is the carrier or systems integrator providing installation personnel? (All relevant parties must be trained in the equipment, frame relay, and the management system. Even if another group is doing the installation, the customer should participate.)

Equipment

- ☐ Are the FRAD, router, and other equipment available and configured (for frame relay and application interfaces) in order to test?
- ☐ Has DSU/CSU, TA, or other line termination and cable been configured?
- ☐ Have relevant loopback or other "sanity check" tests been run?
- ☐ Are datascopes, breakout boxes, or other equipment available to verify frame relay connection and DLC numbering and status?
- ☐ Are equipment management and control procedures ready and working?
- ☐ Has problem escalation been worked out?

Exhibit 3-2-9. Frame Relay Installation Checklist

generally requires only a one-time charge, as opposed to monthly, recurring fees.

PLANNING FOR BACKUP

The most likely backup problems occur in the local loop. Backup scenarios include:

- Having a modem, SW56, or ISDN dial backup scheme that dials around the frame relay network.
- Having an alternate frame relay connection at zero (0) kilobits per second for low cost, if available (this strategy is not recommended, however, because all data may be discarded by the carrier), or at the same CIR for highest availability. From a network design standpoint, use of another carrier (in particular another local loop connection) or perhaps a non-frame relay leased line should be available as an alternate path if the primary one goes down.
- Obtaining switched access or switched backup from the carrier. Carriers

most likely use V.34 modem or ISDN BRI (basic rate interface) service through their frame relay service (as opposed to around it).

- Never putting backup connections over the same conduit or local loop as the primary, if at all possible. If the local loop is affected by the proverbial backhoe, all associated links may be affected adversely. Over the long haul, diverse routing is recommended.

Another problem that may become more visible in coming years is network congestion. It is important for frame relay customers to monitor their service using available statistics and reports. When carriers see on their own that service is not being maintained, or when customers complain, the carriers typically add or upgrade equipment in order to return to appropriate service levels.

SUMMARY

As a high-speed packet-switching protocol, frame relay is an efficient technology designed for today's reliable circuits. Corporate network managers should find that frame relay, when properly applied, offers high value and predictable costs. Nonetheless, there are some items to watch for when designing and implementing a frame relay application. Not all potential pitfalls and suggested solutions are listed in this section (an entire book could be written on the rules and pitfalls of frame relay network design), but after reviewing the applications discussed in this chapter, network managers should be able to let their common sense guide their network design and debugging procedures.

Problem. Some or all frame relay connections are more expensive than a leased line.

Solution. Run frame relay, X.25, or PPP over a leased or switched line for these connections. Private frame relay networking is an option that is especially practical when frame relay makes sense overall. If the network is a small hub-and-spoke configuration, some carriers have a special service in which they groom leased lines into a T1—that is, the carrier routes multiple DDS lines into DS0s on a single T1 and presents the customer with the T1 interface. Exhibit 3-2-10 illustrates this solution using equipment from FastComm Communications Corp. Running frame relay over the user's own leased lines is not a problem. This is essentially how the carriers get frame relay traffic over the local loop to the customer from wherever their closest frame relay switch is.

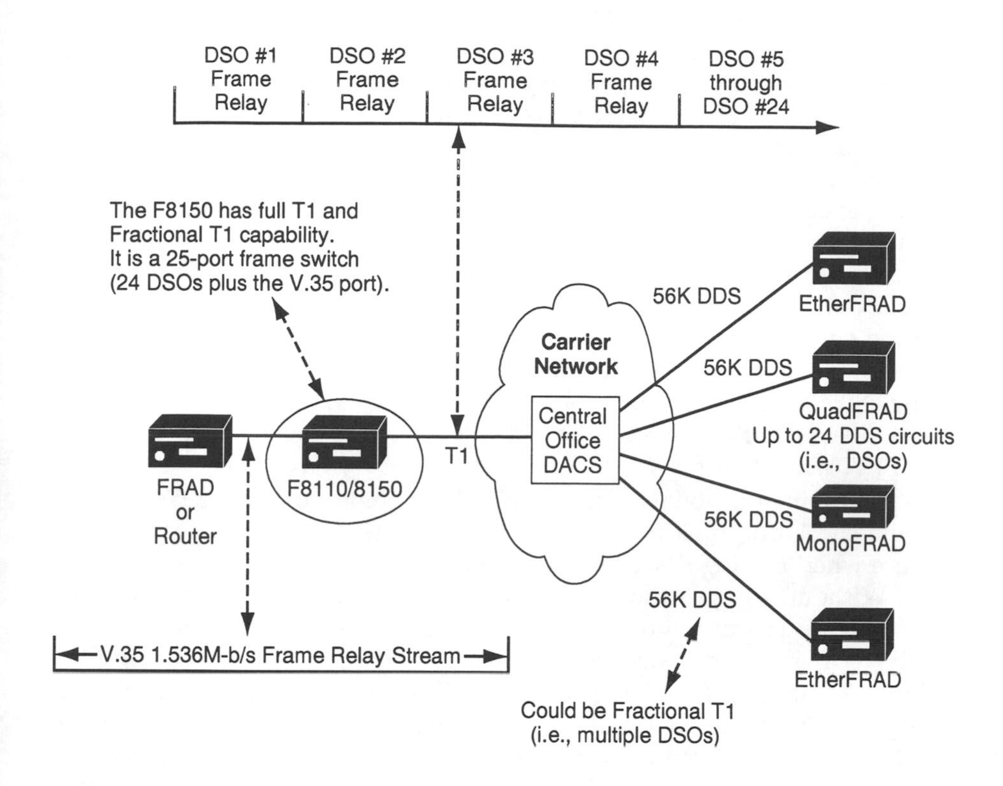

Key:

DACS Digital Access and Crossconnect System
DDS Dataphone Digital Service

SOURCE: FastComm Communications Corp. All rights reserved. Printed with permission.
MonoFRAD, QuadFRAD, and EtherFRAD are trademarks of FastComm Communications Corp.

Exhibit 3-2-10. Running Frame Relay over Multiple DDS Lines from a T1

Problem. Configurations home many remotes (e.g., more than 30) to a single router or FRAD with a T1/E1 port. Some routers and FRADs limit the number of DLCs per port (usually when dynamic-routing updates are turned on), largely because of the need to do more processing associated with port handling and computing routing updates. Customers may assume that they can bring 100+ sites (DLCs) in over a single T1/E1 frame relay connection, but this equipment configuration limit could prevent them from doing so.

Solution. Turn off dynamic routing and use a frame relay switch to break out the T1 into multiple serial connections (thus spreading out the DLCs over several ports), or choose a FRAD or another router that does not have the DLC limitation.

A related situation is the so-called split horizon problem. The problem arose because routers would not send routing updates back along the port from which they arrived (i.e., back over the horizon to avoid broadcast storms). With frame relay private virtual circuits (i.e., DLCs), most routers would not send an update back along the same physical port the update arrived on, even though that port might now have several DLCs on it—one that delivered the routing update and the rest, which should receive their copy of it. This scenario, now solved in most routers, was never a problem in most FRADs and frame relay switches, which were already used to dealing with PVCs and SVCs.

Some tuning may be required for NetWare environments. NetWare users should check out burst mode operation for a significant increase in performance when routing IPX over wide area connections (frame relay and others). Without using all the features available in NetWare 3.X, performance will suffer as compared to IP over frame relay, although it will undoubtedly be better than most dialup connections. The newer NLSP protocol in NetWare 4.X should offer some parity with TCP/IP environments. (Tunneling IPX through IP does not address the windowing and other performance-related issues associated with IPX, originally intended for the LAN environment only. Therefore, use of IPX/IP gateways will not automatically result in increased performance over frame relay or any other wide area link, unless tuning is done.)

Problem. At one end of the spectrum, some carriers offer zero CIR service with attractive pricing. At the other end, the cost exceeds that of a 56K-bps leased line.

Solution. If the data is not mission critical or can tolerate retransmission, customers can use a zero CIR connection. When the frame relay network gets congested, however, all frames are eligible to be dropped. Therefore, this approach is not recommended for inexpensive backup lines.

Customers who want a higher CIR to guarantee that peak bandwidth is available, but who find the price is too high, can try the following approaches:

- Use a lower CIR if the application can tolerate it (most can when they were not using all of the previously available bandwidth on average).
- Negotiate a lower rate with the carrier (frame relay is not tariffed, so customers should negotiate).

- Choose another carrier.
- Use a leased line or switched service for that link, possibly in conjunction with data compression.

If the network is not able to sustain end-to-end traffic at the CIR that was contracted for, or if the contract states that data cannot burst above the CIR, then the carrier's network may be congested. Keep an eye on frame relay statistics and contact the carrier in this situation. The contract should specify CIR and any upper limits (i.e., the burst rate).

Problem. Configurations require a backup connection in the event that the primary frame relay connection goes down.

Solution. Two approaches are readily available. One is to have a modem, SW56, or ISDN dial backup scheme that dials around the frame relay network and connects to the host or server that the original connection homed to. This solution works well for LAN applications where TCP/IP and the dynamic address/routing functions take care of the new, switched connection so that traffic can be resumed with a minimum of fuss.

The second option is to have an alternate frame relay connection (from zero kilobits per second for low cost if available, or at the same CIR for better availability) over another carrier, but preferably over another local loop connection or a non-frame relay leased line, which is available as an alternate path if the primary path goes down. This approach is often more expensive but offers faster resumption of service if the networking equipment is able to automatically use the alternate line through or around the frame relay services. Otherwise, a manual reconfiguration will be required. Treat the frame relay DLCs as logical leased lines in this scenario.

The possibility of dropped frames scares some people, but it need not. Leased line and other networks also drop or damage frames when bit errors or line problems occur. This is why most protocols have a cyclic redundancy check (CRC) or other error detection at the link level and at other protocol levels.

Although some systems, such as the AS/400, are more sensitive to line problems than others, if there is an end-to-end error detection and retransmission mechanism in place in the system (eg., TCP, LLC2), dropped frames will only affect overall throughput, not data integrity. When using frame relay, the equipment chosen (and that the carrier uses) has error statistics at a number of protocol levels, so customers can tell if frames are being lost or damaged. If the error rate appears higher than it should be, contact the carrier (or the equipment vendor, if the cause is not the frame relay network). The frame relay nodes (backbone switches) the carrier uses most likely offer automatic rerouting of DLCs in the event of an internodal link failure.

3-3 Global Frame Relay

Andres Llana, Jr.

MANY ORGANIZATIONS ACCUSTOMED to the convenience of domestic frame relay service are considering international service to expand their markets. Global frame relay services facilitate this process by providing a means of linking remote LANs to international sites. A growing market demand ensures the continued increase of service nodes worldwide.

Based on revenue figures, AT&T (Warren NJ) controls 36% of the domestic frame relay market, Sprint (Westwood KS) maintains a consistent 20%, and MCI Communications Corp. (McLean VA) controls 16%. Domestic frame relay, in many cases, is less expensive than dedicated leased lines. As a result, more users are migrating to frame relay now that it is possible to support voice over frame relay.

To encourage the transition to frame relay, service providers are offering greater discounts and conveniences such Internet and World Wide Web access through gateways. Users must, however, carefully consider their bandwidth requirements for Internet access. Frame relay access requires 56K-bps or higher. The user could be billed for 56K-bps access to the Internet when 9.6K-bps access may be all that is required.

FRAME RELAY-TO-ATM SUPPORT

Although many users may be years away from requiring asynchronous transfer mode (ATM) bandwidth, frame relay service providers plan to offer frame relay-to-ATM support. An ATM backbone will be established that can be accessed through public frame relay to aggregate high-speed traffic between distributed, high-speed host locations. For example, Sprint recently adopted the Frame Relay Forum's standard for frame relay-to-ATM service. Sprint plans to aggregate frame relay nodes running at 56K-bps to 1.544M-bps at an ATM access switch for transport to a host site operating at ATM speeds. However, ATM does not yet support voice-over-frame relay traffic, which may eliminate the business case for early ATM deployment.

FRAME RELAY-TO-ISDN SUPPORT

Although service providers support primary rate integrated services digital network (ISDN) service, some have not yet offered to interface ISDN B service. (ISDN B is a channel for transmitting voice, data, or video.) In some major market areas, the regional Bell operating companies (RBOCs) offer ISDN anywhere, which means they will bear the cost of tie lines from the customer site to the site that supports ISDN B. In many areas, ISDN B service is offered at about $40 per month, which is a significant savings over the typical cost of 56K-bps access to a frame relay switch (i.e., $150–$160 or more per month).

FRAME RELAY DISASTER RECOVERY SERVICES

One of frame relay's greatest advantages is that it works on the basis of a permanent virtual circuit (PVC). Therefore, if a link or circuit is cut or disrupted, the network simply selects another available link. This automatic rerouting ensures the reliability of a user's network. This would not hold true in the case of a multidropped dedicated SNA network, in which a break in a circuit could collapse major portions of the network. For this reason, disaster recovery is one of frame relay's strongest selling points.

Because there are many rerouting options within a public frame relay network, network service can be restored before users realize that it was down. AT&T, Sprint, MCI, and LDDS Communications, Inc. (Jackson MS) offer support for disaster recovery through four basic configurations:

- Frame relay switched 56K-bps dial backup.
- Rerouting to a backup host or backup site.
- Frame relay switch diversity.
- Local access diversity.

Frame Relay Switched 56K-bps Dial Backup

This service is provided in the event of failure of the primary frame relay access level. A switched 56K-bps call can be routed to a different frame relay switch using the switched 56K-bps lines from both a local exchange carrier (LEC) and interexchange carrier (IXC). This service requires the additional monthly expense of a local dedicated switched 56K-bps line, a frame relay switched 56K-bps backup port charge, backup PVC charges, plus usage charges at the LEC and IXC level. These services may not be available if international locations are involved.

Rerouting to a Backup Site

Some users maintain an alternate host or hot site in case of disaster (i.e., fire or flood) at the principal computer center. The user is responsible for maintaining a frame relay local loop access facility at the alternate site. The user must also establish and maintain backup PVCs at each location. In the event of disaster at the host site, PVCs reroute traffic at the local level to the backup site. The service provider establishes network routes to the alternative host site from the central control facility. One-time rerouting service charges apply.

Frame Relay Switch Diversity

This service requires two separate serial connections on the user's local router. The serial interface connections are linked through separate frame relay access facilities to the same local exchange and frame relay point of presence (POP). At the POP, the user's connections are routed to separate frame relay switches in separate switching centers.

Local Access Diversity

In some areas in the US, service providers support alternate routing from the user's premises to the local exchange. These alternate facilities are linked to the frame relay service provider's POP. The user pays for the alternate service provider's line, a second PVC, and the committed information rate (CIR) monthly access charge.

SWITCHING VOICE AND DATA OVER FRAME RELAY

Several vendors, including Hypercom, Inc. (Phoenix AZ) and Motorola, Inc. (Schaumburg IL) have announced products that support voice over frame relay in various forms. However, ACT Networks Inc. (Camarillo CA), Micom Communications Corp. (Simi Valley CA), and Memotec Communications Corp. (North Andover MA), who have mature multiplexer and frame relay assembler/dissembler (FRAD) product lines, were among the earliest to support voice over frame relay. These three manufacturers have offered voice compression as a module on their multiplexers for some time, and voice over frame relay was a natural extension. This new breed of frame relay switches coupled with the global reach of frame relay services has changed the entire concept of network architectures.

Memotec and Micom have taken the voice-over-frame relay process a step further by developing a comprehensive software suite that supports routing of voice and data in much the same way that digitized voice traffic is routed over the public switched telephone network. For example, the Memotec CX 1000 frame relay switch allows the user to send multiprotocol traffic over a single dedicated link. IBM Systems Network Architecture/synchronous data link control (SNA/SDLC) traffic is transported transparently through the ability of the CX 1000 to conduct local polling (i.e., spoofing) in the manner of a local controller.

The frame relay switches on the market all use a modular architecture that provides a common backplane for integrating a number of specialized circuit cards. Each of these cards allows the switches to operate as a FRAD, a multiplexer, data service unit/channel service unit (DSU/CSU), bit voice/ data compressor, high-speed interface, or bridge/router. These modularized frame relay switches can be tailored to meet specific user requirements.

MAINTAINING DATA AND VOICE QUALITY

Frame relay switches have several design characteristics that support the integration of voice, data, and fax with consistent reliability.

Throughput Management

Each network node on a frame relay network is configured with a local access loop and a CIR. The CIR establishes a threshold at which data traffic may be transmitted into the frame relay network. For example, a process within the CX 1000 switch's operating system invokes a network management process that carefully monitors committed traffic against the established CIR. This process ensures that traffic does not exceed the CIR level, which could cause sensitive traffic to be discarded within the network while in transit. In this way, sensitive fax and voice traffic packets are protected against discard eligibility once they enter the network.

Jitter Control

Jitter can occur in the public network when an intermediate switch is busy with another packet. In this case, the second packet is held at the switch until the transmission of the first packet is complete. Because frame relay packets can vary in length, the amount of delay is unpredictable. If the jitter exceeds the ability of the receiving device to compensate through buffering, voice quality will suffer or the signal will be completely obliterated.

To address this problem, switches such as the CX 1000 have an elastic buffering process built into the operating software that manages buffering in real time. By reconfiguring its assigned buffer space, the frame relay switch can compensate for network jitter. Most frame relay switches use some form of buffering as a means of controlling network jitter. In this way, the continuity of a telephone conversation is ensured because jitter can be smoothed out before the digital voice packets are converted back to their analog form.

System Tuning

Frame relay switches such as the ACT SDM-FP, Micom Netrunner, and Memotec CX 1000 can be tuned as the network manager monitors performance under a variety of circumstances. Both the CX 1000 and the SDM-FP have default parameters that define the frame sizes for fax, voice, asynchronous, or synchronous data. However, fragmentation is built into the CX 1000, which allows the network administrator to change the size of the frame relay packets in which fax, voice, or data is encapsulated. This control process is based on network monitoring procedures that determine optimal levels of traffic.

The ability to reconfigure the discard eligibility (DE) status of individual channels is unique to the CX 1000. Channels with the DE indicator set indicate that the information that passes through the marked channels are discard eligible. Through this process, channels can be managed to ensure that critical information can be presented to the network without being discarded.

Traffic Prioritization

Frame relay switches use schemes to set up priorities for each voice/fax channel defined on a data channel link identifier (DCLI). The highest default priority is automatically assigned to any channel configured to support either fax or voice traffic. The network administrator can also assign individual priority codes to selected data channels as well as establish a prioritization scheme within DCLIs. This prioritization facility allows data traffic to be buffered until all higher-priority fax and voice packets have been presented to the network.

Optimizing Bandwidth

Frame relay switches deploy various techniques to optimize bandwidth. One popular technique takes advantage of pauses during a conversation or periods of silence to further reduce bandwidth requirements. This process, known as *digital speech interpolation,* eliminates much of the bandwidth that might otherwise be wasted through the creation of empty packets.

Another technique is to deploy various forms of data compression. For example, the CX 1000 and Netrunner can support data compression on all data channels, which allows these switches to increase overall data throughput fourfold. This process supports the movement of much higher volumes of traffic over the public frame relay network without a corresponding increase in CIR levels.

Both the ACT SDM-FP and the Memotec CX 1000 use a multiprotocol encapsulation scheme (i.e., RFC 1490). This process supports the optimization of the frame relay access link supporting the packing of all outbound traffic over a single DCLI.

ROUTING VOICE OVER FRAME RELAY

Originally, voice over frame relay involved the point-to-point transmission of digitized voice. In early FRAD devices, the transmission of voice traffic was accomplished in much the same manner as a dedicated circuit. However, these devices supported the routing of voice over the public frame relay network according to specific routing instructions.

Comprehensive software suites have been developed to support switching strategies across the frame relay network. For example, CX 1000 software emulates the switching process used in the public switched telephone network (PSTN). The CX 1000 achieves this by setting up a number of telephone hunt groups that are usually associated with hunt groups of other company nodes on the network.

As a number is dialed for a specific extension located on a hunt group, the CX 1000 listens to the dual-tone multifrequency (DTMF) signals, which are mapped to a specific location on the network based on hunt group listings. This information is used to support a routing scheme for sending a call to its ultimate destination.

Hunt groups can be programmed on the switch so that outbound calls emanating from a single location are routed to any location across the network. Several users might have their calls routed to many different locations across the company's internal network. There are a variety of routing schemes for sending faxes, voice, or data to virtually any location on the network.

Voice Digitization

The digitization of voice coupled with the application of efficient compression algorithms makes it possible to route voice traffic over frame relay. However, digitized voice is not new. Telephone companies have long agreed on the pulse code modulation (PCM) standard, which runs at 64K-bps. To advance this standard, various enhancements were developed, such as adaptive pulse code modulation (ADPCM), which compresses the voice component down to 32K-bps and enables T1 multiplexers to make better use of T1 bandwidth, allowing the integration of additional voice traffic.

A T1 link can support up to 48 voice channels. As more companies began to employ T1 backbone networks, the T1 multiplexer manufacturers developed better digitization algorithms that provided for voice compression at 16K-bps, 8K-bps, and 4.8K-bps, down from 32K-bps.

These early attempts at reducing voice information down to the lowest bit level met with mixed results. A standard was needed to measure these new proprietary algorithms against the already established ADPCM standard. Although somewhat subjective, one method that evolved for judging the differences in voice quality produced by the various proprietary algorithms was the mean opinion score (MOS) . This method assigned an MOS rating for a given compression level when compared against the 64K-bps standard. For example, PCM is rated at 4.4, and ADPCM might be rated as 4.1. Another algorithm, adaptive transform coding (ATC), supports variable-rate schemes and may have an MOS of between 2.0 to 3.8, depending on the digitization rate.

Although improved multiband excitation (IMBE), another hybrid algorithm that uses either 2.4 or 8K-bps, may be rated at a lower MOS, it can still achieve an acceptable level of communications quality.

Algorithms for Voice Compression. Algebraic code-excited linear prediction (ACELP) has followed as the result of earlier attempts to develop algorithms that made better use of bandwidth. ACELP is a comprehensive algorithm that involves modeling of the vocal track, a sophisticated technique for voice pitch extraction and coding, and a special excitation modeling and coding process.

This algorithm has produced the best-quality voice levels at 16K-bps, 8.0K-bps, and in some cases 4.8K-bps, achieving an MOS rating of 4.2 out of a possible 5.0. The Memotec CX 1000 uses the ACELP algorithm in conjunction with its network switching matrix to support advanced frame relay switching.

ANATOMY OF THE NEW FRAME RELAY SWITCH

In the case of the CX 1000 frame relay switch, a special module (i.e., MC 600) serves as a data compressor that interfaces with the multiplexer on the network interface side. The AC 600 is the voice compression module. This module contains a computer chip that drives the ACELP algorithm software to achieve the bit-compression level suitable for transmitting voice over frame relay. The ACELP algorithm is the software model used by Micom and Memotec to support voice compression.

The Memotec CX 1000, Micom Netrunner, and the ACT SDM-FP chassis can be configured with various multiples of specialized function modules. For example, the CX 1000 could be configured with two AC 600 voice compression modules to support up to 16 compressed voice channels. Another specialized module, PX 670, supports asynchronous, X.25, SNA, or SDLC traffic. This module can be used to poll remote SNA terminals in the same manner as an IBM remote controller. In this way, SNA traffic can be transported over frame relay transparent to the SNA network monitoring process. This module links to the X.25 interface on the network side.

The CX 1000 is configured with an SNMP (system network management protocol) controller that allows it to be managed remotely by an SNMP network manager. The ACT and Micom switches use proprietary net managers.

A special frame relay module (i.e., FR 600) for the CX 1000 supports the packetizing process associated with frame relay, switched multimegabit data services (SMDS), and ATM networks. This module links to the frame relay SMDS/ATM interface on the network side of the switch.

Another special circuit card, the CL 600, supports a LAN bridge/router function. This card houses a special computer chip that contains the routing software that supports a comprehensive routing scheme. It is this routing system that emulates the network switching function similar to the PSTN. The Micom Netrunner uses a special router card to provide a similar function as well as support for ATM technologies.

The CX 1000 has a built-in DSU/CSU for 56K-bps access as part of the standard configuration. However, a special module (i.e., ISU 5600 and IDM 1500) is available that can serve as a multi-rate or high-speed CSU/DSU supporting the interface of fractional T1 or full T1 service. Both the Memotec and Micom switches provide an interface for the integration of ISDN access.

PLANNING AN INTERNATIONAL NETWORK

It is now possible to configure international networks through the application of global frame relay service providers such as AT&T, MCI, LDDS, and Sprint. In Europe, international service providers include Unisource Business Networks and Scitor International Telecomm Services. Exhibit 3-3-1 provides a list of frame relay switch vendors and service carriers.

Exhibit 3-3-2 includes some of the countries around the world that are provisioned with frame relay switches to support Sprint global frame relay services. AT&T, MCI, and LDDS all have similar network arrangements. If a service provider does not have a frame relay switch in an area convenient to the user's location, an extended access line can be used to link the user to the frame relay network. However, because each service provider has a different distribution of frame relay switches, the cost of installing an extended access varies depending on the distance the user is located from the nearest frame relay switch. This is usually the cost of a dedicated international circuit. Planning an international network should include consultation with several vendors.

Frame Relay Service Carriers

MCI Corp., 1801 Pennsylvania Avenue, Washington DC 20006, (202) 887-3325

AT&T, 295 North Maple Avenue, Basking Ridge, NJ 07920, (908) 221-9000

Sprint, 3100 Cumberland Circle, Atlanta, GA 30339, (800) 733-2287

LDDS WorldCom, Cherry Tree Corporate Center, Cherry Hill, NJ 08002, (800) 929-0722

Scitor International Telecom Services, 26020 Acero Street, Suite 200, Mission Viejo, CA 92691, (714) 470-6300

Frame Relay Switch Vendors

ACT Networks, Inc., 188 Camino Ruiz, Camarillo, CA 93012, (805) 388-2474

Memotec Communications Corp., 1 High Street, North Andover, MA 01845, (508) 681-0600

Micom Communications Corp., 4100 Los Angeles Avenue, Simi Valley, CA 93063, (805) 583-8600

Exhibit 3-3-1. Resources for Frame Relay Equipment and Service

---

Australia
Austria
Belgium
Canada
Denmark
Finland
France
Germany
Hong Kong
Ireland
Italy
Japan
Luxembourg
Netherlands
New Zealand
Norway
Portugal
Puerto Rico
Singapore
South Korea
Spain
Switzerland
United Kingdom

Exhibit 3-3-2. International Locations for Sprint Frame Relay Switches

---

International Service Costs

As carriers establish more frame relay switching centers overseas, the cost for service will drop. Costs vary for installation of service, local loop access, and monthly access, and depend on the user's location relative to the carrier's frame relay switch.

Location	Local Loop Per Month	PVC/CIR Per Month
Buffalo NY	$ 412	$133
Potenkill NY	$ 460	$133
Bennington VT	$ 731	$133
Nashua NH (US hub)	$1,200	$133
Ireland (European hub)	$ 850	$850
Manchester UK	$ 850	$850
Winchester UK	$ 850	$850
Madrid Spain	$ 850	$850
Iyla Denmark	$ 850	$850

Exhibit 3-3-3. Sample Two-Year Pricing for Global Frame Relay

Overseas, the issue of network-to-network interface (NNI) may not be uniform between competing carriers. For example, domestically, NNI between the LECs and IXCs has been clarified, but the same is not true for overseas service providers. However, some domestic service providers do offer a level of NNI between their domestic network and their overseas facilities. For example, MCI, LDDS, and Sprint all support NNI between their domestic and common market facilities. The advantage is that these networks offer relatively low intercontinental transmission costs.

Exhibit 3-3-3 shows the monthly costs for the international network initially proposed by Sprint. Although the overall monthly costs are higher than the US locations, the monthly local loop and PVC/CIR are fairly uniform for all of the European locations. These prices are subject to final negotiations that could result in lower costs if a long-term contract is established.

Other price quotes might include higher installation costs because of the need to install an international dedicated circuit at one location to connect to the service provider's frame relay switch. In this situation, the local loop access price would be inordinately high because of the monthly cost for a dedicated international circuit.

European Service Providers. International European frame relay service providers have different pricing plans than US domestic service providers. European pricing plans are aimed at supporting networks with high monthly traffic volumes. As a result, if a user is planning to increase the level of traffic, the European service provider will offer dial-up access over ISDN. This arrangement does not facilitate linking an integrated network with a North American affiliate because typical NNI would not be possible. A dial-up arrangement over ISDN would not support voice over frame relay, either.

Support in the Host Country. Because frame relay switching has matured into a more comprehensive product, the user would be remiss if strong consideration was not given to voice over frame relay. Domestically, there would be no problem finding suitable equipment to support network nodes; however, it is quite different overseas.

Each country's PTT—the government agency that handles postal, telegraph, and telephone services—may have its own special requirements for customer premises equipment (CPE) that interfaces directly with the public network. For this reason, it is important to establish that the network equipment chosen has been approved by the host country's PTT. Furthermore, the manufacturer of the equipment may also have in-country representatives that the network planner should contact. The user cannot simply show up with equipment and set up shop, because only vendors licensed by the PTT will be allowed to connect equipment to the public network.

Configuring an International Network

It is possible to configure an international network using frame relay services in which users have no responsibility for network management. All of the arrangements for local loop access in the overseas locations can be handled by the domestic carrier's overseas offices and the host country can assume responsibility for services.

In a network configuration in which the user has deployed Memotec CX 1000 as the frame relay access and switching solution, network costs are reduced to a single high-speed local loop access line that carries voice, fax, and data traffic for one monthly charge.

The CX 1000 switch, for example, has features that allow the user to switch both voice and data across domestic and international public frame relay networks. Global frame relay service eliminates the high costs for what might be associated with more traditional international transmission services.

The global reach of this network also allows the user to support a worldwide customer base and provide for the timely management of all company resources. The advantage is that users can leverage the attractive pricing of a global frame relay service through the application of frame relay switches. These same switches might also be deployed in a hybrid integrated network supporting a combination of frame relay and leased lines, where the combination of facilities would produce the most cost-effective arrangement.

Rules of Thumb for Buying Global Services

There is not a great deal of technical support available in the global services arena yet. It is important to obtain specific installation costs for each network node and for monthly local loop access, port costs, and CIR levels. Carriers may want to lump some costs together, which makes it difficult to determine the best network arrangement. The possibility of cost negotiation is not out of the question with most vendors, especially when a long-term service contract is concerned.

Levels of service for all nodes on the network should be established in the beginning. Provisioning ISDN at the local loop level is still somewhat mysterious to many users and therefore should be explicitly established at the start of negotiations. Where ISDN service is planned as the local loop access to the frame relay network, some vendors offer rebates for installation and conversion to traditional 56K-bps access if the ISDN service does not interface to the frame relay network.

Some service providers offer zero-level CIR, which lowers costs for the customer. However, traffic levels should be carefully evaluated before the rates are negotiated for bursting above an agreed upon CIR level. If users plan to integrate voice over frame relay they should plan on a CIR of at least 32 K-bps.

The frame relay premises equipment should work on the service provider's network at all of the user's locations. This is particularly important for overseas locations.

SUMMARY

Global frame relay service can be a cost-effective, high-performance solution for international networking of voice, LANs, and host systems. Like legacy X.25 networks, frame relay uses transmission facilities efficiently—only when needed. Like a dedicated leased line service, frame relay transports information quickly with high reliability and few delays in network processing.

Global frame relay service simplifies interconnectivity because all service can be transported over a single composite link with each network node fully accessible. As a result, all of the benefits and product services of the domestic service provider can be extended to the user's international locations.

3-4
Traffic Control Functions in ATM Networks

Byung G. Kim

ASYNCHRONOUS TRANSFER MODE (ATM) is the universal transport vehicle for multimedia traffic in the context of the broadband integrated services digital network (B-ISDN) specification. ATM is a connection-oriented technology that transports data in 53-byte cells. To accommodate different types of connections, four service classes are currently defined in ATM networks:

- Constant bit rate (CBR).
- Variable bit rate (VBR).
- Available bit rate (ABR).
- Unspecified bit rate (UBR).

ATM SERVICES

Constant Bit Rate Service

CBR service is designed for connections that require a fixed amount of bandwidth for the entire connection time. The amount of bandwidth is equivalent to the peak rate of the connection. CBR service is intended to support real-time applications with stringent requirements for transfer delays and delay variations (e.g., digitized voice and Px64 video).

Variable Bit Rate Service

VBR service, on the other hand, is designed for those sources that generate cells at time-varying rates. Real-time VBR service provides tight bounds on cell delays and delay variations, whereas non-real-time VBR service provides a low cell loss ratio. Examples of VBR connections are an MPEG full-motion video source producing heavier traffic at scene changes and a bursty file transfer connection, occasionally sending a large amount of data.

VBR service uses statistical multiplexing, by which the available bandwidth not used by light-traffic connections is taken advantage of by connections with temporarily heavy traffic. However, there is always a risk that the given VBR capacity may be saturated when many sources generate heavier traffic than expected. The VBR capacity and the buffer space should be carefully designed and managed so that occurrences of such capacity saturation are strictly controlled. Because of the statistical nature of traffic fluctuation in VBR sources, the management of network resources is usually difficult for VBR service.

Available Bit Rate Service

The ABR service class is designed primarily for LAN traffic across an ATM network. The bandwidth left unallocated to CBR and VBR sources is made available to ABR sources for sharing.

An ABR source is required to regulate its transmission rate according to the prevailing congestion condition in the network. Control cells that carry information about congestion, called resource management (RM) cells, are periodically inserted into the cell stream so that the network condition can be acquired from returning RM cells. A network guarantees a low cell loss ratio and a fair sharing of ABR bandwidth for those end systems that adapt their transmission rate according to the feedback information in RM cells.

Unspecified Bit Rate Service

UBR service demands no specific quality of service from the network and is thus handled only after service requirements of CBR, VBR, and ABR connections are satisfied.

CBR AND VBR TRAFFIC CONTROL

An ATM source declares its expected traffic pattern in the source traffic descriptor. The source traffic descriptor includes the peak cell rate (PCR), the sustainable cell rate (SCR), and burst tolerance (BT). The source then requests a connection by submitting connection traffic parameters to the network, which consist of the source traffic descriptor, the cell delay variation (CDV) tolerance, and the conformance definition.

Conformance Definition Using GCRA

As soon as a connection is established, the ATM network starts the usage parameter control function to monitor the actual traffic generated from the source against the source traffic parameters declared for the connection request. This function enforces the service contract so that network resources are protected from malicious or unintentional misbehavior of the source. For an unambiguous specification of the conformance definition, the generic cell rate algorithm (GCRA) is devised.

For each cell arrival, the GCRA determines whether the cell conforms to the declared source traffic parameters. The GCRA can thus be used as a formal definition of traffic conformance test for a connection. However, the network is not obliged to use this algorithm for usage parameter control (UPC) as long as the operation of the UPC does not violate the quality of service objectives of compliant connections.

The GCRA defines the relationship between the peak cell rate and the cell delay variation tolerance, as well as the relationship between the sustainable cell rate and the burst tolerance. The GCRA is based on two parameters, the increment I and the limit L, and is denoted by GCRA(I,L).

For example, let ta(k) denote the arrival time of the k-th cell in a connection. The basic idea of the GCRA is to have a variable—the theoretical arrival time (TAT)—keep track of the expected arrival time of the next subsequent cell. When TAT is updated for the k-th cell, three cases are handled:

- *Case 1:* $<$ *or* $=$ *TAT* $t_a(k)$. The cell arrival time is on or after the expected arrival time. This cell is arriving at a slower pace than expected and is thus conforming to the source traffic parameters. TAT is updated to ta(k) + I.
- *Case 2:* $TAT\text{-}L$ $<$ *or* $=$ $t_a(k) < TAT$. The cell is arriving before the expected arrival time but within the limit of L. Although the cell is generated slightly ahead of the schedule, the cell is considered to be conforming. TAT is updated to TAT + I.
- *Case 3:* $t_a(k) < TAT - L$. The cell arrived too early and is not conforming. TAT remains unchanged. The nonconforming cell may be either discarded or tagged to a lower priority.

Traffic Policing

This section explains how the generic cell rate algorithm is used to police cells from a CBR source. For simplicity, the time to transmit an ATM cell is set to 1. The GCRA is specified by GCRA(T_c, τ), where T_c and τ denote the contracted cell interarrival times and the cell delay variation tolerance, respectively.

Suppose the CBR source violates the contract by generating cells at the fixed intervals T_i, though slightly faster than the contracted interval T_c ($T_i < T_c$). If $\tau = 0$, then Exhibit 3-4-1 illustrates that TAT is updated by Tc and that cells are arriving at every T_i. One extra cell appears within the increment interval of T_c. Because the CDV tolerance (τ) is 0, this extra cell is policed,

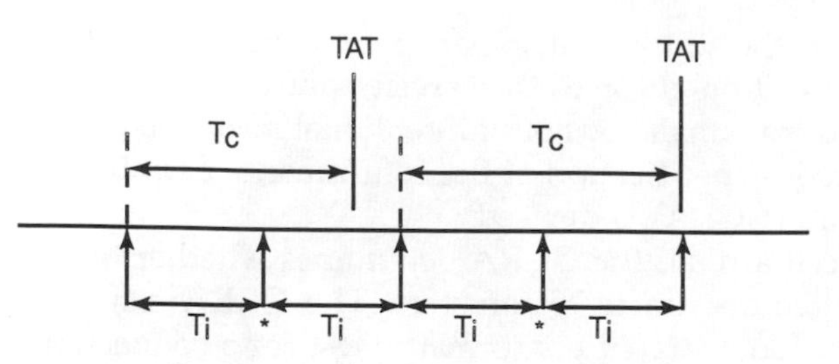

Note:
Asterisk (*) denotes the cell to be policed.

Exhibit 3-4-1. Interval Updating with GCRA

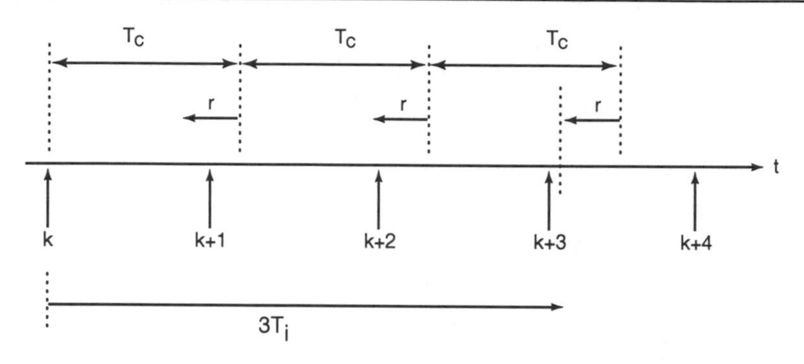

Exhibit 3-4-2. Interval Updating for a CBR Source With GCRA

resulting in policing of every other cell. When policed cells are discarded, the cell loss ratio (CLR) becomes 0.5.

For $\tau > 0$, suppose TAT is updated from the arrival time of the k-th cell. If the (k+d)-th cell is to be policed by GCRA (T_c, τ), the arrival time of the (k+d)-th cell should be within the margin of CDV tolerance τ of TAT for the (k+d)-th cell. An example with d = 3 is depicted in Exhibit 3–4–2. Cell k+3 is shown to be policed because its arrival time is earlier than the GCRA limit τ . At an arbitrary value of d, the (k+d)-th cell is discarded when $d \times T_i < d \times T_c - \tau$. When policed cells are discarded, this amounts to a single cell loss in (d+1) cell arrivals. CLR changes in a stepwise fashion as T_i decreases, as shown in Exhibit 3–4–3.

Policing of a VBR source is performed on the basis of the sustainable cell rate (SCR). The SCR specifies the upper bound on the average cell rate and can be significantly lower than the peak cell rate. The time intervals between two successive cells arriving at the average and the peak rates are denoted by T_s and T_p, respectively. They are inverses of sustainable and peak rates. A

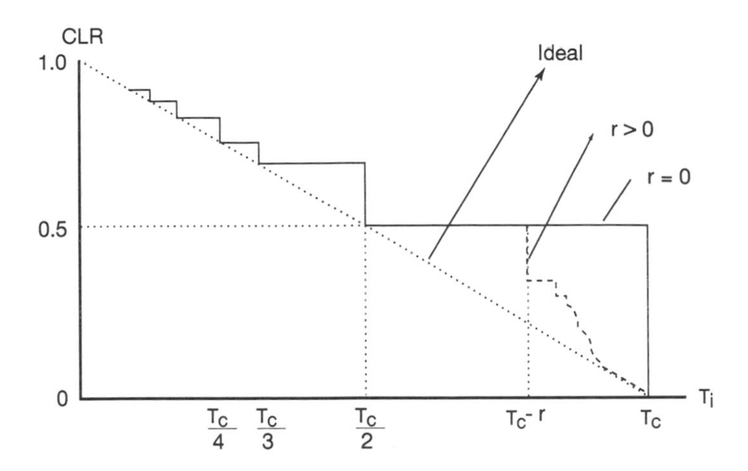

Exhibit 3-4-3. Cell Loss Rate of a CBR Source by GCRA

burst is defined as a group of cells generated at the peak cell rate. The burst tolerance determines the maximum burst size, b, that may be transmitted at the peak rate.

Assessment of Monitoring Techniques

During a connection establishment process, the ATM network has to reserve sufficient resources to meet the quality of service (QOS) demand for the connection. Once admitted, the actual incoming traffic is policed to keep the traffic entering the network in conformance to the connection traffic parameters. The usage parameter control, or UPC, procedure is a preventive control mechanism so that a potential overload from a source is not permitted beyond the preestablished limits set by the connection parameters. However, a cell stream from a source may be altered along the path as it is mixed with other cell streams. The UPC procedure has to be flexible enough to allow for some deviations from the connection parameters by including cell delay variation tolerance and burst tolerance parameters.

Because a cell stream conforming to the admission contract can be altered after a multiplexing, traffic shaping may be necessary to preserve the characteristics of the original cell stream. For example, a nonzero CDV tolerance allows the UPC to accept some cells from a CBR source, although they are spaced closer than the contracted interval.

After the UPC checks for the conformance, it may be desirable to restore cell stream into the one matching the admission contract. A need for traffic shaping is strengthened from some results, which indicate that individual cell streams become burstier after a series of GCRA control actions.

Although a usage parameter control algorithm such as the GCRA may be easily implemented, the traffic shaping is a costly operation since some history of a cell stream has to be maintained if a cell stream is to be reconstructed. Furthermore, some buffering may be needed as the cell stream is restored. The buffering per virtual channel (VC) is exactly what an ATM node wanted to avoid, however.

ABR TRAFFIC CONTROL

ABR traffic control relies on cooperative interworking among three components:

- The source end system (SES).
- The ATM switch.
- The destination end system (DES).

Briefly, their control actions are as follows: An SES keeps track of the best estimate for its cell transmission rate, or allowed cell rate (ACR), according to the perceived congestion status in the network. To obtain the congestion status in the network, the SES periodically inserts a resource management cell in its data cell stream.

The DES returns the RM cell back to the SES so that the network status is conveyed back to the SES. In the backward RM cell, a single congestion indication (CI) bit may be set for the SES to adjust its cell transmission rate. Optionally, an ATM switch may determine the best cell rate for an SES and convey it as the explicit rate in the backward resource management cell.

Source and Destination Behavior

When a source sends the first cell after connection setup, or after not sending any cells for T_{tm} seconds or longer, it sends a forward RM cell to obtain the network congestion status. After every N_{rm}-1 data cells, the source inserts a forward RM cell according to the prevailing ACR. When a backward RM cell is received with CI = 1, then ACR (i.e., the allowed cell rate) is reduced at least to ACR$\times$RDF, where RDF is the rate reduction factor and is a constant (1/16, for example).

On the other hand, if CI = 0 in the RM cell, ACR is increased by a constant additive increase rate. After the ACR is adjusted according to the CI bit, the new ACR is taken as the smaller of the updated ACR and the explicit rate stored in the RM cell.

Destination behavior is relatively simple. Every ATM cell has an explicit forward congestion indication (EFCI) bit in its header. An ATM switch may set the EFCI bit if its buffer occupancy exceeds a certain threshold. The destination end system saves the EFCI state stored in the ATM header (i.e., the payload type field) of an incoming data cell. Upon receiving a forward RM cell, the DES retransmits the RM cell back to the SES after setting the direction of the cell from forward to backward and copying the saved EFCI state to the CI bit field in the RM cell. Namely, the CI bit from DES is set by the congestion status experienced by the most recent data cell.

Switch Behavior

A switch may use at least one feedback mechanism to control congestion. It must be pointed out that this is not an area of standardization and that it is purely up to the switch manufacturer or service provider to decide which scheme to implement.

Binary Feedback. The binary feedback scheme is based on the EFCI bit. A switch monitors the queue length and marks the EFCI bits of passing ATM cells, as long as the queue length exceeds a predefined queue threshold value. The binary feedback scheme is known to suffer from a potential unfairness when all connections share a common queueing buffer. Typically, a connection with more hops has a better chance of running into a congested switch than those with a smaller number of hops.

Explicit Rate Feedback. The explicit rate feedback scheme can provide the fairness by having each switch determine the suitable rate for an SES not to overload the switch. This rate is sent to the SES as the explicit rate. At the same time, each SES declared its prevailing cell rate in the current cell rate (CCR) field of an RM cell.

The enhanced proportional rate control algorithm (EPRCA) computes the mean ACR (MACR) among all connections as a running exponential weighted average ($MACR = (1- \alpha)MACR + \alpha CCR$). A typical value of α is chosen to be 1/16 and CCR is taken from the passing RM cell. The fair share is then taken as a fraction (e.g., 7/8) of MACR, and any SES sending more than the fair share is asked to reduce its rate by setting the explicit rate field in the backward RM cell to the fair share.

Congestion Avoidance. Congestion avoidance schemes monitor the prevailing load according to the load factor z = Input_ Rate/Target_ Rate. The input rate is measured over a fixed averaging interval, whereas the target rate is set slightly below the ABR bandwidth.

There are two variations of this scheme that differ in the manner by which the fair share or the explicit rate is computed. In the explicit rate indication for congestion avoidance (ERICA) algorithm, the fair share is given as the target rate divided by the number of active connections. A switch determines the explicit rate as ER = max(CCR/z, Fair-Share). ER is updated periodically using current cell rate information from the forward RM cells and the load during the averaging interval.

The ERICA algorithm attempts to guarantee at least the fair share amount of capacity to each active connection. Any excess capacity is distributed to active sources in proportion to their rates.

The congestion avoidance using proportional control (CAPC) algorithm, on the other hand, does not update ER for each connection, but instead uses the same ER for every connection. ER in CAPC is computed and updated from the fair share. If $z < 1$, Fair-Share = Fair-Share $*(1 + (1-z)*R_{up})$. Otherwise, Fair-Share $*(1 - (z-1)*R_{dn})$. In these cases, R_{up} and R_{dn} are constant slope parameters to increase and decrease the rate (the amount of changes allowed in each update is limited as well).

Assessment of ABR Control

In general, it is difficult to draw a general comparison among a number of ABR control schemes, since source and switch behaviors are affected by many factors, including the network topology. In order to have some idea about the effectiveness of different control strategies, three control algorithms are considered:

- Binary feedback control.
- EPRCA.
- ERICA.

Three ABR sources send cells to the switch buffer. They are located at 1, 10, and 100 km from the switch. Exhibits 3–4–4, 3–4–5, and 3–4–6 plot changes in ACRs in three sources.

Clearly, the simple binary feedback algorithm shows a significant oscillation, whereas the oscillation is eliminated in ERICA. The cause of oscillation is that in a high-speed network such as an ATM network, the latency for feedback information can be significant. By the time, the source recognizes congestion in the network, a few megabytes of data may already have been sent out.

When in congestion, every source attempts to reduce its rate, producing an excess capacity a short time later. When not in congestion, every source attempts to transmit a little bit more, producing a congestion a short time later. In general, however, oscillation of cell rates does not produce an ad-

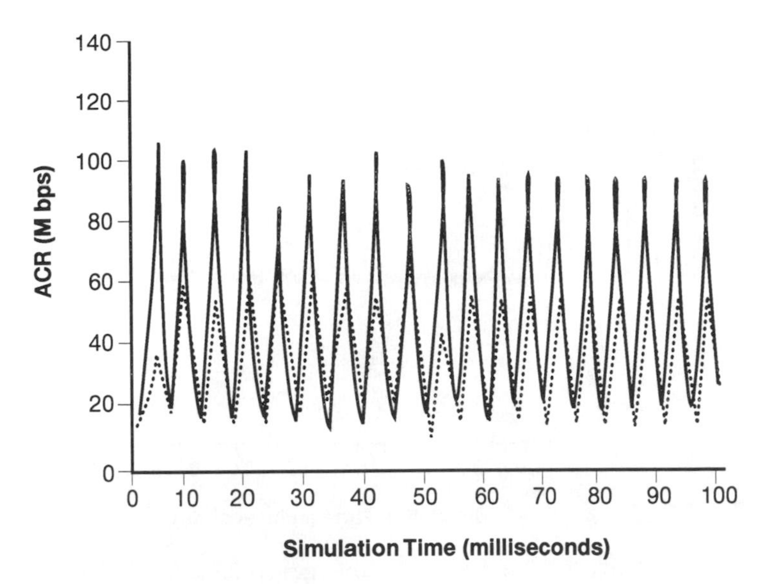

Exhibit 3-4-4. ACR According to the Binary Feedback Control

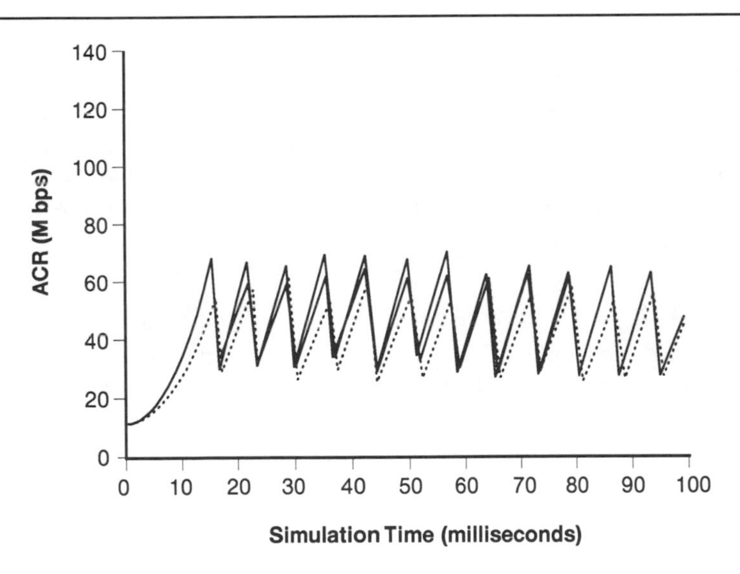

Exhibit 3-4-5. ACR Changes in EPRCA

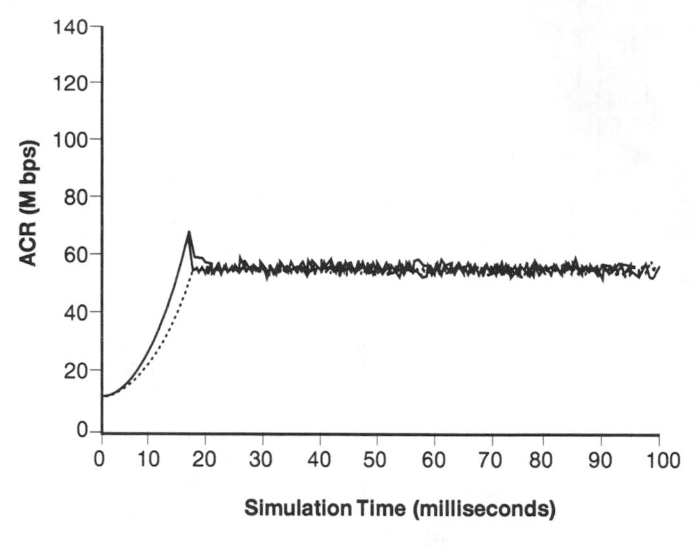

Exhibit 3-4-6. ACR Changes in ERICA

verse effect. Oscillation simply means that a source cannot send at a constant rate for an extended period of time.

Both the latency and the rate oscillation will not be of any concern in a LAN environment where the delay and reaction time are considerably short. As soon as LANs are connected over a long distance, however, control parameters should be tuned to account for potential oscillatory behavior.

SUMMARY

In summary, network techniques to recover from congestion and limit it, such as usage parameter control and traffic policing, are successfully applied to ATM constant bit rate and variable bit rate service. The generic cell rate algorithm, as specified by the ATM Forum, achieves traffic shaping, ensuring that traffic matches the service negotiated between users and the network during connection establishment.

Techniques to avoid congestion, recover from congestion, and ensure flow control in ATM available bit rate service include the use of resource management cells and switch behavior.

3-5 A Guide to Conferencing Technologies

Nathan J. Muller

COMMUNICATIONS TECHNOLOGIES are changing the way organizations do business. In particular, videoconferencing (video) and teleconferencing (voice) technologies are enabling executives and employees to conduct business once typically reserved for in-person gatherings. This capability results in significant cost savings in terms of travel and work time while enhancing the speed and quality of decision making—all of which can translate into a strategic business advantage for the organization. This chapter helps IS managers ensure that the two technologies are a valued and cost-effective part of the corporate communications toolbox.

BUSINESS BENEFITS OF VIDEOCONFERENCING

Videoconferencing is the capability to broadcast audio and full-motion, slow-scan, and freeze-frame images over a closed-circuit to one or more locations. Images, text, and graphics from a variety of sources can be multiplexed over the video circuit to permit interactivity among conference participants without the need (or cost) of additional communications links.

Technology advances have made videoconferencing a strategically significant communications tool. Some of the business benefits videoconferencing offers include:

- *Faster time to market.* Videoconferencing can speed product development by facilitating collaboration between engineering, manufacturing, marketing, and distribution units within the organization—regardless of their location.
- *Reduced costs.* Cost savings result from reduced travel expenditures as well as from faster time to market. The latter promotes the more effi-

cient use of organizational resources, which in turn can produce even more cost savings.

- *Increased presence.* Videoconferencing can be used to establish and maintain relations with external constituencies such as key customers, suppliers, distributors, and strategic partners.

Videoconferencing's growing popularity also results from the increase in available transmission options, which have improved video communications and greatly reduced costs. Today, these transmission alternatives include fiber-optic and Fractional T1 lines, integrated services digital network (ISDN) facilities, and satellite links over inexpensive, rapidly deployable very small aperture terminals (VSATs). If companies prefer not to operate their own videoconferencing facilities, IS managers should evaluate package deals offered by regional and long-distance carriers that include all the necessary lines and equipment.

The falling cost of equipment is another contributor to the increased use of videoconferencing. For example, a $50,000 midrange videoconferencing system that is depreciated over five years costs $10,000 a year or only $833 per month. Additional costs of $200 per month for maintenance and $60 a month for an ISDN BRI (basic rate interface) line bring the monthly fixed cost to $1,093. When usage charges of about 50 cents per minute for the ISDN line are factored in, the cost of videoconferencing falls well below $100 per hour. Assuming an average of five participants at each end (for a total of ten) per conference, the cost of videoconferencing can be as low as $10 an hour per participant.

TYPES OF VIDEOCONFERENCING SYSTEMS

Videoconferencing systems fall into four categories: room-based systems, midrange or rollabout systems, desktop systems, and videophones.

Room-Based Systems

Room-based systems provide high-quality video and synchronized audio generally through the use of one or more large screens in a dedicated meeting room with environmental controls. Because the system components—screens, cameras, microphones, and auxiliary equipment—will not be moved to another room or building, they can even be permanently installed. Prices for room-based systems start at around $150,000 but can go much higher as more sophisticated features are added.

An important consideration for determining the practicality of this solution is frequency of use. A room-based system that is used throughout the business day may be more cost-effective per person than midrange or desktop solutions that are used only occasionally. Equipping every desktop with $2,000 to $5,000 worth of add-in hardware and software can become prohibitively expensive, especially if the number of PCs exceeds 50 or is continually growing. For example, equipping only 50 PCs for videoconferencing can cost between $100,000 and $250,000—making a room-based system that can be used by many more people cost-effective.

IS managers whose organizations do not rely extensively on videoconferencing but consider it an important capability to have may find that a midrange system offers comparable features and flexibility at a reduced cost.

Midrange Systems

Midrange or rollabout videoconferencing systems can be moved to any desired location and plugged in for immediate use, thereby eliminating the need for specially equipped rooms. Typically, such systems use one screen and no more than two cameras and three microphones. Prices range from $20,000 to $50,000, depending on options.

Despite the availability of lower-cost desktop videoconferencing systems, the demand for midrange systems is still high. There are several reasons for this:

- Desktop systems have exerted downward price pressure on midrange systems, making the latter systems more attractive to organizations that might have dismissed them before.
- The picture quality of midrange systems is still much better than that offered by most desktop systems, an important consideration if videoconferencing is used frequently.
- More and better collaborative computing tools are available for midrange systems than for desktop systems.
- Midrange systems have more available options than do desktop systems, including dual monitors, document camera, wide area network (WAN) interfaces, online graphics, and whiteboard capabilities.
- Unlike desktop systems, which are generally limited to a head-and-shoulders view of the participants, midrange systems can show an entire group. In addition, optional far-end camera control lets viewers at one location enhance viewing by moving the cameras at other locations.

Desktop Systems

Desktop videoconferencing is becoming popular because it allows organizations to leverage existing desktop assets. Videoconferencing becomes just another application running on the PC, which can then be used for video mail over LANs as well as for videoconferencing over WANs. The ability to use widely available Ethernet networks makes videoconferencing technology more accessible, less costly, and easy to deploy. Through the use of OLE (object linking and embedding) technology available through the Microsoft Windows environment, multipoint data sharing can also be implemented over Ethernet. Additionally, data sharing can be accomplished through an optional whiteboard capability.

Desktop videoconferencing systems that support the Transmission Control Protocol/Internet Protocol (TCP/IP), the most commonly used WAN protocol, are particularly useful in campus-style settings and for corporations with remote offices. TCP/IP support ensures that the system can be used in conjunction with standard bridges, routers, and dialup lines, giving users convenient, cost-effective access to videoconferencing capabilities, particularly in areas where ISDN service is not available. It also eliminates the need for expensive on-premises switching systems and potentially costly telephone company surcharges for routing services required of ISDN networks.

Desktop systems based on PC platforms are more affordable—in the $5,000 to $15,000 per unit range, depending on options. At the low end, the videoconferencing capability is added to an existing PC with appropriate hardware and software. At the high end, the vendor provides the PC already configured for videoconferencing.

The difficulty in installing and configuring add-in components makes the purchase of a configured system appealing. It can take a whole day to disassemble a PC, add new hardware, and get the software working properly. When a problem arises, the vendor often blames it on lack of memory or a conflict with an existing application, leaving users to fend for themselves. The hidden cost of configuring, testing, and debugging videoconferencing systems can be greatly reduced or eliminated by purchasing an integrated system with all hardware and application software installed by the vendor.

Because vendors each use a slightly different packetization scheme for video, different desktop systems cannot communicate with one another over the LAN. Individual vendors also may use proprietary video compression/decompression algorithms optimized for use on a LAN. However, efforts are under way by Intel and other vendors to develop a Personal Conferencing Specification that will make desktop videoconferencing systems compatible with one another.

Videophones

The latest in videoconferencing equipment is the videophone, used for one-on-one communications and particularly for impulse videoconferencing. The unit includes a small screen, built-in camera, video coder/decoder, audio system, and keypad. The handset lets the unit work as an ordinary phone as well as a videoconferencing system. Prices start at about $1,000 for models that work over ordinary phone lines. Picture quality is generally poor but adequate for brief conferences.

High-end models that work over switched 56K-bps service or ISDN BRI can cost as much as $10,000. The use of digital facilities results in picture quality of up to 30 frames per second, which is comparable to larger systems. These videophones typically come with communications software that provides a number-storage directory and on-screen menus that guide users through videoconferencing call setup or preprogrammed dialing processes.

Some videophone vendors offer additional interface ports that allow plug-in of auxiliary devices such as a camera, slide-to-video converter, personal computer, printer, document camera, or a large monitor. Some videophones even support picture-in-picture capability.

MULTIPOINT CONFERENCING

Videoconferencing between more than two locations requires a special device called a multipoint control unit (MCU). The device is essentially a switch that connects video signals among all locations, enabling participants to see each other and converse and to work simultaneously on the same computer document or view the same graphic. The multipoint conference is set up and controlled from a management console connected to the MCU.

A common approach to multipoint conferencing is to let all conference participants see the person who is speaking. The MCU operator can switch between images manually or the MCU will do it automatically as speakers change. Another option is for the speaker to appear on the screen in full size, and for other participants to appear in smaller windows on the screen.

Conference Management Features

The MCU makes it relatively easy to set up and manage conference calls between multiple sites. Following are some of the features that facilitate multipoint conferencing:

- *Meet-me.* The meet-me feature enables participants to enter a conference by dialing an assigned number at a prearranged time.
- *Dial-out.* This feature is used to automatically dial out to other locations and add them to the conference at prearranged times and dates.
- *Audio add-on.* Participants use the audio add-on feature to hear or speak to others who do not have video equipment (or compatible video equipment) at their locations.
- *Tone notification.* Through this feature, special tones alert participants when a person is joining or leaving the conference and when the conference is about to end.

- *Dynamic resizing and tone extension.* This feature allows locations to be added or deleted and the duration of a bandwidth reservation to be extended during a conference without having to restart the conference.
- *Integrated scheduling.* Through this feature, videoconferences are set up and scheduled days, months, or a year in advance using a Windows-based graphical user interface. The MCU automatically configures itself at the scheduled time and dials out to participating sites.

Conference Control Features

The MCU also provides the means to precisely control the videoconference in terms of who is seeing what at any given time. Some of the advanced conference control features of MCUs include:

- *Voice-activated switching.* This mode allows all participants to see the person speaking, while that person sees the last person who spoke.
- *Contributor mode.* The contributor mode works with voice-activated switching to allow a single presenter to be exclusively shown on a conference.
- *Chair control.* The chair-control mode lets a person request or relinquish control, choose the broadcaster, and drop a site or the conference.
- *Presentation or lecture mode.* Through this mode, a speaker makes a presentation and questions participants at several locations. Participants can see the presenter at all times, but the presenter sees whomever is speaking.
- *Moderator control.* This mode allows a moderator to select which person or site appears on-screen at any given time.
- *Broadcast with automatic scan.* This mode lets participants see the presenter at all times. To gauge audience reaction, the speaker sees participants in each location on a timed, predetermined basis.
- *Continuous presence.* The continuous presence mode divides the screen into quadrants so that up to four locations can be shown on the screen.

MCUs are priced according to the number of ports and can easily cost more than the videoconferencing equipment itself. An eight-port MCU, for example, can cost $100,000 or more. For this reason, MCUs are most likely to appeal to heavy users of videoconferencing who want to expand the connectivity options between their sites. When purchase of a standalone MCU is not cost-effective, IS managers should consider carrier-provided multipoint videoconferencing services.

Network Connectivity

Depending on the choice of MCU, several network connectivity options are available. Generally, the MCU can be connected to the network either directly through T1 leased lines or ISDN PRI (primary rate interface) trunks or indirectly through a digital PBX. Some MCUs can be connected to the network using dual 56K lines or ISDN BRI. Others connect videoconferencing systems over the WAN using any mix of private- and carrier-provided facilities or any mix of switched services regardless of the carrier. This capability offers the most flexibility in setting up multipoint conferences.

IS managers whose organizations lack high-speed links should investigate MCUs that include an inverse multiplexing capability that combines multiple 56/64K bps channels into a single higher-speed 384K bps channel on a demand basis, thus improving video quality. Most MCUs use the inverse multiplexing method standardized by the Bandwidth On Demand Interoperability Group (BONDING).

Although most MCUs are designed for use on the WAN, some MCUs are available for LANs. These devices are useful for providing multipoint desktop videoconferencing over local networks within a campus environment or many floors in a high-rise building.

STANDARDS

Standards for videoconferencing and other transmission technologies are established by the International Telecommunications Union-Telecommunications Standards Section (ITU-TSS), formerly the Consultative Committee on International Telegraphy and Telephony (CCITT). By establishing worldwide videoconferencing standards, the ITU-TSS helps ensure that videoconferencing systems from diverse manufacturers will be able to communicate with one another.

Standards give users greater flexibility in setting up videoconferencing networks, with low-bandwidth and high-bandwidth systems operating on a single network to suit the needs of diverse applications and accommodate various cost constraints. In addition, videoconferencing standards are expected to promote the growth of both public and private videoconferencing networks worldwide.

Video Transmission Standards

Recommendation H.320 is the umbrella standard developed by the ITU that defines the operating modes and transmission speeds for videoconferencing system codecs, including the procedures for call setup, call tear-down, and conference control. Codecs that comply with H.320 are interoperable and deliver a common level of performance.

The H.320 videoconferencing standard includes associated specifications that define how the MCUs of different vendors interoperate. MCUs that comply with H.320 can connect any videoconferencing system that is also compliant with H.320.

A particularly important component of H.320 is the H.261 video compression specification, which defines how digital information is coded and decoded. The standard also permits the signals to be transmitted at a variety of data rates from 64K bps to 2.048M bps in increments of 64K bps. Finally, it defines two resolutions:

1. The Common Intermediate Format (CIF), which is usually used in high-end room systems and provides the highest resolution at 352x288 pixels.
2. The Quarter Common Intermediate Format (QCIF), which is used by most desktop videoconferencing systems and videophones and provides lower resolution at 176x144 pixels.

Just because a codec vendor claims H.261 compatibility does not mean that the product will provide the best possible picture. Buyers should ask the following questions relating to H.261 compatibility:

- Is the codec compliant with H.261? Is it available now, and if so, who is actually using it?
- Does the codec provide high resolution (i.e., CIF) operation, or is it limited to low resolution (i.e., QCIF) modes?
- Does the codec provide pre- and postpicture processing? (These enhancements greatly improve image quality by removing unwanted noise from the picture.)
- Does the codec compensate for motion? If so, is it full search, which provides superior performance, or limited search?
- Is the codec designed for twin-channel operation? (Twin channel is a means for increasing bandwidth, allowing two dialup lines to support a videoconference for higher picture quality.) If so, is channel synchronization automatic, or must the user redial the lines until synchronization is achieved?
- How is the audio transmitted? Although the CCITT has standardized Adaptive Differential Pulse Code Modulation—ADPCM—to compress audio down to 32K bps, some vendors use proprietary schemes to transmit audio at 16K bps or lower, limiting the reach of videoconferencing to those who use the same scheme.

Another recommendation, H.230, describes the signals that enable conferencing systems and MCUs to exchange instructions and status information during the initiation of a conference and while the conference is in progress. A related standard, Recommendation H.243, defines the basic MCU procedures for establishing and controlling communication between three or more videoconferencing systems using digital channels up to 2M bps.

Audio Compression Standards

A set of ITU recommendations standardizes audio compression for videoconferencing equipment. Compression is important because it squeezes the audio component of the videoconference into a smaller increment of bandwidth, freeing more of the available bandwidth for the video component. This results in higher quality video, without appreciably diminishing audio. The three key standards in this area are:

- *G.711.* This standard defines the requirements for 64K-bps audio—the least compressed and highest quality audio.
- *G.722.* This standard defines 2:1 audio compression at 32K-bps.
- *G.728.* This standard defines 4:1 audio compression at 16K-bps.

Video Input/Output Standards

Many videoconferencing systems have ports for such auxiliary devices as televisions, cameras, and VCRs. The two most pervasive standards for the quality of video input/output of such devices are NTSC (National Television Standards Committee), a North American standard; and PAL (Phase Alternating by Line), which is used in Europe. NTSC specifies 320x240 resolution at 27 to 30 frames per second. PAL specifies 384x288 resolution at 22 to 25 frames per second. Most video equipment on the market supports both NTSC and PAL. Other lesser used standards that are country-specific may be optionally supported by some vendors.

TRANSPORT ISSUES

Several transport issues must be considered when implementing videoconferencing over the WAN. The following sections discuss issues that determine video quality.

Bandwidth. There must be enough bandwidth available to support the videoconference. The amount of bandwidth is largely determined by the type of facilities or services selected for videoconferencing. Dialup lines offer the lowest amount of bandwidth and, consequently, the poorest quality video. Cell-switched asynchronous transfer mode (ATM) services offer the highest amount of bandwidth, which results in the highest quality video. The bandwidth available on a LAN varies with the number of users and types of applications but could be high if it is dedicated to videoconferencing.

Latency. Latency is the amount of delay inherent in the transmission of voice and video over a particular type of line or service. Dialup lines, including ISDN, are relatively high in latency. ATM-based services provide the lowest latency, making them better suited to supporting videoconferencing.

Isochrony. Isochrony refers to timing. Voice and video require that signals be sent at regular intervals. Data can be sent in bursts at irregular intervals. All switched services are isochronous, making them better suited to videoconferencing than native LANs, which are nonswitched. Vendors get around this limitation with proprietary schemes that give priority to voice and video and by packaging these types of data into frames of uniform length.

Resource Contention. Users of videoconferencing vie with other users for the available resources, which can be ports or bandwidth. With dialup lines, ISDN, and switched 56K bps services, users contend for available ports at the local central office. Once a port is secured, it is held for the duration of the call. Ethernet users, however, contend for a chance to send information over the LAN; the delays inherent in this contention scheme are unsuitable for voice and video.

Availability. Availability refers to whether the required bandwidth will be there when a call is placed. Because dialup lines are ubiquitous, the bandwidth will most likely be there to support the call once an idle port is seized. The trouble is dialup lines do not offer enough bandwidth to support videoconferencing without serious degradation in picture quality. An Ethernet LAN has high availability because the entire amount of bandwidth can be used by only one node at a time.

MANAGEMENT SERVICES

Because managing videoconferencing communications is often expensive and poses staffing challenges, IS managers may want to consider outsourcing this function through local and interexchange carriers that offer videoconferencing network management services. By outsourcing management responsibility for videoconferencing, IS managers need not divert employees from other important duties to arrange videoconferences. Instead, the carrier's trained staff take care of all the details, including network implementation, administration, end-to-end management, asset management, and accounting management.

Depending on the carrier, such services may also cover reservation and scheduling of videoconferencing rooms, maintenance and management of video equipment, and installation of inside wiring and cabling. Other services can include the design of the physical network as well as of the videoconferencing room(s). Detailed summary and billing statements for chargeback to corporate departments are also available.

Under these management services, the carrier may also offer:

- Multipoint bridging for connecting three or more sites.
- Codec conversion for connecting locations that have dissimilar video equipment.
- Speed conversion for connecting locations using different digital transmission speeds.
- Cross-network connectivity for bringing together conference sites that use different network transport services.
- Inverse multiplexing for enhancing video quality by combining channels for increased bandwidth.

As videoconferencing technology continues to improve and the market for it continues to grow, carrier-provided management plans will become more popular, especially for smaller businesses that want the benefits of videoconferencing but cannot afford the high start-up costs associated with ownership.

TELECONFERENCING BASICS

Teleconferencing can be implemented in a variety of ways. For a basic telephone conference, all the users need is a telephone set with either three-way calling (a custom-calling feature available from the phone company) on the line or a conferencing feature on the set. For a teleconference with more than three parties, the attendant console operator can set up the connections through the PBX and add more participants into the call than can be accommodated from a single set. Alternatively, the phone company can implement a conference call through the conference operator.

If an organization frequently conducts meetings by phone, IS managers should consider installing a group conference unit, which enables 10 or more people in a meeting room to participate in a conference with a remote location. This equipment is positioned in the center of the table and uses an omnidirectional microphone. Sometimes individual microphones for each participant are connected to the unit.

Like videoconferencing systems, teleconferencing systems are available in various levels of sophistication. Digital conferencing and switching systems that can be linked directly to a network or installed behind a PBX can link up to 96 callers on the same conference call, with various operations controlled by both participants and attendants. Optional capabilities include data transmission so users can send computer text and graphics and slow-scan video as they converse.

Bridging Technology

The most economical way to teleconference is to call the parties on separate phone lines, then join the phone lines by bridging them. Teleconferencing devices amplify and balance all the conversations, allowing everyone to hear everyone else as though they were talking one-on-one.

These devices, called bridges, connect through ports off the PBX—either line side ports or trunk side ports. The conference can be implemented in several ways. First, each participant is called and then transferred to one of the PBX ports assigned to support the conference. Another method entails participants' dialing a designated phone number at a preset time and automatically being placed on the bridge. Still another method of access involves participants' dialing a main number and being manually transferred onto the bridge by an operator. In either case, whenever a new person joins the party, a tone indicates that person's presence. The more conversations brought into the conference, the more free extensions are required.

An optional moderator phone may be used to set up conferences. The moderator phone allows the operator to initiate conferences and actively participate as well as terminate participants or have an isolated conversation with one person and then either readmit or disconnect that person. Among the useful features of such bridges is security lockout, which bars unauthorized persons or intruders from accessing the conference call.

Another feature is automatic disconnect. Usually, when a conference has ended and callers hang up, a disconnect signal is sent from the Centrex switch or PBX to the bridge to let it know that the ports are free and may be used for new calls. However, not all Centrex switches or PBXs provide this. To compensate, some external bridging devices detect silence on the line and, after two minutes of silence, automatically disconnect the calls.

ENSURING SUCCESSFUL TELECONFERENCES

IS managers can ensure successful teleconferences by considering the type and size of the room, the transmission link, and the audio system. The criteria to consider in the selection of a room include the acoustics and ambient noise. The transmission links that can be used include a telephone line, a PBX bridging device, and an outside bridge. Managers should consider clarity, volume, interactivity, and capacity when selecting a transmission link. All the audio systems that will be in the room with the transmission link should be thoroughly tested. The criteria to consider when testing audio systems include clarity, volume, interactivity, portability, ease of use, and aesthetics. Other criteria (e.g., the environment, transmission facilities, and the audio system), if addressed properly and in order, can guarantee acceptable audio during the teleconference.

The Environment

The acoustics of a room is the first environmental element that must be addressed. Hard reflective surfaces cause sound waves to bounce back and forth, resulting in a hollow, reverberant, bottom-of-the-barrel effect. Any soft absorptive material that covers hard surfaces will stop the sound waves from reverberating and reduce that hollow effect.

The level of ambient noise in a room, inherent in every meeting place, is another potential obstacle to quality audio. Ambient noise is any sound except human speech bouncing around in the room. There are two types of ambient noise:

- Steady-state noise, which does not change in volume and comes from something like heating, ventilating, and air conditioning (HVAC) systems.
- Transient noise, which is not constant and is caused by rustling paper, finger tapping, or coughing.

The best method of dealing with ambient noise is to try to avoid it. IS managers should consider moving the teleconference site to a quieter environment rather than redesigning the HVAC to suit a particular room. Although some teleconferencing systems can adapt to poor room acoustics or high levels of ambient noise, even the best audio systems will perform better in more efficient environments.

Transmission Facilities

The second criterion for a successful teleconference is the transmission facilities. For two-wire teleconferencing, this can be a standard telephone line, the bridging device contained in a PBX, or a carrier-provided bridging service. When selecting the transmission link, IS managers should address the following questions:

- *Clarity.* Can the transmission link conduct sound accurately with little or no distortion?

- *Volume.* Is the connection strong and stable enough so that every participant can be heard at an acceptable level?
- *Interactivity.* Can all participants interrupt each other and speak as if they were all present in the same room?
- *Capacity.* Can the bridge or bridging service provide enough ports to connect all participants?

In most cases, digital facilities will support the most demanding requirements for teleconferencing, whereas analog lines often exhibit poor or variable quality.

The Audio System

At the heart of teleconferencing is the audio system. The best way to choose an audio system is to test various audio systems in the room where the teleconferences will be held, using the preferred transmission link. Different acoustic environments and transmission links will cause audio systems to produce varying results. The performance issues for an audio system begin with the same criteria used to rate a transmission link: clarity, volume, and interactivity, and capacity. Additionally, three other factors should be considered:

- *Portability.* If the system must be moved from room to room, is setting up the system an easy procedure?
- *Ease of use.* Is the system easy enough for even the most impatient executives to use?
- *Aesthetics.* Does the system blend into the decor of the conference room so that it will not intimidate the participants?

Quality audio communication is the key component of integrated communications solutions. Previously, teleconferences used suppression technology (i.e., gated or push-to-talk microphones) and therefore were restricted to a half-duplex ability (one-way talking at one time). Echo cancellation allows dynamic, two-way conversation, without the clipping and distortion inherent in older systems. Echo cancellation's successful treatment of acoustic problems, including noise and echo, makes the technology suitable for multipoint and roll-out conferencing systems.

Manufacturers differentiate their systems by performance, ease of use, and price. Most teleconference systems accommodate only two of the three characteristics. Therefore, IS managers considering the purchase of a system should make an informed assessment of their communication needs and available solutions.

Audio-only teleconferencing may be inadequate for meetings that involve distributing documents, the ability to see participants' reactions to ideas being discussed (e.g., during labor negotiations between remote locations), or training sessions in which a higher level of involvement is required. For this type of meeting, a videoconference provides a more effective solution.

CONFERENCING OVER THE INTERNET

The most economical conferencing is accomplished through software that works over the Internet. A variety of vendors offer software for text-based chatting, audio conferencing, and videoconferencing on a point-to-point and multipoint basis. Some products, such as White Pine Software's Enhanced CU-SeeMe, combine audio and videoconferencing. Users who want to transmit images can purchase a camera for as little as $99, but the camera is not needed for audio conferencing. To reduce demand on the network, video images over CU-SeeMe are transmitted in black and white at a slower frame rate than full-motion video. One of the main advantages of CU-SeeMe for audio conferencing is that users can either connect directly to each other or they can enter a conference of up to 400 people at a reflector (i.e., a server-like component). A 14.4K bps modem is adequate for audio conferencing, but a 28.8K bps modem is recommended for conveying images. Although the video component may appear grainy and choppy, the fact that conferencing over the Internet is free (i.e., it does not add to the cost of an Internet connection) may make it worthwhile for occasional use.

In addition to the algorithms that compress/decompress sampled voice and video, Internet-based conferencing products may include optimization techniques to deal with the inherent delay of the Internet. The packets may take different paths to their destination and may not all arrive in time to be reassembled in the proper sequence. In the case of ordinary data, late or bad packets would simply be dropped and the host's error checking protocols would request a retransmission of those packets. But this concept cannot be applied to packets containing compressed audio or video without causing major disruption to the session, which is supposed to be conducted in real time. If only 2% to 5% of the packets are dropped, users at each end may not notice the gaps in their conversation. When packet loss approaches 20%, however, the quality of the conversation begins to deteriorate. Some products minimize this problem by using predictive analysis techniques to reconstruct lost packets.

Standards

A pair of TCP/IP standards developed within the Internet Engineering Task Force (IETF) will greatly improve the quality of real-time voice and video over the Internet. They are RSVP (resource reservation protocol) and RTP (real-time transport protocol).

RSVP. As an Internet control protocol, RSVP runs on top of IP to provide receiver-initiated setup of resource reservations on behalf of a multimedia application data stream. When an application requests a specific quality of service for its data stream, RSVP is used to deliver the request to each router along the path(s) of the data stream and to maintain router and host states to support the requested level of service. In this way, RSVP essentially allows a router-based network to mimic a circuit-switched network on a best-efforts basis. This means that a permanent virtual circuit will be set up between the sender and receiver to support multimedia applications, including audio and videoconferencing. With such a circuit, bandwidth is dedicated to the session, eliminating variations in delay. The result is a much smoother flow of audio and video.

RTP. RTP runs over TCP and is typically implemented as part of the application. RTP provides functionality suited for carrying real-time content; specifically, a timestamp and control mechanism for synchronizing different streams with timing properties. RTP inserts timing and sequencing information into each packet. Although RTP does not enhance the reliability of a transmission, applications make use of the timing and sequencing information to enable audio and video streams to be run smoothly, despite occasional packet loss.

SUMMARY

Conferencing technologies, both voice and video, are finally taking their place as strategically significant corporate communications tools. Improvements in conferencing technologies are pushing costs down. With lower costs comes increased demand, which will, in turn, fuel competition and encourage further innovation.

3-6 Choosing a Remote Access Strategy

Scot McLeod
Mark Monday
Bill Heerman

NETWORKS HAVE HAD more to do with recent changes in corporations than any other factor. The corporation is no longer confined to a building; rather, it is the sum total of its stored information. Thus, leveraging information, or knowledge assets, is the key strategic focus of the future for most enterprises.

Already corporate networks contain customer histories, inventories, accounting information, engineering designs, E-mail systems, planning instruments, market encyclopedias, documents, and applications—information that can be leveraged to make better business decisions. To make even better use of network services already in place and better manage the allocation of such systems, an organization may create a remote computing strategy to allow mobile sales forces or corporate telecommuters to access enterprise LANs.

The components of remote computing are similar, regardless of the technology used. For example, each technology starts with a remote PC and can transmit over various devices and carriers (see Exhibit 3-6-1). The ultimate destination of the transmitted data determines the benefits and limits of each technology. The following sections show what different remote technologies have in common and what makes them unique.

THE REMOTE PC

The remote PC can be an IBM-compatible running DOS, Windows, or OS/2, or a Macintosh computer. It can be a laptop or a desk model. It can

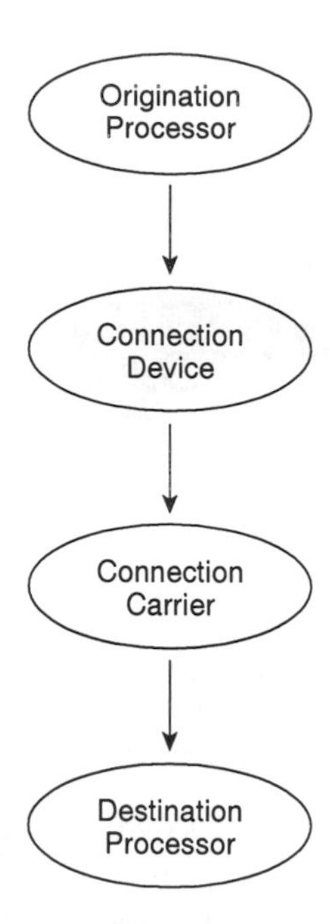

Exhibit 3-6-1. Components of Remote Computing

incorporate the latest, fastest chip or operate with slower, older technology, depending on the demands of the application or destination computer.

CONNECTION DEVICES AND CARRIERS

Modems

A modem converts digital information that circulates in a computer to analog information that can cross a phone line, and then back to digital information again on the other end of the connection.

The modem is usually the slowest link in the remote chain, determining the speed limit to the remote network. For instance, a typical Ethernet LAN connection is 160 times faster than a 14.4 modem. (See Exhibit 3-6-2 for a speed comparison of different technologies.)

Modem vendors are constantly increasing throughput on their modems by adopting new standards and by using their own proprietary technology. Updating a modem can be the easiest way to increase throughput for the remote connection. The 14,400-bps modem, now commonplace, is rapidly being replaced by a new standard that operates at 28,800 bps. (To realize

Link	Speed in K bps × Compression Factor	Effective Throughput Speed in K bps	Speed Relative to the Network*
Network*	3,000	3,000	1
ISDN, 2 Lines with Compression	128×4	512	1/6
ISDN, 1 Line with Compression	64×4	256	1/12
ISDN, 2 Lines	128	128	1/18
ISDN, 1 Line	64	64	1/36
Modem** with Compression	14.4×4	57.6	1/52
Telephone Line/Modem	14.4	14.4	1/60
X.25 Cloud/PAD	9.6	9.6	1/260

* Network speed varies depending on the type of network, number of users, and overhead factors. Speed in this example is what a typical Ethernet user could expect.

** Fastest modem in general use at highest achievable compression; not possible with all file types.

Exhibit 3-6-2. Connection Speed Comparison

maximum throughput, a 16,550 or better universal asynchronous receiver transmitter is required in the PC.) Each speed gain increases network efficiency on a logarithmic scale.

Telephone Lines

Today, traditional analog (i.e., asynchronous) lines are the most popular communication method for remote users because:

- Widespread telephone service is readily available.
- Analog public switched telephone service (PSTN) tariffs are inexpensive.
- The technology is mature and well understood by applications developers, security vendors, and Internet service providers.

Nevertheless, analog transmissions also suffer from several drawbacks because:

- Transmission of data is slow—currently the maximum speed is V.fast (i.e., 28.8K bps).

- The connections are occasionally unreliable or of poor quality.
- The system offers limited scalability for building large networks.

X.25 Networks

Generally an X.25 network is a set of private lines dedicated to one company's use. X.25 refers to the protocol, or agreed-on system of transfer, on such a network. However, the term X.25 is often applied to the whole network.

An X.25 network can be accessed two ways: through a modem and a telephone line or directly through an X.25 card in the PC. The first method is common in North America where the price and quality of telephone lines are consistent. Elsewhere the X.25 card is common.

When data leaves the computer via a modem, it travels a relatively short distance as analog information en route to the X.25 network. At the gateway to the X.25 network, a packet assembler/dissembler (PAD) converts this analog signal to a digital one and sends it over the X.25 network. This same result is gained directly with the X.25 card in typical European applications of X.25. In either case, the data reaches the X.25 network cloud and travels to its destination, where a server with an X.25 card unloads the data to a LAN.

The X.25 approach is similar to walking to the car (PC to modem), driving to the airport (modem to PAD), flying across the country (X.25 network), getting in a taxi (server with X.25 card), and arriving at a hotel (the LAN). Exhibit 3-6-3 illustrates an X.25 network.

X.25 networks are generally implemented as a solution to cost or quality constraints. Typically, parts of the X.25 network operate at 9,600 bps, the same speed as many older modems. The entire network is only as fast as its slowest link. Thus, X.25 does not offer speed or data throughput as advantages. It does, however, offer cost and reliability advantages because the X.25 network is entirely within the auspices of the corporation. In Europe, X.25 networks can span several national telephone systems, saving switching costs and improving quality. In North America, online services such as CompuServe offer X.25 value-added networks to corporate clients for competitive flat rates plus usage costs.

ISDN Connections

An integrated services digital network (ISDN) does not need a modem. An ISDN terminal adapter on the PC sends data out in digital form to travel across phone, fiber-optic, or high-bandwidth lines to another computer without the need for modulation to analog signals. Thus the transmission runs as fast and error-free as possible.

ISDN offers affordable, fast, high-quality transmission of data, voice, and

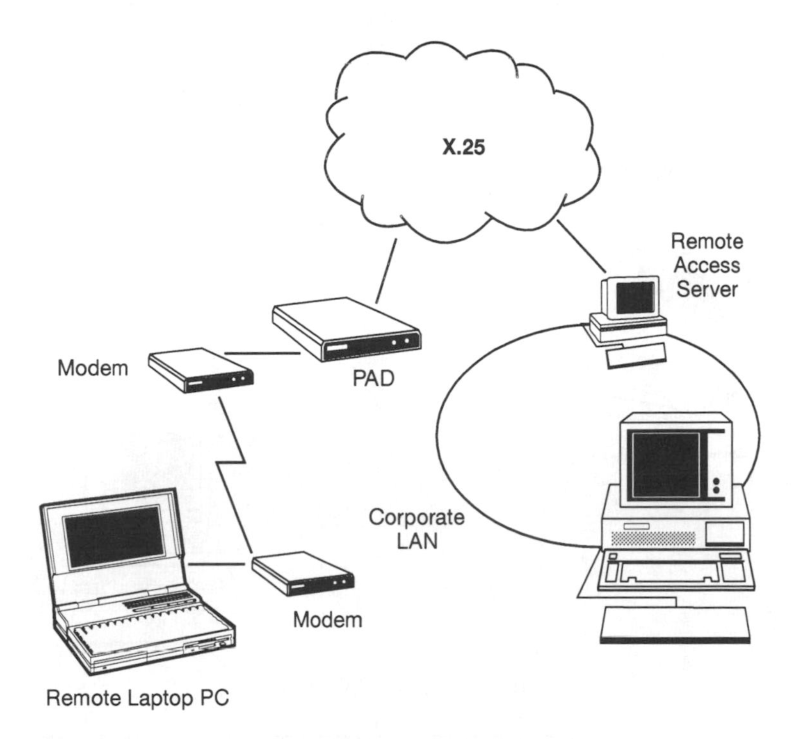

Exhibit 3-6-3. An X.25 Network

video. For example, standard voice lines today can carry up to 28K bps. ISDN increases throughput up to 128K bps. ISDN also compresses the data using one of many standard compression systems, which effectively increases speed by sending more information in the same amount of time. By coupling ISDN with data compression, remote users can achieve exceptional data throughputs—up to 384K bps—making ISDN one of the most common connection methods of the future. ISDN lines are still unavailable in parts of the US. Exhibit 3-6-4 illustrates an ISDN network.

Cellular Transmission

A cellular transmission simply adds another link to the communications chain. Any cellular connection is simply a radio call to a transmission station, referred to as the cell, where that signal is picked up and relayed to another station, eventually arriving at a hard wire and continuing via the network of choice (e.g., a leased line or X.25).

A cellular modem converts the digital information to analog information for it to be carried on the radio waves to the cell, much like a standard modem

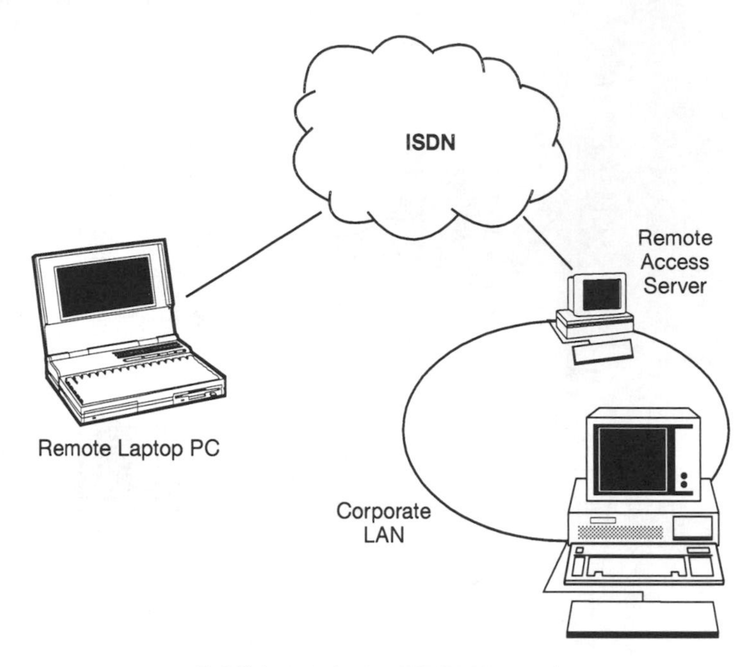

Exhibit 3-6-4. An ISDN Network

converts digital data from a computer to analog data to travel over phone lines. In either case, the data is carried on waves, whether through the air or on a wire.

Cellular data calls work exactly like cellular voice calls except they cannot reliably switch cells, which happens when traveling. Data cellular calls are best completed when the user is stationary. Exhibit 3-6-5 shows a cellular transmission.

CDPD. Cellular calls can also be digital. This technology, called cellular digital packet data (CDPD), breaks data files into digital units, called packets, and sends them on upgraded cellular voice networks. Within the next decade, CDPD will emerge as the standard for cellular transmissions.

THE DESTINATION

Remote PCs can access a variety of destinations including other PCs, mainframe computers, and LANs. Although the capabilities of PCs and mainframes are more obvious, the LAN is somewhat analogous to a street—streets lead to and provide access to individual homes. The network provides

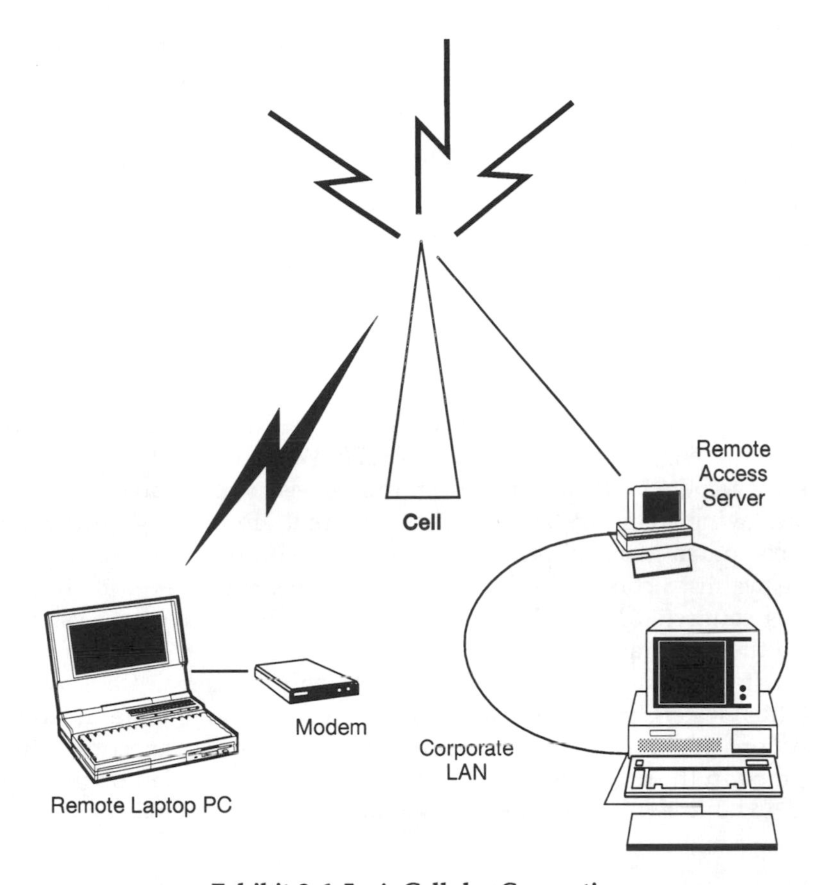

Exhibit 3-6-5. A Cellular Connection

"streets" or venues to files, printers, applications such as E-mail, and access to host computers. To reach the network is to reach all of the devices connected to that network.

When connected to the LAN the remote user can do anything locally attached PCs do. A mainframe computer, for example, could be accessible via a gateway on the LAN. Any network-mapped resource can be accessed, including special PCs dedicated to running network applications, such as file servers or application servers.

CONNECTION TECHNOLOGIES

Mainframe remote terminal and remote control are an outgrowth of the original way computing was conceived. Both of these technologies treat the remote PC as a dumb device. Using these techniques, the remote PC is stripped of any processing obligation and functions only as a control and display device similar to the way a remote control changes channels on a television set.

The third technology, remote node, treats the remote PC differently, as an extension of the LAN. Remote node treats both the remote PC and whatever it accesses as a processor and depends on each, to a certain extent, to do the work of computing. Exhibit 3-6-6 illustrates the differences between mainframe-centric, PC-centric, and LAN-centric access methods.

Mainframe Remote Terminal

Essentially this technology connects a remote PC to a mainframe over a phone line, using a protocol such as TN3270. When connected to the mainframe, the remote PC then emulates a terminal—issuing keyboard commands and displaying character-based applications that are running on the mainframe computer. Minimal processing takes place on the remote computer; it only relays instructions and displays results. This is true even if the mainframe is connected in another way to a LAN. The remote user cannot go beyond the mainframe.

For a small number of remote users who only need mainframe applications, this system proves efficient. However, on a large scale, mainframe connections are limiting and can be redundant to other connections already configured in the organization. They also make no use of laptop processing power. Exhibit 3-6-7 illustrates a mainframe remote terminal.

Remote Control Technology

Remote control technology allows a PC to link to another PC using a modem and a telephone line. Under remote control, the destination PC processes the application receiving input from the remote keyboard and mouse and sending screen output to the remote monitor. The remote computer handles no processing. This system is somewhat like stretching the keyboard and monitor cable from the calling point to the calling destination. The user must work on the hard drive of the destination PC. All files and applications reside there and remain there, as does access to the network.

Remote control technology was invented to handle character-based applications with relatively small input and output demands. With Windows-based applications, the technology bogs down somewhat because graphical-based input and output requires more information to be passed over the modem and phone line. Although recent advances in reducing the necessary load of data for Windows application input and output have improved its workability, several problems still persist in the practical use of remote control, including:

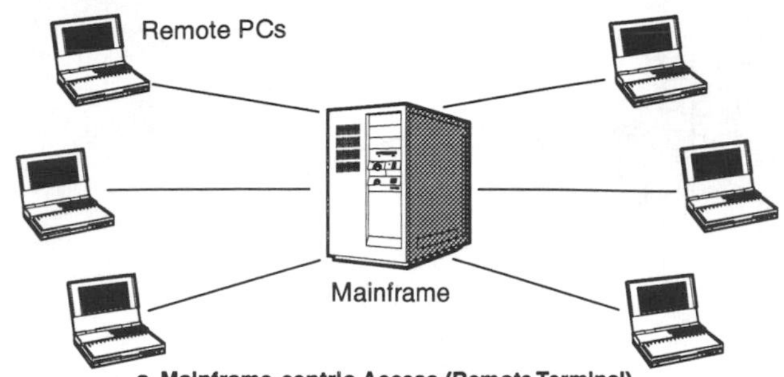

a. Mainframe-centric Access (Remote Terminal)
This technology lets the user get to the mainframe but not beyond it.
It makes no use of existing LAN resources.

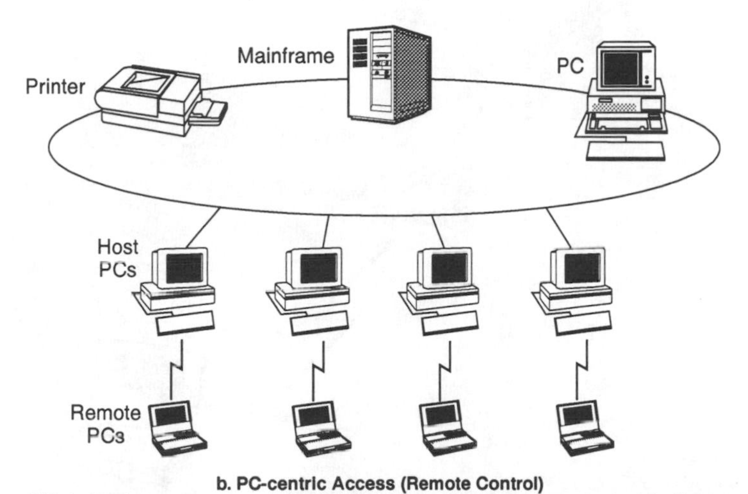

b. PC-centric Access (Remote Control)
This technology gets users onto the LAN but wastes the processing power of the laptop.
Users must buy a dedicated PC for each laptop in use; the potential for a security breach is multiplied by each access point.

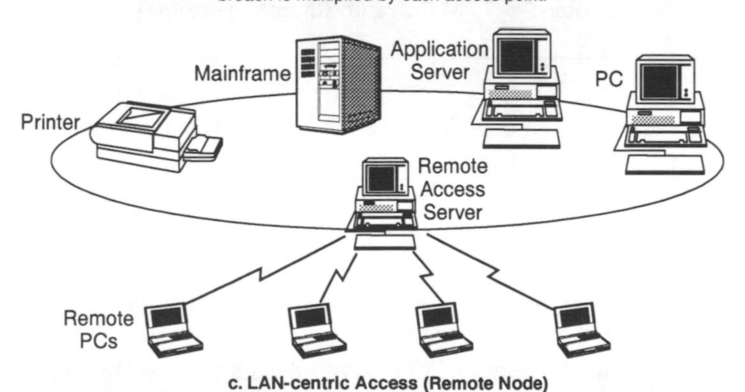

c. LAN-centric Access (Remote Node)
LAN-centric access allows users to access everything on the LAN, control security through one server, and save costs by eliminating dedicated PCs with this technology.
For data-intensive applications, an application server can be added.

Exhibit 3-6-6. Comparison of Access Methods

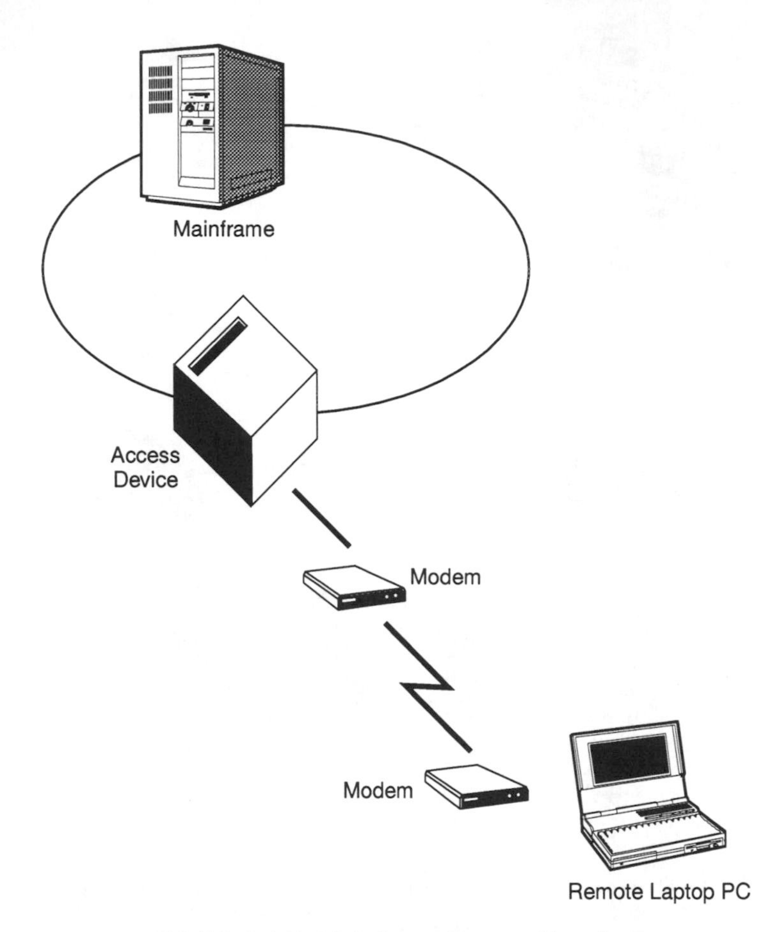

Exhibit 3-6-7. Mainframe Remote Terminal

- There typically has to be a dedicated PC for each user dialing in, making equipment costs very high.
- The user must be connected to run an application; users have no independent functional capabilities despite having applications at the remote PC.
- Access is not transparent (i.e., applications run within applications, adding to complexity and confusion).
- Performance of the remote PC is always dictated by the PC it is connected to.
- User entry into corporate computers is on an individual, unmonitored, and unsecured basis.

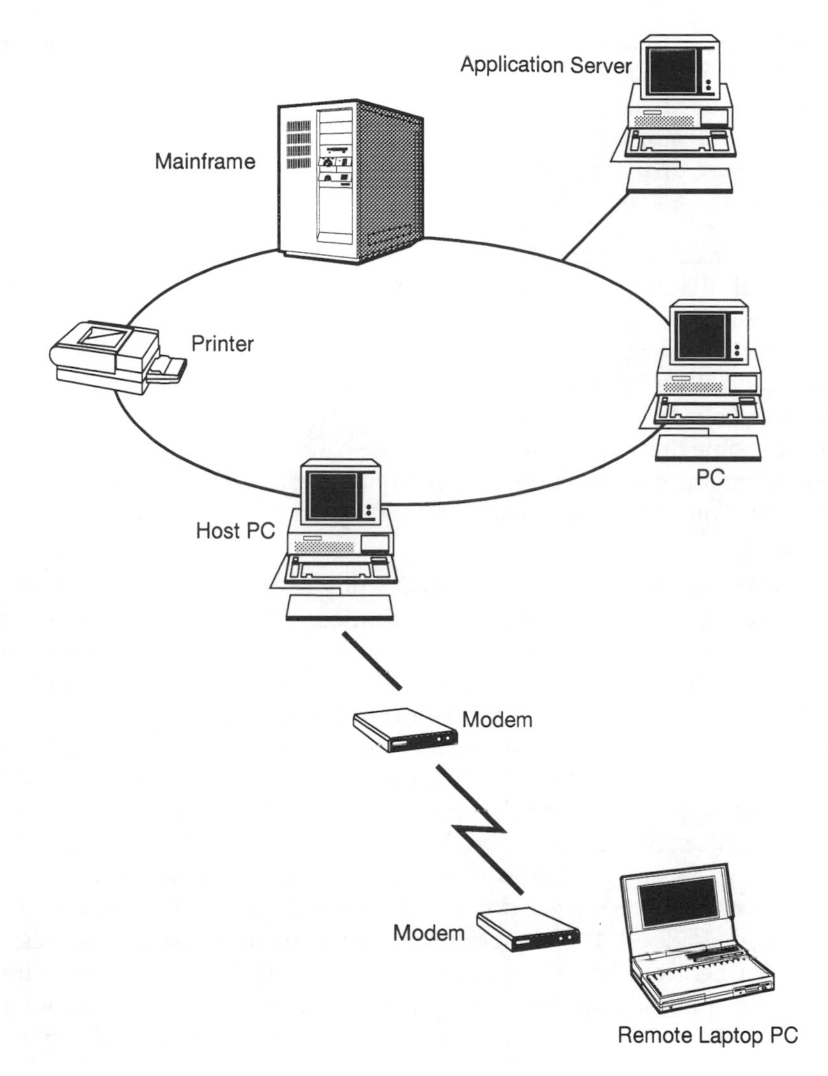

Exhibit 3-6-8. Remote Control Network

With remote control technology, each user multiplies the potential for security and management breaches. In fact, once the remote user reaches a destination PC on a network, remote users can direct their hard drives to all networked devices, gateways, and ultimately to corporate hosts. With so many potential points of access to the network, security is unmanageable. Exhibit 3–6–8 illustrates a remote control network.

Hybrid Remote Control Technologies. Multiprocessing is one way to address the cost/performance concerns of remote control. Essentially, multiprocessing is a way of dividing one PC into many sub-PCs, each of which can run separate versions of remote control software concurrently. This can be done with hardware or by tricking the PC with special software to think it has many different CPUs.

Remote Control Servers. Another hybrid is the remote control server, a PC on the network that is preloaded with programs remote users need to use. The server sends screen information and receives keyboard and mouse information from the remote user. Actual processing takes place on the LAN-based remote control server, so it is ideal for data-intensive applications such as data base programs.

One possible drawback is that the remote control server usually must have all the programs each user needs, already loaded. Users will not be able to process files on their remote computers; they actually work on the server.

Application Servers. Another way to efficiently process data-intensive applications is an application server. The application server directs a remote computer's demand for an application to a file server where that application is stored. Then the file server sends the program to the application server and, in turn, it returns the screen information to the remote computer.

Although this technique may sound more complicated than the remote control server, this system actually makes better use of existing LAN resources and is not limited by the memory capabilities of the remote control server. The file server is already on the LAN with the applications needed for those LAN users and can be accessed by the remote user as well.

Because application servers do the processing, they can give remote PCs a CPU-like upgrade. A 286 remote PC could get the speed and processing power of a Pentium-based application server by dialing in. Multiple users can execute application programs on the server. A multitasking environment supports multiple remote users from a single microcomputer directly attached to the LAN.

Application server technology has become a key component in solving many of the problems previously inhibiting the deployment of network remote access solutions. However, the remote PC still must be connected to the network to run applications.

Application-Specific Technology. Many software vendors have developed ways to access their own products remotely. A common example of such application-specific remote technology is found in E-mail software. Unfortunately, this kind of remote access technology cannot be used to access anything except a specific vendor's software product. Such a narrowly defined solution to remote access makes incomplete use of the network infrastructure. However, application-specific products are precursors to a more open, remote node technology.

Remote Node Technology

Using remote node technology, a user dials into the LAN instead of into a specific PC. This extends the LAN to the remote PC, enabling the user to take advantage of anything offered by that LAN, including E-mail, mainframe access, groupware applications, and network printing.

With remote node technology, the remote user runs applications and saves files at the remote PC. Input commands from the keyboard and mouse do not travel over remote lines; they work locally, so there is no potential for a bottleneck. However, by extending the LAN across remote lines, it no longer works at Ethernet or Token Ring speeds; it works at modem or ISDN line speeds. This is not a problem until large files must be transferred across the remote connection. Consequently, passing large files, like data base files, across this link can make remote node technology unsuitable unless used in conjunction with an application server.

In terms of feasibility, remote node makes best use of the network infrastructure already in place. There is no longer the need to maintain a 1:1 ratio of field to office computers because the remote users can log into the network on a server, a device that extends entry to the LAN to multiple users.

Currently, node access products provided by some network and modem vendors favor a particular and proprietary network operating system. Using such a system for remote access can preclude access to environments where several operating environments are expected to coexist. Banyan, Novell, and Microsoft all provide remote solutions specifically tuned to their own preferred network operating system.

Conversely, a node solution that is transparent and nonproprietary can access different networks and their distinct protocols and operating systems. This solution offers the best opportunity to manage and secure access because all entry to corporate computers is through a server located on the LAN. Exhibit 3-6-9 illustrates a remote node solution.

HARDWARE AND SOFTWARE OPENNESS

Whether using a Windows-based computer at home, an Apple PowerBook laptop in the field, or a UNIX workstation at the corporate LAN, users expect the same procedures and level of network service. In turn, a small branch office will demand high performance, dial-in and dial-out security, ease of administration, central network management, and a future pathway to client/server applications. A successful remote LAN access architecture must

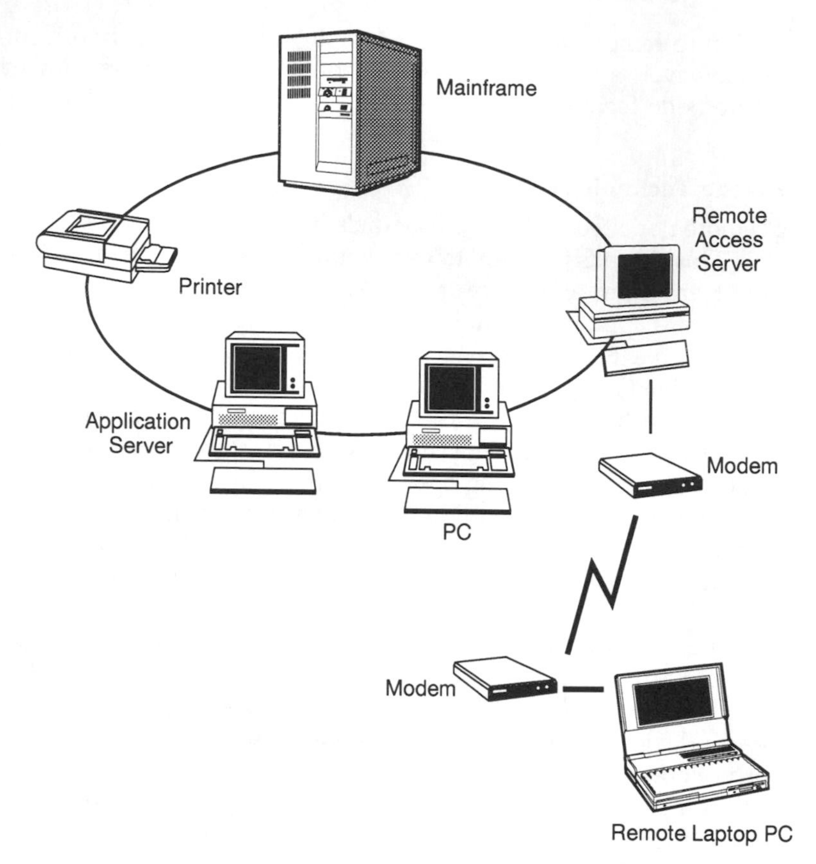

Exhibit 3-6-9. A Remote Node Solution

deliver client-level access to users even when the network computing infrastructure is unpredictable.

The typical network computing infrastructure combines multiple platforms from different vendors and employs differing network operating systems, physical-layer access methods (i.e., Ethernet, Token Ring, and all their emerging bandwidth variations), and network protocols (i.e., TCP/IP, IPX, NetBIOS, NetBEUI, SNA, AppleTalk).

Solutions that conform to industry-specified, nonproprietary standards stand the best chance of surviving technological changes. Some examples of such open standards are point-to-point protocol, simple network management protocol (SNMP), network driver interface specification (NDIS), and open data link interface (ODI) programming interfaces.

Openness also pertains to applications. The optimal solution should be

Equipment	Quantity Needed	Unit Price*	Remote Control Extended Price	Remote Node Extended Price
Laptop PCs**	32	$3,000	$96,000	$96,000
14.4 Models	32	$200	$6,400	$64,000
Phone Lines***	32	$70/Month	$2,240/Month	$2,240/Month
32 Node Server	1	$12,995	NA	$12,995
Remote Software	32	$200	$6,400	$0
Dedicated PCs	32	$1,500	$48,000	NA
Total			$159,040	$117,635

* Estimated prices. Prices vary by vendor and quantity.

** Assumes purchase of a new unit for each remote user.

*** Ongoing monthly fee.

Exhibit 3-6-10. Simple Cost Analysis—Setup and First Month for 32 Users

able to handle all corporate computing needs—even some that are difficult to anticipate. This means a nonproprietary system that can handle both data-intensive and graphical applications, word processing and spreadsheet software, E-mail, groupware, and mission-critical client/server applications.

CHOOSING A REMOTE LAN ACCESS SOLUTION

Costs

If cost were no object, anyone could put together a perfectly functioning system. But often cost is the main object, though there are many additional factors to consider. For example, will added telephone lines put users over their current PBX limit? Such factors vary individually. For a more tailored analysis, Exhibit 3-6-10 provides a remote access cost worksheet.

Security

After access, security is the foremost consideration for a remote computing solution. Having all systems accessible from a single point of entry is a major security benefit. Although it may be unrealistic to expect a totally impregnable environment, the solution should extend multiple layers of security. Users should be allowed to select as much or as little security as their organizations' needs dictate.

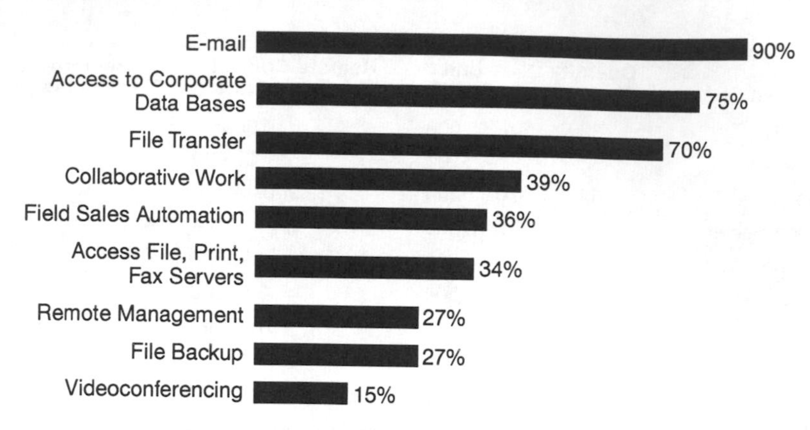

Exhibit 3-6-11. Leading Applications for Remote Access to the Network

Theft, disruption, and malicious intrusion by hackers are obvious problems for any corporation, but even unintentional intrusion can cause an enterprisewide network crash. Network administration of remote branch offices, employees with offices at home, customer/supplier guest accounts, or mobile computing users add to the burden. Although bulletproof dial-in security is desirable, given the potential for an unpredictable network access, it is also imperative that network security operates within each site without complicating the work of legitimate users and network administration.

Network Management

Most network managers want a remote solution that lets them configure, monitor, and troubleshoot from any PC on the LAN; includes a single access server as a primary entry point for all remote user information; provides an easy way to organize their users into logical groups; and provides a way to track remote users and their LAN use.

Reliability. Of the utmost importance is the reliability of a product. Any remote computing solution must include a hot backup feature in case of a remote access server failure, to preserve access to the LAN and valuable data.

Support. Putting together a corporatewide solution can be complicated. When evaluating a solution, users should be sure to evaluate the company behind the product. An organization should be knowledgeable as well as willing and able to help users through any early growth and implementation stages. Once in place, the solution should be easy to manage.

Requirement	RT	RC	AS	RN
Character-Based, Mainframe Application Access	x	x		x
DOS/Character-Based LAN-Centric Applications		x	x	
Access to a Specific LAN-Attached Desktop Platform		x		
Windows or GUI-Based LAN-Centric Applications			x	
Satisfactory Access Using a Low-Performance Remote PC			x	
Data-Intensive Applications		x	x	
Windows or GUI-Based Client/Server Applications				x
Transparent Access Across Different Networking Environments				x
Single Point of Access to Information and Services Anywhere on the Network				x
Consistent User Interface for File Copying and Transfer, Execution of Applications, and Connections to Other Networks				x
No Dedicated PC Requirement				x
Robust and Easy-to-Manage Network Security				x
Automated and Centralized Administration of Network Management Tasks				x
Efficient and Effective Use of Switched Digital Services (e.g., ISDN)				x

KEY :
RT Remote Terminal
RC Remote Control
AS Application Server
RN Remote Node

Exhibit 3-6-12. Remote Access Solutions

Understanding the Remote Workplace

Controlling expectations can be the key to successful implementation of the remote workplace. Often, legitimate purposes are confused with more general goals that cannot be addressed by remote products.

Remote access is not a panacea for all organizations. It is a tool that solves problems and should be applied in the situations that merit it. Exhibit 3-6-11 lists the leading applications for remote access to the network are, for example, E-mail, access to corporate data bases, and file transfer. To succeed, however, remote access must be part of an overall plan to accomplish specific goals. To be accepted, remote access must benefit both employees and corporations, producing more income and reducing workloads.

Demand for a remote workplace typically starts with mobile employees.

Sales and other client-intensive occupations greatly benefit from the transfer of activities from the office to the field. The kind of employees who operate best in remote computing situations are those who have been on the job long enough to solve problems independently, who are self-disciplined and self-motivated.

Groupware. Teamwork is another part of work that is altered by the remote paradigm. Although day-to-day contact with coworkers is lost, a new type of collaborative work has emerged to replace it. This networked electronic form of collaboration is a new software genre known as groupware.

Groupware products permit remote and traditional workers to log on to a server. Through groupware, workers separated by distances can still work together on documents and have discussions threaded like conversations. Groupware products save travel time as well as adding collaboration to the list of tasks that can take place outside of traditional office walls.

SUMMARY

Ultimately it is people who perform work, not technology. Different remote access solutions optimize different functions. Choosing or combining the right solutions can optimize remote access and the efficiency of remote workers. Using the chart in Exhibit 3-6-12, network managers can evaluate some common requirements and the remote schemes that fulfill them best.

3-7

Considerations for Implementing Corporate Intranets

Nathan J. Muller

A CORPORATE INTRANET ENTAILS THE DEPLOYMENT AND USE of Internet technologies such as the Web, E-mail, and Transmission Control Protocol/Internet Protocol (TCP/IP) on a closed private network within one organization or within a group of organizations that share common relationships. Because intranets are based on TCP/IP networking standards, they can include anything that uses this protocol suite, including existing client/server technology and connectivity to legacy host systems. Companies can benefit from Internet technology and avoid its drawbacks—particularly, its lethargic performance and lack of security.

Intranets support communication and collaboration, information retrieval, sharing and management, and access to data bases and applications. None of these functions is new, but the promise of an intranet is that it can use Internet and World Wide Web technologies to do things better than before.

For example, according to Microsoft Corp., Netscape Communications Corp., Oracle Corp., and Sun Microsystems, Inc., a Web browser could become the standard interface used to access data bases, legacy applications, and data warehouses throughout the enterprise. In this scenario, the thin client (i.e., the browser) can make applications easier to maintain, desktops easier to manage, and substantially trim the IT budget.

A company's customers, suppliers, and strategic partners in turn can benefit from the improved communication, greater collaboration, and reduced IT expenditure associated with implementing an intranet. They can even access each other's intranet services directly, which would speed decision making as well as save time and money.

Achieving these benefits comes from properly implementing an intranet, which is far from straightforward. First, resources must be available to establish the service, to establish the TCP/IP network over which it runs, and to train users. Second, the impact on existing systems must be considered. This includes, for example, the capacity of the current network to support an intranet, the future usefulness of existing legacy systems, and the availability of hardware to run multimedia applications.

More difficult to resolve is the issue of intranet content. It is necessary to decide what information will be presented, where it will come from, how its accuracy will be assured, and how often it will be updated. The resources must be available to do this extra work.

"FAT" VERSUS "THIN" CLIENTS

Corporate intranets provide an opportunity to ensure universal access to applications and data bases while increasing the speed of applications development, improving the security and reliability of the applications, and reducing the cost of computing and ongoing administration.

"Fat" and "thin" refer primarily to the amount of processing being performed. Terminals are the ultimate thin clients because they rely exclusively on the server for applications and processing. Standalone PCS are the ultimate fat clients because they have the resources to run all applications locally and handle the processing themselves. Spanning the continuum from all-server processing to all-client processing is the client/server environment, where there is a distribution of work between the different processors.

Client/Server. A few years ago, client/server was thought to be the ideal computing solution. Despite the initial promises of client/server solutions, today there is much dissatisfaction with their implementation. Client/server solutions are too complex, desktops are too expensive to administer and upgrade, and the applications are still not secure and reliable enough.

Furthermore, client/server applications take too long to develop and deploy, and incompatible desktops prevent universal access. However, all this is about to change as companies discover the benefits of private intranets and new development tools like Java and ActiveX, as well as various scripting languages such as Javascript and VBScript. All of these tools are helping to redefine the traditional models of computing and forcing companies to reassess their IT infrastructure.

JAVA-ENABLED BROWSERS

Browsers that are used to navigate the World Wide Web are usually thin clients when they render documents sent by a server. The special tags used throughout these documents, known as the hypertext markup language (HTML), tell the browser how to render their contents on a computer screen.

However, browsers can get very fat when other components are sent from the server for execution within the browser. These components could be specialized files like audio or video that are interpreted by plug-ins registered with the browser. When the browser comes across an HTML tag that specifies a file type that is associated with one of these plug-ins, the application is automatically opened within the browser. This permits an audio or video stream to be played instantly without the user having to download the file to disk and open it with an external player.

Applets

Another way that the browser can become fat is by absorbing Java applets that are downloaded from the server with the HTML documents. Applets are small applications designed to be distributed over the network and are always hosted by another program such as Netscape's Navigator or Microsoft's Internet Explorer, both of which contain a "virtual machine" (VM) that runs the Java code. Because the Java code is written for the virtual machine rather than for a particular computer or operating system, by default all Java programs are cross-platform applications.

Java applications are fast because today's processors can provide efficient virtual machine execution. The performance of graphical user interface (GUI) functions and graphical applications are enhanced through Java's integral multithreading capability and just-in-time (JIT) compilation. The applications are also more secure than those running native code because the Java runtime system—part of the virtual machine—checks all code for viruses and tampering before running it.

Application development is facilitated through code reuse, making it easier to deploy them on the Internet or corporate intranet. Code reuse also makes the applications more reliable because many of the components have already been tested.

ActiveX and Java

Another way the browser can be fattened up is by bulking up on components written in ActiveX, Microsoft's answer to Sun's Java. Like Java, ActiveX is an object-oriented development tool that can be used to build such components as Excel spreadsheet interpreters and data entry programs. Functionally, the two development tools are headed for increasing levels of convergence.

For example, the Microsoft Java VM is an ActiveX control that allows Microsoft Internet Explorer 3.0 users to run Java applets. The control is installed as a component of Internet Explorer 3.0. The Java VM supports integration between other ActiveX controls and a Java applet. In addition, the Java VM understands the component object model (COM) and can load COM classes and expose COM interfaces. This means that developers can write ActiveX controls using Java.

Scripting Languages

Browsers can also fatten up by running functions written in scripting languages like Netscape's Javascript and Microsoft's VBScript. VBScript is a Web-adapted subset of Visual Basic for Applications (VBA), Microsoft's standard Basic syntax. Both Javascript and VBScript are used to manipulate HTML from objects like check boxes and radio buttons, as well as add pop-up windows, scroll bars, prompts, digital clocks, and simple animations to Web pages.

The important thing to remember about these tools is that the features they create rely on scripts that are embedded within the HTML document itself, initiating extensive local processing. Browsers are becoming "universal clients," so much so that Microsoft's next release of Windows 95 will even have the look and feel of a browser.

With the popularity of the Internet continuing to grow, most PCS today come bundled with a browser. Now several vendors, including Microsoft, have latched onto the idea of offering a new breed of computer that relies on a browser as the graphical user interface, Java or ActiveX as the operating system, and servers for the applications. With Java and ActiveX, a net-centric computing solution is emerging that can potentially offer major improvements in simplicity, expense, security, and reliability versus many of the enterprise computing environments in place today.

FEEDING CLIENT APPLICATIONS

How fat the client is may be less important than how the code is delivered and executed on the client machine. Because Java applications originate at the server, clients only get the code when they need to run the application. If there are changes to the applications, they are made at the server. Programmers and network administrators do not have to worry about distributing all the changes to every client. The next time the client logs onto the server and accesses the application, it automatically gets the most current code. This method of delivering applications also reduces support costs.

"Fat" may be interpreted as how much the client application has to be fed in order to use it. For example, a locally installed emulator may have the same capabilities as a network-delivered, Java-based emulator, but there is more work to be done in installing and configuring the local emulator than the Java-based emulator that is delivered each time it is needed. The traditional emulator takes up local disk space whether it is being used or not. The Java-based emulator, in contrast, takes no local disk space.

ActiveX components are a cross between locally installed applications and network delivered applications. They are not only sent to the client when initially needed, but are also installed on the local disk for future use. This means that local disk space is used even if the component was only used once and never used again. Updates are easy to get because they can be sent over the network when required. With Java, the component is sent each time it is needed unless it is already in the browser's cache. This makes Java components instantly updatable.

Because Java is platform-independent, a Java-based T27 emulator for Unisys hosts or a 3270 emulator for IBM hosts, for example, can run on any hardware or software architecture that supports the Java virtual machine. This includes Windows, Macintosh, and UNIX platforms as well as new network computers. This gives any Java-enabled browser access to legacy data and applications.

Which Is Better, Fat or Thin Clients? As with most issues, the answer is "it depends." There is no right answer for all applications and all environments. Each has advantages and disadvantages, so it is necessary to do a cost-benefit analysis first. Even if a significant number of desktops must stay with the fat-client approach, there still may be enough incentive to move the others to the thin-client approach.

According to The Gartner Group (Stamford CT), the annual cost of supporting fat clients—Windows 95/NT, UNIX, OS/2, and Macintosh—is about $11,900 per seat. Substantial savings could be realized for as many as 90% of an enterprise's clients, with only 10% of users needing to continue with a fat client for highly intense applications. Thus, the support costs for moving from a fat-client to a thin-client architecture could be as much as $84.6 million annually for a company with 10,000 clients.

IMPROVING NETWORK PERFORMANCE

Intranets are becoming pervasive because they allow network users to easily access information through standard Web browsers and other World Wide Web technologies and tools to provide a simple, reliable, universal, and low-cost way to exchange information among enterprise network users. However, the resulting changes in network traffic patterns require upgrading the network infrastructure to improve performance and prevent slow network response times.

The network will have to be upgraded to accommodate the graphical nature of Web-based information, which dramatically increases network traffic and demands greater network bandwidth; the integration of the Internet Protocol (IP) throughout the network; easier access to data across the campus or across the globe, which leads to increased inter-subnetwork traffic that must be routed; and new, real-time multimedia feeds that require intelligent multicast control.

LAN switches traditionally operate at Layer 2, or the data link layer, providing high performance segmentation for workgroup-based client/server networks. Routing operates at Layer 3, or the network layer, providing broadcast controls, WAN access, and bandwidth management vital to intranets. Most networks do not contain sufficient routing resources to handle the new inter-subnetwork traffic demands of enterprise intranets.

The optimal solution—intranet switching—is to add Layer 3 switching, the portion of routing functionality required to forward intranet information between subnetworks, to existing Layer 2 switches. This enables network managers to cost-effectively upgrade the Layer 3 performance in their networks. This is the approach being taken by new intranet switches and software upgrades to existing switches.

Intranet Switching

Intranets are increasingly being used to support real-time information, such as live audio and video feeds, over the network. These multimedia feeds are sent to all subscribers in a subnetwork, creating increased multicast traffic and impeding network performance by consuming ever greater amounts of bandwidth. Intelligent multicast control provided by intranet switches helps organizations conserve network bandwidth by eliminating the propagation of multicast traffic to all end stations in a subnetwork. The intranet switches monitor multicast requests and forward multicast frames only to the ports hosting members of a multicast group.

Most enterprise networks use multiple protocols. Intranets are IP-based, requiring IP on all intranet access systems throughout the network. To ease IP integration, intranet switching supports protocol-sensitive virtual local area networks (VLANs), which allows the addition of IP without changing the logical network structure for other protocols.

By combining IP and asynchronous transfer mode (ATM) routing through integrated private network-to-network interface (I-PNNI) signaling, network management is simplified because only one protocol is managed rather than two. Providing this unified view of the network by implementing a single protocol leads to better path selection and improved network performance.

To accommodate intranet traffic demands, increased switching capabilities must be added to both the edge of the network and to the backbone network. Many organizations are using intranets for mission-critical applications, so the backbone technology must deliver superior performance, scalability, and a high degree of resiliency. For these reasons, ATM is the optimal technology for the core technology for intranet switches.

INTRANET OPERATING SYSTEM

As today's networks assimilate additional services originally developed for the global Internet, they are gaining new flexibility in the ways they provide access to computing resources and information. Network operating systems make this easier to accomplish by providing integral access to intranet resources such as Web servers, file transfer protocol (FTP) servers, and WAN connections to the Internet. Novell has done this with its IntranetWare offering, which is built on the NetWare 4 network operating system. IntranetWare provides both IP and IPX access to intranet resources.

INTRANETWARE

IntranetWare incorporates all of the networking services of NetWare 4.11 such as Novell Directory Services (NDS), symmetric multiprocessing (SMP), and core file and print services with new intranet and Internet capabilities, creating a comprehensive solution for building networks and corporate intranets.

These solutions include a high-performance NetWare Web Server 2.5; FTP services, the Internet-standard method for allowing users to download files on remote servers via the Internet; Netscape Navigator; an IPX-to-IP gateway to provide IPX users with access to all IP resources, including World Wide Web pages; and integrated wide-area routing to connect geographically dispersed LANs to a corporate intranet or to the greater Internet.

At the heart of IntranetWare's management is NDS, which enables administrators to manage a network from any workstation and provides sophisticated access controls for all the resources on the intranet. With the centralized administration enabled by NDS, organizations can dramatically reduce management and administration expenses, which are the primary costs of operating a network.

IntranetWare also qualifies for C2 network security certification, enabling the complete network—server, client, and connecting media—to be completely secure.

IntranetWare offers an IPX/IP gateway to enable IPX networks to connect to IP resources. Its routing capabilities enable corporations to extend their intranets to their branch offices and to connect to the Internet via integrated services digital network (ISDN), frame relay, ATM, or leased-line connections. Add-on software from Novell enables mainframe and mid-range computers to become a part of the corporate intranet.

IntranetWare provides comprehensive client support for DOS, Windows, Windows 95, Windows NT, Macintosh, OS/2, and UNIX workstations.

FIREWALL SECURITY

A firewall is server software that protects TCP/IP networks from unwanted external access to corporate resources. With a firewall, companies can connect their private TCP/IP networks to the global Internet or to other external TCP/IP networks and be assured that unauthorized users cannot obtain access to systems or files on their private network. Firewalls can also work in the opposite direction by controlling internal access to external services that are deemed inappropriate to accomplishing the company's business.

Firewalls come in three types: packet filters, circuit-level gateways, and application gateways. Some firewall products combine all three into one firewall server, offering organizations more flexibility in meeting their security needs.

Packet Filtering

With packet filtering, all IP packets traveling between the internal network and the external network must pass through the firewall. User-definable rules allow or disallow packets to be passed. The firewall's graphical user interface allows systems administrators to implement packet filter rules easily and accurately.

Circuit-Level Gateway

All of the firewall's incoming and outgoing connections are circuit-level connections that are made automatically and transparently. The firewall can be configured to enable a variety of outgoing connections such as Telnet, FTP, WWW, Gopher, America Online, and user-defined applications such as Mail and News. Incoming circuit-level connections include Telnet and FTP. Incoming connections are only permitted with authenticated inbound access using one-time password tokens.

Applications Servers

Some firewalls include support for several standard application servers, including Mail, News, WWW, FTP, and Domain Name Service (DNS). Security is enhanced by compartmentalizing these applications from other firewall software, so that if an individual server is under attack, other servers/functions are not affected.

To aid security, firewalls offer logging capabilities as well as alarms that are activated when probing is detected. Log files are kept for all connection requests and server activity. The files can be viewed from the console displaying the most recent entries. The log scrolls in real time as new entries come in. The log files include:

- Connection requests.
- Mail log files.
- News log files.
- Other servers.
- Outbound FTP sessions.
- Alarm conditions.
- Administrative logs.
- Kernel messages.

An alarm system watches for network probes. The alarm system can be configured to watch for the TCP or user datagram protocol (UDP) probes from either the external or internal networks. Alarms can be configured to trigger E-mail, pop-up windows, messages sent to a local printer, or halt the system upon detection of a security breach.

Another important function of firewalls is to remap and hide all internal IP addresses. The source IP addresses are written so that outgoing packets originate from the firewall. The result is that all of the organization's internal IP addresses are hidden from users on the greater Internet. This provides organizations with the important option of being able to use non-registered IP addresses on their internal network. In not having to assign every computer a unique IP address and not having to register them for use over the greater Internet, which would result in conflicts, administrators can save hundreds of hours of work.

INTRANET MANAGEMENT

Intranets bring together yet another set of technologies that need to be managed. Instead of using different management systems, organizations should strive to monitor and administer intranet applications from the same console used to manage their underlying operating system software and server hardware. This is a distinct advantage when it comes to ensuring end-to-end availability of intranet resources to users.

For example, the hierarchical storage management capabilities of the Unicenter platform from Computer Associates can be extended to HTML pages on a Web server. HTML pages that are not accessed from the server for a given period of time can be migrated to less costly near-line storage. If a user then tries to access such a page, storage management will direct the query to the appropriate location.

Some enterprise management vendors are turning to partnerships to provide users of their management platforms with data on intranet server performance. For example, Hewlett-Packard Co. and Cabletron Systems, Inc. have joined with BMC Software Inc. to provide application management software that monitors Web-server performance and use. The software forwards the data it collects to management consoles, such as HP's OpenView and Cabletron's Spectrum, in the platforms' native format or as basic simple network management protocol (SNMP) traps. Instead of looking at their internal Web sites in an isolated way, this integrated method allows full-fledged enterprisewide applications management.

IBM's Tivoli Systems unit provides Web server management through a combination of its internally developed applications and software from net.Genesis Corp. Tivoli is also working with IBM and SunSoft, Inc. to develop the Internet Management Specification (IMS) for submission to the Desktop Management Task Force. IMS would provide a standard interface for monitoring and controlling all types of Internet and intranet resources.

IP Administration

Managing Web servers is only one aspect of keeping an intranet up and running. IP administration can also become unwieldy as intranets lead to a proliferation of devices and addresses. Intranet-driven IP administration can be facilitated by dynamic host configuration protocol (DHCP) software, which streamlines the allocation and distribution of IP addresses. This insulates network operators from the complexity of assigning addresses across multiple subnetworks and platforms. Because intranets depend on the accurate assignment of IP addresses throughout a company, such tools are invaluable to ensuring the availability of resources.

Managing Bandwidth

Intranets also have the potential to significantly increase traffic, causing bandwidth problems. This has some technology managers concerned that bandwidth for vital business applications is being consumed by less-than-vital intranet data. Users access files that may contain huge graphics, and that has created a tremendous bandwidth issue. As Web servers across an enterprise entice users with new content, intranets also can alter the distribution patterns of network traffic as users hop from one business unit's intranet server to another's and as companies make it easier to access information and applications no matter where they may be located.

A Policy-Based Solution

More servers and bandwidth can be added and the network itself can be partitioned into more subnetworks to help confine bandwidth-intensive applications to various communities of interest. But these are expensive solutions. A policy-based solution can be just as effective, if not more economical.

To prevent these applications from wreaking too much havoc on the network infrastructure, companies can issue standards that establish limits to document size and the use of graphics so that bandwidth is not consumed unnecessarily. These policies can even be applied to E-mail servers, where the server can be instructed to reject messages that are too long or which contain attachments that exceed a given file size.

SUMMARY

Companies may be tempted to jump on the intranet bandwagon using the fastest means possible. This tactic may meet basic requirements, but it often does not take into account future network growth, the advantages gained by leveraging existing data and resources, nor how to add new intranet-enhancing products as they become available. These considerations demand that intranets be flexible, open, and integrated.

Any time a company makes information accessible to a wide group of people or extends an intranet to suppliers or vendors, it must establish appropriate security mechanisms, ranging from firewalls to access control to authentication and encryption.

Despite the allure of corporate intranets, companies will not be able to move rapidly towards the kind of full-fledged intranet being predicted by some vendors, with a single browser-type interface and thin clients that download applications and data all at once. For some considerable time to come, intranets, as defined by the browser suppliers, will be distinct from and complementary to existing systems. For one thing, investment in non-intranet solutions is too great, and for another, organizations have become more wary of the bandwagon effect.

Despite these concerns, intranets will undoubtedly benefit many companies. This is as long as, according to the Gartner Group, they are not seen as the next "hammer" for every "nail" within the enterprise. Companies that have implemented intranets are gradually finding that they are able to use Internet technologies to communicate and link information—internally and externally—in ways that often were not possible before.

Section 4
Interoperability and Standards Issues

THE NEW MODEL OF TELECOMMUNICATIONS is open, competing aggressively for business, and global in scope. Communications systems managers that are to succeed in deploying, controlling, and managing a worldwide network in this environment must be aware of one factor that can influence the degree of success—interoperability.

Interoperability saves money, reduces frustration levels, and makes users happy by hiding the details of the networking infrastructure. Interoperability implies out-of-the-box, plug-to-socket compatibilty between network elements such as workstations, modems, access devices, routers, and switches. Interoperability applies equally to software and hardware and encompasses all levels of architecture from the physical wire to the user's application. It is important to understand that "interoperable" does not mean identically implemented.

From a communications point of view, what needs to be interoperable is the protocol that operates to control the flow of user information across an interface. The implementations of the protocol need not be identical, but the behavior of the two interfacing elements must be predictable. Provided that a system's behavior follows the protocol rules, the user does not need to know how it is implemented. When a frame is delivered across the interface to a frame relay network, for example, the user does not know or care how the service provider carries that frame through its network. Users are only are concerned with the interfaces to and from the network.

The three primary forces working toward interoperability are the national and international standards bodies, the various industry forums, and the dominant manufacturers. The official and quasi-official standards bodies, of which the American National Standards Institute (ANSI), the International Telecommunication Union (ITU), and the International Standards Organization (ISO) are the foremost examples, promulgate de jure standards that promote interoperability across national boundaries, between industry groups, and among trade groups. Their standards have the advantage of being internationally recognized and accepted.

These organizations, however, are often criticized for being slow to develop standards and making them overly complex with too many options. Problems often arise because of the difficulty of achieving a consensus among contributors from many different companies, countries, and cultures, each with their own ideas about what a standard should be.

Industry forums have emerged to play an important role in interoperability. They bring together developers, manufacturers, and sometimes users. Their goals are to accelerate the standards process, to sort out and select those options that make the most sense from a market point of view, and to define the next generation of requirements surrounding a particular technology. Examples of industry fora include the Frame Relay Forum and the ATM Forum. These organizations usually work closely with the national and international standard bodies.

The third major player in establishing interoperabilty standards is the dominant technology supplier in a particular aspect of the technology. Examples include the dominant role played by IBM in establishing early computer, network architecture, and communications protocol standards and the current position of Microsoft in operating systems. These types of interoperability standards are often *de facto* rather than *de jure*, but no less effective in the real world. Because they were developed by a single vendor, they are often very well-focused and specific.

If there is a single force preventing total interoperability it is the vendors, who strive for product differentiation. It is absolutely essential for vendors to somehow make their products unique in the marketplace. There must be a reason for customers to buy this product rather than a competitor's product. As a result, vendors build in features designed to attract and lock in customers. These features are by nature proprietary, nonstandard and, often, innovative. However, they make it difficult to achieve true interoperability among products from different vendors.

To ensure interoperability, the systems manager can recognize and understand the forces that work for interoperability as well as the forces that tend to prevent interoperability. The combination of standard and proprietary seems to be the most prevalent approach because it works toward the standard but also recognizes the difficulty of achieving true plug-and-play compatibility. Managers should understand that vendor self-interest is not a bad thing, but often leads to the next generation of technology. The manager can then work with service suppliers and equipment and software manufacturers to promote practices that will make the procurement and operations management tasks easier.

A very important issue for managers working toward achieving interoperability is conformance testing. Chapter 4-1, "Worldwide Activities in Standards Conformance Testing," reviews the need to make sure that a manufacturer's implementation of a standard actually conforms to the requirements of the standard. Someone must first define what conformance means, describe thc necessary test bed, create test scenarios, describe actual test procedures, document the results, and certify the tested product. Methods of ensuring conformance are discussed in this chapter.

Every network element and every user of a network must be uniquely identified—this is the purpose of an addressing scheme. Joe User should be able to ask for a connection to Mary User on the corporate local network in Tampa. Joe should not have to know the twenty octets of binary addressing information that may be necessary to uniquely identify Mary. Neither should Joe worry about the route that his data will take on its way to Mary. All of these considerations are explored in Chapter 4-2, "Open Network Addressing."

An example of a standard designed to enhance interoperability is discussed in "HSSI Standard for High-Speed Networking," Chapter 4-3. This standard is for a high-speed serial interface between computers, routers, and other terminal equipment and the wide area communications access equipment such as multiplexers and channel service units. The chapter delineates the need for the 50M-bps serial interfaces offered by HSSI and describes the operation and implementation of the HSSI protocol.

As applications become more distributed in the client/server environment, it becomes crucial for a network manager to be able to remotely analyze and diagnose network abnormalities. "RMON: The SNMP Remote Monitoring Standard," Chapter 4-4, provides managers with this capability. RMON can monitor packets in the network watching for and reporting the occurrence of any abnormal conditions defined by the network manager. RMON is an Internet standard that is being implemented in a wide variety of products.

4-1 Worldwide Activities in Standards Conformance Testing

Keith G. Knightson

THE INFRASTRUCTURE BEING CREATED TO SUPPORT conformance testing is of considerable importance to the purchasers of open systems equipment. Conformance testing gives purchasers more power over vendors, simplifies procurement, and significantly reduces the risk of incompatibility between different vendor equipment (and in some cases, even guarantees appropriate redress). Vendors will need to adjust their practices accordingly.

It is also fair to say that these activities provide an incentive to apply Open Systems Interconnection (OSI) standards. Users are more likely to be attracted to standards that carry guarantees of compliance and interoperability than to standards that do not.

IMPORTANCE OF CONFORMANCE TESTING

Vendors frequently make claims that their products exactly suit customers' needs. Sometimes they exaggerate, and sometimes sales representatives are not entirely familiar with their own products. Data sheets and specifications documents are often not sufficiently detailed to cover all operating circumstances. The problem for the purchaser is to know what exactly is being offered before making the purchase (all too often the facts become apparent only after the purchase). Standards can be very complicated, and glib assurances of compliance with various standards are insufficient. It is important to know in what way equipment complies with a given standard and that an implementation of a standard is free of errors.

The goals of OSI conformance testing are to increase the probability that different OSI implementations interwork and to ensure that an implementation has the required capabilities and that its behavior conforms to the relevant specification. Conformance testing thus can eliminate the problems often encountered in the user's procurement process and the vendor's design process. The difficulty lies in achieving a universally consistent environment in which to conduct conformance testing.

To test whether equipment from two manufacturers can be interconnected, simple pairwise testing suffices. However, even in this simplest case there is the problem of which is the conforming and which is the nonconforming equipment in the event that something needs to be fixed. If more than two manufacturers are involved in a given system, things begin to get far more difficult.

To advertise a product as generically interoperable, a manufacturer would have to conduct pairwise tests with equipment from all other manufacturers. For a group of 6 manufacturers, 15 pairwise tests would be necessary. For 100 manufacturers, 4,950 pairwise tests would be necessary. The number of pairwise tests for a group of n manufacturers is $n(n-1)/2$. Even this would not provide consistency; the result might be 4,950 variations of a given implementation. Clearly, in an open systems environment consistency with the standards is obligatory.

The answer lies in testing every manufacturer against a single reference, say X (for a given standard), as shown in Exhibit 4-1-1. If A works with X, and B works with X, and so on, A will work with B, and so on. The problem is what is X, and can X be made consistent given that there will not be just one version of X in practice. There will be a number of testing laboratories, in different countries, and test equipment from a number of different vendors as well. In addition, testing conditions, sets of tests, and other elements must be made equal.

NECESSARY ENABLERS AND INFRASTRUCTURES

Conformance testing would be of little use if a given item of equipment could pass a conformance test at one testing laboratory and then fail at another. Similarly, if equipment from two manufacturers could not be made to interoperate, despite both having been successfully conformance tested, the exercise would be pointless.

The issues can be resolved only by adequate standardization of the conformance testing process itself. In this complex process, provision must be made for the following:

- The availability of an unambiguous base standard.
- Unambiguous identification of exactly what is to be tested for a given implementation.
- Standardization of the test environment (test harness).

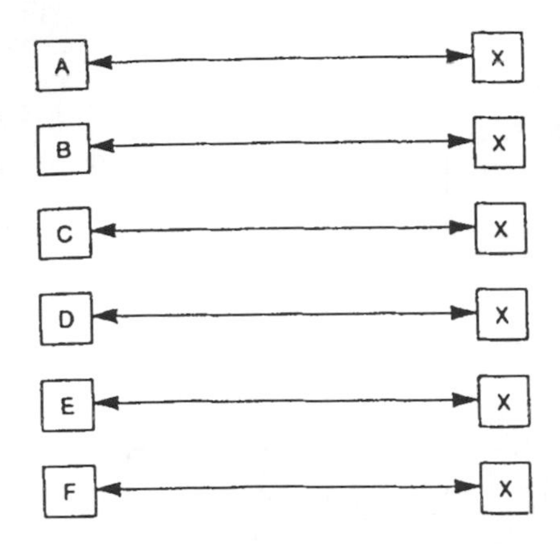

Exhibit 4-1-1. Testing Against a Single Reference

- Adequate standardization of test equipment.
- Standardization of testing laboratory procedures.
- Establishment of criteria for examining testing laboratories.
- Establishment of authorities to accredit testing laboratories.
- Availability of challenge and arbitration procedures.
- Public access to registers of unbiased results.

In other words, another orthogonal set of standards is required to deal specifically with conformance testing (of the base standards).

ISO Standards for Conformance Testing

To cover these conformance testing criteria, the International Standards Organization (ISO) has produced a multipart standard specifically related to testing OSI standards. It is IS 9646, "OSI—Conformance Testing Methodology and Framework."

This standard has seven parts:

- Part 1: General Concepts.
- Part 2: Abstract Test Suite Specification.
- Part 3: Tree and Tabular Combined Notation.
- Part 4: Test Realization.
- Part 5: Requirements on Test Laboratories and Clients.
- Part 6: Profile Test Specification.
- Part 7: Implementation Conformance Statements.

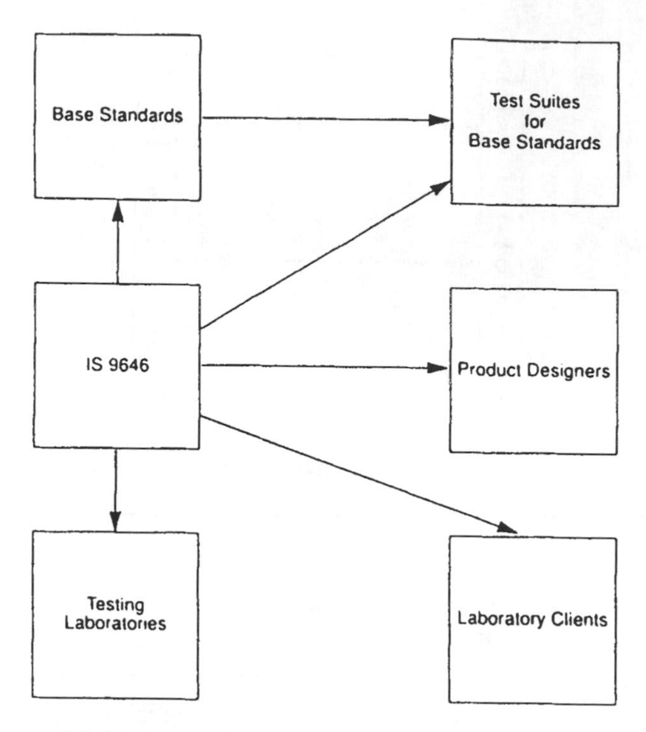

Exhibit 4-1-2. Scope and Application of IS 9646

Briefly, these standards fall into three categories:

- Those concerned with identifying requirements intrinsic to the base standards and ensuring that the base standards themselves are adequately written.
- Those concerned with standardizing the testing machinery.
- Those concerned with the process of conducting conformance testing.

The scope of IS 9646 is shown in Exhibit 4-1-2. A brief review of each part of IS 9646 follows.

Part I: General Concepts. Conformance requirements are classified as static or dynamic elements. Static elements are checklist items, and dynamic elements are those concerned with real-time behavior. Each of these elements can be further classified as mandatory, optional, or conditional. These classifications are applied to all elements of the base standard to produce a protocol implementation conformance statement (PICS) template. When an implementation is submitted for testing, the submitter must complete this

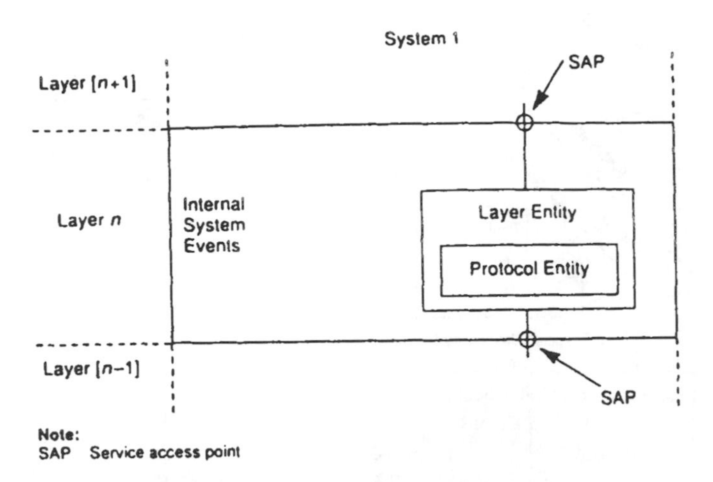

Exhibit 4-1-3. End System Architecture

template for the implementation to be tested. The answers in the PICS are used to select the tests to be run.

Three testing architectures are defined: remote, distributed, and coordinated. All the architectures are based on the layering principles defined in the OSI seven-layer model. Within a layer, a protocol's behavior (this is in addition to the sequence of protocol exchanges) can be analyzed if events at the lower- and upper-service access points can be controlled and observed. Exhibit 4-1-3 illustrates the basis of the testing architectures.

The remote method is intended for black-box testing where there are no useful internal interfaces within the equipment to assist in the testing process. The distributed method is designed for use where there are interfaces within the equipment that can be used to carry out control and observation. The coordinated method is used when the entire testing process can be controlled remotely by the test equipment through the use of another protocol called the test management protocol (TMP).

Part 2: Abstract Test Suite Specification. This part is concerned with the nature and content of test suites. Test suites are structured as shown in Exhibit 4-1-4. A test case is the element that assigns verdicts of pass, fail, or inconclusive, according to the behavior of the implementation under test (IUT). Test events are lower-level elements (e.g., the sending or receiving of a particular protocol data unit), whereas a test case may be something like verifying that a connection can be established. Part 2 specifies what events must be contained in the test specification for the various testing architectures defined.

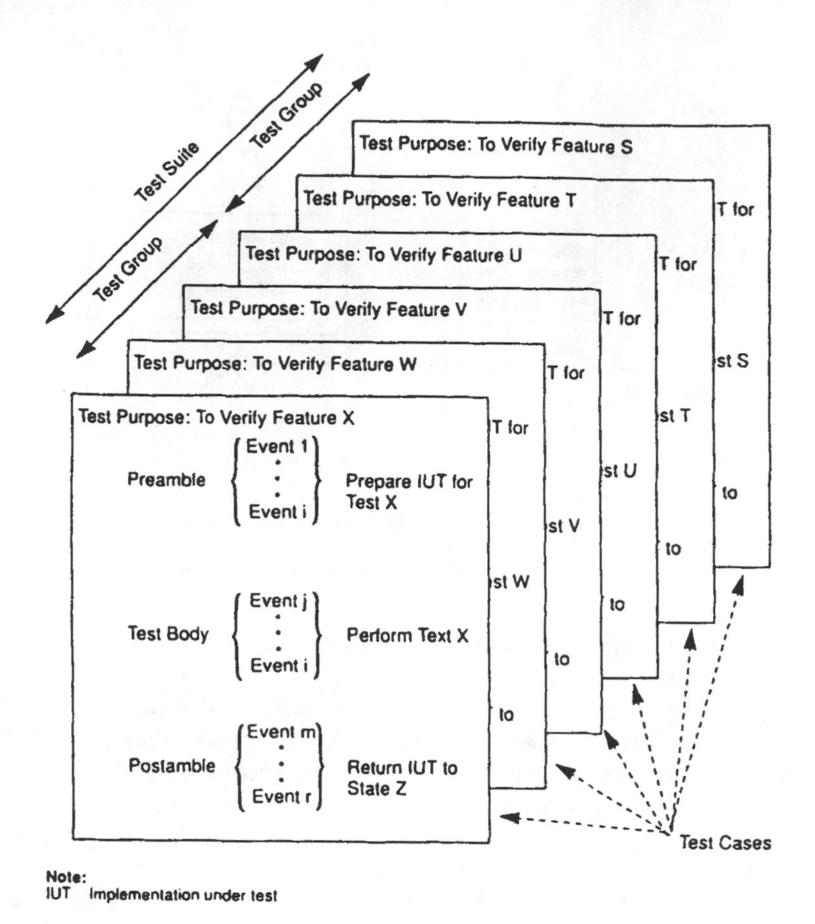

Exhibit 4-1-4. Components of a Test Suite

Part 3: Tree and Tabular Combined Notation. The tree and tabular combined notation (TTCN) is a special language designed for writing test suites. This has two purposes. It permits the universal understanding of a test specification and it allows automation of the testing process. This latter aspect is important in ensuring that test equipment from different vendors can execute the same test suite and yield the same results.

The basis of TTCN is relatively simple. It uses a tree of alternatives, based on actions and allowed reactions. Exhibit 4-1-5 shows how various sequences of events can be represented by a tree structure. The TTCN statements are programmatic in form, and thus constructs that are similar in form to a computer program can be made. These can contain variables

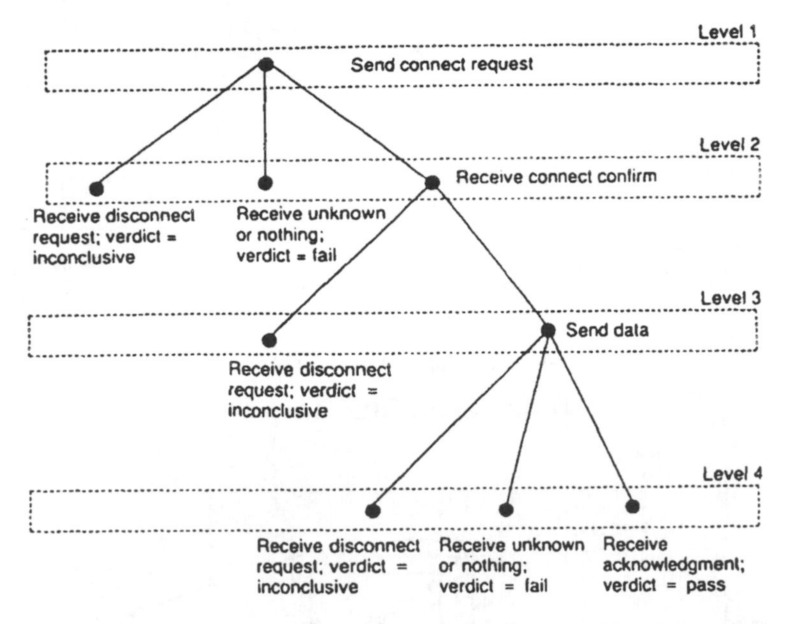

Exhibit 4-1-5. Tree and Tabular Combined Notation Tree Structure

and expressions as well as the send, receive, and timer events associated with protocols.

Part 4: Test Realization. This part specifies the requirements that manufacturers of test equipment must meet. The test equipment must be able to derive an executable test suite from the standardized test suite and the information about the particular implementation under test (e.g., from standard PICS templates). The test equipment must then execute the tests and produce a log of the test results in a specific form.

Part 5: Requirements on Test Laboratories and Clients. This part specifies the process the testing laboratory must use and the relative roles and responsibilities of the laboratory and the client. A simplification of the process is shown in Exhibit 4-1-6.

A test campaign is the execution of a block of tests. (In many cases, this may be the entire test suite.) Tests are not conducted one at a time. This is not simply a question of saving time through a process of automation. It guards against changes made for one test affecting previously run tests. Thus, if a particular problem is found and then fixed by the client, the entire set of tests (i.e., the campaign) is rerun.

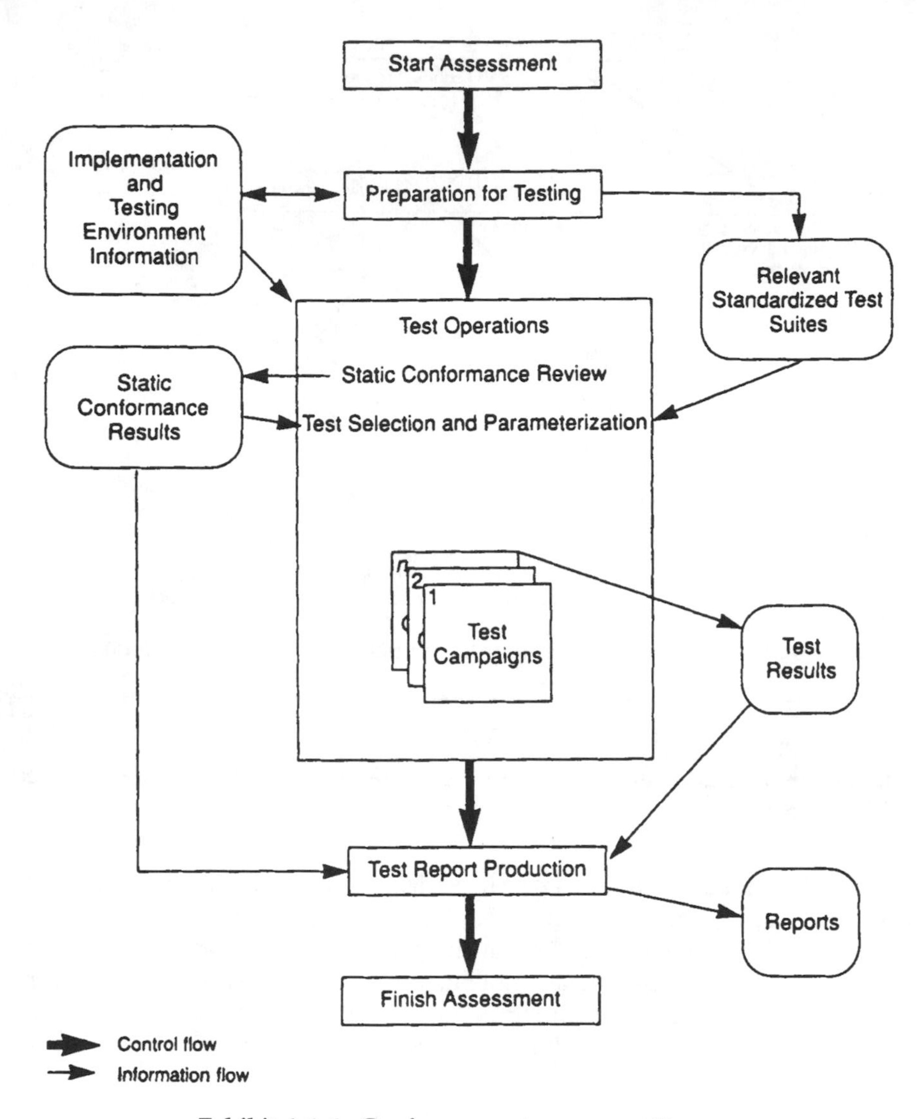

Exhibit 4-1-6. Conformance Assessment Process

Part 6: Profile Test Specifications. Part 6 is concerned with how to specify the differences in the case in which a profile is being tested rather than the base standard.

Part 7: Implementation Conformance Statements. Part 7 specifies the notation and content of the templates (PICSs) used to gather the necessary information about the implementation to be tested.

National and Organizational Procurement and Testing Programs

Many large organizations mandate the use of information standards. General Motors produced a set of standards-based specifications for factory floor automation, the Manufacturing Automation Protocol (MAP). Boeing produced a companion set of specifications for office automation, the Technical and Office Protocol (TOP). As interest in these specifications grew, a more widely based organization of users known as the MAP/TOP Users Group emerged to take over the development of the MAP/TOP specifications.

Similarly, the National Institute of Standards and Technology (NIST) has produced specifications for US government procurement known as the Government OSI Profile (GOSIP). The NIST activities extend well beyond simply mandating the use of OSI standards.

To understand an organization's participation in standard setting entails more than a synopsis of its activities. Therefore, several organizations are examined individually. Most of these organizations are concerned with establishing national or international policies and infrastructures for conformance testing and in some cases, for interoperability testing. The need for the real world activities under way—and the issues related to them—are summed up in the goals of the policy document produced by the North American OSI Testing and Certification Policy Council:

- Establishment of accreditation systems for assessment of test laboratory competence.
- Establishment of mutual recognition agreements.
- Establishment of the criteria on which certificates of conformity are based.
- Establishment of the criteria on which supplier declarations of conformity are based.
- Establishment of neutral, publicly available information on the availability of tested OSI products and the testing and certification processes to which such products have been subjected.
- Promotion of harmonized OSI conformance testing specifications and their worldwide adoption.
- Promotion of harmonized conformance testing services.
- Promotion of a harmonized approach to interoperability testing.

Generally, these activities are outside the scope of the standards process, because they are primarily concerned with the application of IS 9646 and the commercial issues related to application of IS 9646.

NIST and the US GOSIP Program. NIST organizes regular workshops aimed at producing implementation agreements. These workshops are totally open and comprise a mix of vendors, users, and other organizations. Implementation agreements provide further clarification of the base standard, where options exist, where some element is deemed insufficient or lacking for the purposes of implementation, or where implementation guidance would help to provide interoperability. Generally, the aim is to avoid individual interpretations of a base standard that would cause incompatibilities. The workshop process leads to use of the term "profile." Profiles are documents that propose a particular version of a given standard, agreed to by a specific community of interest.

In addition, NIST is responsible for Federal Information Processing Standards (FIPS) and formulating the US GOSIP requirements and related program elements. The US GOSIP requirements are specified in FIPS 146, which specifies the major elements required.

US GOSIP has placed a high value on conformance testing, and so conformance testing has been made a mandatory prerequisite for procurement. NIST has compiled and maintains a register of OSI products that meet US government procurement requirements and have been conformance tested. This is available on a publicly accessible data base.

NIST itself is not a testing laboratory, although it has performed some research in this area. External testing laboratories perform the required testing. Because testing must be related to NIST's specific requirements, the capability, integrity, and reliability of a testing laboratory in relation to the GOSIP program must be ensured. NIST has also specified requirements for the assessment of testing laboratories. The process of assessing a testing laboratory and giving it some seal of approval is known as accreditation. Accreditation should not be confused with certification. Certification is the result of a product being successfully tested by an accredited testing laboratory.

NIST intends to make interoperability testing, in addition to conformance testing, mandatory. This involves testing a product against a reference implementation provided by NIST (or against a number of other implementations). To assist this process, NIST ran an interoperability test network, known as OSINET, on which suppliers could test interoperation between their products. NIST is establishing a publicly accessible data base of information about interoperable products.

National Voluntary Laboratory Accreditation Process. The National Voluntary Laboratory Accreditation Process (NV/LAP) accredits laboratories for the US GOSIP program, in accordance with accreditation criteria developed by NIST. NVLAP is administered and staffed by NIST.

NVLAP accreditation criteria are included in the Code of Federal Regulations (in the process of being harmonized with ISO Guide 25, "General Requirements for the Competence of Calibration and Testing Laboratories").

NVLAP is mandated to accredit those laboratories in the US wishing to provide testing for the US GOSIP program.

Corporation for Open Systems. The Corporation for Open Systems (COS) International is a US-based, not-for-profit organization with an international membership of computer and telecommunications suppliers and government and corporate information technology users. The primary function of COS is that of a testing laboratory.

The COS mission is: "To provide an international vehicle for accelerating the introduction of interoperable, multivendor products and services under agreement to OSI, ISDN, and related standards to ensure acceptance of an open network architecture in world markets."

The major way in which COS achieves its mission is through the provision of conformance testing facilities. The background activities involved in this approach include:

- Managing a user-driven requirements definition process.
- Accelerating internationally recognized profiles, tests, and laboratory accreditation.
- Providing promotion, education, and training and support products and services.

COS has been one of the pioneers in making test equipment available. COS has developed and sponsored the development of an array of test equipment. The test development program is driven by the user groups; they identify and assign priority to the various OSI protocols they feel are important to their corporate needs. Test equipment for these protocols is then used by COS itself within its own testing laboratory and is also made available for purchase to vendors and suppliers that wish to perform first-party testing.

COS was the first organization to institute licensing for conformance testing. The supplier is allowed to display the COS Mark after successfully completing conformance testing at the COS laboratory, provided that the supplier also enters into a legally binding agreement with COS regarding the liabilities and assurances associated with the Mark.

The COS Mark program is a commitment between COS and the supplier for the product in question. It ensures that the purchaser has recourse in the event of problems, in which case COS acts as an independent arbitrator. COS can revoke the COS Mark if the supplier does not cooperate in solving the problem, or has violated the original agreement in some way. Such violations might include not maintaining the quality of the product or, perhaps, secretly introducing changes that affect the validity of the conformance tests originally performed. COS reserves the right to perform spot checks on products that carry the Mark. COS provides an interoperability analysis service to assist in the resolution of problems.

COS relies on NIST/NVLAP procedures to accredit vendors using COS-supplied test equipment for first-party testing. This ensures that the suppliers maintain the same standards as if COS itself were performing the testing. Such suppliers still enter into the COS Mark agreements after the self-certification process and are subject to the regular COS Mark liabilities and commitments required by COS.

Standards Promotion and Application Group. The Standards Promotion and Application Group (SPAG) is a European organization, originally formed by a number of suppliers. SPAG covers ground similar to that covered by NIST and COS but has also sponsored the development of test tools and more recently of a testing program, similar to that of the COS Mark, called the Process to Support Interoperability (PSI) trademark.

SPAG is famous for its Guide to the Use of Standards (GUS), which represents some of the earliest work on functional standards and profiles. SPAG plays a major role in the European workshop on Open Systems (EWOS) to which the functional standards work has very largely been transferred. SPAG continues to play a major role in conformance testing with its PSI trademark and its testing center in Belgium.

The PSI trademark is awarded to suppliers whose equipment has been successfully conformance tested and that have agreed to abide by the agreements of the trademark license. The PSI trademark combines both conformance testing and interoperability. To be certified, products must be successfully interoperated with products from at least three other vendors.

PSI includes a conciliation mechanism that can be invoked if interoperability problems arise after testing. The licensing agreement requires vendors to participate in the conciliation process and abide by its recommendations. The trademark can be withdrawn if the vendor does not meet its obligations.

Canadian Interest Group on Open Systems. The goal of the Canadian Interest Group on Open Systems (CIGOS) is "the accelerated development, incorporation, and exploitation of OSI-based information technology products and services to enhance the competitiveness of Canadian goods and services in domestic and foreign markets, and to protect the information technology investment of Canadian users."

With regard to conformance testing, CIGOS has been active in two main areas: the establishment of a Canadian conformance testing center and a campaign for the establishment of harmonized accreditation criteria.

The problem highlighted by CIGOS concerned the inadequacy of existing laboratory accreditation criteria for protocol conformance testing. The existing accreditation material contained in ISO Guide 25 applies mainly to administrative procedures for laboratories engaged in all types of testing. The criteria used to determine whether a laboratory can adequately test crash helmets hardly suits protocol testing.

The Canadian government (through CIGOS) proposed that projects be initiated within the International Laboratory Accreditation Conference (ILAC) and within the Conformity Assessment Council (CASCO) Committee of ISO, to augment the existing guides to cover OSI testing. The Canadian government also submitted a draft document defining laboratory accreditation criteria, to get things started. This draft was based on a comparison of various documents on accreditation, from NAMAS, from the US GOSIP program, and with additional Canadian input.

National Measurement Accreditation Service. The National Measurement Accreditation Service (NAMAS) is the UK accreditation service and is part of the National Physical Laboratory (NPL), which has also been a pioneer in conformance testing research and development. NAMAS is the accreditation authority that has accredited the SPAG OSI Test Center for the various testing services that SPAG offers.

Reseau Nationale d'Essais. The Reseau Nationale d'Essais (RNE) is a French accreditation organization performing duties similar to those of NVLAP in the US and of NAMAS in the UK.

European Workshop for Open Systems. EWOS is the European regional workshop responsible for the development of functional profiles. Functional profiles developed by EWOS are submitted to CEN/CENLEC for progression as European Norms (ENs). EWOS would also likely submit such documents, or derivatives thereof, to ISO as candidate ISPs for inclusion in the relevant taxonomy group.

Asia-Oceania Workshop for Open Systems (AOW) andPromoting Conference for Open System Interconnection (POSI Japan). These two organizations are equivalent to EWOS but cover the regional interests of Asia/Oceania and Japan, respectively.

European Procurement Handbook for Open Systems. EPHOS is equivalent to the US GOSIP program and is designed to provide the governments of member states with procurement specifications for open systems.

OSI Network Management Forum. The OSI NMF is an international organization, composed of the major communications vendors and computer manufacturers, whose mission is to accelerate the availability of OSI-based network management products.

The OSI NMF is similar in nature, in many respects, to COS except that it concentrates solely on the OSI stack for network management (the Common Management Information Protocol (CMIP) and the definition of managed objects for various types of communications equipment. The OSI NMF has delegated the testing of OSI NMF compliant products to COS, among others.

International Laboratory Accreditation Conference. ILAC is a group of laboratory accreditation organizations. This is not a standardization body in the normal sense but more a body of practitioners.

Conformity Assessment Committee. CASCO is a committee of ISO that produces standards like documents (Guides) on testing, inspection, and product certification. CASCO also deals with mutual recognition schemes and promotes the appropriate use of other ISO standards related to testing.

Harmonization Efforts

Not surprisingly, the involvement in conformance testing of so many organizations created the need for harmonization, and thus a whole new set of organizations.

ISO Regional Workshops Coordinating Committee. This committee of ISO forms the umbrella under which the harmonization efforts on profile and test procedures are performed.

North American OSI Testing and Certification Policy Council. The OSI Testing and Certification Policy Council was formed as a result of an initiative by the Corporation for Open Systems. The mission of the Policy Council is, "To provide the North American focal point for the development, coordination and harmonization of policy as it pertains to worldwide OSI testing and certification."

Only consensus organizations may participate in the Council, and not individual companies of interests. Current members include COS (which also acts as secretariat), the Computer and Business Equipment Manufacturers Association (CBEMA), the Canadian Interest Group on Open Systems, NIST, and the American National Standards Institute (ANSI).

The Council issues policy statements in the following areas:

- Standards.
- Functional specifications.

- Abstract test suites.
- Means of testing.
- Test laboratories.
- Laboratory accreditation.
- Recognition arrangements for conformance testing.
- Recognition arrangements for interoperability testing.
- Certification and supplier declarations.
- Registers of conformant and interoperable products.
- Procurement profiles.
- Change management.
- Market access issues.

For each area, a set of objectives, rationale, set of policy statements, action plan, and current status summary are given.

The work of the Policy Council highlights the issues to be solved in establishing a worldwide harmonized and consistent framework for conformance and interoperability testing, enabling fair and equitable market opportunity for both manufacturers and purchasers.

The current document is incomplete but already more than 100 pages in length. Copies can be obtained directly from the Corporation for Open Systems.

TEST TOOL DEVELOPMENT

Test tool development is a research-intensive process. The market for test tools is limited to a few (third-party) testing laboratories and a few large manufacturers (that can afford to perform first-party testing). The outright purchase of test tools is very expensive, costing as much as $500,000 or more, depending on, for example, the number of layers and the applications to be tested.

Because of the high cost of development, a collegial approach has been adopted in many cases, with test tool developers concentrating on non-overlapping, specific areas to create a synergy within the industry and to eliminate the costs of duplicate development.

Recently, there has been an initiative to harmonize tester design: the Open Integrated Tool Set (OITS). The basic idea is to break down a tester into a set of functional components and design a set of corresponding interfaces to exchange data in standard ways. This would make it possible to build a tester by interfacing the set of independently developed components. In addition, a number of standardized commands could be defined for the human interface for the operator. A number of organizations are participating in this effort, including COS, SPAG, POSI, and test tool manufacturers.

STRATEGIC USE OF CONFORMANCE TESTING

Conformance testing eliminates the need for individual procurement specifications and it enables buyers to skip a great deal of product research. Because conformance testing ensures compliance, a potential buyer does not have to prepare individual performance specifications; only the procurement specification that the product must meet needs to be stated. In the business relationship between users and vendors, conformance provides a rigorous and legalistic frame of reference. The associated testing will reveal any discrepancies between the sales information and the reality of the implementation.

The conformance test process produces extensive test reports. These reports provide an objective, unbiased evaluation of the equipment under test, amounting to a wealth of detail that previously would not have been available. These reports not only enable the technically aware customer to assess the equipment but also provide enough material to assess and compare the different manufacturers' equipment. This should take much of the risk out of the procurement exercise.

Pairwise testing is an expensive and tedious business and, in the event of a discovered incompatibility, it does not resolve the question of whose equipment is not in compliance. Industrywide pairwise testing is clearly not practical, and limited pairwise testing would probably favor certain groups of manufacturers. Conformance testing by a third (impartial) party is in the best commercial interests of many vendors.

Customers are more likely to purchase equipment that has been subjected to and has passed conformance tests than equipment that has not. The endorsement of conformance on a piece of equipment can be a significant selling point and can provide a market acceptance edge.

The availability of globally identical testing standards should minimize the amount of testing required for the international market. Testing could be performed anywhere, and once performed, the results should be acceptable everywhere (without further testing).

SUMMARY

Conformance testing is the key to simplifying and solving the problems associated with building multivendor systems. It takes the risk out of procurement and can provide a means for redress. This is what makes conformance testing so important to the user.

To be equally fair to buyers, manufacturers, and those performing the testing, it is essential that the process be understood by all parties. Vendors must be satisfied that they are being treated fairly and must trust and accept the results (in particular because test results may reveal flaws in their products). Arriving at this understanding is not an easy task. The standards process can and does provide some of the rules, but others are subject to commercial considerations outside the scope of standards. In these cases, other means are necessary to establish an infrastructure within which the standards can operate.

A tremendous amount of activity has been undertaken by many organizations to establish this infrastructure. This infrastructure will ensure that conformance testing continues, to the advantage of both purchasers and manufacturers, and generally enhance and secure the future of the information technology environment.

4-2
Open Network Addressing

Howard C. Berkowitz

PROPRIETARY NETWORKING ARCHITECTURES such as SNA and NetWare were designed to suit single organizations. The Novell identifier assigned to a workstation in one company can therefore duplicate that of a workstation in another company. But as business requirements continue to encourage internal and external networking among organizations, the duplication problem becomes much more significant. Proprietary address schemes also limit users' abilities to select alternative vendors' solutions when those alternatives use different yet incompatible addressing schemes.

BASICS OF NAMING AND ADDRESSING

Names and addresses are the attributes of network objects that enable them to be found and identified by other network objects. A name can exist at different locations; an address is a specific logical or physical place at which names can be located. The systems that accomplish internetworking make use of addresses, not names.

Names, on the other hand, are more easily understood by the human users of networks, so there must be a provision for translating names for the use of automated network elements. A subscriber's initial business relationship to a telephone company, for example, is based on a human name, which is then associated with a unique account number. The account number is then associated with one or more software-defined unique telephone numbers, and these telephone numbers are then associated with real wires at real locations. The business office deals with names, which may not be unique, and logical telephone number assignments. Plant personnel, however, deal with wires and other physical facilities that map to the telephone number.

Most people accept that it would be impractical to label wires in a wire closet with the subscriber's name; a mapping to a telephone number would be all that is reasonably expected. Unfortunately, many network users assume that name and address registration is a simple process in which an address is associated with a hardware address. Realistically, it is at least a two-step process, where a name is associated with a network layer address and a network address is associated with a hardware address.

MAPPING NAMES TO ADDRESSES

The names visible to users map to upper-layer addresses. In X.400 this is done precisely, but not simply. The name Howard Berkowitz, for example, might map to an X.400 messaging address of C=CA, ADMD=DATAPAC, PRMD=PSC, SN=BERKOWITZ. More complex mappings are possible, such as expansion of the name "Washington Office" to several individuals' X.400 addresses. Name-to-address mappings may involve organization policy decisions: If Mary Jones is promoted and transferred, should business messages addressed to her be routed to her at her new location, or to her successor in her old job?

People now put X.400 addresses on business cards, but X.400 may not become more popular until better directory tools free users from the details of addresses. If a relatively simple Internet name of hcb@world.std.com is too complex for many people, an X.400 address with fields for country, administrative management domain, private management domain, organization, organization unit, subunit, and personal name can be overwhelming.

X.400 addresses meet real technical needs; the problem is to improve the tools for using these addresses rather than changing the address structure.

Below the routing functions performed by the messaging service, which are geared to delivering the message to the end user, are packet-oriented network layer routers and switches. Either a directory service or manually created table is necessary to map addresses between the upper layers and lower layers. If only a single address applies to upper and lower layers, a user's electronic mail address would have to change whenever that user moved to a new desk or used a laptop to read mail while traveling.

Addressing conventions need to be more complex than, for example, the relatively simple rules that apply to human names. Humans can use context to decide which of several people with identical names they wish to speak with. Current computers, however, need to be told explicitly that "Mary Smith, the CEO," is different from "Mary Smith in Accounting."

The most common and practical means of ensuring name and address uniqueness is to manage them in a hierarchy. A precedent for this exists in international telephony. An organization having international authority assigns country codes and then delegates the next level of addresses to national organizations. National organizations in turn assign area codes and delegate actual telephone number assignment to a subordinate organization.

The scope in which an address authority can assign unique addresses is called a domain. A complete real-world address consists of segments that correspond to nested domains and subdomains; in practice, a real-world globally unique address usually contains a domain identifier as well as the addresses specific to those domains.

The problem is not so much the theoretical maximum of network addresses as the number of unusable addresses created by an addressing scheme. If the country identifier forms the top level of the address, enough addresses have to be assigned to meet the needs of each country. If an addressing scheme supplies a country like the US with enough addresses and the same number of addresses are reserved for a country with much more modest requirements, many addresses will be wasted. In the current Internet addressing plan, the basic quanta of address assignments are blocks of approximately 65,000 or 250 addresses. The former is too large for most organizations to fill; the latter is too small for many requirements.

MOBILE ADDRESSING CONSIDERATIONS

An evolving and not completely solved addressing problem becomes more apparent as cellular telephony and other mobile technologies become more common. As a car phone moves, it travels through a series of cells, or small radio service areas that connect the phone to the telephone network. Each cell has a number, and while a cellular phone is in a cell, the phone uses a specific frequency between itself and the cellular telephone switch. Cellular switches are designed to keep track of the changing relationship between the persistently assigned telephone number and the transient cell number/frequency pair. While conventional network layer protocols can deal with an address (e.g., a LAN MAC address) below the network layer, these protocols do not assume this hardware address will be dynamic.

APPLICATION-LEVEL ADDRESSING

If real networked applications are to work, system administrators must deal with both application-level and network-level naming and addressing. For example, if an organization wanted to be addressable for Internet electronic messaging, it would need both a domain name (e.g., whitehouse.gov) and an IP address for that domain. Users sometimes incorrectly assume that directory or other management tools can generate one given the other. The main global addressing need is for globally unique messaging addresses. The two main global addressing systems for messaging are those of the OSI X.400 standard and the Internet's Domain Naming systems (DNS). Both schemes are hierarchical and can interwork by using a gateway to convert addresses from one scheme to the other.

High-level DNS addresses are obtained from the Internet's Network Information Center; subordinate addresses are managed by the user or user organization. Address assignment for X.400 is more complex.

The Telecommunications Standardization Sector of the International Telecommunications Union (ITU-TSS) authorizes certain organizations to act as X.400 national naming authorities, which can be national telecommunications monopolies, other national addressing authorities, or carriers offering public X.400 services. They are assigned address components at the Administrative Management Domain (ADMD) level. They, as national authorities, or ADMD owners, can in turn assign Private Management Domain (PRMD) address components.

PRMD addresses directly assigned to user organizations by a national telecommunications monopoly may duplicate PRMDs assigned by other administrations. This does not create a practical problem, because an ADMD is not necessary if communication is directly between PRMDs not subordinate to ADMDs. For example, if airline A connects its X.400-based reservation system directly to that of airline B, there is no need for ADMDs as long as both airlines' PRMDs are assigned by the national body or by the national bodies of different countries.

Such directory services as X.500 accept either name or high-level address arguments and return the appropriate network address. They can also do pure high-level name-to-address translation, distribution list expansion, and mappings based on local policies.

BETWEEN APPLICATION AND NETWORK

Both Open Systems Interconnection (OSI) and the Internet have a concept of something on the network to which datagrams are routed. In Internet protocol (IP), this is the Internet address. In OSI this is technically the network entity title (NET), but the term network service access point (NSAP) address is commonly, if not quite precisely, used.

To link an actual transport or higher protocol in an IP host, a supplemental protocol identifier further qualifies the IP address; the combination of IP address and protocol identifier uniquely identifies a user of IP. In OSI, the NSAP address actually is composed of the NET and a selector field. If a selector field has a value of zero, the NSAP address identifies the target of routing (i.e., the specific host interface). Non-zero selector field values are similar to the IP protocol identifier; they identify a user of the network service entity rather than the entity itself.

Whereas application addresses identify service access points that are meaningful to users, and network addresses identify points in an intelligent network, there is often an additional need for identifiers between the application and network functions. OSI-compliant systems and the Internet, for example, both concatenate an end system service identifier to the network address of end systems.

The Internet transport protocols (TCP and UDP) call this identifier the port number. In OSI, it is the Transport Service Access Point Identifier (TSAP-ID). The function of these transport-level identifiers may be more easily understood if they are viewed as analogous to extension numbers associated with a telephone switchboard. An Internet application may be reached using the IP address, protocol identifier, and an identifier of the user of the transport service above IP. The latter identifier is called the port number. OSI also has an identifier for the transport entity that uses the network service (i.e., the transport service access point identifier), but presentation and session service identifiers, as well as an application entity title, are necessary to reach the application.

When these identifiers are used in a particular way by a vendor they are often complex and obscurely documented. Customers should demand that vendors clearly demonstrate address configuration procedures.

INTERNETWORK ADDRESSING

Functionally, both OSI network layer protocols and the Internet protocol are equivalent. They differ primarily in address formats and parameter encodings, and it is feasible to build a single piece of software that processes both the IP and OSI connectionless network protocol (CLNP). The OSI reference model definition of the network service applies well to both. Indeed, there is an active proposal called TUBA (TCP and UDP over bigger addresses) intended to deal with the IP address shortage by substituting CLNP for IP in existing TCP/IP stacks. In both architectures the job of the network layer is to forward traffic one logical step closer to the destination. These logical steps can be routers or can be the actual destination host. The same addresses are used for routers and for hosts, because upper-layer addresses only are examined when the forwarded traffic reaches its destination. Common addressing for hosts and routes makes network administration far more consistent.

Different types of intelligent systems cooperate at this architectural layer. End systems are hosts (or gateways to other architectures); intermediate systems are routers and packet switches. Intermediate systems do not contain application-level management functions, so management and control of intermediate systems must be distributed to other entities, including end systems. This entails problems with network layer communications between end systems, communications among intermediate systems that are in a domain under a single administrative authority, and communications between intermediate systems that are in different administrative domains.

To participate in network layer interaction, an entity must know its address. Real network entities often have two levels of "network" address.

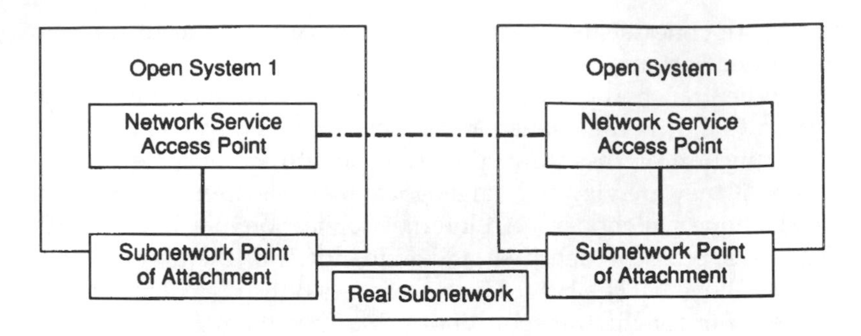

Exhibit 4-2-1. Open System NSAPs and SNPAs

They have one that belongs to the global, medium-independent addressing context defined by OSI or IP. They have another, hardware-oriented address to which the global address is mapped. Examples of such hardware addresses include LAN MAC addresses and X.25 addresses in the X.121 format. The lower-level addressing environment, such as X.121, is called the subnetwork in relation to the higher-level true network address. OSI formalizes two address constructs here, the NSAP address at the global level and the subnetwork point of attachment (SNPA) address at the hardware level (as shown in Exhibit 4-2-1). NSAP and IP addresses are logical rather than physical and always must map to an actual physical address.

The word "network" usually refers to collections of wires and switches that carry bits of data. A network does not mean the same thing in OSI. In OSI, the network service defines a set of capabilities of the network layer of a collection of open systems, which does not necessarily have to form a real-world network. The OSI term for a real-world network is subnetwork. In a similar manner, IP operates in catenets—that is, collections of intelligent networks that may not actually constitute a simple network.

Subnetworks have SNPAs where OSI and non-OSI systems can access their services. In contrast with SNPAs, NSAPs identify conceptual access points for accessing open system capabilities, not for accessing subnetwork services. NSAPs exist within a conceptual global address space used by OSI and are implemented inside open systems, not as addressable parts of communication networks.

The IP uses a similar address model: IP addresses are defined for the global IP address space, and IP addresses are mapped to addresses defined by interface protocols of the subnetworks over which IP runs.

OSI has an additional identifier, the network entity title, which identifies the set of NSAPs in an end system. Additional identifiers may be needed in the future, to deal with mobile communications.

OSI network layer architecture includes three main functions common to modern network architecture. The subnetwork independent function defines the view of the pure network service within the architecture's global address space, the subnetwork access function defines the specific procedures for interfacing to a real subnetwork technology, and the subnetwork dependent convergence function maps between the preceding two. Understanding differences between subnetwork dependent and subnetwork access requirements, as well as the convergence mechanisms, is the key to integrating OSI and Internet network services with existing subnetworks.

OPEN NETWORK ADDRESS STRUCTURES

The Internet addressing scheme has been in use long enough for there to be extensive experience with it. IP addressing has been effective, but is now showing some weaknesses as technologies and network sizes advance.

The NSAP addresses used in OSI are based on ideas first used in creating IP addresses. IP addresses are 32 bits long. They contain three fields of varying length, the first identifying the format of the rest of the address, the next identifying the network domain, and the third identifying a system within that domain.

The structure of the 32-bit address—that is, the use of a format identification field and the division of the address into domain identifiers—means that all possible 32-bit addresses (i.e., 232, approximately 5.6 billion addresses) are not available for assignment to network entities. While other proposals are still being examined, the two main long-term solutions to IP address exhaustion are called the Simple Internet Protocol (SIP) and the TUBA proposal. SIP extends IP addresses to 64 bits, whereas TUBA uses variable-length OSI NSAP addresses.

An interim solution called classless interdomain (CIDR) has been adopted to provide better use of the existing 32-bit addresses. Its basic principle provides a mechanism to subdivide the approximately 65,000 address class B blocks into smaller units, still larger than the approximately 250 address class C blocks. Class B addresses were previously associated with single-user organizations; they will now be shared. Although CIDR does not require a completely new set of protocols, it does restrict the routing protocols usable to those that support variable length subnet masks. It also introduces new administrative procedures for address assignment.

OSI NSAP addresses have a variable length and can be up to 32 octets long. Like IP addresses, they begin with a field that identifies the address format. This field, the initial domain part (IDP), contains fields identifying both an addressing authority and the format of the domain-specific address. The remainder of the NSAP address, the domain specific part (DSP), is unique within an addressing domain.

The DSP is large enough to contain complete subnetwork addresses, such as those used by X.25 or Integrated Services Digital Network (ISDN). As discussed next, LAN devices that have a hardware-defined address can automatically create their own NSAP address using their hardware address and higher-level information that is broadcast by OSI routers.

SUBNETWORK ADDRESSING

Subnetwork attachment protocols map OSI network services onto an underlying real subnetwork—a subnetwork that may use an architectural model completely different from that of OSI. The subnetwork itself may have its own internal intelligence. Careful development of a subnetwork address structure and choice of switching logic are important in building private networks or in interfacing to externally provided subnetworks. Important subnetwork address structures include:

- X.121 addresses used by X.25 and X.75.
- ISDN circuit switching addresses.
- IEEE medium access control (MAC) addresses.
- D channel addresses.
- Frame relay addresses.

OSI provides a mechanism for compatibility with different subnetwork technologies. Network entities often can dynamically learn their NSAP addresses by fixing a subnetwork address with an appropriate IDP or by combining low-order address information with information obtained from broadcasts of layer management entities.

X.25 AND OPEN NETWORK ADDRESSES

X.25 implementations can directly support NSAP addresses of the OSI connection-oriented network service (CONS) by means of address extension fields. X.25 does not have built-in support for IP addressing.

Because there can be interoperability problems between connectionless (e.g., IP and CLNP) and connection-oriented network services, X.25 is more often used as a subnetwork rather than a full network service. When X.25 is used as a subnetwork, IP or CLNP is layered on top of it.

There are two texts for X.25 addressing: pure X.25, and X.25 within the context of such larger networking standards as OSI and the Internet protocol suite. ITU-TSS's X.121 Recommendation defines the structure of X.25 addresses, which are split into an international unique prefix called the data network identification code (DNIC) and a national network part. The latter

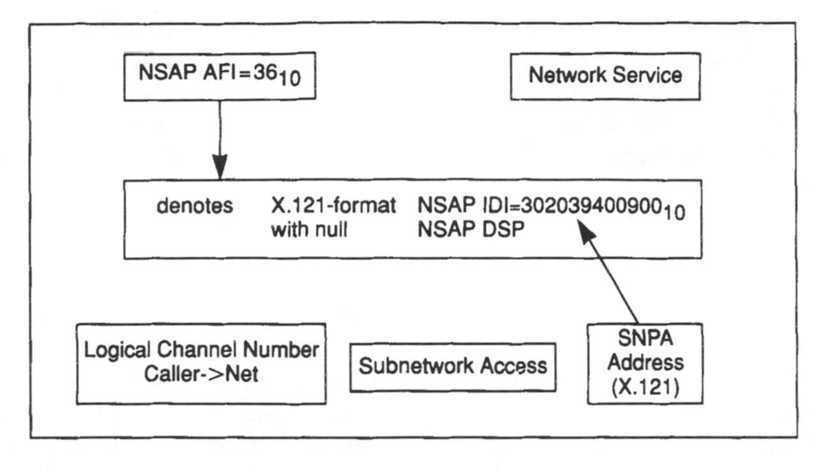

Notes:
DSP Domain Specific Part
NSAP Network Service Access Point
SNPA Subnetwork Point of Attachment

Exhibit 4-2-2. Distinction Between Persistent and Nonpersistent Connection Identifiers

part is not internationally standardized but prescribed by each network. It often takes the general form of a telephone number, that is, including an area code, city exchange, and subscriber (e.g., host) number.

X.25 packet layer standards contain both persistent X.121 addresses and nonpersistent connection identifiers (i.e., logical channel numbers). Logical channel numbers are not part of the larger OSI NSAP or IP addressing plans (as shown in Exhibit 4-2-2).

Configuring an X.25 Address

Using the X.121 address as part of the NSAP address has the attraction of making the NSAP address self-configuring. There are advantages and disadvantages, however, to self-configuration. Self-configuring reduces system administrator workload and the chance of addressing errors caused by incorrect entry of addresses. But if individual human users (i.e., names) are bound to self-configured NSAP addresses, the situation is equivalent to permanently associating a telephone number with a human being. This situation has distinct operational problems if, for example, the individual moves to a location served by a different X.121 address or replaces X.25 service with LAN service.

An alternative to self-configuration is to separate the NSAP and SNPA addresses. A user's NSAP address will then reflect an organizational structure, and the current SNPA address for the user—whether X.25/X.121, LAN, or other service—is obtained from a directory service. As mobile communications become more prevalent, such separation of NSAP and SNPA will become necessary. The telephone equivalent of not doing such a separation requires first trying an office number, then a car phone, then a residence number, and so on.

Self-configuring OSI NSAP addresses, however, is a reasonable short-term mechanism. Self-configuration by including the X.121 address is not a practical alternative for the Internet, because Internet addresses are too small to contain X.121 addresses. There is no single way to map between X.121 and Internet addresses, although RFC 1236 recommends one method. Standards bodies did recognize the need to interconnect public networks such as those operated by Administrations and registered private operating agencies (RPOAs). They defined the X.75 protocol as the means of interconnection. Where X.25 uses two addresses, calling and called, X.75 uses three addresses: calling, gateway, and called. X.75 has not been implemented widely in packet switches intended for users as it is intended for intercarrier use.

Address Software Translation. X.75 is generally reserved for interconnecting user X.25 networks with multiple public X.25 networks, each with their own X.121 addressing plans. A more practical and popular method of interconnecting is translating X.25 addresses from those of one network to those of another. Software to do this translation is switch- and user-specific.

When translating between public and private X.25 addresses, the data switching equipment (DSE) maps "public" X.25/X.121 addresses into addresses meaningful within the private network. Such mapping is not part of ITU-T or ISO standards.

As an example of user-specific address translation, a user organization established a convention that a zero in the DNIC part of an X.121 address meant that the address was part of a local address space. In this address space, the first three digits of the X.121 address identified a DSE and the last four digits identified a port on that switch.

If the DNIC part of an address was nonzero, switches were programmed to route that call to a gateway serving the external network identified by the DNIC. Address translation software would translate external references (e.g., 3020-888-001) to internal references (e.g., port 999 on switch 999). While an internal network user at 997-0002 would see a direct circuit to 3020-888-0001, that virtual circuit was actually composed of circuits between 997-0002 and 999-9998, and between 999-9998 at 3020-888-0001.

Most newer public network technologies, such as ISDN and frame relay, can accommodate existing applications that have an X.25 interface. The underlying technologies, however, are quite different than those traditionally used with X.25.

ISDN

Integrated Services Digital Network, or ISDN, generally provides subnetwork technologies over which network services (i.e., OSI or Internet services) run. Subnetwork facilities can be provided on a semipermanent "dedicated" basis, on dynamically switched circuits, or through packet handling. OSI NSAPs can include ISDN subnetwork addresses as the DSP.

Semipermanent modes may provide a virtual wire to a network entity for conventional LAPB access. Alternatively, they may provide a mechanism for encapsulating X.25 packets and delivering them to a packet network. ISDN global addresses are not significant in semipermanent access using LAPB, because connectivity between end systems is established automatically by the ISDN provider. When semipermanent access is at the packet level, the ISDN provider defines the mapping of LAPD terminal endpoint identifiers to external X.25 destinations.

Circuit switching, using the Q.931 protocol, can set up virtual wires on an as-needed basis. In this case, ISDN addresses are used to set up "data telephone calls" between X.25 entities.

LAN Addressing

IEEE standards for LANs use a scheme that assigns a unique hardware address to every LAN interface. This relieves the network administrator of assigning addresses to LAN interfaces, but adds the complication that a MAC-sublayer address will change whenever the hardware is changed.

Correspondences need to be established between MAC-sublayer addresses and network addresses. Internet Protocol suite systems use the address resolution protocol to discover the network address assigned to them. OSI systems include the MAC address with their NSAP address, which is constructed using prefixes obtained from local router broadcasts.

COLLAPSED LAYERED ADDRESSING

While the OSI reference model broke valuable ground in subdividing network technologies into a set of layered functions, this model did not anticipate all transmission technologies. Especially where communications functions are built into such hardware as LANs and new high-speed transmission technologies, addresses associated with individual layers may be "collapsed" into addresses mappable to multiple layers. These collapsed addresses, which are used for transmission efficiency, are mapped into standard addresses at endpoints or internetwork gateways.

Technologies using collapsed addresses include frame relay, ISDN, and new transmission systems such as asynchronous transfer mode (ATM). In both frame relay and ISDN's nailed D channel operation, the destination address into which the layer 2 address maps can be the access point for a standard X.25 network or an IP address. X.25 packets run unchanged over these services, so multiple X.25 switched virtual circuits can be run over a single ISDN or frame relay address pair.

In frame relay, a short layer 2 identifier is associated, by the network provider, to an endpoint of the frame relay subnetwork. Frame-relay endpoint addresses have only local significance. Permanent virtual circuits are defined between these endpoints. The frame relay protocol is not aware of higher-level addresses, the protocol data units, that flow over it.

SUMMARY

Open network architectures, such as OSI and Internet, have mechanisms that support truly global networking over both new and old transmission systems. These powerful mechanisms can lead to interoperability problems if the associated configuration rules are not well understood. Users need both to understand general address mechanisms and to obtain clear configuration procedures from their vendors.

4-3
The HSSI Standard for High-Speed Networking

Nathan J. Muller

HIGH SPEED SERIAL INTERFACE (HSSI) is a full-duplex synchronous serial interface capable of transmitting and receiving data at up to 52M-bps between data terminal equipment (DTE) and data communications equipment (DCE). Jointly developed by OnStream Networks, Inc. (Santa Clara CA) and Cisco Systems, Inc. (Menlo Park CA), HSSI provides the physical and electrical interface between DTE (e.g., computers, routers, channel extenders, and peripherals) and DCE (e.g., multiplexers and channel service units/data service units (CSU/DSUs) that, in turn, establish links over the wide area network (WAN) (see Exhibit 4-3-1).

This arrangement supports full-duplex transmission of digital data at rates of up to 52M-bps. For example, HSSI can connect through a data service unit (DSU) to carrier-provided full or fractional T3 at 44.736M-bps or E3 at 34.368M-bps. HSSI also supports switched multimegabit data services (SMDS) and frame relay at T3 rates, as well as asynchronous transfer mode (ATM)-based services and point-to-point T3 interconnection between remote fiber distributed data interface (FDDI) local area networks (LANs) using a router and a DSU. HSSI also provides for the interconnection of multiple 16M-bps Token Ring and 10M-bps Ethernet LANs over the WAN.

Working with a bandwidth manager or multiplexer, HSSI can run at any speed to support bandwidth allocation from 1M to 52M-bps. This capability makes HSSI a particularly valuable tool for efficiently managing transmission facilities. Bandwidth managers can dynamically allocate bandwidth to many devices and applications at varying speeds across T3 wide area connections.

SHORTCOMINGS OF V.35 AND EIA-422/449

Before HSSI, standard input to communications devices (i.e., routers, multiplexers, CSU/DSUs) generally supported only V.35, EIA-422/449, and T1

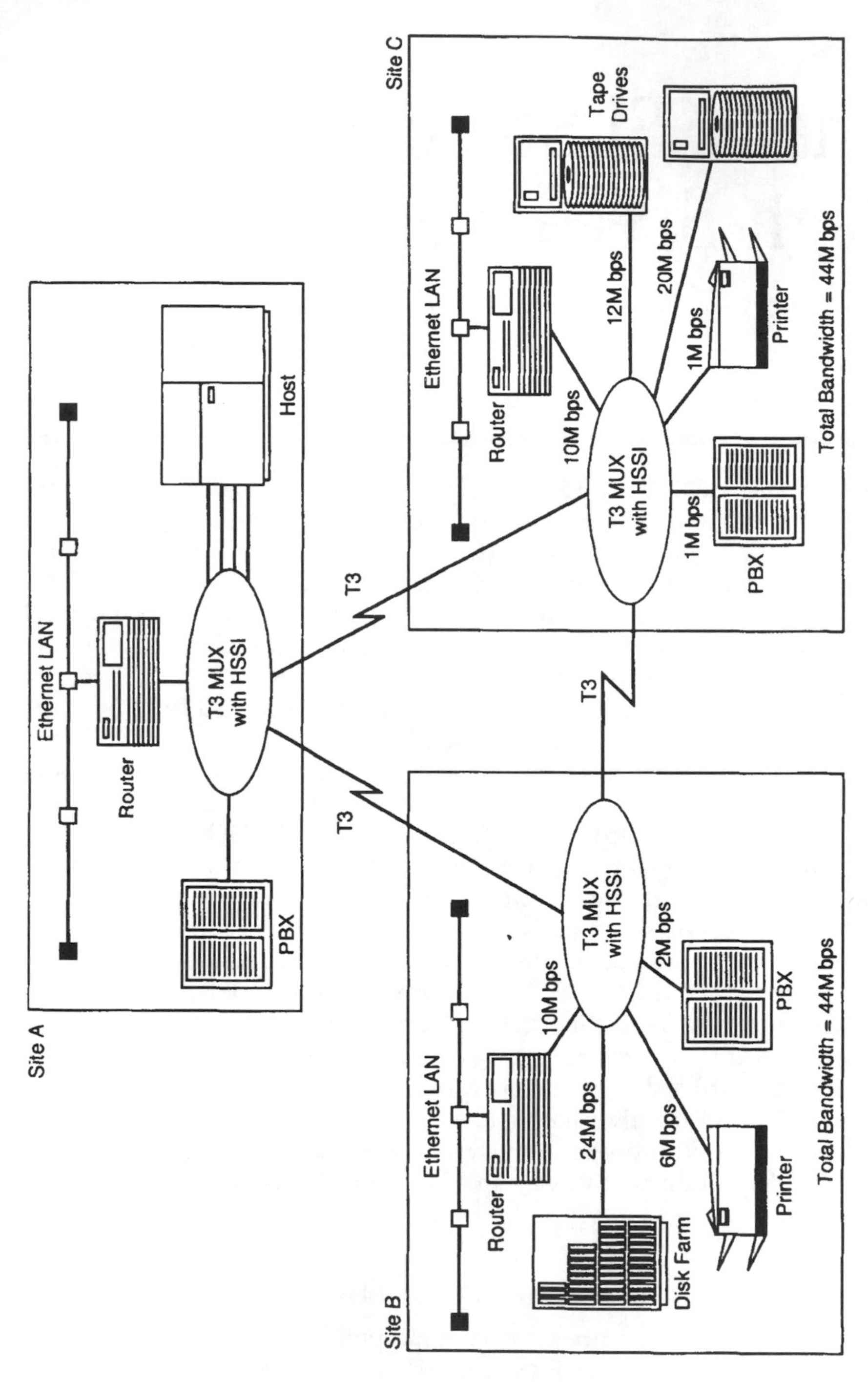

Exhibit 4-3-1. Use of HSSI in a High-Bandwidth Network

connections. Although T1 specifies a maximum data rate of 1.544M-bps, the upper limits of V.35 and EIA-422/449 are more difficult to define because the data rate varies with distance: the shorter the connections between DTE and DCE, the higher the data rate.

V.35 was originally intended to support no more than 6M-bps; the ceiling for EIA-422/449 is 10M-bps. Although there are ways to increase the data rates of these interfaces to as high as 45M-bps, the drawback is that they become, in essence, proprietary interfaces. Furthermore, increasing the speed of V.35 and EIA-422/449 can also increase electromagnetic emissions, which can cause interference problems with other nearby equipment and connections.

Although relatively few single devices need the full bandwidth of a T3 line, HSSI makes it possible to allocate the 44.736M-bps among different DTE. V.35 and EIA-422/449 are not designed to do that. With HSSI, a single-port can support six router connections, for example. In contrast, six V.35 interfaces would be needed to link the same six routers.

THE HSSI SPECIFICATION

HSSI is not really a technology in the usual sense, but a set of specifications that define various electrical and physical characteristics that, together, permit data transmission to 52M-bps over the relatively short distances (up to 50 feet) between DTE and DCE.

Signal Definitions

There are 12-signal definitions in the HSSI electrical specification. Depending on the function, the signals may travel to or from the DCE (Exhibit 4-3-2). The signals and their functions are as follows.

RT: Receive Timing. RT is a gapped clock with a minimum bit rate of 52M-bps, providing receive signal element timing information for RD.

RD: Receive Data. The data signals generated by the DCE, in response to data channel line signals received from a remote data station, are transferred on this circuit to the DTE.

ST: Send Timing. ST is a gapped clock with a maximum bit rate of 52M-bps, providing transmit signal element timing information to the DTE.

TT: Terminal Timing. TT provides transmit signal element timing information to the DCE. TT is the send timing signal echoed back to the DCE by the DTE.

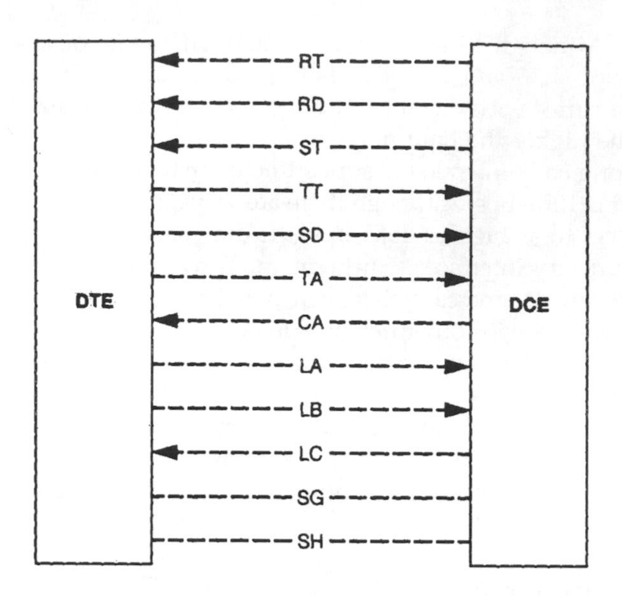

Exhibit 4-3-2. HSSI Signaling Between DTE and DCE

SD: Send Data. The data signals originated by the DTE are transmitted via the data channel to a far-end data station.

TA: Data Terminal Equipment Available. TA will be asserted by the DTE, independently of data communication equipment available, when the DTE is prepared to both send and receive data to and from the DCE. Data transmission does not commence until data communications equipment available has also been asserted by the DCE. (If the data communications channel requires a keep-alive data pattern when the DTE is disconnected, the DCE will supply this pattern while TA is de-asserted.)

CA: Data Communications Equipment Available. CA will be asserted by the DCE, independently of the data terminal equipment available, when the DCE is prepared to both send and receive data to and from the DTE. This indicates that the DCE has obtained a valid data communications channel. Data transmission does not commence until the data terminal equipment available has also been asserted by the DTE.

LA: Loopback Circuit A, LB: Loopback Circuit B, and LC: Loopback Circuit C. LA and LB are asserted by the DTE to cause the DCE and its associated data communications channel to provide one of three diagnostic loopback modes

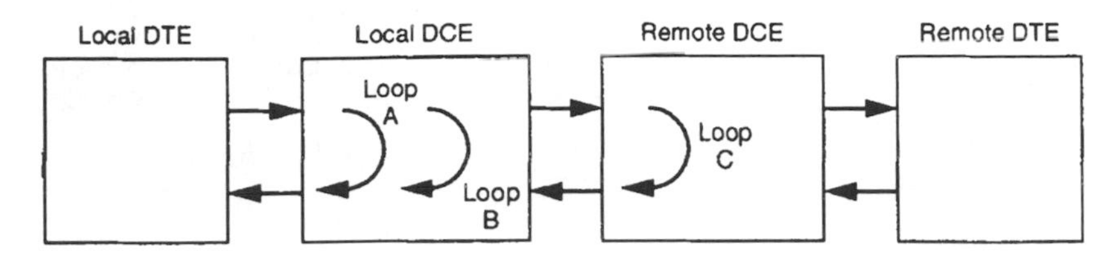

Exhibit 4-3-3. Use of Circuits for Loop-back Testing

for testing purposes (Exhibit 4-3-3). LC is an optional loopback request signal from the DCE to DTE, which requests that the DTE provide a loop-back path to the DCE. Specifically, the DTE would set TT=RT and SD=RD. ST would not be used, and could not be relied on as a valid clock source under these circumstances. This would then allow the DCE/DSU network management diagnostics to test the DCE/DTE interface independent of the DTE. This follows the HSSI philosophy that both the DCE and the DTE are intelligent, independent peers, and that the DCE is capable of and responsible for maintaining its own data communications channel. In the event that both the DTE and DCE assert loopback requests, the DTE will be given preference.

SG: Signal Ground. SG is connected to circuit ground at both ends. SG ensures that the transmit signal levels stay within the common mode input range of the receivers.

SH: Shield. The shield encapsulates the cable for the purpose of limiting electromagnetic interference (EMI) and is not intended to carry signal return currents.

Pin Assignments

Included in the physical specifications of HSSI are the pin assignments for the 2-row, 50-pin plug connector and receptacle that are used with all HSSI-compliant equipment. The pin assignments are described in Exhibit 4-3-4.

The type of cable specified by the HSSI standard consists of 25 twisted pairs (28 AWG) cabled together with an overall double shield and polyvinyl chloride jacket. Whereas V.35 and EIA-422/449 use 34-pin connectors, HSSI uses a 50-pin connector that is identical to that used by the small computer systems interface-2 (SCSI-2), which is becoming a common interface for microcomputer peripherals.

Signal Name	Direction	+ Side Pin No.	- Side Pin No.
SG—Signal Ground	N/A	1	26
RT—Receive Timing	to DTE	2	27
CA—DCE Available	to DTE	3	28
RD—Receive Data	to DTE	4	29
LC—Loopback Circuit C (optional)	to DTE	5	30
ST—Sending Timing	to DTE	6	31
SG—Signal Ground	N/A	7	32
TA—DTE Available	to DCE	8	33
TT—Terminal Timing	to DCE	9	34
LA—Loopback Circuit A	to DCE	10	35
SD—Send Data	to DCE	11	36
LB—Loopback Circuit B	to DCE	12	37
SG—Signal Ground	N/A	13	38
Reserved for future use	to DCE	14-18	39-43
SG—Signal Ground	N/A	19	44
Reserved for future use	to DTE	20-24	45-49
SG—Signal Ground	N/A	25	50

Note:
Pin pairs 18, 43, 20 to 24, and 45 to 49 are reserved for future use. To allow future backward compatibility, no signals or receivers of any kind should be connected to these pins.

Exhibit 4-3-4. HSSI Pin Assignments

HSSI IMPLEMENTATION

HSSI is being used in conjunction with such network devices as T3 multiplexers, bandwidth controllers, routers, channel extenders, and DSU/CSUs.

Multiplexers

The latest generation of T3 multiplexers can divide a T3 pipe into multiple T1 circuits; however, they only accept T1 input. This means that if the T3 multiplexer is being used for point-to-point LAN interconnection, among other applications, there will be a bottleneck on the input side of the multiplexer. This is exactly the kind of problem that HSSI addresses.

When HSSI modules are added to T3 multiplexers, users can pool T1 circuit bandwidth or segment T3 circuits for transmissions requiring bandwidth greater than 1.544M-bps. This allows the T3 multiplexer to accept inputs ranging in speed from 1.544M-bps to 44.736M-bps. For example, to support a 4M-bps router connected to a Token Ring LAN, the T3 multiplexer's HSSI module would segment the incoming channel into multiple T1 increments for transmission, those channels would be combined at the remote location.

HSSI may also be used to connect various DTE to synchronous optical network (SONET)-compliant DCE (Exhibit 4-3-5). Such devices allow corporations with many data centers to view their geographically remote data centers as a unified operation, with bandwidth allocation managed from a single point of control. Most providers of T3 multiplexers now have models that support HSSI.

Bandwidth Controllers (Imuxers)

HSSI is also being used in conjunction with bandwidth controllers, also known as inverse multiplexers. Inverse multiplexing is an economical way to access the high-speed digital services of the interexchange carriers to provide bandwidth on demand.

With this technique, users dial up the increment of bandwidth needed to support a given application and pay for these 56K-bps or 384K-bps local-access channels only when they are set up to transmit voice, data, or video traffic. Upon completion of the transmission, the channels are taken down. No expensive private T1 access lines are required to support temporary or constantly changing applications.

An inverse multiplexer that supports HSSI can support multiple devices on the input side and build the required aggregate bandwidth on the output side from switched 56K-bps or 384K-bps services.

As new carrier-provided digital services, such as switched multimegabit data services, become more widely available, HSSI will be built into new T3 bandwidth controllers. With a T3 bandwidth controller that supports HSSI, the user will be able to assemble local-access T1 channels to achieve the bandwidth required to support multiple high-speed data applications. In the case of SMDS, the access channels for T3 are 4M-bps, 10M-bps, 16M-bps, 25M-bps, 34M-bps and 44M-bps. Through the HSSI, any number of T1 channels may be supported by the T3 bandwidth controller up to the maximum 44M-bps, each carrying a different application.

Routers

Routers are also available with HSSI ports to support LAN internetworking over high-speed digital facilities. Cisco Systems, for example, offers HSSI in its AGS+ router, a router that also supports interfaces to FDDI. The HSSI board links the router to an external DS-3 DSU that is required for connection to carrier-provided T3 lines. Via HSSI, the router also supports SMDS.

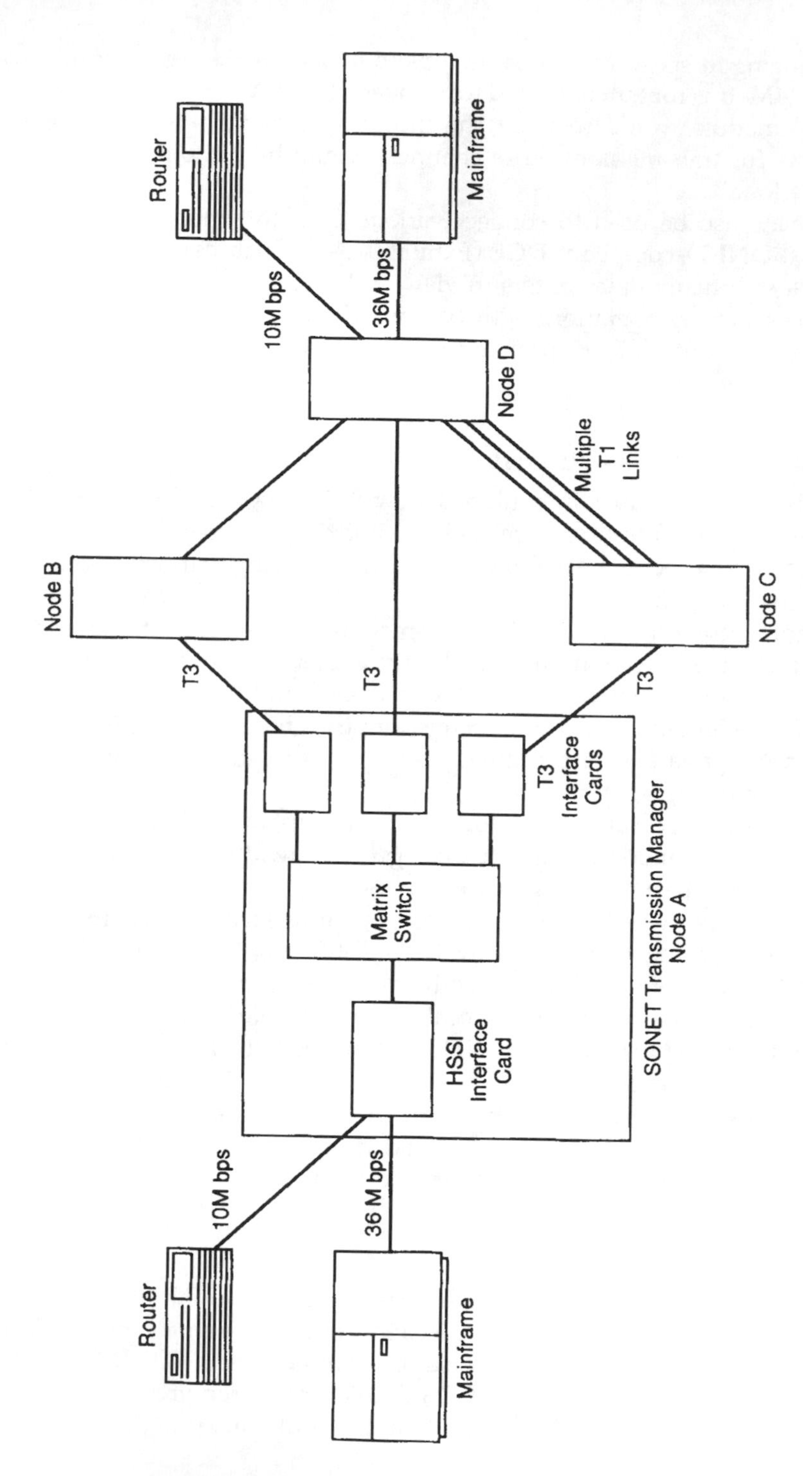

Exhibit 4-3-5. DTE Connected Through HSSI to SONET Compliant DCE

HSSI has also found its way into channel extenders, which are typically used for connectivity between mainframes or between the mainframe and a LAN server. With HSSI, multiple mainframe channels can share a single T3 link. If the channel extender accepts multiple HSSI connections, other devices such as bridges and routers can share the T3 bandwidth as well.

DSU/ CSUs

Customer-provided CSUs or DSUs encode serial data from terminals or computers and, to ensure an acceptable level of network performance, do wave-shaping of the transmit signal before it is sent over the digital facility. The CSU is positioned at the front end of a circuit to equalize the received signal, filter both the transmitted and received waveforms, and interact with the carrier's test facilities. Part 68 registration rules from the Federal Communications Commission (FCC) require that every digital circuit be terminated by a CSU. The DSU element transforms the encoded waveform from alternate mark inversion (AMI) to a standard required business equipment interface, such as V.35. It also performs data regeneration, control signaling, synchronous sampling, and timing.

Both the CSU and DSU fall into the category of customer-premises equipment (CPE). Because of the complementary functional relationship of the CSU and DSU, most vendors have integrated the two devices into a single unit, which is sometimes referred to as a digital data set.

A T3 DSU/CSU equipped with an HSSI port can accommodate a connection from a router that is also equipped with an HSSI port. The T3 DSU/CSU provides clocking and frame synchronization functions for data sent over or received from the T3 lines.

SUMMARY

Designed to meet the growing demands of high-speed data applications, HSSI eliminates the throughput bottlenecks of current V.35 and EIA-422/449 serial interfaces. Today's growing need to interconnect LANs and support bandwidth-intensive applications requires a standardized interface that can support the higher speeds offered by carrier-provided facilities and services. HSSI addresses this need, not only by providing economical and efficient LAN-to-WAN connectivity, but by providing increased configuration flexibility as well.

4-4 RMON: The SNMP Remote Monitoring Standard

Nathan J. Muller

REMOTE MONITORING (RMON) enhances the management and control capabilities of simple network management protocol (SNMP)-compliant network management systems and local area network (LAN) analyzers. RMON-equipped probes can view every packet and produce summary information on various types of packets, such as undersized packets, and events, such as packet collisions. The probes may also store information for further analysis by capturing packets according to predefined criteria set by the network manager or test technician.

For example, the network manager might only be interested in examining NetWare packets to track down the source of an intermittent problem that occurs when opening files at a remote server. According to an established threshold, an alarm can be generated if there is more than one file-open error per 100 file opens. Packets that match the filter criteria are stored in the probe's memory and those that do not match are discarded. At any time, the SNMP-based management console can query the RMON probe for this information so that detailed analysis can be performed. Once stored, a remote manager can play back the error history of any station to pinpoint where and why an error occurred.

RMON agents are becoming available for a variety of products, including network interface cards (NICs), the network modules or intelligent hubs, and the boards of bridges and routers. RMON has been so widely accepted by the vendor community that it will not be long before just about any device that is attached to the network—regardless of media—and is manageable with SNMP can be optionally equipped with RMON agent.

Initially, RMON defined media-specific objects for Ethernet only. Media-specific objects for Token Ring have recently been added and those for the fiber distributed data interface (FDDI) are anticipated in the near future.

The RMON MIB is organized into optional object groups with the following associated variables:

- *Statistics Group.* Maintains low-level use and error statistics (e.g., number of packets sent, broadcasts, and collisions).
- *History Group.* Provides user-defined trend analysis based on information in the statistics group.
- *Host Table Group.* For each host, contains counters for broadcasts, multicasts, errored packets, bytes sent and received, and packets sent and received.
- *Host Top N Group.* Contains sorted host statistics (e.g., a complete table of activity for the busiest nodes communicating with each host). N indicates the number of nodes specified by the network manager.
- *Alarms Group.* Allows a sampling interval and alarm threshold to be set for any counter or integer recorded by the RMON agent.
- *Filters Group.* Provides a buffer for incoming packets and user-defined filters.
- *Packet Capture Group.* Allows buffers to be specified for packet capture, buffer sizing, and the conditions for starting and stopping packet capture.
- *Events Group.* Logs events (e.g., packet matches or values that rise or fall to user-defined thresholds).
- *Traffic Matrix Group.* Arranges use and error information in matrix form to permit the retrieval and comparison of information for any pair of network addresses.

Compliance with the RMON MIB standard (Request for Comment 1213) requires that the vendor provide support for every object within a selected group. Each group is optional, so when selecting RMON MIB agents users should determine the features they require and verify that those features are included in actual products. What follows is a more detailed discussion of the features provided by each group.

STATISTICS GROUP

The Statistics Group provides segment-level statistics. For an Ethernet segment, for example, these statistics show packets, octets (or bytes), broadcasts, multicasts, and collisions on the local segment, as well as the number of occurrences of dropped packets by the agent. Each statistic is maintained in its own 32-bit cumulative. Real-time packet size distribution is also provided.

The number of collisions detected by the agent depends on the capability of its internal or externally attached transceiver, or media access unit (MAU). The MAU should be receiver-based, meaning that it can detect all collisions on the segment.

The RMON MIB includes error counters for five different types of packets. For Ethernet, for example, the following types of errors are counted:

- *Undersizes (runts).* Ethernet packets of less than 64 bytes, which are usually caused by a collision on the network. A normal Ethernet packet is 1,518 bytes. Excessive runt packets may indicate that the transmitting station is not configured properly.
- *Fragments.* A packet whose total length is less than 64 bytes (excluding framing bits) and is not an integral number of octets in length or which has a bad frame check sequence.
- *CRC and alignment errors.* Cyclic redundancy check (CRC) is the last 32 bits of information contained in a packet or frame and is used for detecting transmission errors. When the CRC value of an incoming frame is not identical to the CRC value of an outgoing frame, a bit flop is said to occur, which generates a CRC error. This is usually caused by faulty cable, as when an impedance mismatch occurs, causing a reflection on the cable, which in turn causes the bit flop. An alignment error is a packet that is not an integer number or bytes in length and is between 64 and 1,518 bytes in length (including the frame check sequence but excluding the framing bits) or which has a bad frame check sequence. An alignment error is most often caused by a frame collision on the network, although loose connections or noisy cable can also be the cause.
- *Collisions.* Data sent by a device on the network without regard for any other devices that may also be trying to transmit. When two or more devices try to transmit at the same time, a collision occurs, a situation that causes the signals to collide and the data to become garbled.
- *Oversizes (giants).* A packet that exceeds the maximum 1,518 bytes (excluding framing bits but including the frame check sequence), and which has a good frame check sequence. This type of packet can be caused by a node that is not configured properly.

These counters provide useful network management information beyond that provided by typical network interface cards, for example. Industry-standard cards usually provide only two separate counts of CRC and alignment errors, and will not count packets that are either too small or too large. These runt and giant packets are counted by the RMON MIB agent because they usually indicate configuration problems in the transmitting station. Such packets usually are not passed from the receiving card driver, resulting in failed transmissions.

History Group

With the exception of packet size distribution, which is provided only on a real-time basis, the History Group provides historical views of the statistics provided in the Statistics Group. The History Group responds to user-defined sampling intervals and bucket counters, allowing for the complete customization of trend analysis.

The RMON MIB comes with two defaults for trend analysis. The first provides for 50 buckets (or samples) of 30-second sampling intervals over a period of 25 minutes. The second default provides for 50 buckets of 30-minute sampling intervals over a period of 25 hours. Users can modify either of these or add additional intervals to meet their specific historical analysis requirements. For example, the sampling interval can range from 1 second to 1 hour.

Host Table Group

A host table is a standard feature of most current monitoring devices. The RMON MIB specifies a host table that includes node traffic statistics: packets sent and received, octets sent and received, broadcasts, multicasts, and error packets sent. In the host table, the classification "errors sent" is the combination of undersizes, fragments, CRC and alignment errors, jabbers and oversizes sent by each node. The RMON MIB also includes a host timetable that shows the relative order in which each host was discovered by the agent. This data is not only useful for network management purposes, but also assists in uploading to the management station only those nodes of which it is not aware. This index entry improves performance and reduces unnecessary SNMP traffic on the network.

Host Top N Group

The Host Top N Group extends the host table by providing sorted host statistics (e.g., the top 10 nodes sending packets or an ordered list of all nodes according to the errors sent over the last 24 hours). Both the data selected and the duration of the study are defined by the user at the network management station, and the number of studies is limited only by the resources of the monitoring device.

When a set of statistics is selected for study, only the selected statistics are maintained in the Host Top N counters; other statistics over the same time intervals are not available for later study. This processing—performed remotely in the RMON MIB agent—reduces SNMP traffic on the network and the processing load on the management station. The station would otherwise need to use SNMP to get the entire host table and sort it locally.

Alarms Group

The Alarms Group provides a general mechanism for setting thresholds and sampling intervals to generate events on any counter or integer maintained by the agent (e.g., segment statistics, node traffic statistics defined in the host table, or any user-defined packet match counter defined in the Filters Group). Both rising and falling thresholds may be set, because both can indicate network faults. For example, crossing a high threshold can indicate network performance problems; crossing a low threshold may signal the failure of scheduled network backup.

Thresholds can be established on both the absolute value of a statistic or its delta value, so the manager is notified of rapid spikes or drops in a monitored value. Rising and falling thresholds work together in an alternating fashion. After a rising threshold is crossed, another rising event is not generated until the matching falling threshold is crossed.

Filters Group

The Filters Group provides a generic filter engine that implements all packet capture functions and events. The filter engine fills the packet capture buffer with packets that match the user-specified filtering criteria. Any individual packet match filter can serve as a start or stop trace trigger.

The conditions within a single filter can be combined using the Boolean parameters AND or NOT. Multiple filters are combined with the Boolean OR parameter. Users can choose to capture packets that are valid or invalid, or are one of the five error packet types (discussed previously). If the proper protocol decoding capability is in place at the management station, this filtering essentially provides distributed protocol analysis to supplement the use of the technicians equipped with portable protocol analyzers.

The monitor also maintains counters of each packet match for statistical analysis. Either an individual packet match, or a multiple number of packet matches through Alarms, can trigger an event to the log or the network management system using an SNMP trap. Although these counters are not available to the History Group for trend analysis, a management station may request these counters through regular polling of the monitor so that trend analysis can be performed.

Packet Capture Group

What packets are collected is dependent on the Filter Group. The Packet Capture Group allows users to create multiple capture buffers and to control whether the trace buffers will wrap (i.e., overwrite) when full or stop capturing. Depending on the agent, the user may expand or contract the size of the buffer to fit immediate needs for packet capturing without permanently committing memory that will not be needed later.

The network manager can specify a packet match as a start trigger for a trace and depend on the monitor to collect the trace without further manager involvement. The RMON MIB includes configurable capture slice sizes to store either the first few bytes of a packet—where the protocol header is located—or to store the entire packet. The default slice setting specified by the RMON MIB is the first 100 octets.

Events Group

The Events Group is used to create entries in the monitor log or to create SNMP traps, from the agent to the management station, on any specified event. An event can be a crossed threshold on any integer or counter or from any packet match count. Vendors may add other notification capabilities, including administrative events from the agent (e.g., a power failure or reset).

The log includes the time of day for each event and a description of the event. The log wraps (overwrites) when full, so events may be lost if they are not uploaded to the management station periodically. The rate at which the log fills depends on the resources the monitor dedicates to the log and the number of notifications the user has sent to the log.

Traps can be delivered by the agent to multiple management stations if each station matches the single community name destination specified for the trap. An RMON MIB agent will support each of the five traps required by SNMP: link up; link down; warm start; cold start; and authentication failure. Three additional traps are specified in the RMON MIB: rising threshold; falling threshold; and packet match.

Traffic Matrix Group

The RMON MIB includes a traffic matrix at the medial access control (MAC) layer. A traffic matrix shows the amount of traffic and number of errors between pairs of nodes—one source and one destination address per pair. For each pair, the RMON MIB maintains counters for the number of packets, number of octets, and error packets between the nodes. Users can sort this data either by source or destination address.

BENEFITS OF RMON

RMON is typically offered as an optional expansion package to vendors' SNMP-compliant network management systems, facilitating the gathering of information from network devices, which can be used for fault diagnosis, performance tuning and network planning.

Although the standard MIB II is designed to provide real-time protocol management for devices that use the Exterior Gateway Protocol (EGP), Internet Protocol (IP), Internet Control Message Protocol (ICMP), Transmission Control Protocol (TCP), and SNMP, the accumulation of historical performance data requires that the network management station constantly poll every network device and store the data itself. In many cases, however, historical data is better gathered at the network device itself, and only retrieved occasionally by the network management station. This is exactly what the RMON MIB is designed to do. It reduces network overhead caused by SNMP polling and reduces network management system processing overhead.

RMON also can be used to provide "proxy" management, either intercepting SNMP requests for other agents, or acting as an agent for network devices that are not SNMP-compliant. In the case where multiple management stations are trying to manage a single agent, a proxy agent can shield the real agents from redundant requests, and can respond for them. Proxies also can send a single agent's trap to multiple management stations.

Proxies can reduce the amount of SNMP traffic hitting the network, enhance SNMP security, and be used to manage proprietary network devices or devices such as MAC-level bridges, wiring hubs, or modems that work below the network layer of the OSI reference model.

IN-BAND AND OUT-OF-BAND ACTIVITIES

All monitoring and control interactions between agent modules and network management stations occur using SNMP. The information collected by RMON-compliant devices can be sent to the central management station in either of two ways: in-band through the data stream or out-of-band through an RS-232 serial link and modem.

In-band transport protocol options include standard User Datagram Protocol/Internet Protocol (UDP/IP), and UDP/IP encapsulated within source-routed frames (which allows the SNMP traffic to pass through internetworks joined by source-routing bridges). In-band management can also be conducted over a Telnet session running over the LAN.

Out-of-band management may be accomplished through modem connection to an RS-232 port, or by using the Serial Line Internet Protocol (SLIP) or the Point-to-Point Protocol (PPP) over router-based WAN connections when the network management system is remotely located. The modem connection can be used as a backup, allowing out-of-band communications should in-band communications fail.

SUMMARY

RMON provides the benefits of standardization, including multivendor support and potential commodity pricing. As the vendor community continues to offer RMON-compliant products, users will be able to mix and match them and manage them from a single network management system.

The RMON MIB is more a flexible and economical solution to monitoring network performance than the proprietary systems currently on the market. Many of these vendors are expected to support RMON, so products already purchased may only have to receive software upgrades to be of continued use.

Section 5
Communications Services

VERY FEW CORPORATE COMMUNICATIONS SYSTEMS are totally self-contained. Except for local networks, the network infrastructure is rarely owned and maintained by the enterprise. Usually a very significant portion of the infrastructure is supplied by a telecommunications service provider, especially the switching and transport portions of a wide area network. Consequently, communications managers must contract with a telephone company or some other alternative provider to acquire what is known as "services."

A service is composed of a set of the network technical attributes observed by the user combined with a set of other attributes associated with the network's provision of the service. The former includes attributes such as capacity, error rates, and protocols. The latter includes operational, maintenance, and billing attributes. As the name "service" implies, the user contracts for a set of characteristics visible at the network interface. The user does not contract for the use of a specific trunk or other network element. A contract for T1 service obligates the network to provide the technical capability to transfer 1,536K-bps between two points with, for example, a minimum of so many error-free seconds and a maximum of so many minutes of outage per year. How this service is provisioned by the supplier is of no direct concern to the user.

Services are packaged by the service providers in the hope that this particular combination of attributes will attract buyers. Carriers generally try to deploy services in response to market demand, which requires that they try to determine whether a proposed service will improve customer performance or satisfy a new requirement. Is there demand for this service? Can the service be provisioned at a profit? What is the competitive environment? Are there alternatives to this service?

Because service deployment can be an expensive proposition with a multiyear time frame, it is essential that the service provider develop a strategy and a carefully designed launch plan. Communications managers can be easily frustrated by the sometimes slow launch or limited availability of a service, which is usually the result of the service provider trying to "test the water" to see if a service is viable before investing in widespread deployment. One thing a manager can do to help the process is to have a firm grasp, or at least a good understanding, of future requirements. Because of the long lead times to service deployment, the providers are trying to anticipate future customer demand. Yet when providers survey users about requirements five years down the road, users often answer that they have no idea except that they will need more capacity.

Another service related difficulty is the need to remain backward compatible with current services. The infrastructure must be designed to not preclude the continued customer use of existing services. The IS department in an organization could never persuade users to do a wholesale replacement of their existing software and hardware because IS deployed a new infrastructure. This means, for example, that as the networks deploy ATM in the network core, that core must continue to support the frame relay service, the dedicated T1 services, and of course, voice telephony.

This section is intended to provide the communications systems manager with information about the telecommunications service environment. The section begins with "An Overview of Switched Digital Services," Chapter 5-1. Dedicated, point-to-point leased lines are an excellent choice where the amount of traffic between two locations is relatively high or where the transfers are of long duration. In many cases, however, where the monthly connect-times are measured in hours, or where small amounts of data are exchanged sporadically with many different locations, switched services may prove to be very advantageous. This chapter discusses the basics of the switched digital services now available. The service offerings of the many providers are described and cost comparisons are included.

The proliferation of local networks has been a leading market driver for high-speed transmission services. As corporate intranets are deployed to allow information to be accessed from anywhere within the organization, there is an increasing demand for interconnection of formerly isolated facilities. Chapter 5-2, "High-Speed Services for LAN Interconnection," provides a perspective of the carrier services available to support this demand. This chapter contrasts with Chapter 5-1 in that the emphasis here is on the non-switched services. T1, Fractional T1, SMDS, SONET, and ATM are described in this useful chapter.

Organizations frequently use service provider facilities for a significant portion of a corporate network. In response, service providers have developed the concept of a virtual private network (VPN). A VPN maps the characteristics of a private network onto the facilities of a public network. The user gains all the benefits of a private network, such as unified, corporatewide numbering plans, without the investment and ongoing expense of deploying private facilities. VPNs also solve problems of dealing with continuous changes in network size and scope and of maintaining a staff of network designers and traffic engineers. These responsibilities are devolved to the VPN service provider. Chapter 5-3, "A Management Briefing on Virtual Private Networks," details how VPNs work and offers descriptions of the services offered by the leading providers.

The handbook also addresses the issue of access technologies and pricing. As discussed earlier, services are based on characteristics of the network. These are implemented in switches, trunking facilities, and a whole spectrum of intelligent processors and peripherals. To use these services, the user must first gain access to them, which is the subject of Chapter 5-4 "A Guide to Network Access and Pricing Options." This chapter reviews the various types of access, discusses different pricing plans, and offers advice on contracting for access.

Because all communications managers are concerned with controlling costs, the handbook includes Chapter 5-5, "Auditing Telecommunications Costs for Data and Video Communications." This chapter examines the cost structure of different technologies. ISDN, microwave, frame relay cellular, and the Internet are all explained from the point of view of cost control. The objective is to provide readers with guidelines to assure efficient use of facilities and to minimize communications costs.

The worldwide movement toward deregulation of telecommunications should provide communications managers with a vastly increased choice of services and service providers. Deregulation, defined as the elimination of a monopoly in a public market, is easily legislated but very difficult to implement. Chapter 5-6, "The 1996 US Telecommunications Act and Worldwide Deregulation," details the difficulties of implementing deregulation. The chapter emphasizes Federal Communications Commission's struggle with interpreting the intent of Congress and creating the rules under which competitors may enter the local and long-distance service markets. Deregulation in the United Kingdom, Japan, and Europe is also discussed in this very timely and important chapter.

5-1
An Overview of Switched Digital Services

Steven E. Turner

THE NEED FOR HIGH-SPEED DATA TRANSFER has grown exponentially, driven largely by increased computer processing rates and distributed processing. The growth of local and wide area networks (WANs) has also fostered an increased need for quick and reliable data communications. Fortunately, various digital communications services and products are available to help users meet these demands and use communications resources in a cost-effective manner.

No single solution can satisfy the demand for data communications products and services. For example, analog modems now operate quite reliably at basic data rates of 33.6K-bps and modem information rates now exceed 112K-bps using advanced data compression and error control. Leased digital services and basic-rate integrated services digital networks (ISDNs) operate at similar rates. Another solution, switched digital services, offers the flexibility of the dial-up network and the cost-effectiveness of a pay-per-use fee structure.

DIGITIZING THE NETWORK

To understand the nature of these all-digital techniques and the products that support them, it is important to review the evolution of the telephone network during the past few decades. Digitization of the public switched telephone network (PSTN) began during the early 1960s, when pulse-code modulation was introduced to encode voice-band analog signals into 64K-bps digital bit streams. By packing these high-speed digital signals into the formerly all-analog channels, the telephone company significantly increased its network capacity. Pulse-code modulation also improved the quality of long-distance connections, because the digital bit streams could be transmitted over longer distances without the inherent degradations that accompanied analog signals.

As pulse-code modulation encoding spread throughout the network, T-carrier lines were introduced to take advantage of these digital signaling techniques. T-carrier systems (e.g., T1 networks) pack multiple, digitized 64K-bps pulse-code modulation data streams onto a single-telephone line, again increasing the telephone network's signal capacity. Almost all long-distance telephone lines are digital, T carrier-based connections, and analog lines are becoming a thing of the past.

After T-carrier systems were installed, the only remaining analog portion of the dial-up telephone network was the switching network. This is also changing, however, and analog switches are largely being replaced by digital switches that can support many digital transmission techniques. More than 90% of the analog telephone switches in the US have been replaced with digital switches. Most of the remaining analog switches are located in rural areas less likely to need high-speed digital access just yet; these switches are scheduled for replacement with digital devices in the near future. Europe and Asia are also moving rapidly to incorporate digital switches into their telephone networks. The end result is the ability to achieve end-to-end dial-up digital connectivity on almost all telephone calls throughout the world.

Because most analog telephone switches have been replaced by digital switches, switched digital services are becoming popular. Whereas DDS involves leased telephone lines connecting two specific locations, switched digital services use the standard, dial-up network. This allows users to send high-speed digital data to any location served by a digital telephone switch, much as users of dial-up voice or modem equipment have done over analog telephone channels. It also brings unparalleled flexibility to the high-speed data world, combining the speed and reliability advantages of digital transport media with the flexibility of the ubiquitous public telephone network.

SWITCHED DIGITAL SERVICES OPERATION BASICS

Several switched digital services offerings are available, and the inner workings of each are unique. All, however, use the basic encoding technique known as alternate mark inversion (AMI) to transmit data. This is the same technique used in leased-line DDS service; in fact, AMI is the universal transmission method for the digital portion of the wire communications network. Unlike analog modulation, in which each data symbol is multiplied by a carrier signal for transport at a different frequency and data rate, digital products send encoded ones and zeros at the base-line frequency of the service.

With AMI, every one bit in the data stream corresponds to a pulse in the outgoing digital system, and every zero bit in the data stream is represented by the absence of a pulse. During normal operation, positive and negative pulses alternate on successive one bits, giving AMI its name. All zero bits result in no pulse at the transmitter's output. Together, this series of positive, negative, and absent pulses represents the information being transmitted over the link.

AMI is used instead of simple ones and zeros because of a requirement by the telephone network that a significant amount of activity take place in the network during transmission. Otherwise, some network repeaters fail to operate properly. The alternations inherent in AMI guarantee the required network activity level.

To further ensure correct operation over the digital portion of some networks, switched digital services (as well as other AMI-based techniques, such as DDS) use part of their encoded bandwidth to guarantee an adequate density of ones. Specifically, some network repeaters have a tendency to fail if more than seven consecutive zeros appear in the data stream. To accommodate this, some switched digital systems reserve the last bit of each data octet for maintaining the density of ones, and this bit is always set to one. This requirement uses 8K-bps of the 64K-bps capacity of these switched digital systems, reducing the maximum data rate from 64K-bps to 56K-bps.

SERVICE OFFERINGS

The particular switched digital service offered in any area is determined by the Regional Bell Operating Co. (RBOC) that serves that area and the digital switches used in its system. Four switched digital services dominate the US market: Northern Telecom, Inc.'s Datapath; AT&T's Accunet; Sprint, Inc.'s Virtual Private Network; and the Circuit Switched Digital Capability (CSDC). Collectively, these four form the core of a technology that many experts believe points the way to the future of data communications.

The customer premises products used in switched digital services are similar to the data service units used in DDS; however, they include the dialing and control capabilities needed to access the switched telephone network and communicate with dial-up network equipment. They also contain special circuitry that allows them to accommodate different types of digital telephone switches and use any of the four switched digital services.

Accunet

Originally offered by AT&T as a simple extension of DDS, Accunet now includes services ranging from 56K-bps to T1. To use the service, subscribers must first obtain digital lines from their business location to the nearest Accunet switching node. For the 56K-bps service, for example, the digital lines are four-wire lines approved for carrying 56K-bps data using AMI encoding. Unlike DDS, which can connect the customer only to a specific, previously selected destination, Accunet uses a dialing mechanism to select the destination of calls.

The Circuit-Switched Digital Capability

The CSDC is a Bellcore-defined service provided by the individual RBOCs and can operate only through AT&T's 5ESS central office switches. This limits the availability of this service to specific locations and connections. The basic concept of the CSDC is similar to that of the all-digital local loop, which is the two-wire circuit that connects business locations to the telephone company's central office. The CSDC allows subscribers to use the local-loop wire pair already connected to their voice telephone to gain access to the digital data network.

The only drawback is that to allow high-speed data operation, the band-shaping loading coils attached to local loops that are used exclusively for voice transmission must be removed before data transmission begins. For adequate data reliability, the distance between the customer's location and the central office must also be limited. This is generally not a problem, because high-density business areas usually require a nearby central office facility to serve its local customer base. As these requirements are met, the local loop can be used interchangeably for the CSDC switched digital service and for normal analog voice traffic.

Call setup in the CSDC service is accomplished with an ordinary analog telephone; after the call has been established, the circuit is switched to data mode. Although data over the long-distance network is passed at the nominal 56K-bps rate, the CSDC sends data over the local loop (between users and their local central office switch) using time-compression multiplexing. The primary benefit of time-compression multiplexing is that it simulates full-duplex operation for users at both ends of the CSDC link. This is accomplished by buffering the 56K-bps data received from the long-distance network at the central office's line interface equipment and transmitting the data to the customer in short bursts clocked at 144K-bps. In turn, the customer's CSDC transmitter sends data to the local central office in 144K-bps data bursts. Each burst contains 199 bits, including 168 data bits, 4 control bits, 3 synchronization bits, and 24 bits currently reserved for future network expansion. The bursts are synchronized so that each end transmits a burst and then enables its receiver to accept a corresponding burst from the other end. Each end sends and receives data in such rapid succession that the link appears to be full duplex, though in reality it is a high-speed, half-duplex operation.

Datapath

The Datapath service was defined by Northern Telecom, Inc., for use with its own DMS-100 central office switches. It is also offered selectively by different RBOCs throughout the US. Like the CSDC, Datapath operates on the concept of the digital local loop, but the implementation is somewhat different. Unlike the other services, the maximum data rate for Datapath is 64K-bps; this rate can be achieved, however, only on calls that traverse a single central office switch (i.e., local calls). In these instances, a separate 8K-bps signaling channel provides control information, and all 64K-bps of the information channel are dedicated to data transfer. Although the out-of-band signaling protocol used in this scheme is not the same as the ISDN signaling protocol, Datapath offers a 64K-bps B-channel and an 8K-bps D-channel corresponding to encoding, but because it must accommodate a 64K-bps data rate, the format is somewhat different from the CSDC format. In Datapath, the burst clock rate is 160K-bps, and each burst is only 74 bits in length—64 data bits, 8 signaling bits, 1 start bit, and 1 stop bit.

A unique feature of Datapath is its proprietary rate adaptation protocol, called T-Link. T-Link allows the user's terminal equipment to send data at a variety of synchronous and asynchronous data rates. This rate adaptation method permits asynchronous data rates to 19.2K-bps and several synchronous rates between 1.2K-bps and 64K-bps. T-Link accomplishes this by converting the data rate output of the users' terminal into 64K-bps bit streams compatible with the base Datapath service rate. At rates of 19.2K-bps and below, Datapath uses T-Link to provide data redundancy, creating a built-in error correction technique to identify and correct bit errors. It is expected that eventually, the CSDC and Accunet will have similar rate adaptation schemes.

SERVICE INTERWORKING

Despite the differences in technique, Accunet, the CSDC, and Datapath can interwork. That is, calls from any of the services can be made to equipment supported by the other services. The data rate is fixed at 56K-bps for interswitch and interservice calls between the CSDC and Accunet, and Datapath can also be configured to interoperate with these services at that rate. The biggest difficulty in this type of interworking is that Datapath cannot use its proprietary T-Link rate adaptation protocol. Compatibility is maintained in such circumstances by having the Datapath unit use the inband signaling technique used by the other services.

Although all three services can interwork with one another, none can completely guarantee end-to-end connectivity on a switched digital link unless the call remains entirely within a single switch. This is because digital telephone switches must be installed at both ends of the link to make switched digital access possible. Because almost all analog switches have now been replaced with digital switches, the probability of encountering this difficulty is remote.

SERVICE AVAILABILITY

As stated earlier, the availability of each of the switched digital services in any particular area is determined by the RBOC and the type of switches used to serve that area. Of the three established services, Accunet is the most widely available. The switching nodes required for Accunet's four-wire connections are located primarily in major metropolitan areas, and leased digital circuits are widely available into these nodes. Accunet service can be offered in more remote locations as well, but the required leased lines cost more with increased distance from the switching nodes. Therefore, Accunet is sometimes impractical in areas too far removed from an Accunet node.

Datapath is also popular with a large installed base in Canada and a growing customer base in the US. The limitation in Datapath availability is its dependence on the DMS-100 central office switch. Fortunately, most major metropolitan areas are served by several of these switches, and customers served by other types of switches can access this service with an auxiliary channel bank to the switch hardware. The channel bank provides a crossover point for Datapath customers to a nearby switch that has Datapath capability.

Although the CSDC is the least accessible switched digital service, it has strong pockets of availability around the US and its use is steadily growing. The CSDC uses 5ESS central office switches in much the same way that Datapath uses DMS-100 switches. Crossover auxiliary channel banks are available for CSDC service customers without AT&T switch access, so end-to-end CSDC service can usually be achieved in most areas.

Accunet is currently marketed nationally by AT&T under a single Accunet name, but Datapath and the CSDC are offered regionally and independently by the RBOCs under various names. A great deal of confusion has resulted, because potential customers often cannot determine which service is available in their local area. To eliminate some confusion, Exhibit 5-1-1 lists some of the names for Datapath and the CSDC used by the RBOCs in the US.

APPLICATIONS ISSUES

Switched digital services have definite advantages over such leased digital services as DDS when the application is appropriate. Because every appli-

RBOC	Service Name	Service Type
Ameritech	Public Switched Digital Service	Datapath, CSDC—Public
	Centrex Digital Service	Datapat—Centrex
Bell South	Accupulse	Datapath—Public
	Digital Essex	Datapath—Centrex
NYNEX	Switchway 56	Datapath—Public
	Intellipath II	CSDC—Centrex
Pacific Telsis	Public Switched Digital Service	Datapath, CSDC—Public
Southwestern Bell	Microlink II	Datapath—Public
	Essex Custom Digital Service	Datapath—Centrex
US WEST	Public Switched Digital Service	Datapath, CSDC—Public
	Switched Net 56	CSDC—Public

Note:
Centrex is a tariffed telephone company service offering features similar to those of a PBX, plus additional call ID accounting features. Many service providers offer switched digital access through both Centrex and the public dial-up network.

Exhibit 5-1-1. RBOC's Switched Digital Service Offerings

cation is different, the choice of a data transfer service must be based on the needs, goals, and constraints of the user. Several factors must be considered (e.g., data rate, switch availability, tariffs, use, and reliability). To aid in the analysis of these considerations, Exhibit 5-1-2 lists the most important factors for both analog and digital telephone services.

If all of these telephone services are available to a user, the most obvious consideration in selecting a service is the data rate required by the application. For example, analog leased lines and PSTN services are geared toward the individual user access, whereas digital services are better suited to applications such as LAN-to-LAN communications.

Another important consideration in service selection is the number of locations that must be served. When high-speed data service is required, switched digital services have an advantage over DDS if the flexibility of access to multiple locations is important.

SERVICE COSTS

A key factor in selecting a digital service is the cost of use. Digital services and channels typically cost more than analog, but the difference is steadily narrowing. The additional cost of the digital service, however, is often offset

Service	Data Rates	Switched Access	Configuration	Analog Phone Available	Data Integrity	Availability	Basic Tariff	Use Charge	Typical Applications
PSTN	≤14.4K bps	Yes	Point-to-Point	Yes	Moderate to Very Good	Virtually Universal	Low	Yes—Long Distance	Personal Computer to Personal Computer and Terminal to Mainframe: • Typically Low Data Rate, Low Use, Flexible Connections
Analog Leased Line	≤19.2K bps	No	Point-to-Point or Multipoint	No	Good to Very Good	Generally Available in US	Moderate to High (distance dependent)	No	Personal Computer to Personal Computer and Terminal to Mainframe: • Multipoint Networks • Typically Low Data Rate, High Use, Fixed Connections
DDS	2.4K, 4.8K, 9.6K, and 56K bps	No	Point-to-Point or Multipoint	No	Excellent	Generally Available in Metropolitan Areas	Moderate to High (distance dependent)	No	Mainframe to Mainframe and LAN to LAN: • Typically High Data Rate, High Use, Fixed Connections
Accunet	56K bps	Yes	Point-to-Point	No	Very Good	Available in Most Metropolitan Areas	Low to Moderate	Yes	Mainframe to Mainframe, Workstation to Workstation, and Workstation to Mainframe: • Typically High Data Rate, Low Use, Flexible Connections
CSDC	56K bps	Yes	Point-to-Point	Yes	Very Good	Available in Some Major Metropolitan Areas	Low	Yes	Mainframe to Mainframe, Workstation to Workstation, and Workstation to Mainframe: • Typically High Data Rate, Low Use, Flexible Connections
Datapath	300 bps to 19K bps (asynchronous) and 1.2K bps to 64K bps (synchronous)	Yes	Point-to-Point	No	Very Good	Available in Most Major Metropolitan Areas	Low	Yes	Mainframe to Mainframe, Workstation to Workstation, and Workstation to Mainframe: • Typically High Data Rate, Low Use, Flexible Connections

Exhibit 5-1-2. A Functional Comparison of Tariffed Telephone Line Data Transport Services

by the higher-data rates achieved with these links. Another important cost issue is whether to choose dial-up or leased-line service.

From a cost standpoint, the optimal service depends on use rate. Leased lines are rented at a flat monthly rate, regardless of use, and are preferred by customers who use communications links often. If the links are used sparingly, the dial-up rate paid on each use is lower on a monthly basis than the flat-rate cost of the leased line. Users must evaluate their needs to determine which service best meets their requirements.

SUMMARY

Switched digital services, combining the speed and reliability of digital transmission with the flexibility of the dial-up network, are now a widely available, competitively priced option for providing high-speed, general-purpose data transmission. Significant growth in switched digital services has occurred in recent years, and the service is expected to retain its popularity into the next decade. As with all data communications services, the choices available will continue to evolve, eventually being replaced by the next generation of technology. Until that time, switched digital services are likely to remain one of the most practical and economical means of data transport for many users.

In the future, the switched digital services currently available will be supplemented by similar services, all leading to the eventual implementation of a worldwide ISDN. Until that time, switched digital services are likely to remain the most practical and economical means of data transport for most users.

5-2

High-Speed Services for LAN Interconnection

Nathan J. Muller

IN TODAY'S DISTRIBUTED COMPUTING ENVIRONMENTS, with so much information traversing wide area internets, it is imperative that the right transmission service be selected to ensure the uninterrupted flow of information between interconnected local area networks (LANs). With the types of applications running over LANs, relatively high bandwidth is required for relatively low-duty cycles. The transmissions onto the wide area networks (WAN) consist of bursts of data set at intermittent intervals.

A variety of carrier-provided services are available to support bursty LAN traffic. Understanding the range of choices and their advantages and disadvantages allows LAN managers to select the best facilities for LAN internetworking, perhaps cost-justifying replacement of existing narrowband networks with new technologies that offer a clear migration path to broadband networks. This chapter provides an overview of available carrier-provided transmission services and assesses their suitability for supporting LAN traffic.

Transmission services are typically categorized as either switched or non-switched. These, in turn, may be separated into low-speed (narrowband) and high-speed (broadband) services. Exhibit 5-2-1 summarizes these options.

NARROWBAND SERVICES

The technologies and transmission media employed to implement switched services determine the speed at which data is routed through the public network. For the purposes of this discussion, narrowband services are those that use data rates of DS-2 (6.312M bps) and below.

Analog Dialup and Leased Lines

Conventional switches such as AT&T's 5ESS and Northern Telecom's DMS series, which are in widespread use in local central offices, are currently

Nonswitched	Switched
Analog, 4.8K to 19.2K bps	Dialup with modem, 2,4K to 38.4K bps
Digital Data Services, 2.4K to 56K bps	ISDN, 64 K bps to 1.544M bps
Fractional T1, $N \times$ 64K bps	Packet-switched, 2.4K to 56K bps
T1, 1.544M bps	Virtual Private Networks, 2.4K bps to 1.544M bps
Frame Relay, 1.544M to 44.736M bps	Frame Relay, 1.544M to 44.736M bps
Fractional T3, $N \times$ 1.544M bps	SMDS, 1.544M and 44.736M bps
T3, 44.736M bps	Broadband ISDN, 155M and 600M bps
SONET, 51.84M bps to 2.488G bps	

Note:
FDDI at 100M bps is not included in this chart because it is a private network solution, not a carrier offering.

Exhibit 5-2-1. Summary of Carrier-Provided Communications Services

limited to switching in 64K-bps increments. AT&T is upgrading its switches to accommodate higher speeds, but most are suitable now only for voice and low-speed data. The transmission medium is typically copper wire, but because transmission over copper wire may be affected by various impairments, data transmission over modems is currently limited to 33.6K-bps, or 115.2K-bps when using MNP 5 compression. It is expected that modem speeds of 56K-bps will become common during 1997.

Because line quality may vary by location and even from one moment to the next, many modem manufactures offer a capability known as automatic downspeeding. When line quality deteriorates, a modem operating at 33.6K-bps could downspeed to 28.8K-bps. At the lower transmission rates, the data is less susceptible to corruption. Correspondingly, a 14.4K-bps modem will downspeed to 9.6K-bps. Sensing better line quality, some modems even return to the higher speed automatically. The advantage of automatic downspeeding is that it keeps channels of communication open; the drawback is that much of the time the modem will transmit at speeds below advertised line rates, which curtails cost savings on line charges and prolongs the payback on the hardware investment. To minimize the effects of line impairments on copper-based voicegrade leased lines, extra-cost line conditioning may be requested from the carrier. But there are no performance guarantees with line conditioning; the carrier promises higher-quality lines on a best-effort basis.

The low speed and uncertain line quality of analog dialup and leased lines renders them suspect for carrying LAN traffic. They may be used as a temporary backup for failed digital lines and only when the volume of LAN traffic is very low. They can also be used to communicate with isolated bridges and other interconnection devices for diagnostic purposes when digital links go down. LAN interconnection requires that higher-quality digital offerings be explored, starting with digital data services (DDS).

DDS

DDS, the first digital service for private-line communications, was introduced by AT&T during the mid-1970s. It offers a range of speeds from 2.4K-bps to 56K-bps. Being a digital service, DDS does not require a modem but a digital termination device called a digital service unit that is required at each end of the circuit. The appeal of DDS for data is the higher quality of digital transmission. Until recently, DDS at 56K-bps was a popular way to connect LANs by remote bridges. However, the 56K-bps line rate is a potential bottleneck to LANs that operate at speeds in the 10M bps to 100M bps range.

For this and other reasons, DDS is rapidly being supplanted by newer offerings such as 64K-bps generic digital services and fractional T1. GDS does away with hub architecture—and high cost of DDS—because it can be implemented from any of the local exchange carrier's serving wire centers. The growing popularity of fractional T1, and the fact that AT&T plans to enhance it with management and diagnostic capabilities, has fueled industry speculation that AT&T might be phasing out DDS. Moreover, new 128K-bps digital services are becoming available, implemented by the recently adopted two binary, one quaternary (2B1Q) line code standard.

The 2B1Q line code standard provides 128K-bps of bandwidth over the same pair of wires that now support 56K/64K-bps digital services from the local central office to the customer premises. The standard has been adopted by the American National Standards Institute (ANSI) and major European postal, telephone, and telegraph administrations and others as well, including the European Telephone Standards Institute and the International Telecommunications Union-Telecommunications Standards Sector (ITU-TSS, formerly the CCITT). The 2B1Q scheme is viewed as essential to the success of basic rate ISDN (2B + D).

PACKET-SWITCHED SERVICES

Packet-switching networks thrived during the 1970s and 1980s as a way for asynchronous terminals to access remote computers. The driving force behind the acceptance of packet-switching networks was the ITU-TSS X.25 standards. Packet networks based on the ITU-TSS X.25 standard are generally optimized to run only at 56K-bps, mostly because of the X.25 protocol's overhead burden. This protocol divides the data stream into manageable segments and encapulates them into envelopes referred to as packets. A single data stream may be segmented into many packets, each containing the address, sequence, and error-control information required to allow a network switch to identify the destination of the packet.

Packet switches may be divided into low, midrange, and high-end categories, depending on the number of ports and the packet throughput. These two measures are interrelated. As the number of switch ports increases, the throughput, or capacity, must also increase or overall performance will degrade. Packet switches designed during the 1970s employed single 8-bit microprocessors, which limited them to a throughput maximum of 100 packets per second (pps), usually measured by a packet size of 128 bytes. During the 1980s, the introduction of more powerful 16-bit and 32-bit microprocessors increased the throughput of packet switches to between 300 and 500 pps.

More elaborate multiprocessor design can increase throughput to 1,000 to 10,000 pps, which is required for support of T1 line rates. To increase throughput beyond this value requires the use of technologies that increase the switch processing speed through either hardware or protocol innovation. Frame relay represents a protocol innovation. Such evolution is worthwhile because public networks offer very powerful network management capabilities.

Packet-switched networks employ in-band management with most of the routing information embedded in the individual packets. For a packet-switched network, a management packet is just another type of packet that is routed along the same routes as data. But reading the packet information presents a considerable software processing burden for the nodes. Circuit-switching devices are faster than packet-switching devices because they are hardware based rather than software intensive. As the core microprocessor technology continues to evolve to deliver higher performance, so do the packet-switched devices.

X.25. Public X.25 packet networks offer users convenience and instant worldwide connectivity. In the private networking environment, some LAN interconnect vendors offer a software interface option that enables a single X.25 connection to support both bridged and routed low-volume traffic, instead of requiring separate connections for bridges and routers.

Frame Relay. For the immediate future, however, frame relay holds the most promise. Although equally applicable to public and private networking, frame relay's advantages are quite compelling in the private networking environment:

- It provides mutiplexing with high throughput and can be readily managed.

- It offers an optimized protocol for networking bursts of data—for example, LAN bridging at T1 rates.
- It establishes a standard interface and transport for the many differing data communications requirements of an organization.
- It offers an open architecture for using public-switching services.
- Its implementation is vendor independent.

Frame relay will ultimately provide the benefit of vendor-independent implementation. This is certainly not the case today, because the carriers use either Cisco (formerly StrataCom) or their own proprietary fast packet switches to implement frame relay services. Users must therefore buy frame relay access devices that are certified to work on the selected carrier's frame relay network.

The reasons for using frame relay in the public network are no less compelling:

- It allows user access to public services through ISDN's D, B, and H channels.
- It implements control access through ISDN's D channel at 64K-bps, using Q.931 call-setup procedures.
- It allows existing X.25 packet switches to be upgraded to faster frame relay switches.

Frame relay is discussed in further detail later in this chapter.

Fractional T1

Fractional T1 entails the provision and use of incremental bandwidth between 56K-bps and 768K-bps without having to pay for an entire T1 facility. Instead of paying for a full T1 pipe of 24 DS-0s (at 56K/64K-bps each), fractional T1 allows users to order individual DS-0s or bundles of DS-0s according to the requirements of their applications. Thus, a user may order six contiguous DS-0s, which constitutes a single 384K-bps channel, for the purpose of linking geographically separate LANs over a pair of bridges. This saves the user the expense of a partially filled T1 line while easing the WAN bottleneck between LANs, which was a problem with DDS at 56K-bps.

Because fractional T1 is widely available among interexchange carriers, there is no back-haul problem to contend with, as in DDS. Fractional T1 is an offering of interexchange carriers. To date, among local carriers, only New England Telephone, Ameritech, and PacBell offer tariffed fractional-T1 access (some independent telephone companies offer fractional-T1 access on an individual case basis). For the most part, users must still pay for a full T1 for access, even through they may need only one-quarter of one-half of the bandwidth for LAN traffic.

Apparently, with the falling price of T1, many local telephone companies have come to believe that fractional T1 is unnecessary. For the foreseeable future, this means that the cost savings associated with fractional T1 will accrue only on the long-haul portion of the circuit; the longer the circuit, the greater the cost savings over a full T1.

Integrated Services Digital Network

Primary rate integrated services digital network (ISDN) (23B + D) is another wide area internetworking option. ISDN is a switched digital service that is billed for on a time-and-distance basis, just like an ordinary phone call. However, the 64K-bps B (bearer) channels cannot be strung together the way DS-0s can under fractional T1. Like DDS at 56K-bps, ISDN B channels pose a potential bottleneck for the traffic bursts between LANs.

A number of high-capacity ISDN are channels are included with the primary rate interface, two of which are now available from AT&T: the 384K-bps H0 channel and the 1.536M-bps H11 channel. These channels are more suited for interconnecting LANs.

Tariffs for ISDN services vary from region to region. Some local telephone companies offer flat-rate service with unlimited local calling, while others charge fees according to a combination of time and distance of use. Interexchange services are time and distance sensitive. Data communications managers should carefully evaluate the available pricing options before determining the suitability of ISDN to a particular situation.

Dedicated T1 Lines

T1 digital lines are an excellent medium for interconnecting LANs from point to point. They offer excellent reliability and availability, in addition to high capacity. An increasing number of bridges and routers offer T1 interfaces. Some T1 vendors have integrated the bridge function into their multiplexers.

The advantage of this hybrid arrangement lies with the nature of the traffic. All of the capacity of the T1 line may not be required for LAN traffic. Therefore, voice and low-speed data from multiple sources can be integrated on the same facility for economical long-haul transport. Each type of traffic can even be separately managed. If line impairments threaten to disrupt high-speed data, these channels can be rerouted to other lines having spare capacity, or to bandwidth-on-demand services like ISDN. Channels carrying voice traffic and low-speed data, which are more tolerant of bit error rates that affect high-speed data channels, can remain where they are.

Another advantage of using T1 multiplexers in conjunction with bridges is that it allows the logical integration of voice and data into a single, centrally managed network. This arrangement becomes all the more important as

LANs evolve from workgroup solutions to general-purpose transport vehicles. The higher-order management system of the T1 multiplexer can be used to provide an end-to-end view of the network for performance monitoring, diagnostics, and status reporting. The network elements of both LANs and WANs show up on the same screen of the operator's console to simplify performance monitoring and expedite trouble handling.

Emerging applications are already putting a strain on today's Ethernet and Token Ring LANs, so it should come as no surprise that linking LANs over packet-switched and T1 networks is rapidly becoming inadequate. Other LAN interconnection strategies are called for. Another bandwidth-on-demand service—frame relay—offers an alternative to traditional T1 services. Other alternatives fall into the category of broadband services. There are three transitional service offerings: fractional T3, T3, and switched multimegabit data service (SMDS). They are transitional in the sense that they straddle the narrowband and broadband service categories; that is, they operate in the 1.544M-bps to 45M-bps range and provide the transition to ATM cell switching provided over the emerging synchronous optical network (SONET).

Frame Relay Services

An outgrowth of ISDN, frame relay is a bearer service that has become a high-performance alternative to X.25 packet-switched networks for LAN interconnection. The technical concept behind frame relay is simple: eliminate protocol sensitivity and unnecessary overhead—as well as associated processing at each node—to speed up network throughput. Error correction and flow control already exist at the upper layers of most modern computer communications protocol stacks and thus may be relegated to the edges of the network rather than performed as every node along a path.

Virtually all carriers offer frame relay services, including AT&T, MCI, US Sprint, British Telecom, and Worldcom (formerly WilTel). These networks offer one standard high-speed interface. Initially, frame relay networks will use permanent virtual circuits for connecting a source and destination. The routes are defined in fixed tables located in the network switch, rather than established during a call-setup phase by reading the frame address. Eventually there will be switched frame relay calls, using switched virtual circuits. The pricing strategies for both permanent and switched virtual frame relay differ among the carriers. Although some carriers have published their pricing schedules, flexibility is often achieved through negotiation on an individual basis.

With frame relay, available bandwidth is used in the most efficient manner possible. A frame relay application will momentarily seize the entire amount of available bandwidth to transmit information in bursts. Upon completion of a duty cycle, it relinquishes the bandwidth to other voice and data applications.

A management issue with frame relay networks is congestion control. At any given time, several users might want to access the full transmission bandwidth simultaneously. The network must have a means to detect such situations and initiate procedures to prevent an overload condition. Although congestion control is an integral function of X.25 and LANs, the problem is usually more pronounced in frame relay networks. This is because the speed of the access links may be much higher than the frame relay backbone. A LAN operating at 10M bps, for instance, can easily overwhelm a frame relay link operating at the T1 speed of 1.544M bps.

Typically, frame relay carriers manage the congestion issue by establishing for each subscriber a committed information rate (CIR), measured in bits per second. By establishing the CIR for a subscriber, the carrier guarantees to route traffic that is submitted at or below that rate. Above that rate, the carrier will accept traffic for transmissions, but will set the discard eligibility (DE) bit on the excess frames, indicating to the network that the frames may be discarded if needed, to alleviate congestion problems. The CIR guarantee is a contractual guarantee, rather than a technical guarantee. It is up to the carrier to ensure that sufficient bandwidth is in place to assure delivery of the aggregate CIR of its subscriber base.

Currently, the alternatives to frame relay include private digital connections, leased lines, dialup modem links, and other packet services. Of these, private digital links are relatively expensive, and the others are often too slow and prone to creating bottlenecks.

Virtual Private Networks

In recent years, virtual private networks (VPNs) have become a workable method for obtaining private network functions. Under the VPN concept, an organization's PBXs are linked together over the public network. The intelligence embedded in the virtual network—at the carriers' serving offices—provides digit translation and other conversion services. The carrier assumes the responsibility for translating digits from a customer-specific numbering plan to and from the carrier's own numbering plan. All routing and any failures are transparent to the customer and consequently to each individual user on the network. PBXs are connected to the carrier's point of presence (POP) through various local access arrangements. The private network exists as a virtual entity on the carrier's backbone network.

Until recently, virtual private networks could handle only voice and low-speed data traffic. New high-speed data capabilities are available that are suited to such applications as LAN interconnection and disaster recovery. Switched 384K-bps and switched 1536 bps are two of the services that are being phased in by AT&T, for example, under its software-defined data network (SDDN). These services offer user the performance of private lines with a bit error rate of 10^{-6}, which equates to 95% error-free seconds on 95% of premises-to-premises calls, 95% error-free seconds on 99% of POP-to-POP calls, and 95% error-free seconds on 98% of premises-to-POP calls.

Fractional T3

As LAN interconnectivity fuels the need for additional bandwidth, an increasing number of users are asking: What is the right solution for connecting LANs at speeds greater than T1 but less than T3? Some vendors are trying to capitalize on the appeal of fractional T1 by extending the fractional concept to T3. Although fractional T3 might constitute one solution, it is by no means the most efficient or economical. In fact, it is very unlikely that fractional T3 with the contiguous bandwidth characteristics of fractional T1 will be provided within asynchronous T3 infrastructure.

To obtain an analogous service using DS-3 signaling, which would offer bandwidth in $N\times$DS-1 increments, the DS-3 signal would have to ensure frame integrity between individual DS-1s. Although this is possible, it is highly inefficient because of the asynchronous multiplexing structure of DS-3.

In constructing the DS-3 frame, DS-1 goes through two stages of multiplexing with bit stuffing and destuffing at each stage. The first stage combines four DS-1s into a DS-2 (6.312M bps), and the second stage combines seven DS-2s into the DS-3 frame, which equals 28 DS-1s. The DS-1s are not synchronized within DS-3, have differing frame alignment, and do not allow frame integrity that would be required for a fractional T3 offering. Thus, the only practical fractional T3 bandwidth increment afforded by DS-3, other than DS-1, is the 6.312M-bps DS-2 intermediate rate.

The migration to SONET will change this situation and truly open up incremental bandwidth beyond DS-1 (T1). Being a synchronous network, SONET will enforce both byte and frame integrity, which can easily lead to new $N\times$DS-0 and $N\times$DS-1 services. It would be far better to look toward fractional SONET services that can be deployed today than toward fractional T3. As SONET evolves it is conceivable that a fractional version of the first level of SONET service (fractional OC-1) could offer services with bandwidths of up to $28\times$DS-1 or $672\times$DS-0.

SONET is the foundation for broadband ISDN with access lines in the future operating at 155.52M bps (OC-3) or 622.08M bps (C-12). Both ANSI and ITU-TSS standards for broadband ISDN require that the broadband ISDN structure be capable of providing bandwidth in increments of any 64K-bps (DS-0) integer. This sets the stage for fractional services up to $2{,}016\times$DS-0 (C-3) and up to $8{,}064\times$DS-0 (OC-12).

Current discussions of fractional T3 among service providers acknowledge the fact that these are really DS-1 channelized services implemented through M13 multiplexers. AT&T provides this services as an option for its T45 service, which includes an element called M28, an M13 offering. This option is available under AT&T's FCC Number 11 for individual case basis T45.

Another example of this M13-type service is the multiplexing option found in New York Telephone's T3 individual case basis, per FCC Tariff Number 41. The M13 option, however, has a number of drawbacks that prevent it from implementing fractional T3 services that are truly analogous to fractional T1. Among them, M13 cannot be remotely controlled to permit customers to add bandwidth incrementally; this must be done manually by technicians who rearrange patch panels, a process that can take weeks to implement.

One implementation of fractional T3 entails breaking down the local access DS-3 circuit into DS-1 circuits for routing over the service provider's network (POP-to-POP). However, there is confusion about the way to access these T1 channelized services. All T3 channelized T1 services require strict compliance with the M13 format. Some T3 products are not compatible with M13, such as those that adhere to the Syntran structure or use an incompatible proprietary multiplexing scheme for T3. Today, users must choose either M13 or proprietary T3, which are mutually exclusive.

Most service providers do not consider M13-type options to be true implementations of fractional T3. They are looking instead at 1/3 crossconnects and SONET in pursuit of true fractional services above T1. With SONET products currently being deployed and being introduced by more vendors, the next step beyond fractional T1 is not fractional T3, but fractional SONET.

BROADBAND SERVICES

Broadband networks are analogous to multilane highways. Large bandwidth applications require them, but the need may be only temporary and smaller roads may suffice most of the time. The prospect of bandwidth on demand is being viewed with increasing interest among LAN users as well as carriers. An intelligent broadband network that spans public and private domains would be capable of managing multiple high-capacity LAN channels plus video and other bandwidth-intensive applications.

T3 Service

T3 service is offered at the DS-3 rate of 44.736M bps, typically over fiber facilities. The applications touted by T3 advocates include:

- LAN interconnection.

- Multiple T1 line replacement.
- High-speed backbones integrating voice, data, video, and image.

However, T3 requires the use of proprietary optical interfaces and so entails the construction of special individual access lines from the customer premises to the carrier's serving office. Special construction costs at each end differ widely from region to region, from a low of about $8,000 to a high of about $150,000. Because these costs are almost never factored into the crossover comparisons with T1, the true cost of T3 is so high that it is difficult for even the largest companies to justify it.

The majority of T3 devices sold are M13 multiplexers. The M13 is a simple T1 concentrator, which lacks network management capabilities. Although DS-3 is a standard electrical interface that transports DS-1, the optical interface is proprietary. Also, many M13s lack support for such features as binary eight zero substitution (B8ZS), used for clear channel transmission, and extended super frame, used for maintenance, essentially because these formats were developed after the introduction of DS-3/M13. With DS-3, bipolar violations of DS-1 are always lost. Further, DS-3 was rate-adapted long before network management (over T1 networking multiplexers) became widely used. Consequently, DS-3 services generally do not provide the network management features T1 users have come to expect. This situation is changing through the implementation of C-bit parity on T3 facilities. This alternative T3 framing format is roughly equivalent to T1's extended superframe format.

C-bit parity can help ensure reliable service through end-to-end performance monitoring, remote maintenance and control, performance history, error detection, and trouble notification. Most new T3 equipment supports both C-bit parity and M13. Users of M13s must add new interface cards and reconfigure their software to work with the new framing format. AT&T, MCI, and Worldcom are among the carriers that support C-bit parity.

DS-3 is often transmitted over fiber, which requires an interface for electrical-to-optical signal conversion. The lack of optical standards for DS-3 has led to a proliferation of proprietary interfaces, which restrict the ability of users to mix and match different manufacturers' equipment end-to-end. T3 services, like AT&T's T45, require the customer to negotiate the type of optical interfaces to be placed in the interexchange carrier's serving offices. In contrast to a ubiquitous service such as DS-1, T45 requires expensive special construction for the local channel, from customer premises equipment to POP.

Ideally, it would be preferable for large organizations to use transmission facilities operating at the 45M-bps rate or higher, rather than to lease multiple T1 lines to obtain the equivalent bandwidth capacity. After all, AT&T's T45 service, which provides about 45M bps of bandwidth between its serving offices, may be cost-justified at five to eight T1 lines, depending on distance.

Because the FCC has ruled against the use of individual case contracts, T3 access connections must be tariffed like any other service. Preliminary tariff filings indicate that it will now take 8 to 15—and perhaps more—T1 lines to cost-justify the move to T3. The individual case contracts require that users commit to the service for three to five years, during which they could lock themselves out of upgrades to SONET.

Users who cannot afford to wait for SONET can opt for T3 now but should not sign long-term contracts with carriers, unless they are prepared to pay hefty penalties to terminate service in favor of SONET.

New York Telephone, for example, charges $12,192 a month for a T3 access line but offers discounts that range from 10% to 30%, depending on the length of service commitment. At a commitment level of 3 years, the user qualifies for a 10% discount; at 5 years, 20%; and at 7 years, 30%. These service discount plans do not protect the user from rate increases; they lock in only the discount percentage. If the user terminates the service before the end of the selected commitment period, New York Telephone charges 50% of the applicable monthly rate for each month the service is disconnected before the end of the discount term.

SONET

SONET access with OC-1 customer premises equipment eliminates the cost of fiber-to-electronics multiplexing from the custom construction costs of T3. This is similar to the migration of T1 multiplexing from within the central office to customer premises equipment on T1 private networks. Incidentally, T1 does not drastically change under SONET standards but remains an entry level to SONET, which is defined identically to today's T1/DS-1, per D4 or extended superframe structures.

SONET is a worldwide standard that includes the fiber-optic interface SONET transmission begins at a rate of 51.84M bps and reaches 2.488G bps with provision to go to 13G bps. Most SONET standard work has been completed, including standardization of the basic transmission formats and network maintenance and protection, as well as internetworking between different vendors' equipment through midspan meets on the public network.

The enormous amounts of bandwidth available with SONET and its powerful management capabilities will permit carriers to create global intelligent networks that will spawn bandwidth-on-demand services of the kind needed to support new LAN applications, including three-dimensional computer-aided design (CAD), CAD conferencing, high-resolution imaging, virtual reality stimulation, and videoconferencing. All of these require the transfer of large blocks of data in bursts.

Such a transport network, which can be used as simply as the current voice telephone network, will emerge from the electrooptical SONET standards that have been winding their way through North American and ITU-TSS standards groups since 1985.

SONET offers advantages over today's WAN solutions that are too competitive to ignore. For example, because T3 is proprietary, it limits product selection and configuration flexibility while increasing dependence on a single vendor. SONET standards change all that, offering seamless interconnectivity among carriers. Under SONET, bandwidth may be managed to the DS-0 level to provide maximum control and support of bandwidth-on-demand services, regardless of carrier. SONET also provides central control of all network elements and permits sophisticated self-diagnostics and fault analysis to be performed in real time, making it possible to identify problems before they disrupt service. Intelligent network elements automatically restore service in the event of failure—a vital capability for large, complex LAN internets.

Eventually, SONET will dominate the transport and switching fabrics of the public network, and will support a wide range of services, including SMDS and broadband ISDN. SONET can even transport the fiber distributed data interface (FDDI) and metropolitan area networks (MANs) to extend the reach of high-speed LANs.

SMDS

SMDS is a public high-speed packet-switched transport service for connecting LANs, host computers, image data bases, and other high-speed applications. SMDS is ideal for bursty data applications, including the interconnection of geographically dispersed LANs, and is capable of supporting such applications as video and imaging.

The Bellcore SMDS standards were finalized in late 1989. To the original standard was added 802.6, a MAN standard approved by the Institute of Electrical and Electronic Engineers (IEEE) after eight years of research and development. The Bellcore standard specifies how customer premises equipment can access an SMDS switch using twisted-pair wiring at the T1 rate or optical fiber at the T3 rate. The IEEE standard incorporates the connectionless data part of the distributed queue dial bus architecture for media access control. This allows up to 16 user devices in a LAN arrangement to share a single access link to an SMDS switch.

SMDS employs a packet-switched, dual counter-rotating ring architecture. The two rings each transmit data in only one direction. If one ring fails, the other handles packet transmission. This protection mechanism safeguards users' data against loss due to a network fault.

Although SMDS shows promise as a LAN interconnection service, it is important to note that current standards have yet to specify network management and billing interfaces. Bellcore is expected to address billing interfaces, and the IEEE 802.6 Committee is expected to broaden its MAN standard to address such areas as network management, high-speed interfaces (e.g., T1, T3, and SONET), and isochronous service. Isochronous service would require that cell-relay switches, which form the switching fabric for SMDS, be expanded to switch voice and video traffic in addition to data.

Broadband ISDN and Asynchronous Transfer Mode

By the time narrowband ISDN emerged from the laboratory during the late 1980s and began to be deployed in the real-time business environment, it was already apparent that ISDN's B (64K-bps) channels would be inadequate to support emerging high-speed data applications. Specifically, users wanted to interconnect LANs and achieve high-resolution images and video—all of which requires considerably more bandwidth. Consequently, ISDN came under increasing attack as being too little, too late.

Standards bodies recognized early on, however, that the requirement for high-capacity channels within the framework of ISDN could be satisfied sometime in the future under a vaguely defined concept called broadband ISDN. Broadband ISDN increases the channel capacity to 600M bps and beyond for high-capacity connectivity and bandwidth on demand. But broadband ISDN requires a transport network of tremendous capacity. With no such network in place at that time, interest in broadband ISDN waned in favor of other broadband technologies and services, such as FDDI and SMDS, which were projected for near-term deployment.

Separately, but at about the same time, SONET was conceived to provide carriers with a standard optical transmission medium for interoffice trunking. Later, in consideration of private network interest in T3 transmission, the SONET concept was extended to include the local loop. More recently, the concept has made its way into the customer premises.

With the advent of SONET, it became apparent that the standardized, high-capacity intelligent transport medium offered by SONET provided the infrastructure that would make broadband ISDN a reality much sooner than originally thought. Thus, originally conceived and developed for different reasons, SONET and broadband ISDN subsequently became interrelated through the work of standards bodies with the idea of creating advanced voice, data, and video networks.

In support of these concepts, the technology now known as Asynchronous Transfer Mode (ATM), based on the use of cell-relay switching, emerged. ATM implements the services of broadband ISDN over SONET-compatible transport media. ATM is being deployed more slowly than some experts predicted, but is making steady inroads nonetheless.

Because the development of ATM and cell relay switching parallels that of the SONET and broadband ISDN standards, many design aspects have been developed to accommodate high capacity services on a worldwide basis.

Taking advantage of the widespread availability and capacity of single-mode optical fiber, of which more than 2 million miles have been installed in the US alone, this technology brings the ability to reliably transport enormous quantities of information. ATM services are available at speeds ranging from 45M bps to 622M bps, using cell-switching technology. Not coincidentally, these rates correspond to speeds that have been standardized for SONET fiber-optic systems.

With this much capacity, ATM efficiently carries a multitude of high-speed services that can support such applications as LAN interconnection, videoconferencing, and imaging, as well as bursts of asynchronous data and voice. Although most services will require only a portion of the bandwidth, others may take the entire channel.

Broadband ISDN standards are introducing such maintenance features as trouble indicators, fault locators, and performance monitors. These will interface with any variety of customer equipment—PBX, LAN, T1 multiplexer—and provide transport to the central office where a SONET cross connect or switch will further concentrate or segment the information. The network topology between central offices is most likely to be star or mesh, while the topology between the customer premises and central office could range from star, ring, or point-to-point.

The Internet

Commercialization of the Internet has added new choices for LAN interconnection. From practically any given location, an organization can establish a connection to the Internet using a variety of interconnection speeds and technologies. Numerous Internet service providers (ISPs) offer complete LAN-to-LAN interconnection services as bundled solutions. Yet, even without such services, it is possible to establish LAN-to-LAN connectivity across the Internet.

Typical configuration scenarios involve the establishment of dedicated Internet connections at each location that has a LAN. These connections are made locally using a variety of the services and technologies described in previous sections of this chapter. Low-end configurations are often serviced with 56K-bps digital circuits or with ISDN connections operating at either 64K-bps or 128K-bps. (Note that ISDN connections used in this manner are generally cost-effective only in those areas with unlimited local ISDN calling.) Moderate amounts of traffic are effectively serviced with T1 connections, while high-end users require options such as multiple T1s or even T3 connections. The use of ATM enhances the choices even further. For instance, some ISPs offer 10M-bps service by establishing an ATM point of presence on the customer premises and providing that customer with a 10Base-T connection to the Internet.

Despite the range of high-end solutions available, even the smallest organizations can benefit from use of the Internet to interconnect their sites. For instance, a company with many small offices scattered around the globe can cost-effectively provide interoffice electronic mail with as little as dial-up Internet connections from each site.

Regardless whether the Internet-based LAN-to-LAN connection is established as part of a bundled ISP service offering or is custom built by the organization, security considerations are of paramount importance. Two aspects of security should be considered in this equation.

First, all traffic traversing the Internet should be considered vulnerable unless steps are taken to encrypt the data. Second, most organizations with dedicated Internet connections implement firewalls to protect their internal networks from unauthorized access. If using the Internet to establish connectivity between protected internal LANs, a careful analysis of the organization's firewall policies and implementation practices is imperative.

Solutions to both these challenges are available, varying widely in cost and complexity, depending on the specific needs of the organization. Thus, it is possible for most organizations to implement adequate safeguards for this type of connection. Despite the added cost of employing such safeguards, use of the Internet in this fashion can be a very attractive, cost-effective alternative.

SUMMARY

With computing resources becoming increasingly distributed over vast distances, the future portends a buildup of LAN traffic between corporate locations. The means to interconnect LANs is provided by myriad current and emerging carrier services that, together, constitute the wide area network. When choosing from among the various transmission services, consideration must be given to the quality and speed of the links for supporting bursty LAN traffic. Choosing a low-speed service limited to 56K/64K-bps may not only create bottlenecks between LANs but entail costly upgrades if the service cannot handle incremental increases in LAN traffic. For this reason, such bandwidth-on-demand services as frame relay, ISDN, and SMDS that run over T1 and T3 links may be more appropriate.

Another factor to consider is whether the organization's needs for LAN interconnection are more effectively met by carrier-provided services via the Internet or by existing private networks. The number of choices and the variety of pricing options are in a constant state of flux, and thus require substantial analysis on the part of the data communications manager. Understanding the available choices and the advantages and disadvantages of each service will enable LAN managers to select the best facilities for LAN internetworking over the wide area network in terms of cost and efficiency as well as reliability and availability.

5-3
A Management Briefing on Virtual Private Networks

Nathan J. Muller

DURING THE PAST 10 YEARS, organizations have eliminated 25% of middle management, resulting in a flatter, leaner bureaucratic structure. With more decision-making and transaction authority now delegated to the lowest common point in the organization—the individual employee—organizations are more effectively positioned to respond to the needs of customers and to exploit emerging business opportunities. Along with changes in organizational structure have come vast changes in the distribution of corporate resources and in the way networks are being set up and managed.

The changes in corporate structure reveal the emergence of several forces shaping the corporate networking landscape in the 1990s:

- Increasing reliance on voice processing, voice messaging, and call distribution technologies to support daily business operations.
- Installation of considerable processing power on the desktops of field, branch, and other corporate locations.
- Increasing use of local area networks (LANs) for information and resource sharing at both the field and corporate levels.
- Use of distributed processing architectures to synthesize field and corporate processing into a coherent computing system.
- Merging of voice calling with relational data bases through automatic number identification (ANI) to serve customers faster and more efficiently.
- Use of wide area networks (WANs) to link all parts of the corporation—as well as customers, suppliers, and strategic partners—over long distances.

The cumulative effect of these trends has been to make networking more complex. Faced with a barrage of new services and features, constantly changing tariffs, and new technologies, many organizations are seeking to offload network management responsibilities so they can devote more attention to core business activities. As the new corporate information technology infrastructure becomes more communications intensive, and as communications costs become one of the largest line items on corporate budgets, organizations are more inclined than ever to experiment with alternative ways of implementing networks to minimize costs. One of these alternatives is the use of virtual private networks (VPNs), which rely more on the facilities and features of the public network than on a private network of dedicated leased lines that are typically managed by in-house technical staff.

For these and other reasons, VPN revenues are expected to grow at a 60% annual rate from the $800 million in sales recorded in 1990. The growth of VPNs is largely fueled by organizations that are less willing to make major capital investments in hardware and training for internal networks but still want the high-performance, fixed-cost benefits of private networks.

A sluggish US economy in recent years has caused a dramatic rise in the growth of equipment leasing and a surging demand for virtual private networks; corporations were wrestling with ways to stay afloat without jeopardizing their long-term competitive positions. Restructuring operations, laying off staff, and drastically cutting prices are some of the ways organizations are attempting to ride out the consumer-driven, demand-oriented recession. Leasing equipment and opting for VPN services is seen as a way to save cash and preserve credit lines, because these costs are not reported on corporate balance sheets as liabilities. Other reasons for the explosive growth of VPNs are related to the increasing complexity of network design, continued corporate expansion (i.e., mergers and acquisitions), and organizations' greater need to interconnect with international operations.

Leading VPN providers include AT&T, MCI Communications Corp., and Sprint Communications Corp. Value-added network (VAN) providers are also entering the market for virtual private networks. Unlike the Big Three carriers, however, the VANs specialize in providing data services through X.25 and frame relay facilities. The major VANs are B1 Concert and Infonet, which together have garnered 50% of the market for VAN services.

THE CARRIER PERSPECTIVE

During the past decade, the public telephone network has also undergone profound change. With its digital switching, stored program control, fiber-optic connections, and the high-speed signaling capability provided by Signaling System No. 7 (SS7), the public network is looking more like a vast distributed computing system capable of providing an incredible array of value-added services. Because their networks are now viewed as computer systems, AT&T and its competitors, MCI and Sprint, want to use them as such. One of the ways in which these carriers can leverage their increasingly intelligent networks is by offering a package of services that mimic private networks. The result is virtual private networks, or VPNs.

Such services offer clear benefits to users, mainly in pricing and in features comparable to older electronic tandem networks (ETNs) and Customer Controlled Switched Access (CCSA) networks. However, there are other, less-visible benefits afforded by the virtual private networks:

- They allow the carriers to route more traffic over existing facilities, most of which are underused.
- They enable carriers to offer volume discounts to large customers, luring them away from less profitable, fixed-cost private facilities.
- They provide a platform for more sophisticated services; in essence, they are a competitive tool in a highly competitive business.
- They offer an upgrade path for private network users seeking expansion without incurring capital costs. This permits the carriers to regain account control.

By encouraging customers to use shared facilities, the carriers themselves benefit from the efficiencies and economies of their networks; simultaneously, they regain account control. Volume pricing plans, for instance, are quite aggressive if customers are willing to commit to four years of service. AT&T offers volume pricing discounts of 30% and more on its rates for Software Defined Network (SDN) service; these rates are already considerably lower than direct distance dialing (DDD) rates. If dedicated access lines are used from the customer premises to an SDN serving office, the customer may realize additional savings.

HOW VPNs WORK

During the mid-1970s, AT&T introduced a new concept in wide-area voice networking, the Electronic Tandem Network. The ETN employed software to define various aspects of the network, offering customers such capabilities as:

- *Unified numbering plan.* The same 7- or 10-digit number can be used to reach a person from anywhere on the network, regardless of where the caller may be located on the network.
- *Automatic alternate routing.* A predefined, alternate path is selected automatically, if the most direct path is not available to route the call.
- *Overflow routing.* Calls will be routed over the facilities of the public network, if they cannot be routed over the corporate network for any reason.
- *Traveling class mark.* The class of service assigned to each station will be passed from the originating private branch exchange (PBX) to the remote

PBX to ensure that specified calling privileges are maintained, regardless of where the call terminates on the network.

The ETN concept was later extended to that of virtual private networks (Exhibit 5-3-1), which offer more access options, features and functions, and interconnectivity with other services and networks. Some of the advantages of using VPNs include:

- Consolidated billing, with only one bill for the entire network.
- Ability to have the carrier monitor the network and reroute around failures and points of congestion.
- Reliance on the carrier for network performance, maintenance, and management reduces the requirement for high-priced technical personnel.

AT&T's SOFTWARE DEFINED NETWORK

The SDN architecture consists of several service elements linked through Signaling System No. 7. Users access the network from an AT&T serving office. The action points provide connectivity through the facilities and switches of AT&T's long-distance network. The network control points store customized call processing instructions to control customer-specific SDN/SDDN (Software Defined Data Network) features, and interact with the network services complex to collect user information. The communications manager provides users with access to management and control functions, using the service management system. The SDN control center provides complete performance and reliability assurance.

Although AT&T originally intended to sufficiently lower the cost of its Software Defined Network to discourage the growth of private networks, it has not yet done so. Most companies have preferred not to give up their T1 backbone networks at any cost. Instead, most users of virtual networks view the service as complementary to private networks already in place—for example, to link low-volume sites to the T1 backbone. But virtual networks are rapidly being endowed with the intelligence to finally live up to their potential as a practical alternative to private networks. The result will be more sophisticated call processing and network management capabilities that allow users to truly customize their own networks as if they were private.

AT&T POP. In addition, the intelligence embedded in the virtual network—at the carriers' serving offices—gives users more flexibility in choosing PBXs from different manufacturers. With AT&T's Software Defined Network, for example, PBXs are connected to an AT&T point of pres-

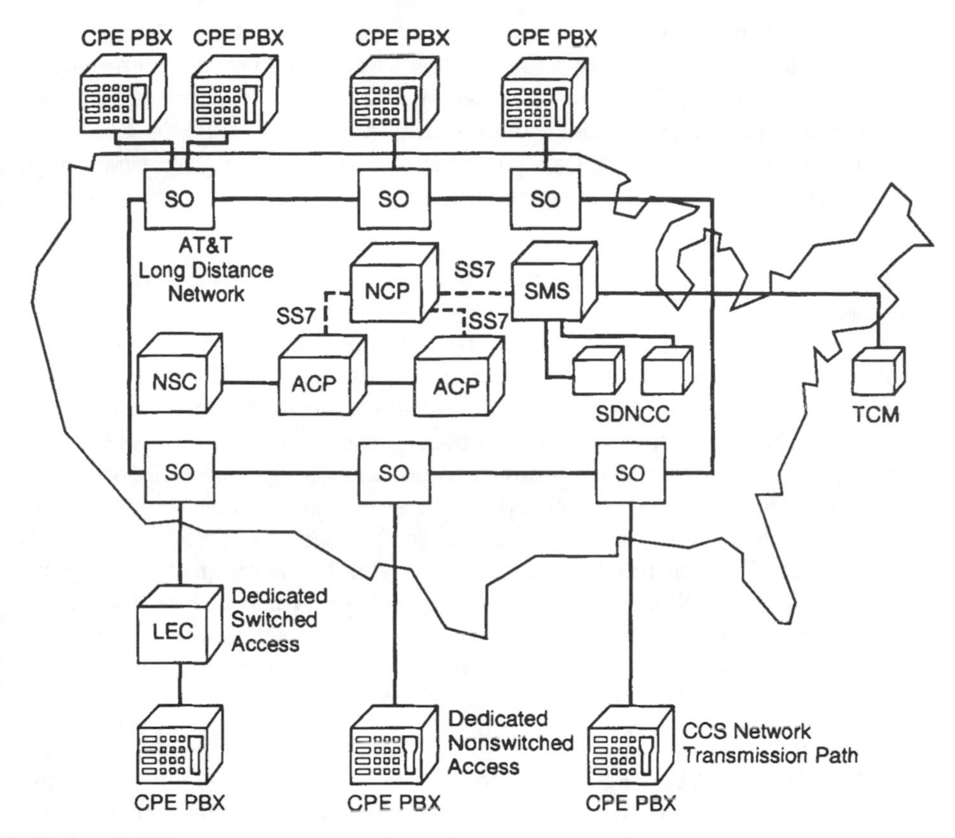

SOURCE: AT&T

Notes:

ACP	Action point
CCS	Custom control signalling
CPE	Customer premises equipment
LEC	Local exchange carrier
NCP	Network control point
NSC	Network services complex
SDNCC	Software Defined Network control center
SMS	Service management system
SO	Serving office
TCM	Telecommunications manager

Exhibit 5-3-1. AT&T's Software-Defined Network

ence (POP) through various local access arrangements. The "private" network exists as a virtual entity on the carrier's backbone network. The carrier assumes the responsibility for translating digits from a customer-specific numbering plan to the carrier's own numbering plan, and vice versa. All routing and any failures are transparent to the customer and, consequently, to each individual user on the network.

ESMS. SDN even provides communications managers with various management and reporting capabilities. For example, with AT&T's Expanded Service Management System (ESMS), the communications manager can control various aspects of the SDN service. This is accomplished with an on-premises terminal and a 9.6K-bps private line connection to AT&T's SDN centralized network information data base, which resides on an AT&T 3B2 minicomputer. There, the communications manager can access information about configuration, use, and equipment status. This enables the user to do such things as set up, change, and delete authorization codes; authorize use of such capabilities as international dialing by caller, work group, or department; and activate SDN's flexible routing feature, which allows the user to direct traffic from one SDN site to another. This last feature would, for example, allow calls to an East Coast sales office to be answered by the West Coast sales office when the East Coast office closes for the day.

NIM. Through AT&T's Network Information and Management (NIM) arrangement, users can obtain SDN reports containing call detail and network use summaries. These reports are similar to those available through PBX and Centrex data collection and reporting systems. The reports help SDN users track costs and bill departments for SDN costs, identify network traffic trends, and review network performance. In addition, users can use the NIM arrangement to request access line status and schedule transmission tests.

Other Customer Benefits

For many companies, virtual private networks offer significant benefits, though these benefits are mostly contingent on geographical proximity to VPN serving offices, traffic characteristics, and enterprise demographics. For example, an organization having much traffic that can qualify as "on-net" from end to end can realize significant savings. Although it will require special engineering studies and significant tariff analyses to design a virtual private network and access to it, the cost and effort can certainly be worthwhile.

VPNs are continually growing in sophistication, offering users such capabilities and features as:

- Different levels of customer network management and reconfiguration.
- The ability to divide a single software defined network into subnetworks, and to manage and obtain billing separately for each.
- Extensive routing capabilities (essentially the routing capabilities of the public switched network) that enables users to bypass offices disabled by emergencies.
- International extensions.
- The ability to switch bandwidth at 384K-bps and 1.536M-bps (and even 44.736M-bps through cross-connect systems) for such applications as videoconferencing and LAN interconnection.
- Sub-20-second network restoration for data, which prevents the automatic time-out of host sessions.
- Private network interface.
- Partitioned data base management.
- Network remote access.
- Multilocation billing.
- Primary rate integration services digital network (ISDN) features and support.

Some organizations searching for ways to reap additional savings on long-distance calling have decided that all calls should be placed over the internal communications system. Branch offices, for example, that rely on the corporate 800 service number to control costs, are, in effect, pushing overall telecommunications costs up. In such situations, limiting all calling to the virtual private network can trim corporate costs 21% to 38%, compared with using the 800 line. An intelligent PBX can be used to route calls in a cost-effective manner. Calls not accepted by the private network are automatically rerouted to the virtual network, where their completion to "off-net" locations is ensured through public facilities.

As more employees engage in off-site work arrangements (e.g., telecommuting), large organizations are expanding their virtual networks into their employees' homes in an effort to reduce communications costs and extend employee work hours. The strategy provides the added advantage of generating more network traffic that, int urn, leads to greater volume discounts for the organization.

INTERNATIONAL VPN SERVICES

A virtual private network gives users many of the benefits available with dedicated private networks. However, until recently only very large companies (i.e., the Fortune 50) could justify the cost of international dedicated networks. Now, even midsized companies (i.e., the Global 1,000 and others) with calling volumes of more than two hours a day can take advantage of communications costs.

Global VPN services allow businesses to design their own private seven-digit dialing plan for international calls within the network. This arrangement features fast call-set-up times, provides high-quality transmission, and offers greater management over international communications. In addition, international VPNs offer the connectivity and interoperability potential of public network services. This is particularly useful for facsimile transmissions and videoconferencing. And, in supporting X.400—the standard for universal message exchange—users of European public electronic mail (E-mail) networks can interconnect directly with the E-mail networks of US service providers.

Businesses that use domestic VPN services can extend the reach of their networks overseas, and do so very easily. For instance, one-stop shopping plans allow clients with heavy international traffic needs in either the US or in international locations to ask either carrier to arrange for and provide dedicated communications services in both countries. This service, designed to streamline network procurement and management, also provides customers with the option of single-end ordering. This allows the customer to be billed in one currency, whether it contacts the US carrier or the participating post, telephone, and telegraph (PTT) administration.

US customers can directly access the services of select PTTs through a US international 800 number. The toll-free number lets callers in the US access a PTT operator and place collect calls to the international location. At the same time, users in international locations can access a VPN operator in the US to call collect or to charge calls to their VPN or local exchange company calling cards.

VPN interconnection arrangements with foreign carriers and PTTs are becoming ever more sophisticated. One such arrangement concerns Sprint and the Canadian carrier, Unitel. Through a technological and business partnership unique in international telephony, Sprint and Unitel will extend the compatibility of their networks beyond the simple interchange of transmission facilities to comprehensive joint network planning. The partnership will jointly develop and implement such network operations as dial administration, signaling, operations methodology, and order procedures.

Sprint and Unitel are accepting full responsibility for service within and between the two countries. This solves problems multinational customers have encountered with traditional bilateral private line service, in which carriers are responsible only for their half of the transmission facility and require customers to deal with two or more entities on the international portion of the link.

Controlled by computer software in Sprint's network, the VPN service allows communications traffic to be carried over shared transmission facilities, providing many of the same services available through dedicated leased channels without the cost of special communications equipment or expensive facility leases. The service includes operationally advanced features such as seven-digit dialing, which is compatible with uniform domestic network numbering for overseas calls.

The expansion of US VPN service providers into the international arena represents a significant step in their global networking plan. Their objective is to interconnect with the wide variety of switching systems in use by the foreign communications companies. Availability of international virtual private network service, or of features within it, will not depend on a single manufacturer or switching system. In pursuing an open international architecture for VPNs, US service providers will give customers a wide range of network options as their businesses become global.

Currently, the network use charges for international VPN services are almost identical among AT&T, MCI, and Sprint. The offerings differ by the number of countries served and the features supported. International VPNs will continually evolve and expand into more countries. Development of new features will depend on advancing technology, both by domestic and foreign communications organizations. Over time, users will see increasing sophistication and more feature-rich services adhering to international standards.

Organization that are now international or that may become so in the foreseeable future should investigate the many advantages offered by international virtual private networks. In today's global markets, the VPN is becoming an essential business tool that can no longer be dismissed lightly.

VPN FOR SPECIALIZED MARKETS

VPN service providers are also looking at ways to offer custom service packages to niche markets. Sprint, for example, has entered the health care marketplace with a comprehensive package of communications services called HANDS (Healthcare Application Network Delivery System) for physicians, hospitals, clinics, and insurers. HANDS includes WATS lines, 800 number services, dial-up video transmission, and an efficient billing system that allows control of costs. Such customers as hospitals and clinics can use the HANDS services themselves, as well as provide direct phone services to patients.

One feature of HANDS is dial-up access to the Sprint Meeting Channel's worldwide videoconferencing network. Sprint's dial-up videoconferencing enables health care professionals to meet face-to-face electronically, using standard voice lines and without leaving their respective cites. Videoconferencing can even be implemented within the operating room, to allow surgeons to converse with peers and experts in remote locations.

Previously, VPN users could connect to videoconferencing facilities only within their own network; additional sites could communicate only through an audio bridge. Sprint claims that VPN videoconferencing costs can be cut by 25%, compared with the costs for non-VPN customers, because VPN users access their own lines to reach the VPN/Meeting Channel Gateway.

A unique aspect of Sprint's HANDS package is the near-instantaneous transmission of medical images anywhere in the country. Physicians can transmit diagnostic quality images from CAT scans, magnetic resonance imaging (MRI) equipment, X-ray machines, and sonograms in seconds, compared with the hours and even days required by other methods of electronic transmission or air-express shipments.

VIRTUAL NETWORKS FOR DATA

Encouraged by their success with virtual private voice networks, some carriers are now offering virtual private data networks, which manage connections between terminals and hosts, administer security, reroute to alternate hosts, and protect connections. AT&T was the first to introduce such a data service, with its Software Defined Data Network, starting with Accunet Switched 56/64K Service and then Switched 384K Service.

AT&T tariffed a switched 1536K-bps digital data service that allows users to dial up T1 bandwidth. This service is ideal for such applications as bulk file transfers, videoconferencing, and imaging. The service also provides an economical alternative to redundant leased lines for backup purposes. The Accunet Switched 1536K Service complements AT&T's existing Accunet Switched 56/64K and Switched 384K services. The Accunet Switched 1536K Service is available as a standalone service or as an option through the SDDN. Both implementations require ISDN PRI-equipped customer premises equipment.

AT&T's SDDN and MCI's switched T1 and T3 virtual private data services are typical of the virtual private networks coming into the market. These new digital services are flexible and cost-effective and will offer many features, but will primarily complement existing private networks instead of replacing them.

Value-added network providers are also getting into the market for data-oriented virtual private networks. Infonet's VPDN service enables users to build custom networks with the functional capability of private networks, but at less cost than standard packet-switching services. There are four components to VPDN: two flat-rate billing plans for packet-switched traffic; the Vstream circuit-switched service; and the Network Control Center-PC, a microcomputer-based network management system. Monthly charges vary according to the number and location of user sites in a configuration. The general price range is between $1,125 and $2,000 per month for communications within a single continent. Network spanning multiple continents cost between $2,500 and $3,100 per month.

Infonet's NetPlus System Reports (Exhibit 5-3-2) are electronically transmitted to customers on a daily, weekly, and monthly basis. These reports apprise users of communications incidents, keep them informed of problem resolution, and provide network availability statistics. A real-time view of the customer's VPDN operations is provided by access to one of Infonet's network control centers. This way, customers can become informed of network problems as they occur and be provided with continuous status updates on the corrective actions taken by Infonet technicians.

FRAME RELAY

To handle LAN interconnection on their VPNs, the interexchange carriers and VANs are planning to enhance their VPNs with frame relay. A faster method of packet switching, frame relay can transmit information at the T1 speed of 1.544M-bps, which helps to explain why it is rapidly emerging as a useful means for data networking, particularly for LAN users who require wide-area connectivity.

Frame relay takes advantage of today's high-quality digital facilities, including optical fiber, to reliably transport high-speed data. Because high-quality lines eliminate the need for embedded error correction in the protocol, this also eliminates the associated processing steps at each network node, which translates into higher throughput and less delay. Frame relay relies upon a low level of functional capability to send frames serially between devices. The frames contain their own control information for source and destination addressing, as well as for error detection. In looking only at destination addresses, the switches are able to pass the frames along very quickly. Errored frames are merely discarded by the network. Error correction, in the form of retransmission requests, is the responsibility of customer premises equipment.

AT&T, Sprint Data Group, MCI, BT North America, Worldcomm, and many other national and regional carriers now offer frame-relay service from selected cities. The technology is expected to make virtual private networks more attractive to customers who would otherwise choose dedicated lease lines for such applications as LAN interconnection.

HYBRID NETWORKS

AT&T, MCI, and Sprint work with customers in designing custom services. Pricing is negotiable even on standard services; AT&T does this through Tariff 12. Referred to as AT&T's Virtual Telecommunications Net-

Net*Plus*
System Weekly Report

Data for week ending Sunday, June 17, 1997
Page 1

Global Electronics
Account: WBZ79W

Node and Port	**126-063**	**007-057**	**053-072**	**019-056**	**010-069**
Location	New York	Singapore	Paris	London	Tokyo
Number of calls	1075	3018	1386	728	1782
Call setup rate	45	145	60	36	84
Total time	330:18:46	1184:12:26	967:07:14	194:10:34	760:42:0
Average call	00:21:08	00:20:48	00:32:49	00:15:34	00:23:4
Kilobytes input	1348	7648	5684	1384	6998
output	18543	68942	54,987	12,780	42,865
Kilobytes per hour	60	64	63	73	66
Link down events	0	0	2	0	1
time	00:00:00	00:00:00	00:32:07	00:00:00	00:21:00
Link availability	100%	100%	99.87%	100%	99.97%
Calls cleared	0	0	0	0	2
Maximum VC used	10	28	14	9	16
Peak b/s	5800	6840	6218	2270	4980
Mean time between failures	44:08:40	63:10:10	33:41:01	93:49:01	07:06:10
Project data volume KB	4378	14454	7150	2926	7546
Project connect	176	594	308	132	308

Part 1: Network Activity Summary

Total Tickets for the period:

	Current	**Previous**
Opened	003	002
Updated	001	000
Closed	001	001

		Opened		**Updated**		**Closed**	
Ticket No.	**Caller**	**Date**	**Time**	**Date**	**Time**	**Date**	**Time**
00100266	J. Dalley	06/13/93	06:18	06/13/93	06:22:00	06/13/93	06:26
	PR: 1 Description: Network access line down in New York						
00100294	B. Dubaugh	06/17/93	18:45				
	PR:1 Description: User held up in Paris						
00100317	F. Wong	06/17/93	19.08				
	PR:1 Description: Network access line down in Singapore						

Part 2: Trouble Ticket Activity

SOURCE: Infonet

Exhibit 5-3-2. Sample Status Report: Weekly

work Service, it is actually a negotiated package deal that includes all of AT&T's regulated services (e.g., Megacom or Accunet T1.5), but not unregulated services (e.g., Accunet Packet Service or VSAT networks). Although in theory AT&T delivers a turnkey network, there is little, if any, customization. Tariff 12 includes the leased lines that the user's private network requires, as well as switched services. In many cases, AT&T has merely sold the same network at bulk discount.

Corporate network managers can realize substantial savings by using virtual networks, as well as customized packages like AT&T's Tariff 12. Hybrid networks, formed by mixing virtual networks with private corporate networks, can generate even greater savings than are offered by each service alone. Hybrid networks capitalize on a key aspect of virtual networks: they cost less when traffic volume between two sites is low, more when traffic is high. The greatest savings in hybrid networks come from voice traffic. These savings come in three components: using lower-cost T1 or fractional T1 trunks instead of virtual networks; using voice compression techniques; and using low-cost, carrier-type circuits to cut off-net calling costs. Voice services now cost as little as $0.04 per minute under Tariff 12.

To support hybrid networks, the major carriers have installed new switching equipment that will streamline their networks, provide more processing power, and allow them to offer frame relay and switched multimegabit data service (SMDS) as well as provision new T1 and T3 services more easily. Once that has been completed, the savings produced will be passed along to customers. In addition to ISDN and Signaling System 7, the new switches will also support calling-card, 800, and VPN services and will speed development of intelligent network services.

The savings expected from the new network will accrue from a reduction in the amount of central office equipment and cabling, and the ability to test the networks and provide services more quickly. In some cases, such as MCI, the equipment will also make it possible for the carriers to eliminate the cross-connect technology on which they have been relying to provide switched T1 and T3 services.

MCI and Sprint, opponents of AT&T's Tariff 12, want the FCC to investigate the legality of offering combined discounts on 800 and other services under the tariff. MCI and Sprint claim that the FCC has already prohibited this in its banning of an AT&T promotion that would have allowed user discounts on the basis of combined use of SDN and 800 services. The FCC disallowed the promotion because of the dominance of AT&T in the 800 services market. Opponents are now pressing the FCC to overturn Tariff 12 or at least to force AT&T to eliminate 800 services from the tariff. Observers contend that if the FCC orders the modification of Tariff 12 to exclude the 800 services, this would make the tariff much less attractive to customers, because the biggest discounts in the tariff are in 800 services.

SUMMARY

As new, higher-speed switched services are developed by the carriers, organizations must accordingly revise the break-even points for the use of private networks as compared with virtual networks. All indications point to the increasing use of virtual private networks as the efficiencies and economies of public network plant and facilities improve. The VPN offerings of the major carriers have already progressed to the point at which they merit serious consideration among companies that rely extensively on public networks. The following factors, if applicable, may provide added justification for considering VPN services:

- A majority of the equipment used on the private network (e.g., modems, multiplexers, and DSUs/CSUs) is nearing full depreciation, or long-term service contracts and equipment leases with carriers and vendors are about to expire.
- The cost savings offered by VPN services outweigh such concerns.
- The VPN service significantly lowers the cost of entry into international markets.
- Competitors are using VPN services, and the savings enable them to compete more effectively because of lower overhead costs.

Many experts agree that virtual private networks are the next big competitive arena within which the big three carriers will do battle. If true, no matter how this scenario plays out, the customer will be the real winner.

5-4 A Guide to Network Access and Pricing Options

Nathan J. Muller

THE RATE AT WHICH NEW NETWORK SERVICES are being introduced and price revisions are issued is due as much to the easing of regulatory constraints as to advances in communications technology. Many of the new ways of accessing services can improve networking efficiency. New pricing plans can provide services more economically to users. For the purposes of this chapter, the different types of access available to users are categorized as basic access, consolidated access, alternate access, and special access.

BASIC ACCESS

Basic access entails the establishment of a physical connection from the customer premises to a local central office port. The connection can be dialed up as needed or provided to the user with a dedicated facility, such as a T1 line.

Dial-Up Connections

In the case of a dial-up connection, once an idle port is seized, the user has access to a variety of services and associated call-handling features provided by the local exchange carrier (LEC) within its local access and transport area (LATA). If the voice or data call is destined for a location outside the LATA, the LEC hands it off to a designated interexchange carrier's point of presence, where the call is transported to the LEC at the destination LATA. The user is billed for each call, based on time and distance. With integrated services digital network (ISDN), billing also takes into account the amount of bandwidth used for the call.

Long-distance or interexchange carriers (IXCs) are required to interface with local telephone companies using points of presence. A point of presence,

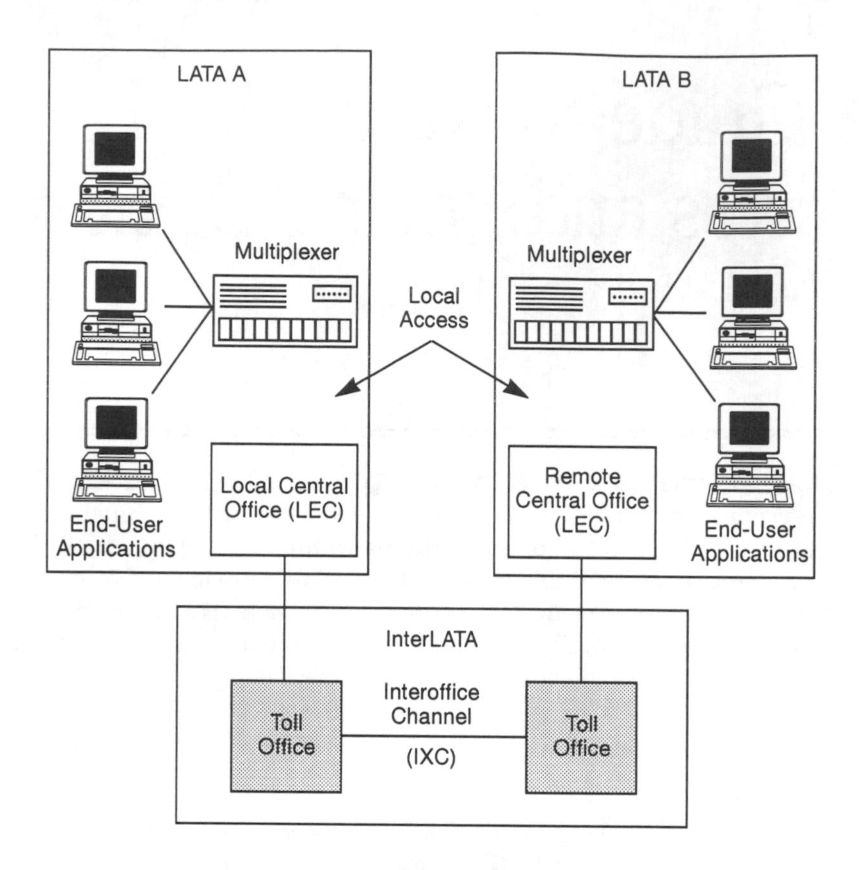

Exhibit 5-4-1. End-to-End InterLATA Circuit

or POP, is a serving office set up in each LATA. It is the point to which the local telephone company connects its customers for long-distance dial-up and leased-line communications between LATAs. The local segment of a dial-up or leased-line circuit is maintained by the LEC at each end, whereas the interLATA portion of the circuit is established and maintained by the IXC (see Exhibit 5–4–1). Billing for dial-up connections is according to time and distance. With a leased line, however, the user has full-time use of the carrier-provided facility and pays a fixed monthly fee based on distance, regardless of how much the line is actually used.

Bandwidth on Demand

One wrinkle in the basic access concept is called "bandwidth on demand," in which the user dials up bandwidth in increments of 56/64K bps to meet specific applications requirements. This can be accomplished with a customer premises device called an inverse multiplexer.

Inverse multiplexing allows users to dial up the appropriate increment of bandwidth needed to support a given application and pay for the number of local access channels only when they are set up to transmit data, image, or video traffic. Upon completion of the transmission, the channels are taken down. This obviates the need (and high cost) for private leased lines to support temporary applications.

Under the bandwidth-on-demand concept, extra switched bandwidth can be dialed up to support temporary applications, accommodate peak traffic periods, or reroute traffic from failed private lines. Advantages to this approach include:

- The immediate availability of bandwidth.
- The need to pay only for bandwidth used, according to time and distance.
- The elimination of the need for standby links that customers are billed for whether fully used or not.

Nynex Enterprise Services. To compete with vendors of inverse multiplexers, the regional Bell operating companies have introduced bandwidth-on-demand services of their own. New York Telephone, for example, offers Nynex Enterprise Services. Customers get up to 100M bps of scalable, managed bandwidth within 24 hours of placing the order. The amount of bandwidth can be adjusted on a time-of-day basis or within one hour of a customer request. The service is priced 10% to 15% less than leased lines of equivalent bandwidth. Discounts are available depending on the length of contract, and there is no installation fee. The drawback is that this service requires fiber optic connections; also, the service is available only in New York State for interconnecting local sites.

Although Nynex Enterprise is the first service of its kind, it is likely to be followed by similar offerings from other local telephone companies. Innovative data service offerings are the most practical way for the LECs to compete with alternative access carrier services.

CONSOLIDATED ACCESS

One access service that offers consolidated multiple services over a single access line is AT&T's Static Integrated Network Access (SINA). SINA is designed to eliminate the cost of maintaining separate access lines for private line services and switched services, because the traffic over the two types of access lines can be combined over the same ISDN access line. This means that a user can access AT&T's Accunet family of private line services, Software

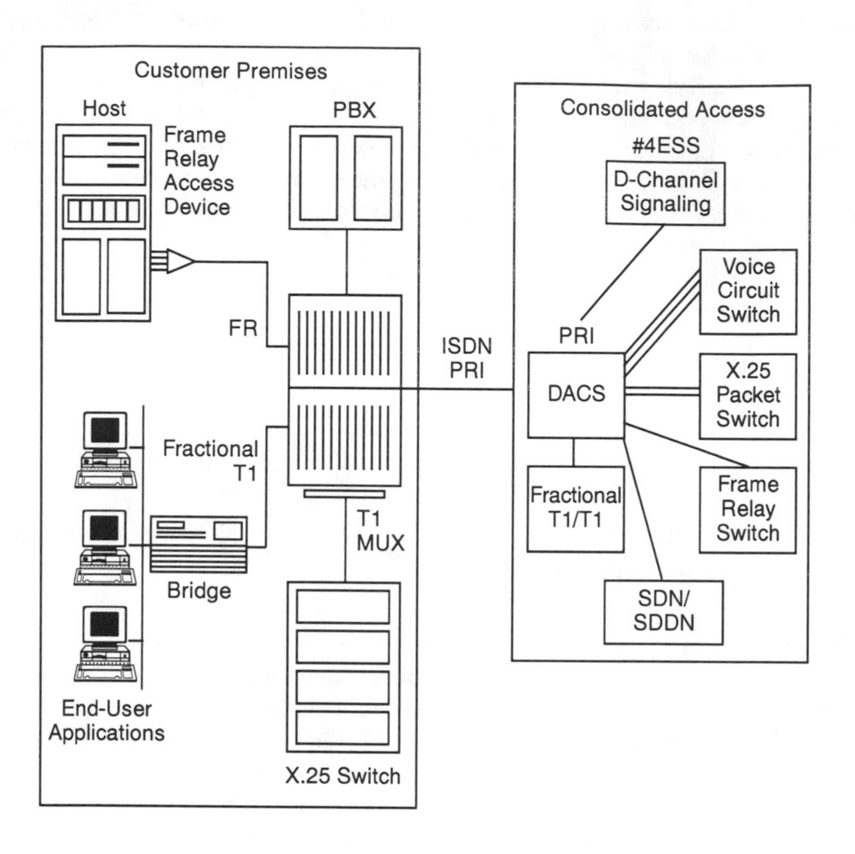

Exhibit 5-4-2. Consolidated Access

Defined Data Network (SDDN), and Accunet Switched Digital Services over a single access line instead of having to pay a separate access charge for each type of service (see Exhibit 5–4–2).

With ISDN primary rate interface (PRI) service, the 23 64K-bps B channels carry voice and data traffic to a digital access and cross-connect system (DACS) at an AT&T central office. The DACS separates switched voice from data and sends it to the carrier's Software Defined Network (SDN) or Megacom services. Switched data traffic is sent to AT&T's SDDN. Other data can be sent over AT&T's Fractional T1 offering. AT&T also provides local access to its InterSpan Frame Relay Service via ISDN PRI trunks. In this way, frame relay traffic can be combined with AT&T's other switched and dedicated network services.

AT&T's Definity G3 PBX is equipped to support SINA. In combination with the PBX's nxDSO capability, customers can size bandwidth and access, through the same local access line, a variety of services that support LAN traffic.

Sprint and MCI offer similar integrated network access services. Sprint uses ISDN PRI to support access to dedicated and switched voice and data services; first, users must purchase an integrated access controller, which is a network hub that includes special software. The service offering is targeted at small and midsize sites, many of which have been unable to cost-justify ISDN primary rate interface for conventional voice and data applications.

ALTERNATE ACCESS

Alternate access, also known as "bypass," entails the use of facilities from a carrier other than an LEC to get direct connections to the long-distance facilities and services of IXCs or value-added network providers. These alternate access carriers include regional teleports and metropolitan fiber companies, some of which compete directly with the local telephone companies by providing businesses with the means to bypass the local exchange and save money on local access lines and usage charges. Typically, alternate access carriers offer service in major cities, where traffic volumes are greatest and, consequently, where businesses are hardest hit with high local exchange charges.

Teleports. A broad array of services, ranging from Fractional T1 to SONET broadband transmission, are offered by alternative access carriers. Teleport Communications-New York, for example, offers Fractional T1 services as well as DS-0 and sub-DS-0 services to business customers in New York City and northern New Jersey. Through the use of intelligent T1 multiplexers and DACS systems, the company provides bandwidth in any multiple of 56/64K bps. It also offers continuous quality monitoring and extensive remote diagnostic capabilities from its network control center, as well as line changes and reconfigurations that are computer-controlled through a DACS. Some teleports offer sophisticated leading-edge services. Teleport Chicago, a subsidiary of the NY-based Teleport Communications, has a SONET-based fiber network that allows long-distance carriers and businesses in Chicago to transmit data at speeds of 150M bps and above.

Advantage Customers. Advantages of alternative access carriers include:

- They offer less expensive and more reliable access service and solve problems faster than the local exchange carriers.

- They offer advanced network protection capabilities, mostly in the form of dual fiber rings. When one ring fails, the other is activated automatically to handle the traffic load without any loss of customer data.
- Some disaster recovery firms have cooperative arrangements with alternative access carriers whereby bandwidth is made available to customers during emergencies when private communication systems may be knocked out.

In some states, the public utility commissions have mandated that the local telephone companies provide local connections to the alternative access carriers, so they can more easily serve business customers. The FCC has ordered the regional Bell operating companies to open up their central offices to the equipment of alternative access carriers seeking to provide leased-line access services, and to allow competitors into the local central office for purposes of routing switched traffic to long-distance networks. These decisions signal that the local exchange monopoly is going to be dismantled sooner than most industry analysts expected.

Although there is still industry disagreement on whether tariffed equipment collocation applies to end users as well as alternative access carriers, some LECs and IXCs provide collocation arrangements for users on an individual case basis. Aside from convenience, these arrangements can allow better quality communications by reducing the potential for impairments in the transport of information over various lines and interfaces that would otherwise be required.

SPECIAL ACCESS

Special access entails the provision of custom connections to LEC or IXC services on an individual case basis. This type of access is used in two situations:

- When proprietary technologies are involved to support requested services.
- When the LEC does not offer a service but will honor a customer request.

T3 Services. T3 is an example of a service that is supported with proprietary technologies. The absence of an optical standard for DS-3 signaling severely restricts users' ability to mix and match different manufacturers' equipment end to end. Therefore, T3 services generally require the customer to negotiate the type of optical interfaces to be placed in the various LEC and IXC serving offices. In contrast to the ubiquitous T1 services, T3 requires special construction of the local channels at each end—from the customer premises equipment to the long-distance carrier's POP. These special construction costs vary, depending on the regional Bell operating company.

Dark Fiber. Another example of special access is the concept of "dark fiber," an arrangement whereby users set up and maintain their own fiber optic networks over unused carrier facilities. Local exchange carriers are required by the FCC to provide—and, in some cases, maintain—fiber links for users that supply the equipment needed to power and send traffic over the links.

Dark fiber, which is less expensive than such fiber-based services as T3 and provides users with greater control, is used in campus environments or clusters of sites within a metropolitan area. An advantage of dark fiber is that the user can determine the level of service simply by the type of equipment installed at each end. For example, dark fiber can be used as a 100M-bps FDDI backbone or a 155M-bps SONET link.

The LECs had been installing dark fiber as the foundation for future high-speed point-to-point links in private user networks and offering it on a contract basis, and they are required to file formal tariffs for the service. However, in not owning the facility, the LECs have no control over what services are delivered over dark fiber pipes; without control, they cannot maximize revenues. As a consequence, the LECs hope to move into the more lucrative T3 market. They must, however, according to the FCC, continue to offer dark fiber, on the grounds that ceasing dark fiber services would discontinue, reduce, or impair service to a community.

Fractional T1. Although offered by the major long-distance carriers, Fractional T1 is an example of an access service that is not widely supported by the LECs but can be obtained on an individual case basis. Fractional T1 entails the provision and use of bandwidth in 64K bps (DS-0) increments up to 768K bps without incurring the expense of an entire T1 facility, much of which may go unused.

Fractional T1 requires the use of customer premises equipment (e.g., an intelligent channel service unit or multiplexer) that is capable of packaging the DS-0s into the bandwidth increment required to support a given application. For example, six DS-0s can be bundled to support viodeconferencing at 384K bps. The LEC routes this bandwidth increment to its destination using a digital cross-connect system located at the local central office.

DISCOUNT PRICING FOR LOCAL ACCESS

Not only do customers have more access options from which to choose, but more attractive pricing plans as well. Historically, the price of local access at each end of a long-distance circuit consumed an inordinately large portion of the total cost. Often the price of local access amounted to half the price of a private line. Today, the IXCs have greatly improved the ways in which they manage their access infrastructures. The resulting lower costs are passed on to customers in the form of discounts on local access. This, in turn, has spawned a new competitive arena for the top three long-distance carriers—MCI, AT&T, and Sprint.

MCI's Access Pricing Plan

MCI, for example, offers its Access Pricing Plan, under which customers receive discounts on local access in exchange for long-term service contracts. The plan offers discounts of 5%, 10% and 20% on each dedicated access line for a one-, two-, and three-year access contract, respectively. The discounts apply to channelized T1 links that access a variety of MCI services, including Vision, Virtual Network, 800 and Prism services, voice-grade private lines, and frame relay.

Customers that pay $550 a month per T1 access line at both ends of a long-haul link can cut that cost by 20% to $440 a month by contracting for the service for three years. A one-year contract will lower the cost to $522.50 a month, which represents a 5% savings.

In response to AT&T offerings (discussed in the nest section), MCI added a four-year contract option with a 22% discount and a five-year option with a 24% discount. MCI's plan also includes a so-called coterminous option, which gives users the flexibility to order additional T1 lines midway through a contract and still receive the maximum discount.

MCI charges users 50% of the nondiscounted monthly rate for access multiplied by the number of months remaining in the contract. The termination charges are:

- One month for a one-year plan.
- Two months for a two-year plan.
- Six months for a three-year plan.
- Eight months for a four-year plan.
- Ten months for a five-year plan.

MCI allows customers to relocate within a LATA and move their T1 access links without penalty. MCI users can upgrade a T1 access line to T3 without an upgrade charge.

AT&T Access Value Plan

Under AT&T's Access Value Plan, customers save as much as 24% on access charges by committing to a five-year contract. The discounts apply to all Accunet T1.5 local access channels, including those used to connect to AT&T's Accunet Switched Digital Services, Digital Data Service. InterSpan Frame Relay Service, Megacom, Megacom 800, and Software Defined Network services.

For example, a one-mile T1 access link from New York Telephone costs $492 a month. Under this plan, the same circuit would be priced as follows:

Length of Contract	*Discount*	*Price*
One year	5%	$467.40
Two years	10%	$442.80
Three years	20%	$393.60
Four years	22%	$383.76
Five years	24%	$373.90

A five-mile dedicated access link from Ameritech in Chicago would cost $506.80 a month. Under the Access Value Plan, it would cost from $481.46 to as little as $385.17, depending on contract length.

AT&T's termination charge is 50% of the nondiscounted monthly rate multiplied by the number of months left on the customer's plan.

THE CUSTOMER PERSPECTIVE

The dilemma for customers is that in committing to long-term contracts with the IXC's they might lock themselves out of greater discounts from the LECs. One advantage for the customer, however, is that purchasing local access lines from the IXCs is simpler with a single point of contact. Moreover, the IXC bears the responsibility for ordering, installing, and troubleshooting the lines. Unless the discount offered by the LEC is large enough to make assuming these responsibilities worthwhile, purchasing local access lines from IXCs may be worth the slightly higher cost.

Enhanced Local Services

Local exchange carriers are also beginning to enhance their private line services to give customers more control. Using software developed by Bellcore called FlexCom/linc, Pacific Bell is providing customers with time-of-day and real-time routing control. Instead of having to give PacBell 24-hour notice for line reconfigurations, customers can do it by themselves at their own discretion or program the reconfiguration to occur by time-of-day.

Southwestern Bell offers a Transport Resource Management (TRM) service with which customers can remotely manage their own private line networks through a workstation linked to a central office-based intelligent D4 channel bank. The service includes a subrate multiplexing feature, so users can send multiple signals at different speeds inside a single 64K-bps circuit, and a voice compression capability that permits transmission of 44 voice calls on a single T1 line. Customers can also set up and control "private" frame relay networks using a central office-based frame relay switch. The service package includes analog, digital data, and voice bridging features and three fault-recovery options.

SUMMARY

From both a technology and service perspective, the public network is undergoing what can only be described as a radical makeover. More access options are available, and local exchange carriers are giving users more control over service provisioning and circuit reconfigurations.

These developments benefit users in a variety of ways, such by reducing the lag time between order requests and order fulfillment for services and associated features. The new capabilities not only add value to local services and help wean customers from dealing with the IXCs for local access lines, but also position the LECs strategically to withstand the growing onslaught of competition from alternative access carriers. The price of access services, which has always comprised an inordinate amount of the end-to-end circuit charge, will be greatly reduced in the near future.

5-5 Auditing Telecommunications Costs for Data and Video Communications

William A. Yarberry, Jr.

DATA AND VIDEO COMMUNICATIONS, as a percentage of total worldwide telecommunications expenditures, have been steadily increasing. As organizations substitute "bits for atoms," the telecommunications infrastructure becomes deeply entwined with the fabric of the business. This chapter identifies areas of potential cost savings and efficiency improvements.

GETTING STARTED

Before examining specific communications systems, the following background information should be obtained:

- The circuit diagrams that show dedicated circuits, private virtual circuits, frame relay, and asynchronous transfer mode (ATM) networks.
- The listings of all video conferencing locations and equipment (e.g., Vtel, Pictel, and CLI).
- The available common carrier data (e.g., circuit listings and sources of raw billing data). This applies to all carriers used by the organization, including long distance, local telephone company, and resellers.
- Any microwave tower information.
- Any information on very small aperture terminals.
- Any information relating to wireless applications, including analog, cellular digital packet service (CDPD), two-way paging applications, and

personal communications systems (PCS). PCS competes with cellular services. It uses less power and higher frequencies (1.850-1.990 GHz). The government has auctioned rights to most of this bandwidth across the country.

- Information about electronic data interchange applications.
- Information about the organizational structure that supports communications within the company.
- All available communications budget information. In some cases, expenditures by vendor can provide the necessary cost information.

COMMUNICATIONS INFRASTRUCTURE

After collecting the preceding background information, the person auditing these systems should examine specific components of the communications infrastructure as discussed in the following sections.

Private Line Versus Dialup

Private (i.e., dedicated) communication lines are remote point-to-point (POP) circuits that permit traffic 24 hours a day at a fixed cost from a common carrier or reseller. These lines are always up and available (except for occasional outages). Dial-up, on the other hand, uses a temporary circuit that is created when the call is made, then eliminated when the calling parties hang up. Dial-up circuits are created on the fly and the quality varies, even when calls are made from and to the same locations. That is why, for example, the quality of voice calls sometimes improves when the parties hang-up and redial—a better circuit is obtained the second time.

The person performing the audit should ask a series of questions to determine if the organization has chosen the most appropriate technology to establish simple point-to-point circuits. A sample of these questions includes:

- Has an analysis of usage been performed to determine if dial-up would cost less than dedicated lines? Exhibit 5-5-1 shows a breakeven analysis based on daily usage. Based on a per minute cost of $0.12 and a dedicated cost of $200 per month for a 56K-bps line, usage of more than 28 hours per month results in higher costs. However, this analysis assumes that volume constraints are not an issue, because a dedicated 56K-bps circuit (i.e., line) has considerably greater bandwidth than a dialup 28.8K-bps line.
- Will the user tolerate dialup delay as modems handshake? A handshake is the process by which two modems attempt to connect by agreeing on the parameters required to pass data.

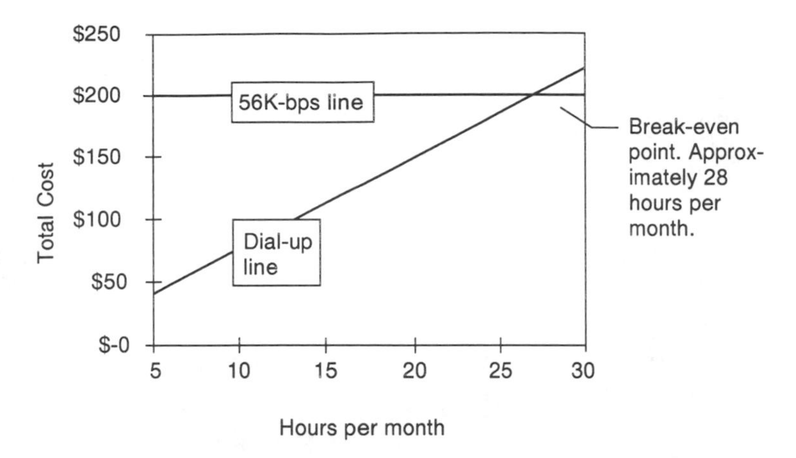

Exhibit 5-5-1. Dial Up Costs Versus Dedicated 56K-bps

- Is the circuit critical to business? Noise and dropped connections are considerably less prevalent for private lines.
- Is the user so remote that private lines are not an option?
- Are bandwidth requirements large? If so, several analog circuits can be "inverse muxed" to allow for dialup lines. Analog lines are the traditional circuits used by the telephone company to carry voice traffic. Although they are used for data, such technology on the user end as high-speed, error-checking modems has been required to enable 28.8K-bps and 33K-bps speeds. An inverse multiplexer (I-mux) takes one large originating circuit and splits it into several lower speed circuits for transmission. For example, an inverse multiplexer may take a video signal at 112 or 384 bps and dial up enough analog (i.e., switched) lines to provide the necessary bandwidth. Thus, the business does not need to lease lines and can have bandwidth on demand—paying only for what time is used.

Integrated Services Digital Network (ISDN)

ISDN is a high-quality, relatively low-cost digital line that can be used for traditional data transfer and videoconferencing. Using ISDN, users enjoy 144K-bps of data transfer with low noise and high availability. It would seem that ISDN would have a throughput advantage of 144/28 = 5 times that of dial-up over voice grade lines. In fact, the advantage is even greater, because voice grade circuits require frequent retransmissions.

Advantages	Disadvantages
1. Instead of seperate lines for voice and data, users can combine them for lower cost and greater functionality.	1. Complex setup.
2. High bandwidth for data, voice and video.	2. Not available everywhere in the US. Prevalent in Europe. Use worldwide is growing.
3. Out-of-band signaling (i.e., provides much more information about who is calling, how to handle certain calls, and auto billing).	3. Fixed cost of equipment is high. For example, an ISDN "modem" (adapter) can cost as much as $800.
4. Ease of desktop videoconferencing.	4. If used with a personal computer, ISDN may cause interrupt problems (although this may be addressed by later releases of various operating system software).
5. Internet access is more practical.	
6. Time to dial a number is greatly reduced.	
7. Caller ID and a large number of application level features.	

Exhibit 5-5-2. ISDN Advantages and Disadvantages

In some areas of the US, ISDN is widely available and, in other areas, deployment by the telephone companies is proceeding rapidly. The disadvantage of ISDN is the bewildering complexity of options and the less than straightforward setup. In addition, competing higher-bandwidth technologies such as asymmetrical digital subscriber lines (ADSL) and cable modems could supersede ISDN.

The auditor should examine the organization's use of ISDN and encourage its use where appropriate. Exhibit 5-5-2 identifies the advantages and disadvantages of ISDN.

The IS manager should ask these specific questions:

- Has ISDN been considered to replace older "Switched 56" lines?
- Does the organization have international locations that could be better served by ISDN rather than dedicated lines that are very expensive to use internationally?
- Has the organization taken advantage of the features of ISDN? For example, complex routing instructions (based on business requirements) to various internal telephone extensions can be developed. This capability is based on availability of caller ID. Links to a customer data base can be used for customer contact applications.

- Is telecommunications equipment used to reduce the number of incoming phone lines?

Internet Communications

With the rise of commercial applications based on the Internet's infrastructure, auditors must include Internet technology in their review of data and video communications costs. The mechanics of Internet implementation should be examined. For example:

- Where does the organization's Internet service provider (ISP) stand in relation to the Internet backbone. If the ISP is a third- or fourth-tier provider, packets will be relatively delayed due to extra router hops. An ISP that is directly on the backbone, on the other hand, will provide a relatively fast response.
- Does the organization use the Internet for non-urgent communications where it is economical? If encryption is needed, token cards and pins provide a practical level of security.
- Does the organization use the Internet for electronic data interchange (EDI)?
- Are firewall products sized appropriately for the volume of traffic?
- Have Internet-based videoconferencing products been considered (far less cost than point-to-point dedicated facilities, although quality may not be as good)?
- Has the organization established guidelines for purchase of Internet-based PC videoconferencing equipment so that standards-based equipment from different vendors can be used together?

COMMUNICATIONS OPTIONS

Equipment selection and configuration affects the efficiency and security of the wide area network. The feature set of routers, Customer Service Unit/ Data Service Units (CSU/DSU), muxes, and other telecommunications equipment should be examined. If advanced features are ignored, the organization may not be properly leveraging its equipment resources.

Although hundreds of options are available for complex communications hardware, the auditor should review the following options:

- *Router compression.* Depending on the traffic, router compression can improve throughput by as much as 40%. Success requires considerable effort for this technology. Hardware should match on both ends (i.e., same vendor's equipment). In addition, the equipment should have a sufficiently fast CPU and backplane that compression does not result in dropped packets, thereby requiring retransmission. If the router cannot keep up with the volume of packets sent to it because it is overwhelmed with the task of compressing data, it will simply skip over a certain number of packets. Those packets will have to be sent through the network at a later time.

- *Routing—not bridging.* Bridging is simple and works well for small networks. However, large networks cannot function with a bridge-only topology. Bridging works as follows: A message is sent from device *A* (e.g., a server) to a bridge. The bridge determines if the destination is "local" (i.e., destined for a station on the local ring or segment) or "not local." If the message is not local, it is shipped to all points on the network. Routers, on the other hand, look at the destination of the message and determine specifically where it is to be sent. The message is then forwarded by the most efficient path to its destination. Most importantly, traffic is not generated for paths that will never use the information. Thus as networks grow, routing, rather than bridging, becomes mandatory.
- *Asynchronous transfer mode (ATM).* In situations where voice, data, and video traffic can be carried on the same circuit, ATM can be an effective means of reducing the number of circuits required (i.e., a separate circuit for each kind of traffic is no longer required). The disadvantage of ATM is that the end equipment is still somewhat expensive, though rapidly coming down in price.
- *Low overhead protocols.* As part of communications standards, the organization should implement low overhead protocols when possible. For example, Novell offers the Netware Link Services Protocol (NLSP) to reduce overhead and provide better throughput for the Internet Packet Exchange (IPX). The NLSP is a new protocol that improves the speed and functionality of IPX traffic in large networks. It reduces the quantity of "ack" and "nack" traffic by sending larger packets through the network. Ack and nack refer to acknowledgment and negative acknowledgment messages, respectively, sent back and forth to determine if the packet was sent successfully. IPX is Novell's standard LAN communication protocol. IPX information is encapsulated and transported by packets used in Ethernet and Token Ring networks.
- *Unnecessary protocols.* Management responsible for maintaining the network should resist "protocol creep," a tendency in organizations to adopt over time a large number of protocols on its network. Individuals within the organization may request support of a new (or older) protocol. Typically, such a request is made because a particular hardware or software vendor requires it to run an application. The costs of supporting a large number of protocols should be evaluated. Is the application's

business benefit worth the increased complexity of troubleshooting, potential negative effects on throughput, and perhaps additional network downtime? Could another protocol be used? Could another vendor be used to run the application?

- *Appropriate filtering.* Devices on a network (e.g., printers, servers, and workstations) often "advertise" their presence on the network by periodically sending out messages that are intercepted by routers. Routers then add information about the devices to their routing tables (i.e., a new piece of equipment is now on the network or, if no message is received for a while, equipment is assumed disconnected). However, when the number of devices on a network grows, advertising traffic can increase rapidly. Particularly for low-capacity wide area links (e.g., 56K-bps circuits), this network overhead can consume up to three fourths of available bandwidth, leaving little capacity for running applications.

Filtering, a software feature of routers, eliminates advertising messages in those parts of the network where they are not needed. For example, a Novell network in a Baltimore office does not need to know about a printer in San Francisco. The auditor should examine router filter tables to determine if the organization is taking advantage of router filtering capabilities. Appropriate filtering can not only improve network responsiveness for users, but can delay the need for bandwidth upgrades as business data volumes increase.

Microwave Private Systems

The use of microwave private systems is declining because of less expensive alternatives available from the public carriers and because of technologies such as frame relay. However, many organizations use microwave private systems for rural areas or for point-to-point communications where the initial cost of the towers, for example, has already been absorbed. If this legacy technology is a significant expense, the auditor should ask the following series of questions to determine if microwave private systems are used efficiently and effectively:

- Does the organization account for the full costs of maintaining a private microwave system? Only by identifying all costs, direct and indirect, can a valid comparison be made between the microwave system and alternatives. For example, costs of tower maintenance (e.g., periodic painting and structural repairs), license management, and personnel should be included in such a comparison. The FCC does not charge (currently) for using a microwave bandwidth in the two or six—GHz bandwidth. This may change in the future. In addition, a significant effort is necessary to keep licenses up to date and to fulfill FCC reporting requirements.
- Are volumes of voice, data, and fax transmissions monitored? If field

offices move or no longer need as much data transmission, alternative means of communications, such as land lines or very small aperature terminals could reduce costs.

- If the organization is committed to microwave private systems, has the technology been converted to digital technology?

Microwave and Radio Bandwidth Are Assets. The Federal Communications Commission (FCC) has shown a strong propensity to auction off sections of the microwave and radio spectrum. The intent is to both raise revenue and ensure that this limited resource is allocated in the most efficient manner. The communications manager should obtain an understanding of what frequencies have value and what licenses and towers the organization owns. The FCC's home page provides detailed background information on spectrum auctions and relevant information. For example, the FCC has implemented legislation that permits personal communications systems (PCS) vendors to obtain rights to certain frequencies currently licensed to organizations around the country. In return, the PCS vendors must either compensate the organization for the use of the frequency or provide microwave equipment that operates at a higher frequency (e.g., 6 GHz).

The communications manager should ensure that the organization recognizes the value (sometimes in the hundreds of thousands of dollars) of licensed frequencies. PCS vendors have an incentive to pay and obtain the airwaves early. By 1998, that incentive will be reduced because they will have the legal right to use the frequency.

In general any licenses for airwaves should be considered assets and monitored. If the organization no longer uses the frequency, it should be sold when a buyer is located.

Use of Microwave and Infrared for Short Distances Between Buildings. Both microwave and infrared technology can provide a shorter start-up time to get communications up and running when office locations are in the same proximity and high volumes of data must be transmitted. In the case of infrared technology, the offices must be within "line of sight" of each other. Communications engineering firms usually offer a turn key solution, in which they provide the design, equipment, and even the license for transmission. The FCC provides the license to the engineering firm, who then permits its customers to operate under its license. Infrared equipment supports transmissions that support Ethernet or Token Ring speeds (i.e., 155 M-bps ATM speeds soon) and requires no government license. In addition, a typical infrared solution costs approximately $15,000 to $25,000, whereas a microwave system between two buildings can cost up to $100,000. The disadvantage of infrared includes its limited distance (one kilometer or less) and susceptibility to interference from heavy fog or rain.

The key question, from a communications manager's perspective, is whether the organization is overly conservative in its choice of communications media. Is "order another T1 or order more leased-line capacity" the knee jerk reaction to the request for more bandwidth?

Frame Relay

Frame relay relies on packet switching. Compared with its older cousin, X.25, frame relay uses smaller packet sizes and requires less error checking. The technology allows customers to sign up for a committed information rate (CIR) of a specified size (e.g., 64K-bps) and burst up to a larger bandwidth (e.g., 256K-bps) as demand requires. Unlike traditional dedicated circuits, the user does not pay 24 hours a day for the maximum bandwidth that is needed, even though most of the time actual usage is far below the maximum. This is sometimes called the "elastic bandwidth" concept—a rubber pipe that can expand temporarily to accommodate an unusually large burst of traffic (typical for LANs) and then shrink back for normal traffic patterns. By combining packets of many customers, carriers can create the illusion of bandwidth on demand, because it is unlikely that all customers will "burst" at the same time. Frame relay pricing is based on CIR, port size (which determines the maximum possible volume of data transmission), local access charges, user distance to the nearest carrier point-of-presence, and other factors.

In many instances, frame relay technology can be substituted for dedicated, point-to-point circuits, resulting in reduced costs and higher reliability. On the other hand, if traffic is constant and always directed to the same location, dedicated circuits make sense.

Although some organizations have implemented voice over frame relay, it is generally considered a marginal use of the technology, requiring significant attention to equipment settings and a higher maintenance effort. The great majority of organizations use frame relay for data traffic only.

The auditor should ensure that frame relay is used where appropriate. Savings vary considerably according to carrier, traffic, and other factors, but generally a range of 10 to 30% over dedicated, point-to-point circuits can be expected. In addition, the mesh topology of frame relay makes it more robust and less likely to suffer interruptions.

Frame Relay Equipment. Routers are the most important feature of the frame relay network. To the extent possible, a single vendor should be used because:

- The communications team can learn and understand how to set up the routing tables more readily if they have only one vendor's equipment to work with.

- Compression algorithms, if implemented, are more straightforward.
- Spares (in the event of a router failure) can be deployed more economically across the network and less inventory is required.
- Troubleshooting is simplified.
- Vendor volume discounts are more readily obtainable.

For small bandwidth networks, frame relay cards inside a LAN server can be used. This eliminates the need to purchase a separate, standalone router for frame relay traffic. The disadvantage of using a card inside a server is the effect on server performance. If data transmission (input and output) steals too many CPU cycles, users will see degradation of service.

The X.25 Packet Switched Network. X.25 is an older standard developed in the 1970s to provide a means for the public networks to communicate with their customers. It has robust error correction algorithms and was designed when communications links were far more noisy than they are today. X.25 networks are nearly ubiquitous. If a country has any data communications facilities at all, it is likely to have X.25.

The communications manager should be aware of the capabilities of X.25, particularly if international communications are required. The service is generally based on a per-minute charge.

Because X.25 uses a packet assembler/disassembler (PAD) at the customer's site, no modems are required. The user dials up the common carrier's "cloud," enters an account number and password, and then is transmitted to the host via a digital line. The PAD is located on the organization's premises and connects to the backbone or mainframe, as needed. For small office or individual users, X.25 is a reliable and cost-effective means to communicate with the office. It is ideal for remote access to intranets, because the traffic does not go over the Internet.

CELLULAR DIGITAL PACKET SERVICE (CDPD)

CDPD takes advantage of pauses of traditional cellular voice conversation to send packets of data at high speeds. It is wireless and uses the existing analog cellular infrastructure. Organizations can use CDPD when wireless transmission of a relatively small quantity of data (e.g., an empty Coke machine sends a message to the warehouse that it needs to be restocked) is practical. Exhibit 5-5-3 identifies the advantages and disadvantages of CDPD.

The manager should evaluate the economics of CDPD deployment. In those areas of the world where it is implemented by the cellular carriers and in those specific applications where it is inconvenient or expensive to have a

Advantages	Disadvantages
1. Efficient, relatively fast means to send wireless packet data.	1. Not uniformly implemented in the US or worldwide. Depends on cellular carriers.
2. Achieves speeds up to 19.2 K-bps.	2. Only small volumes of data are economical to send (cost is per kb).
3. Mobile workers can use combination of voice cellular and CDPD to handle both voice and data needs.	3. CDPD modem rather than standard modem is required.
4. Organization can avoid costs of setting up alternative digital communications network.	
5. Error checking and accuracy are excellent.	

Exhibit 5-5-3. Cellular Digital Packet Data Advantages and Disadvantages

land-based connection, CDPD can be very effective. CDPD has been used to send E-mail messages and for credit card number authorization.

LOW EARTH ORBITAL SATELLITE COMMUNICATIONS

Traditional, geostationary satellite communications systems have two significant weaknesses. First, because the signal has to travel 22,300 miles up to the satellite and the same distance back again, there is a noticeable delay for voice or time-sensitive communications. Second, relatively powerful receiving dishes are required to capture the signal from the satellite. Low earth orbit satellite communications systems, on the other hand, are much closer to the surface of the earth and are able to beam a stronger (albeit narrower) signal to receiving dishes.

Any organization that has remote work areas requiring periodic communications should review low earth orbit (LEO) technology as an option. For example, gas pipelines are required to report corrosion statistics periodically. Rather than typing up a technician's time to visit each site, LEO systems can be used to periodically, and relatively cheaply, transmit current data to a monitoring facility. Orbcom, for example, has developed a number of applications using LEO technology, including remote E-mail, wireless fax, and transmission of cathodic protection data.

INTERNATIONAL COMMUNICATIONS

Crossing international boundaries can involve numerous expenses. If the organization has international operations, the auditor should examine technologies, agreements, and trends to determine if opportunities for improvement exist. The auditor should ask the following series of questions:

- Do traveling employees use all features available from the organization's primary carrier? For example, when using a calling card and dialing several different telephone numbers, do they press the "# " key to dial the second, third, and fourth numbers? By pressing the "# " key, the call setup fee is incurred only once.
- Is the Internet used for noncritical data transmissions with appropriate firewall security and encryption on both ends?
- If Inmarsat satellite communications is used, have volume agreements been examined?
- In areas where new satellite services are available (e.g., American Mobile in the Caribbean), has a comparison been made between traditional Inmarsat and the new service? Competition in the global satellite services, for both data and voice, is intensifying. Costs are expected to drop dramatically, and auditors should be alert to last-minute changes in these services.
- Has a "callback" service been considered? For some countries, the cost to call the US is much higher than the same call from the US to the international location. Callback services automatically switch the call, based on predefined telephone numbers, by hanging up and dialing back the caller from the US. (The dialtone is provided by the callback service.) By using a callback service, an organization can obtain savings of 20 to 60%. However, not all countries permit this service. For example, China and Vietnam prohibit all callback activities—at least the ones that are automated. In general, the largest savings in using a callback service are obtained in countries whose telephone services are provided by state-controlled monopolies.

CELLULAR COMMUNICATIONS

Cellular costs for certain employees can be significant, particularly if they use a laptop to cellular phone modem for LAN access. In conducting a review of cellular communications, the auditor should include these steps:

- *Roaming analysis.* If possible, the auditor should obtain a file from the organization's cellular carrier (preferably on CD-ROM) that identifies roaming charges by employee. Those employees who have very high roaming charges (e.g., 25 to 50% of the total bill) may be able to reduce charges by purchasing service in the multiple areas they cover.
- *Corporate agreements.* The auditor should determine if the organization has used its size (i.e., amount of cellular usage) to negotiate a favorable

rate. Major cellular carriers are anxious to lock in customers—particularly with PCS becoming an option more often.

- *International Cellular.* US cellular carriers offer cellular telephones that can be used internationally. Although users must purchase (or lease) a new telephone, they can use the telephone immediately outside of the US and sometimes avoid high international hotel telephone charges.

FASCIMILE EFFICIENCY

Facsimile (i.e., fax) transmissions, for long documents, can incur significant per-minute costs. An international fax increases the cost due to much higher per-minute charges. For potential cost savings of 10 to 30% the auditor should determine that the following procedures are active:

- *The same fax vendor equipment should be used on both the send and receive ends.* When fax machines from different vendors make a connection, they transmit at the lowest common denominator protocol (i.e., Group III). In contrast, if the originating fax machine recognizes a same-vendor machine, it is free to transmit at a higher speed, proprietary protocol, thus reducing transmission time.
- *Using PCs to transmit long documents internationally should be avoided.* A PC fax is convenient, but slow. Users have the tradeoff of convenience versus cost, but at $3.00 per minute for some countries, it may be worthwhile to use a stand alone fax machine.
- *For lengthy international transmissions, transmissions should be delayed until off-hours.* If the per-minute rates are based on time of day, savings can be obtained by transmitting documents after business hours. Up-to-date fax machines allow the fax to be stored in memory and automatically sent at a specified time (e.g., off hours).
- *For low- to medium-importance documents, the use of an IP/ FaxRouter should be considered.* For example, the Brooktrout IP/FaxRouter allows faxes to be transmitted over the Internet, saving long-distance charges. For security, encryption hardware can be added between the end points. The risk to the organization is potential downtime if the Internet is down; one alternative would be to use public dialup or ISDN lines for backup in case the Internet is overloaded.

VIDEO COMMUNICATIONS

Video is a "bandwidth hog." Choices for transmission include traditional point-to-point dedicated circuits, dial-up using inverse multiplexing, and

ATM. The auditor should ask the following series of questions when reviewing telecommunication costs for video transmissions:

- Is inverse multiplexing (i.e., bandwidth on demand) used for occasional use videoconferencing? If videoconferencing is used only a few hours per week, it may be more economical to dial up the destination and use I-muxes to obtain the necessary bandwidth.
- Has PC videoconferencing been considered? PC videoconferencing, using ISDN , is more economical than the large, dedicated units. In addition, "whiteboard" facilities are available on PC videoconferencing systems that enable users to look at specific applications and annotate them directly on the screen. A whiteboard is a combination of hardware and software that allows participants in different locations to view images, text, and data on a screen simultaneously. For example, engineers in different cities might set up a whiteboard conference to look at a PC screen that displays an engineering drawing and a spreadsheet of related costs.
- Is there a timer on dial-up videoconferencing that shuts the system down after a predetermined number of minutes or hours? It is possible to incur a charge of several thousand dollars because an international videoconferencing session is not terminated at the end of the business session.
- Does all equipment use the H.320 standard or above (or H.324 for low-end products that use a standard analog telephone line)? Using H.320, videoconferencing systems from many different manufacturers can communicate with each other. For example, a PC video unit can communicate with a large, dedicated videoconferencing facility, so long as they are both on the same standard.
- If PC videoconferencing is used, is "throttle" software used to ensure that the LAN does not choke with video traffic? Too many videoconferencing sessions on a LAN can slow down traffic. Software, available from Intel and other manufacturers, should be used to limit the number of simultaneous sessions. Throttle software monitors the quantity of traffic on a network and limits the number of users who can use video applications if the network is becoming slow because of excessive video traffic.
- If only crude (i.e., low resolution, low frames per second) videoconferencing is required, has a low-cost Internet solution, such as CU-SeeMe, been considered? For the price of an ISDN connection and a $100.00 PC camera, Internet videoconferencing can be obtained. CU-SeeMe, for example, allows Macintosh and PC users to send audio and video streams to each other (i.e., point-to-point) or to participate in groups. The qual-

ity is limited, but the system is functional and adequate for some purposes.

SENDING TRANSMISSIONS SAFELY OVER THE INTERNET

The media has reported Internet security problems to such an extent that the general public thinks the Internet and hackers are linked at the neck. Ironically, billions of dollars flow through restaurants and retail stores whose employees have easy access to credit card numbers and expiration dates. Reality is different. Although existing commercial products cannot foil a determined effort by sophisticated espionage agencies or individuals to decrypt messages, the Internet is fully adequate for most file transmissions if sensible security techniques are used. As a result, an organization can forego business opportunities if it does not take advantage of a low-cost file transmission medium because it believes the exaggerated and irrational fears of the reported lack of Internet security.

The auditor should ask the following series of questions to determine if opportunities exist to reduce transmission costs via the Internet:

- Does the organization support small field offices whose need for data exceeds short 28.8K-bps dial-up transmissions? Because Internet providers often charge a flat fee, long duration communications via the Internet are cost-effective, even if a dial-up line is used to connect to the provider.
- Are sensible security measures used for file transmissions? For example, are firewalls used on both sides (i.e., originating and receiving)? E-mail can be protected by inexpensive Pretty Good Privacy (PGP) encryption. PGP encryption is a secure way of endcoding text or binary data. It is a shareware product available on the Internet.
- Does the organization need to transmit to hard-to-reach locations where leased lines may be difficult to obtain? Typically, common carriers such as AT&T, British Telcom, and Sprint, can easily establish circuits from a major metropolitan area to the point of presence (POP) (i.e., the building where a common carrier maintains communications equipment and links to local businesses and residential areas) nearest the desired location. However, delays in establishing circuits frequently occur in getting the circuit from the local POP to the office. Using the Internet for video transmission can be an interim solution.

QUALITY MONITORING

Errors and omissions in the communications network, as elsewhere in the organization, increase costs. The telecommunications function should include self-monitoring procedures to ensure that efficiency is maintained and that costs do not needlessly escalate. The auditor should review the following areas:

- *Network response time.* Are trunk lines load balanced properly? Is there a means to size backbone circuits? Are efficient protocols used?
- *Circuit monitoring.* Are circuit conditions properly monitored?
- *Cabling system.* Do cable lengths and conditions meet specifications. Are cables near high-emission areas or run across fluorescent lighting? Is Category 5, for example, attached to the punch-down block by unwinding the ends, thus taking its properties down to Category 4 or 3 level?
- *Dial blocking.* What percent of incoming calls (i.e., data or voice) are blocked? Generally 1% or less is considered acceptable. Blocked calls should be monitored.
- *Network congestion.* Is overall traffic monitored by hour of day?
- *Protocol enforcement.* Are obsolete, rarely used, or high overhead products allowed on the network? Does peer-to-peer advertising clog the network?

SUMMARY

The growth and diversity of telecommunications presents opportunities for new methods of internal control as well as cost savings and efficiency improvement. By keeping abreast of new technologies, examining communications expenses, identifying obsolete (e.g., high cost/low bandwidth) systems, and recommending efficient, standards-based alternatives, managers can increase the communications technology payback for their organization.

5-6
The 1996 US Telecommunications Act and Worldwide Deregulation

Keith G. Knightson

DEREGULATION OF THE TELECOMMUNICATIONS MARKET is under way in many countries, including the US, the UK, the nations of the European Union, and Japan. This chapter examines various deregulation initiatives and the measures being taken by the 1996 Telecommunications Act to ensure fair competition among telecommunications companies.

DEREGULATION AND COMPETITION

There is a big difference between deregulation and the practical establishment of a competitive market. The term "deregulation" usually means the elimination of a monopoly in a public market. The dominant carrier who held the monopoly before deregulation is often referred to as the incumbent.

The intent of deregulation is to create an open and competitive market. The problem in actually creating such competition is the difficulty that newcomers (i.e., new entrant carriers) have in competing with a well-established incumbent. Unless this problem is addressed, deregulation may not result in competitive supply, and incumbents will retain de facto monopoly positions.

The initiatives being undertaken by the various national regulatory authorities, including the Federal Communications Commission (FCC), are concerned with the practical aspects of establishing an open marketplace subsequent to actual deregulation.

PROBLEMS WITH DEREGULATION

Many countries have been deregulated for years—the US since 1974, the UK since 1983, and Japan since 1985. The European Union has set a target date of 1998 for a fully open European market. Many countries that have been deregulated for some time have, based on their experiences, instituted new telecommunications acts or amendments.

In most cases, the original basis for interconnection between competing organizations depended solely on bilateral negotiations between the parties concerned. In all cases, the various national regulatory bodies usually retain the power to arbitrate and issue orders when negotiations between an incumbent carrier and a new entrant carrier fail to reach an agreement. This hands-off approach has led to:

- Protracted interconnection negotiation periods.
- Anticompetitive practices.
- Lack of free and fair conditions for potential new entrant carriers and their suppliers.

This situation is part of the natural process of developing competition. Drastic changes, such as those involved in moving from a total monopoly to fair competition, are bound to produce unforeseen consequences and associated growing pains.

Incumbents, even if forced by license or statute, do not voluntarily grant interconnection to competing network operators. Without regulatory arbitration, negotiations simply break down. The Ministry of Posts and Telecommunications in Japan and OFTEL, the national regulator in the UK, have learned this lesson from experience. The new 1996 US Telecommunications Act addresses these issues.

THE 1996 US TELECOMMUNICATIONS ACT

The 1996 Telecommunications Act and the accompanying FCC commentary were published in August 1996. The principle goals of the act include:

- Opening the local exchange and exchange access markets to competitive entry.
- Promoting increased competition in telecommunications markets that are already open to competition, including the long-distance services market.

The act directs the FCC and its state colleagues to remove not only statutory and regulatory impediments to competition, but also the operational impediments. Incumbents are mandated to take steps to open their networks to competition, including providing interconnection and offering access to unbundled elements of their networks.

Interconnection

The FCC identifies a minimum set of five technically feasible points at which incumbent local exchange carriers (LECs) must provide interconnection, including:

- The line side of a local switch.
- The trunk side of a local switch.
- The trunk interconnection points of a tandem switch.
- Central office cross-connect points.
- Out-of-band signaling facilities, such as signaling transfer points, necessary to exchange traffic and access call-related data bases.

In addition, the point of access to unbundled elements are also technically feasible points of interconnection. The FCC also anticipates and encourages parties, through open, multilateral negotiation and arbitration, to identify additional points of technically feasible interconnection.

Unbundled Access

Section 251 of the Telecommunications Act also requires LECs to provide access to network elements on an unbundled basis at any technically feasible point, including:

- Local loops.
- Local and tandem switches (including all vertical switching features provided by such switches).
- Interoffice transmission facilities.
- Network interface devices.
- Signaling and call-related data base facilities.
- Operations support systems functions.
- Operator and directory assistance facilities.

Section 251 also states that the commission will consider, at a minimum, whether:

- Access to such network elements is necessary.
- The failure to provide access to such network elements would impair the

ability of the telecommunications carrier seeking access to provide the service that it intends to offer.

The lack of a clear standard is not sufficient reason to deny access. Public standardization at agreed-upon interface points are in the interests of all parties.

Local Loop. A local loop situation can be quite complex because of the variety of transmission techniques and concentration points. However, the FCC requires incumbents to provide access regardless of technology or concentration. Thus, this requirement embraces integrated digital loop carrier (IDLC), two-wire and four-wire voice grade loops, and two-wire and four-wire voice grade loops conditioned to provide such services as integrated services digital networks (ISDN), asynchronous digital subscriber lines (ADSL), high bit-rate digital subscriber lines (HDSL), and DS1-level signals.

IDLC carries aggregated loop traffic from the point of concentration in the LEC's loop facilities directly to the switch via a multiplexed circuit. ADSL provides a high-speed downstream data path, a medium-speed upstream data path, and an analog voice path. HDSL provides 768K-bps data over two-wire and 1.544M bps over four-wire. Therefore, the granularity of unbundling is clearly an important issue.

On the question of this granularity, the FCC declines to define a loop in terms of specific elements or functions, relying on the fallback position of "technically feasible points." However, that raises the subject of sub-loop unbundling, which would provide carriers with access at various points along the loop between the local exchange and the customer premises. Additionally, this would enhance competition because the competitor would only need to purchase loop facilities that they could not provide over their own feeder infrastructure.

COORDINATING INTERCONNECTIVITY

The FCC will monitor the mechanics of achieving the desired level of interconnection and unbundling. Nondiscriminatory access to public telecommunications networks will be established through:

- Planning and design of coordinated public telecommunications networks.
- Ensuring public telecommunications network interconnectivity.
- Ensuring the ability of users and information providers to seamlessly and transparently transmit and receive information between and across telecommunications networks.

The Telecommunications Act states that the FCC can participate, in conjunction with the appropriate industry standards-setting organizations, in the development of interconnectivity standards that promote access to:

- Public telecommunications networks used to provide telecommunications service.
- Network capabilities and services by individuals with disabilities.
- Information services by subscribers of rural telephone companies.

The Network Reliability and Interoperability Committee was established under this mandate.

Network Reliability and Interoperability Committee (NRIC)

The charter of NRIC, set by the FCC, is to ask the council to provide recommendations for the FCC and for the telecommunications industry that will ensure optimal reliability, interoperability, and accessibility to public telecommunications networks. The objective of the recommendations is to ensure that users and information providers can seamlessly transmit and receive information between and across telecommunications networks. The charter asks the council to continue to report on the reliability of public telecommunications networks. The NRIC has organized two working groups to gather and analyze information.

Focus Group 1. The council has organized focus group 1 to identify technical and engineering barriers to network accessibility and interconnectivity and to identify ways to eliminate them. Focus group 1 documents and evaluates the processes by which coordinated network planning and design occur, and will evaluate options for optimizing these processes. The group considers security issues and methods by which the FCC could oversee coordinated network planning. The FCC will provide focus group 1 with conferencing resources, including a Web site, so that it can make work files available electronically.

Focus Group 2. Focus group 2 assesses the effectiveness of the standards-setting process and determines what role is most appropriate for the FCC. Focus group 2 uses conference resources provided by Committee T-1, an ANSI-accredited standards development organization sponsored by the Alliance for Telecommunications Industry Solutions, to make work files available.

Task Groups

Four task groups have been established within focus group 1 to address planning, implementation, operations, and user interoperability.

Task Group 1—Planning. Task group 1 addresses the issue of planning by:

- Identifying the differences between planning for network architectures and network implementations.
- Identifying the differences between the planning of national and regional services.
- Examining the transition of architectures, products, and services from a proprietary to a public status.
- Evaluating the impacts that protecting competitive information has on the planning and design of products and services.
- Examining timing issues relative to matching the availability of network products and services.
- Developing a recommendation on the FCC's role for coordinated network planning.

Task Group 2—Implementation. Task group 2 addresses implementation issues by:

- Monitoring information sharing.
- Monitoring the interconnection environment.
- Acting as an industry liaison to improve implementation processes.

Task Group 3—Operations. Task group 3 addresses issues of operations by:

- Investigating operations systems access (i.e., functionality, interfaces, security, reliability, and measurements).
- Overseeing performance monitoring.
- Determining security requirements (i.e., ID authorization, auditing, access control, partitioning, and measurements).
- Investigating signaling (i.e., congestion control, interoperability, reliability, synchronization, and security).
- Confirmation of interoperability by testing and certification.

Task Group 4—User Interoperability. Task group 4 monitors interoperability by ensuring that:

- There are increased interconnections.
- There are Internet interconnections.
- Asynchronous transfer mode service is available to users.
- There are service interconnection definitions.
- There are adequate standards for vendor compatibility.

WORLDWIDE DEREGULATION ACTIVITIES

Japan's Interconnection Rules

The Telecommunications Council Report indicates that the bilateral negotiations process in Japan has been plagued by problems. Negotiations for frame relay service and virtual private networks service between long-distance new common carriers and Nippon Telegraph and Telephone Corp. (NTT) have taken an unduly long period of time. Thus, interconnection between suppliers with essential facilities, such as NTT's local communications network and other suppliers, has become a very important issue to a telecommunications policy dedicated to fair and effective competition.

As a consequence, the Japanese government created the "Deregulation Action Program" to clarify the basic rules for interconnection to the NTT local communications network. Japan recently established a "set points of interconnection" very similar to those enumerated in the 1996 US Telecommunications Act.

Europe's Open Network Provision

The European community created a directive to establish the internal market for telecommunications services through an open network provision (ONP). The directive states that ONP should include harmonized conditions with regard to:

- Technical interfaces, including the definition and implementation of network termination points.
- Usage conditions, including access to frequencies.
- Tariff principles.

The requirements are published in the ONP standards list. This list is divided into those elements that are formally referenced (i.e., mandatory in the context of ON) and those still considered voluntary. These are known as the reference list and the indicative list, respectively.

The European Telecommunications Standards Institute (ETSI) is the driving force in the development of telecommunications standards for Europe. The ONP mandates ETSI to produce standards to meet its evolving requirements. Most ETSI standards are equivalent to or extensions of international standards.

The UK—A Framework for Action

The UK established two methods of addressing anticompetitive behavior. First, generic interconnection regimes will be set up, as opposed to case-by-case interconnection conditions. Second, all deliberations will take place in the public domain because:

- Publication of cases will promote better understanding of policies and the interpretation of license conditions and legislative provisions, and of what is considered to be conduct or circumstances requiring enforcement or remedial action.
- Publicity of cases in which companies were uncooperative and acted unfairly may discourage others from following suit.

SUMMARY

Countries experiencing deregulation concur that a framework based solely on bilateral negotiations between carriers does not function effectively and that something more needs to be done to level the playing field.

New initiatives are being taken in many regions around the world to address network interconnection in a more generic, consistent, and controlled way. The new provisions for interconnection within the recent 1996 US Telecommunications Act, the European Community's Open Network Provision Directives, and Japan's Basic Rules for Interconnection are ample evidence of worldwide efforts to reintroduce regulation based on collaboration and consensus.

Typically, these initiatives have two major components:

- The establishment of formal points of interconnection.
- The identification of standards to be used at the points of interconnection.

To offset discriminatory practices, the US and Japan enumerated a preliminary minimum set of obligatory interconnection points and accessible network elements selected from an unbundled architecture. Europe appears to be heading in the same direction.

The 1996 US Telecommunications Act and the new activities initiated by the FCC will be of great benefit to the consumer in offering a wider choice of services and suppliers at a lower cost. However, there is still a considerable way to go before a truly open and fully competitive environment will be realized.

Section 6
The Internet and Internetworking

AS COMMUNICATIONS NETWORKING SUPPLANTS THE COMPUTER as the core of the information processing system, internetworking becomes a critical necessity. In the purest sense, internetworking is the process of linking distinct networking domains in a way that is totally transparent to any user connected to any of the domains. The most famous of the public internetworks is, of course, that massive, globe-spanning assemblage of networks called the Internet.

Over the past year, the surge in Internet growth and use has led many organizations to apply the Internet architecture, principles, and protocols to private network versions of the Internet. Because these corporate networks are fundamentally designed to support intracompany communications, they have been dubbed "intranets."

An intranet could be defined as a private enterprise network characterized by the common use of the Internet routing protocol, the use of Web server technology for the storage of information, and the application of browsers as the standard user workstation interface for accessing that information.

Intranets represent the next generation of the corporate network model. The first generation was the host-centric network, which was hierarchical in organization, relatively slow, used low-bandwidth connections, and had tight timing constraints to maintain user associations. The current model is the client/server model, in which the architecture is much more distributed (although still somewhat hierarchical), the traffic is more bursty, the bandwidth is broader, and data rates are 100 times faster.

The new intranet model offers totally distributed, non-hierarchical, peer-to-peer architectures. Bursty and constant-rate traffic share multimedia-capable network facilities. Data rates are increasing into the gigabit-per-second range.

An important aspect of intranetworking is that, in deploying these networks, users are applying technology that has been developed and proven over many years on the Internet. TCP/IP is one example of such technology. These middle layer protocols make the overall intranet independent of the underlying physical networks used at any single location. The individual networks can be local or wide area, Ethernet or Token Ring, and can use cells, frames, or packets as the unit of transfer. This independence means that individual departments or locations or countries can retain a high degree of autonomy in meeting their individual requirements and still remain closely knit with the organization as a whole.

In addition to the use of common networking protocols, intranets offer many benefits to the organization. Because it is estimated that around 70% of US-based organizations have deployed or are in the process of deploying intranets, it is clear that these benefits must be more than mere marketing hyperbola. Intranets make information as easily accessible as people are today via the telephone. In most cases, access can be even easier because intranets are independent of time zones. The servers, as the repositories for the information, are accessible anytime from anywhere.

Intranets can produce significant cost savings in access time as well as in paper and distribution costs. Phone directories, policies and procedures, training schedules, product catalogs, and almost any other kind of document can be accessible and distributed via an intranet.

Intranets also compare favorably with the more traditional corporate networks and with the Internet. The intranet combines the best features of each. An intranet can have the security and management features of the corporate network combined with the Web-based accessibility and standard protocol features of the Internet.

The deployment of an intranet is, however, not a simple task. There can be serious impacts on existing local and wide area networks. For example, because Web server access is unpredictable in the short term, network scalability becomes very important. Bandwidth needs to be flexible. In the wide area case, there may be insufficient capacity in leased lines to meet the surging demands for server access. If access becomes problematic, higher-speed frame relay or cell relay may need to be considered, and location of servers can be critical.

The communications systems manager contemplating a migration to a corporate intranet must first have an accurate picture of the current network. The manger must ask: Who will be using the intranet? Where are they located? What existing resources can be applied? What will be the capacity demand as a function of time? The answers to these and many more questions will provide the foundation for a migration plan.

This section begins with Chapter 6-1, "The Essentials of Enterprise Networking." Chapter 6-1 clarifies the networking problem and offers tools and descriptions that will aid readers in integrating different network types, selecting vendors, dealing with multiple protocols, and understanding addressing and routing. Internetworking, interoperability, and global networking are also discussed.

The subject of network capacity (not to be confused with bandwidth) is often raised as communications managers contemplate the new, higher-speed model of telecommunications. Because of increases in network capabilities, some observers talk about essentially free capacity. Most communications managers realize that capacity will never be free, and that for that reason, careful use of the resource will always save money. As a result, data compression will always be a useful tool. Chapter 6-2, "Data Compression in Routed Internetworks," offers a primer on the subject of compression. Methods of compression, implementation issues, and design criteria are all explored in this chapter.

Chapter 6-3, "Accessing and Using the Internet," reviews the basics of Internet applications including E-mail, file transfer, and remote logon. The World Wide Web is also explained. The chapter emphasizes connection methods and provides a selected list of Internet service providers. The chapter will be useful in selecting an access method appropriate to particular needs.

Internet and intranet security is a common networking topic. Chapter 6-4, "Internet Security Using Firewalls," addresses this important issue. The term "firewall" has come to mean a workstation or other network element designed to screen all incoming traffic for authorized access. The device may also be used to limit traffic leaving the network to that authorized by a network administrator. This chapter looks at the risks of connecting to the Internet and examines different approaches to establishing firewalls.

Networks based on IBM's Systems Network Architecture (SNA) remain the most dominant type of enterprise network. The rapid proliferation of local networks demonstrates the need for Chapter 6-5, "Managing Coexisting SNA and LAN Internetworks." Many scenarios that arise when integrating SNA and LAN traffic are explored in great detail in this chapter. Any manager concerned with SNA and local networking will find this chapter useful.

Most traditional SNA networks use dedicated leased lines. Recently, there has been a trend toward the use of frame relay to replace these leased lines because of the promise of a 20% to 30% cost savings. "SNA Over Frame Relay," Chapter 6-6, reviews how IBM is making frame relay an integral component of its strategy for integrating LANs and SNA systems.

Chapter 6-7, "IPv6: The New Internet Protocol," concludes this section. IPv6 is a major revision of the venerable IP. Many industry experts collaborated in creating this revised networking protocol, which contains features that will bring the protocol up to date and allow it to remain viable as the primary internetworking protocol for the next generation of high-speed networks.

6-1 The Essentials of Enterprise Networking

Keith G. Knightson

ENTERPRISE NETWORKS AND ENTERPRISE networking are buzz-phrases on every salesperson's lips, together of course, with open and open systems. Many products are glowingly described with these phrases. Creating an enterprise network, however, requires more than just a knowledge of the buzzwords. This chapter explains the basic subtleties of an enterprise network and the challenges of establishing one.

THE NEXT GENERATION OF ENTERPRISES

An enterprise is nothing more than a fancy name for a given company or organization. It conveys, however, the notion of a geographically dispersed, multifaceted organization, an organization comprising many branches, departments, and disciplines (e.g., marketing, manufacturing, finance, administration).

In the past, the networking and information technologies deployed in an enterprise were many and disjointed. In some cases, this was because of local departmental or workgroup autonomy, or simply because of ignorance among different parts of the enterprise as to what information systems were being used, or it was an artifact of historical equipment acquisition procedures.

The allegiance of specific departments to particular vendors was also a factor. When acquisition of capital equipment is performed gradually (rather than implemented all at once, across the board), it is difficult to make sure that all equipment is mutually compatible. Finally, the lack of an enterprisewide view, strategy, or policy with respect to networking and information technology—and the possible convergence solutions—are contributing considerations.

Consolidating the Network

In the same sense that the word "enterprise" conveys the totality of an organization's operations, the phrase "enterprise network" means combining all the networking and information technology and applications within a given enterprise into a single, seamless, consolidated, integrated network. The degree of integration and consolidation may vary; total integration and consolidation may not be always achievable.

For example, an organization may have an SNA network from IBM Corp. and a DECnet from Digital Equipment Corp. In all probability, these two networks have their own communications components; there might be one set of leased lines serving the SNA network and another completely independent set of leased lines serving the DECnet.

It would be useful if all the IBM users could intercommunicate with all DEC users, but a first and evolutionary step might be to have both the SNA network and DECnet share the same leased lines. Now, only one physical network has to be managed instead of two separate ones, and more efficient and cost-effective sharing of the physical communications plant can be achieved.

A second step might be to interconnect the mail systems of the two networks to achieve at least the appearance of a single enterprisewide electronic-mail system.

A third step might be to unify the data and information and its representation as used within the organization. This would enable basic forms of data to be operated on by many applications.

The challenges of building an enterprise network fall into two distinct categories: getting the data (i.e., information) from A to B, and enabling B to understand the data when it receives it from A. These two categories are referred to in this chapter as the "networking challenge" and "beyond the networking challenge." In this context, the network is used as it is in the Open Systems Interconnection (OSI) reference model—that is, layer 3 and below.

THE NETWORKING CHALLENGE

The networking part of the problem has three major components:

- Choosing from and integrating the many network technologies.
- Selecting from the many vendor solutions.
- Moving information from a local to a global environment.

Integrating Network Technologies

The first basic problem with networks is that there are so many of them. In this context, networks are taken to mean the raw network technologies—

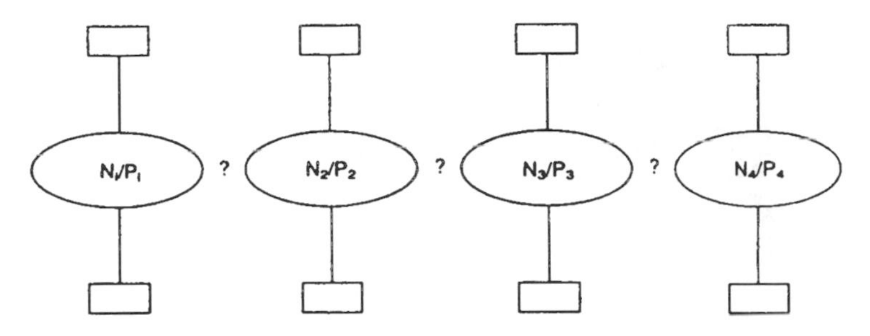

Notes:
- For every (sub)network N_i, there is a network-specific protocol P_i.
- All P_i's are different and thus not directly interconnectable.
- A way to make all the different networks look the same must be found, if the enterprise network is to appear as a single seamless network.

Exhibit 6-1-1. The Interoperability Problem

---

leased lines (i.e., T_1 and T_3), X.25, ISDN, frame relay, asynchronous transfer mode (ATM), and the many and various LAN access methods.

If all the users in an enterprise are connected to the same network technology, there is no problem. Unfortunately, this is not always the case. Communication between users on dissimilar networks (e.g., two different LANs) is where the problem occurs.

Each network technology has its own characteristics and inherent protocols. From an enterprise viewpoint, this is bad news. For example, users connected to an X.25 network cannot easily be connected to those already connected to a LAN. For example, how would the X.25 user indicate the destination's media access control (MAC) address, and vice versa? X.25 networks understand only X.25 addresses and LANs understand only MAC addresses. The differences between network technologies and native protocols almost invariably prevent their direct interconnection.

Differences in addressing schemes present another difficulty. Addressing considerations alone usually dictate the use of a network interconnection device (NID) at the point at which two network technologies come together.

Exhibit 6-1-1 illustrates several network technologies, represented by N_1, N_2, N_3, N_4. Each of these technologies has its own native protocol (i.e., P_1, P_2, P_3, P_4). A way must be found to integrate all these disparate technologies into a single supernetwork, with globally uniform and globally understood characteristics and a single addressing scheme.

This is achieved by operating an integrating, unifying protocol (shown in Exhibit 6-1-2 as Px), sometimes known as an Internet protocol, over the top of all the possible basic communications networks. The Internet protocol (IP) of TCP/IP is one such protocol. The connectionless network layer protocol (CNLP) specified in the OSI International Standard 8473 is another. Propri-

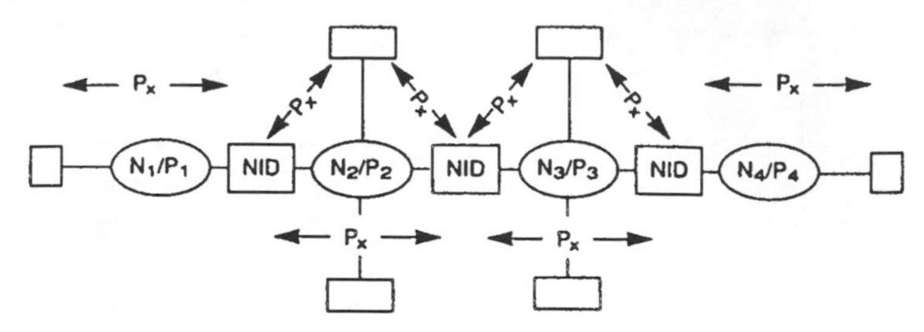

Notes:
NID Network interconnection device
- A neutral protocol P_x is run over the top of every subnetwork.
- Because P_x is now the basis of the enterprise network, the individual networks are called subnetworks.
- P_x operates at the top of the network layer and is usually called an internet protocol.
- Several Px's exist (i.e., IP, CNLP, IPX).

Exhibit 6-1-2. The Interoperability Solution

etary systems have their own Internet protocols—for example, Novell uses its Internetwork Packet Exchange (IPX) and Banyan uses Vines.

From the architectural standpoint, the technical term for such an Internet protocol is subnetwork independent convergence protocol (SNICP). The protocols used on real-world communications networks (e.g., leased lines, X.25, frame relay, LANs) are known as subnetwork access control protocols (SNACP). The basic internetworking architecture is shown in Exhibit 6-1-3.

Unification does not mean simplification. Two protocols operating over a given subnetwork still require two address schemes. Routing tables are then needed in the network interconnection device to map the global enterprise address to the address to be used by the network interconnection device for the next link in the composite path. Exhibit 6-1-4 is a simplification of how the two addresses are used. In practice, the "next" address may be more complex, depending on the internetworking protocols under consideration. A network interconnection device of this type is called a router.

Selecting Vendor Solutions

The second basic problem is that each system vendor has a vendor-specific idea of how to build the supernetwork—the type of supernetwork protocol, the global addressing scheme, and the internal routing protocols to be used. At worst, this leads to a multiprotocol network, which amounts to several separate internets operating in parallel over the same physical communications plant.

Dealing with Multiple Protocols. An alternative to the multiprotocol network is to choose a single protocol for the entire enterprise supernetwork.

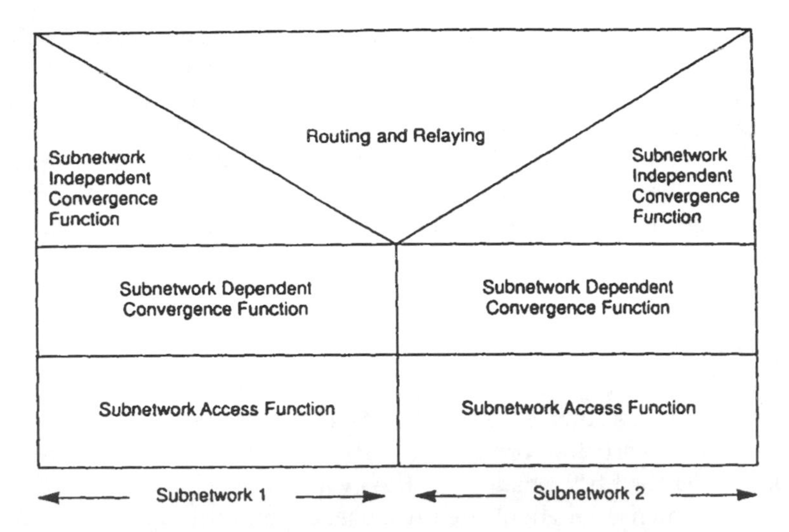

Exhibit 6-1-3. Network Layer Architecture

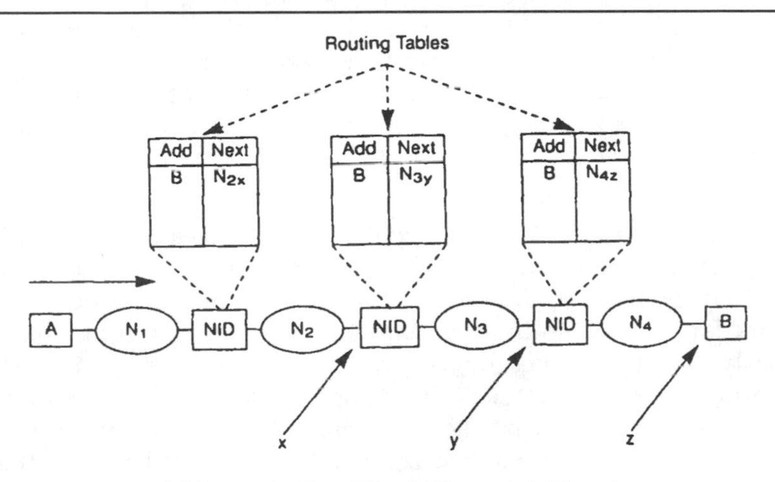

Exhibit 6-1-4. Simplified View of Addressing

This inevitably requires finding techniques to accommodate the systems that do not inherently operate this chosen protocol. Techniques include encapsulation (sometimes called tunneling) at the edges of the single-protocol network, or other techniques such as transport service interfaces and application gateways.

However, even with a single protocol, tunneling permits only the coexistence of incompatible systems; there can be little or no interaction between each of the tunneled applications. The major advantage of tunneling is that the core of the network is unified, optimizing network management and networking skills. The disadvantage is the effort required to set up the tunneling configurations at the edges.

The best solution is for all vendors to use the same Internet protocol. Increasingly, the protocol of choice for this purpose is TCP/IP. Although not a networking panacea, TCP/IP is the protocol of choice for most networking challenges involving multiple protocols.

Going Global

Many LAN-based systems include internal protocols that advertise the existence of various LAN-based servers. Such a protocol is sometimes known as a service advertising protocol (SAP). Protocol exchanges, frequently broadcast over the LAN, ensure that the availability and addresses of various servers are known throughout the LAN user community. This is useful when the geographic area is confined to a work group or a floor of a building; for example, the knowledge of a set of available printers is useful only in the area that has ready access to one of them. Thus, local messages must be constrained to local environments by putting adequate filtering at the point of access to the wide area portion of the enterprise network. There is no point in telling a user on a LAN in New York that there is a printer available on a LAN in Seattle.

WAN Transit Delays. Another global problem relates to the extra transit delay involved in transport over a WAN, especially for nonroutable protocols. Many protocol stacks used in local environments do not contain a network layer protocol—in other words they have no routing layer. Such protocols cannot be routed directly in a router-based enterprise network. Where it is necessary for such an application to be networked outside a particular local environment, the local protocol stack must be encapsulated within an internetworking protocol. Then it can be launched onto the wide area part of the enterprise network.

Many of the local or nonroutable protocols are designed for very rapid acknowledgment. The transfer of these types of protocols across a wide area may cause problems; applications may prematurely time-out or suffer poor throughput because of lack of a windowing mechanism adequate for the wide area transit delay.

To accommodate such applications, it is necessary to "spoof" the acknowledgments. This means that acknowledgments must be generated by the local encapsulation device. This requires the extra complication of adding a reliable transport protocol on top of the internetworking protocol across the wide area portion of the enterprise network. Once a local acknowledgment

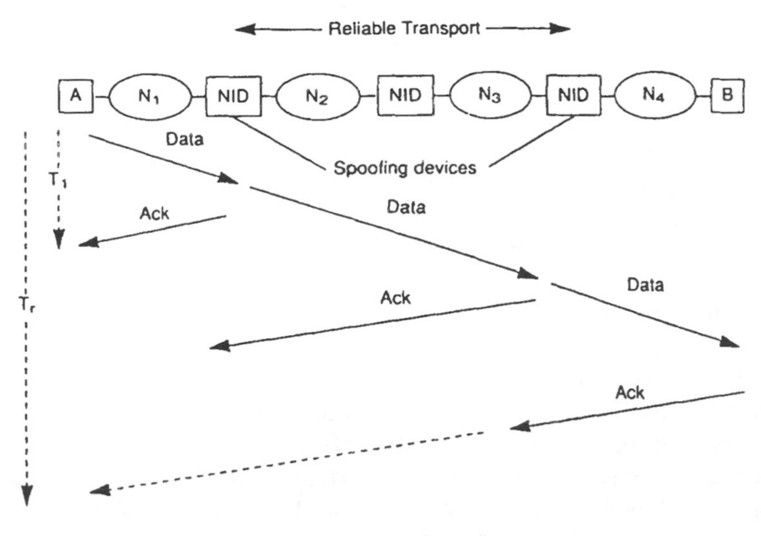

Exhibit 6-1-5. Spoofing

has been given, the originator will discard the original so it is no longer available for retransmission. Having given the local acknowledgment, the spoofing device must ensure reliable delivery to the remote end by employing a transport protocol of some sort (e.g., TCP or OSI Transport Class 4). The scheme, shown in Exhibit 6-1-5, avoids the end-to-end round trip delay T_r for every packet of data by providing an acknowledgment at time T_1.

Addressing. Going global also poses some challenges in the area of network layer addressing, particularly with regard to achieving enterprisewide uniqueness and structuring addresses for scalability and ease of routing.

Usually, addresses of stations within a local workgroup are allocated locally. This can present problems when subsequently the local workgroups must be integrated into a single enterprisewide address scheme. If several workgroup addresses—or parts of an address (e.g., an area or server name)—are the same, some changes will have to be made. From an operational perspective, changing addresses is not a trivial matter. It is best to avoid address allocation clashes right from the outset by having an enterprisewide address registration authority set up within the organization.

Some addressing schemes do have some hierarchy associated with them that can be used to avoid address encoding clashes by ensuring that local addresses are only the low-order part of the total address. Even in this case, however, an enterprisewide perspective is necessary to avoid clashes in the high-order part of the address.

Some vendors achieve uniqueness by allocating unique addresses when the equipment is shipped. However, this usually results in a flat, random address space that makes routing considerably more complex because there is no structure in the address to help "scale" the enterprise network from the routing perspective.

If the enterprise is to be permanently connected to the Internet (as opposed to using a dial-up connection), IP addresses must be obtained from an appropriate addressing authority. Until recently, all addresses were dispensed directly from the Internet Network Information Center (InterNIC). More recently, in response to a number of problems associated with addressing practices in the past, IP addresses have begun to take on a more hierarchical form. As such, the enterprise may need to obtain a block of addresses from its Internet Service Provider (ISP), in effect obtaining a subset of the addresses that ISP has obtained from the InterNIC.

This practice ensures that the appropriate hierarchical relationships are maintained, allowing improved routing, and it has the added benefit of more efficiently allocating the available addresses. The primary drawback from the perspective of the enterprise is that addresses obtained in this fashion are no longer considered permanent. That is, if the enterprise changes ISPs, the addresses may also have to be changed.

Hierarchical Schemes. The most widely documented and hierarchically administered address available today is the OSI address space available for OSI Network Service Access Point (NSAP) addresses. A more recently developed scheme is the next generation of IP, now known as IP version 6 (IPv6), described in RFCs 1883-1886. NSAP addresses can consist of up to 40 digits, and IPv6 addresses can be up to 128 bits, either of which allows good scaling potential and simplified routing.

The reason that a hierarchical (i.e., scaled) address scheme is so important has to do with the way that routers operate and the size of the associated routing tables. If addresses were allocated completely randomly but uniquely from a large address space, every router would need a table with every address in it. Not only would the table be extremely large, but the time needed to find an entry could also be a problem. Routing is thus better arranged on the basis of hierarchical distinctions that are implicit in the address scheme.

To service a local workgroup or other limited geographical area, a local router must know only whether the destination address is internal or external. If it is internal, the router knows how to get the message to the destination; if it is external, the router can pass it on to the next-level router. This leads to the concept of areas, groups of areas, domains, and countries being components of a hierarchical address.

When legacy systems must be accommodated with conflicting address schemes and reallocation of addresses is impossible, tunneling may have to be employed merely to avoid interaction between the conflicting addresses. Because conflicting networks are divided into separate virtual private networks, the protocol under consideration cannot be routed natively even if the backbone routers are capable of doing so.

Routing Protocols. To reduce the amount of time devoted to setting up routing tables manually, and to allow dynamic rerouting and a degree of self-healing, routing protocols are often employed to distribute routing information throughout the enterprise network. These protocols are in addition to the internetworking protocol itself but are related to it.

For every internetwork protocol routed in a multiprotocol network, there may be a specific routing protocol (or set of protocols). This also means in general that there will also be a separate routing table for each internetworking protocol. The situation in which several routing protocols are used simultaneously, but independently, is sometimes known as a "ships in the night" situation, because sets of routing information pass each other and are seemingly oblivious to each other even though there is only one physical network.

Some router manufacturers operate a single proprietary routing protocol between their own routers and convert to individual protocols at the edges of the network. There have been some attempts to define a single standard routing protocol based on the International Standards Organization's intermediate system to intermediate system (IS-IS) standard.

In an enterprise network, end systems (e.g., terminals, workstations, mainframes) usually announce their presence and their own addresses to the nearest local router. The local routers record all the local addresses within their area and inform all neighboring higher-level routers of their own area address. In this way, a router at the next and higher level in the hierarchy only needs to know about areas. Recursive application of these principles to a hierarchical configuration can lead to efficient routing by minimizing the amount of routing information to be transmitted and by keeping the size of routing tables small.

As the process of promulgating routing information proceeds across the network, every router in the enterprise network will obtain a table of reachability that it can then use for choosing optimum routes. Route optimality may be based on a number of independent metrics (e.g., transmit delay, throughput, monetary cost). Invariably, a shortest path first (SPF) algorithm is used to determine the optimal route for any particular metric chosen as the basis for routing. Both the Internet and OSI routing protocols use an SPF algorithm.

ROUTERS

Routers are the key interconnection devices in the enterprise network; subsequently, the router market has been one of the key growth areas during this decade. Some router vendors have grown from small $10 million companies to $1 billion companies.

In most cases, routers are purpose-built communications processor platforms with hardware architectures specifically designed for high-speed switching. Several possible pitfalls await the unwary purchaser of routers. Such a purchase involves four important considerations:

- The capacity and architecture, in terms of the number of ports accommodated and throughput achievable.
- Internetwork protocols supported and their associated routing protocols.
- Support of technologies for the connected subnetworks.
- Interoperability between different vendors.

Capacity and Architecture

The number of ports required determines to a large extent the size of the router required, which in turn affects the architecture and throughput of the router. Physical size of circuit boards dictates how many ports can be placed on a single board. The greater the number of ports, the greater the number of boards required and the more critical the architecture.

Routing between ports on the same board is usually faster than routing between ports on different boards, assuming that there are on-board routing functions. Boards are usually interconnected by means of some kind of backplane. Backplane speeds can vary greatly between vendors. Routing functions and tables may be distributed across all boards or may be centralized. The bottom line is that the architecture affects the performance, and performance figures are sometimes slanted toward some particular facet of the architecture. Thus, some routers may be optimal for certain configurations and not so good for others.

Many of the router manufacturers make several sizes of router, which could be referred to as small, medium, and large. All of one vendor's routers may, regardless of size, offer the same functions, but the circuit boards may not be interchangeable between the different models. This can make a big difference when it comes to stocking an inventory of spare parts. There may also be differences in network management capabilities.

When making comparisons, the data communications manager must carefully analyze vendor throughput and transit delay figures. Although worst cases are helpful for the user and network designer, some vendors specify either the best cases or averages. Other metrics involved in measurement may also be different (e.g., packet size assumed, particular internetwork protocol, particular subnetwork).

Other architectural considerations include extensibility and reliability. For example, is hot-swapping of boards possible? If the router must be powered down and reconfigured to change or add new boards, the disruption to a live network can have severe ripple effects elsewhere in the network. Can additional routing horsepower be added easily as loads increase, by simply inserting an additional routing processor?

The question of using standalone or hub-based routers may also be relevant. This is a difficult problem because of the traditional split between the hub and router manufacturers. Hub vendors tend not to be routing specialists, and router vendors tend not to be experts at hub design. Alliances between some vendors have been made, but the difference in form factors (of circuit boards) can result in some baroque architectures and poor performance. Except in the simple, low-end cases, purpose-built standalone routers usually perform better and are more easily integrated with the rest of the network.

Some standalone routers can directly handle the multiplexed input streams from T1 and T3 links, making voice and data integration possible. This is unlikely to be the case for a hub that has been designed mainly for operation in a LAN.

Internetwork Protocols Supported

Most router vendors claim that they support a large number of internetworking protocols. In some cases, however, there may be restrictions on the number of protocols that can be supported simultaneously. There may also be restrictions on the use of multiple protocols over certain network technologies, or hidden subnetwork requirements. An example of the latter might be the need for a separate X.25 permanent virtual circuit (PVC) for every individual protocol, as opposed to operating all the protocols over a single PVC.

Some vendors may also use a proprietary routing protocol scheme for internal routing, only making the standard protocols available at the periphery of the network. This makes it difficult to mix different vendors' router products on the same backbone or within a single routing domain.

Network Technologies Supported

Most manufacturers provide interfaces to a large number of network technologies (e.g., X.25 ISDN, frame relay, T1, T3, Ethernet, Token Ring). The method of support may also vary. For example, in the case of leased circuits, it may or may not be possible to directly connect the carrier's line to the router. Some routers may accept the carrier's framing mechanism directly;

others may require an external converter to provide a simple serial interface (e.g., V.35) before connection can be achieved. Buyers should remember that the interaction between these interfaces and the multiple internetwork protocols may not be clearly reported by the vendor.

Interoperability

In the not too distant past, there was little interoperability between routers from different vendors. The reason most often cited was lack of standards for operating multiple protocols over a given subnetwork topology. Fortunately, the Internet community has made substantial progress subsequent to its definition of the Point-to-Point Protocol (PPP), which originally defined encapsulation and discrimination methods for multiprotocol operation over leased circuits. More recently, the utility of PPP has been extended with numerous enhancements. For example, it can now be used over switched services, including dial-up, ISDN, and Frame Relay, and it can be used in a multi-link configuration. It can operate with or without authentication, and with or without compression.

This plethora of options has led to the widespread support for PPP, both in terms of the number of protocols standardized for use with PPP, and in terms of the number of vendors building compatible routers. As such, interoperability among routers of different vendors is much more common than it was just a few years ago.

Network Management

It is extremely unlikely that a common set of management features will apply to all vendors' routers. Thus, if several manufacturers' routers are deployed in a given enterprise network, several management systems probably will be required. In the best case, these systems can be run on the same hardware platform. In the worst case, different hardware platforms may be required.

Filtering

The degree of filtering that can be applied—to prevent local traffic uselessly flooding the enterprise network—may vary with the manufacturer. Various parameters can be used as the basis for filtering—for example, source address, destination address, protocol type, and security codes. The disadvantage of using filtering is the labor involved in setting up the filter tables in all the routers.

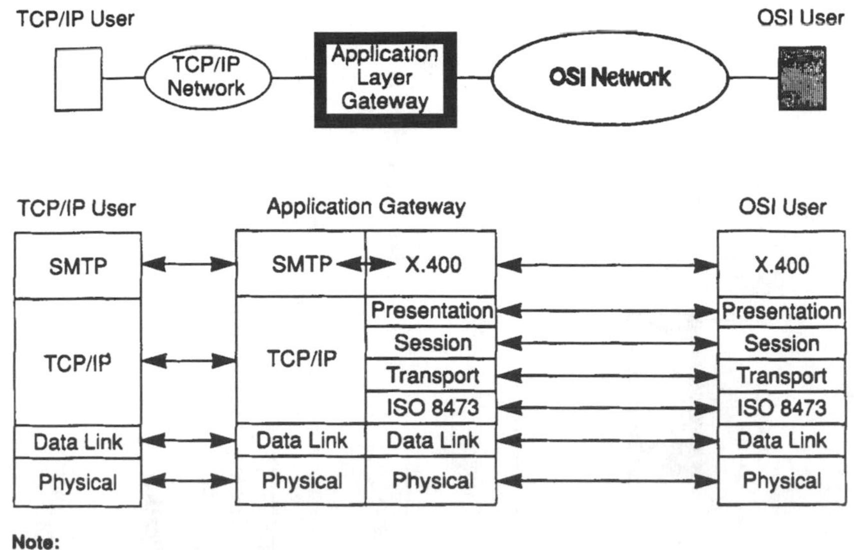

Exhibit 6-1-6. Mail Application Gateway

BEYOND THE NETWORKING CHALLENGE— THE APPLICATIONS

Gateways, Tunneling, and Transport Service Interfaces. All the considerations discussed so far apply to the internetworking protocols. Multiprotocol networks serve only to share bandwidth; they do not allow applications to interoperate. Where that is necessary, with completely different stacks of protocols, an application gateway must be used. Exhibit 6-1-6 shows an OSI-based mail (X.400) application interoperating with a TCP/IP-based mail application over an application gateway.

Such gateways may be sited either centrally or locally. The use of local gateways makes it possible to deploy an application backbone with a single standard application operating over the wide area portion of the enterprise network (e.g., an X.400 mail backbone). This reduces the number of gateways needed for conversion between all the different applications. Only one conversion is necessary for each application (i.e., to the one used on the backbone). A considerable number of different local systems could interoperate through the "standard" backbone application.

The encapsulation technique already mentioned in the context of IP tun-

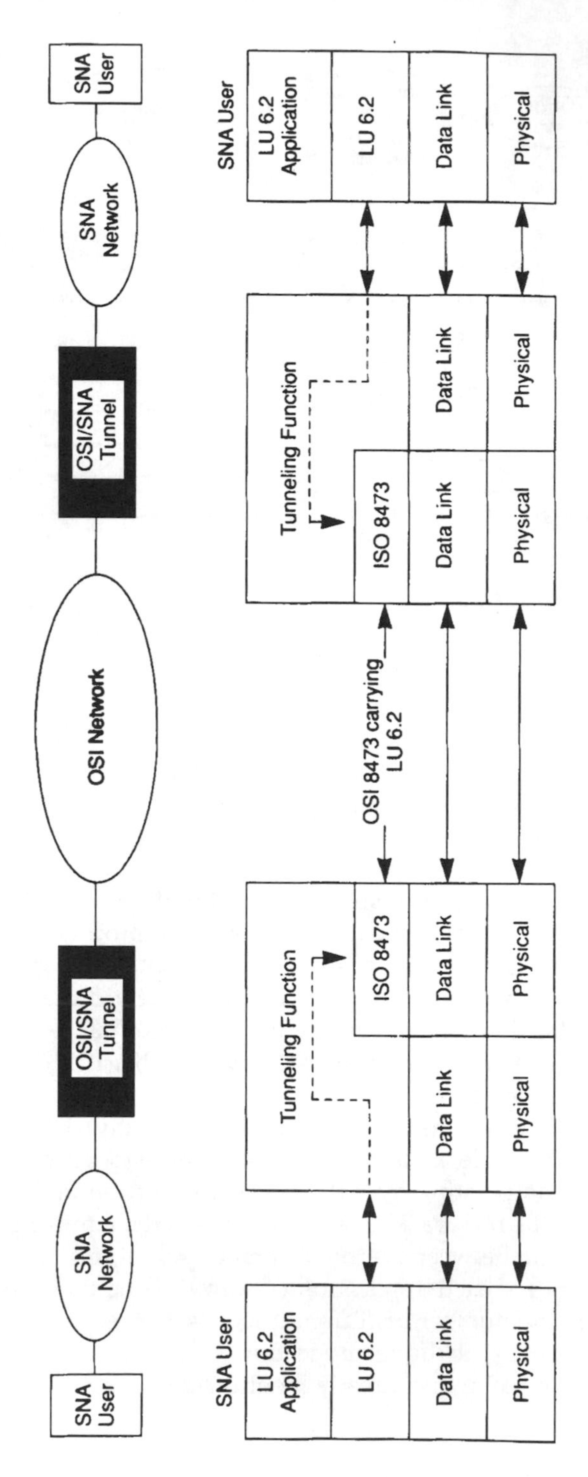

Exhibit 6-1-7. Tunneled SNA Application

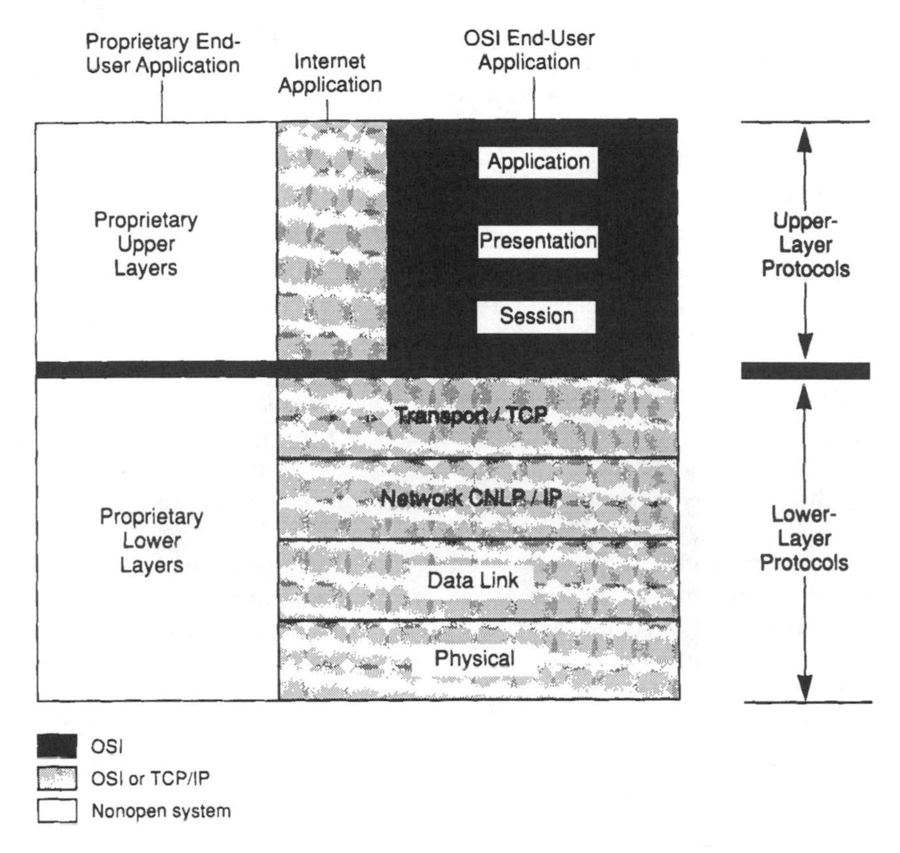

Exhibit 6-1-8. Transport Layer Interface

---

neling allows the applications that can be so configured to operate across the enterprise network. A tunneled SNA application is shown in Exhibit 6-1-7.

Another solution that may help in the future is the availability of transport service interfaces for end systems (e.g., workstations, terminals, servers). A transport server interface allows a given application to be operated over any underlying communications protocol stack. In other words, applications and communications stacks can be mixed and matched as necessary. The so-called open operating systems (e.g., POSIX and X/Open) adopt this approach.

The transport layer is a fundamental dividing line in the system architecture. Network-related functions are separate from application-related functions so that applications work with many communications protocols. Exhibit 6-1-8 shows an end system containing both an open OSI/TCP/IP stack (shaded) and a proprietary stack (unshaded). Within an end system, protocol stacks can generally be separated into the communications-specific lower-layer parts and the application-specific upper-layer parts. The two stacks communicate through a transport layer interface (TLI).

SUMMARY

In practice, legacy systems or other requirements result in the existence of a variety of heterogeneous systems. Several techniques can be applied to at least make the heterogeneous systems networkable over a single physical network. Varying degrees of interoperability between them may also be possible.

TCP/IP is the single common protocol that has made the most tremendous advances toward this objective. With the continued progress in developing Internet protocols, coupled with the impending migration to IPv6, the multivendor networking situation will only improve

Nonetheless, developing an enterprise network architecture continues to pose significant challenges. An overall plan for the network minimizes confusion and puts in place a timed migration strategy toward a completely integrated network. Central control has fallen into disrepute, but without some control over how networking is to be achieved, all the real benefits of an enterprise network will not be realized. Finally, it is probably fair to say that enterprise networking is still something of a black art and is bound to present all data communications managers with some surprises and disappointments.

6-2 Data Compression in Routed Internetworks

Sunil Dhar

CORPORATE INTERNETWORKS CONSIST OF LANS that are connected to each other over WAN transmission links. The WAN links can be very expensive by themselves, and as part of an enterprisewide network, they can consume a significant portion of the operating cost of the network. Data compression enables organizations to put much more data on their existing networks, but for data to continue to move reliably, the data communications manager must understand how compression functions and its effects on routing.

One trade magazine's cost study of a fully meshed six-router internetwork determined the operating costs of the network's hardware, software, and maintenance to be 12.2% of the total operating costs over a five-year period. The balance of the operating costs, 87.8%, was accounted for by the WAN transmission costs.

WAN connections between routers have traditionally been point-to-point leased lines. This is rapidly changing as ISDN, X.25, frame relay, switched multimegabit data services (SMDS), and asynchronous transfer mode (ATM) are made available for use in internetworks. A compression solution for a router must be adaptable to all WAN services. In addition to the many other capabilities developed to make WANs more efficient and cost-effective, compression is targeted to help provide lower cost of internetwork ownership.

WAN-BANDWIDTH BOTTLENECK

The demand for WAN data transport is increasing daily. Driving this growth are the many applications that are being used in the networks and the growth in the size of the networks themselves.

LAN applications can generally be put in two categories:

- Interactive applications that generate transaction-oriented traffic.
- Bulk data transfer applications that generate file transfers and other types of non-transaction-oriented traffic.

Otherwise, applications all have different characteristics and thus demand different services from the WAN transmission facility. Accepted WAN optimization techniques (e.g., header compression and priority queuing) can be applied to the applications only if the different needs of each application are anticipated.

As the terminal-based application model has given way to the LAN-based model, the requirement for wide area network transmission bandwidth has grown. The easiest way to address the problem of additional bandwidth is to add more bandwidth; however, this is an expensive solution. In addition, there are some circumstances in which the bandwidth is not available at any price. Increased bandwidth is nonetheless required as the LAN-based applications become more prolific and cause increased contention for access to the internetwork.

A cost comparison for leased US, European, and transatlantic links is shown in Exhibit 6-2-1 to demonstrate just how expensive bandwidth is in different parts of the world. The figures highlight the monthly and annual costs that are potential targets for such cost reduction techniques as compression.

Cisco Systems, Inc. (Menlo Park CA), for example, offers compression, priority output queuing, custom queuing, access lists, Novell static service advertising protocol (SAP), and Novell IPX SAP filters to ease the WAN-bandwidth bottleneck. The compression features—Transmission Control Protocol/Internet Protocol (TCP/IP) header compression and Digital Equipment Corp. Local Area Transport (LAT) compression—will also be augmented with full data compression.

WHAT IS DATA COMPRESSION?

Data compression involves reducing the amount of time needed to transmit information over a channel of fixed bandwidth by using a special encoding algorithm. The key principle behind the operation of a compression algorithm is removing redundancy in the data patterns. This reduces the size of the message being sent over the WAN. The objective of compression is obvious; however, the use of this technology is not without issues and trade-offs related to the efficiencies achieved.

Data compression algorithms are generally "lossy" or "lossless." Lossy algorithms allow some information to be degraded or discarded—in other words, they do not decompress data 100% back to original. Voice and video

	64K-bps		1.544M-bps	
	Monthly Cost	Annual Cost	Monthly Cost	Annual Cost
New York to Paris	$ 7,906	$ 94,872	$ 79,883	$ 958,596
Paris to Frankfurt	$ 6,068	$ 72,816	$ 62,089	$ 745,068
Paris to London	$ 5,925	$ 71,100	$ 72,770	$ 873,240
London to Brussels	$ 4,849	$ 58,188	$ 63,403	$ 760,836
San Francisco to New Jersey	$ 1,375	$ 16,500	$ 16,360	$ 196,320
Los Angeles to Atlanta	$ 1,085	$ 13,020	$ 13,000	$ 156,000
Chicago to Dallas	$ 570	$ 6,840	$ 6,840	$ 82,080
Total Cost	$27,778	$333,336	$314,345	$3,772,140

Source: Lynx Technologies, *Communications Week*, *Network World*.

Exhibit 6-2-1. Leased-Line Costs

are two applications for which some degradation of quality is acceptable to achieve higher-compression and lower-bandwidth use.

Lossless algorithms (i.e., those that reproduce the original bit stream exactly, with no degradation or loss) are required by routers and other internetworking devices for the transport of data across the WAN. Without the ability to compress the data stream, transport it across the WAN, and then decompress it into its original form, routers would be constantly reversing transmissions, negating the benefits of compression.

Efforts to develop industry standards for voice and video compression are being undertaken by the Telecommunications Subcommittee of the International Telecommunications Union (ITU-TS) and the American National Standards Institute (ANSI). However, these efforts have not translated into a synchronous compression standard for use in routers.

Work is progressing on asynchronous compression techniques for use with modems and communications servers, and may soon start for synchronous techniques as well. The Internet Engineering Task Force (IETF) is actively working, through the Point-to-Point Protocol (PPP) Working Group, to define a compression negotiation mechanism for use by routers and other devices using PPP. This would not define a compression standard, but it would allow users to achieve interoperability between devices using the same compression algorithm and the PPP compression negotiation mechanism.

Two types of algorithms have been developed for lossless compression:

- Statistical encoding.
- Dictionary encoding.

Statistical Encoding. A statistical encoding algorithm (e.g., the Huffman encoding algorithm) essentially takes each symbol (i.e., a character or a number) in a data stream and assigns it a code based on the probability that it will occur in the data stream. Highly probable symbols are assigned short codes; highly improbable symbols are assigned long codes.

Statistical compression can be extremely fast, with low latency, particularly if the code chosen is nonadaptive (that is, the lookup table is predetermined and fixed). However, the very nature of fixed coding implies a tradeoff between the speed at which compression is performed and the optimization of compression.

To obtain acceptable compression ratios with statistical compression requires both adaptation (meaning significant processing) and large amounts of memory. Compression based on nonadaptive statistical encoding is best applied to a single application, in which the data is relatively consistent and predictable. This is not the case on internetworks; hence, this type of algorithm is unsuitable for routers.

Dictionary Encoding. Alternatively, a compression algorithm based on dynamically encoded dictionary (e.g., Lempel-Ziv algorithms) replaces a continuous stream of characters with codes; symbols represented by given codes are stored in memory in a dictionary. Because the relationship between a code and the original symbol varies as the data varies, this approach is more responsive to the changing needs of the data. This is especially important for LAN data, because many applications can be transmitting over the WAN at any one time. As the data varies, the dictionary changes to accommodate and adapt to the varying needs of the traffic. Small dictionaries (i.e., 2K bytes to 32K bytes) are typical, but compression performance may be optimized through the use of larger dictionaries.

Continuous Mode or Packet Mode of Operation. Dictionary-based algorithms may be used in the continuous mode or packet mode of operation. In continuous mode, a continuous stream of characters in which there is no distinction between packets is used to create and maintain a dictionary. The continuous stream of data may consist of packets from many network protocols (e.g., IP and DECnet).

Continuous mode operation requires that the dictionaries on the compression and the decompression side be kept synchronized with each other though the use of a reliable data link mechanism (e.g., X.25 or reliable mode PPP). This ensures that packet errors and loss do not cause the dictionaries to diverge. Without a reliable link mechanism, dictionaries can diverge, and it becomes impossible to accurately decompress subsequent packets. In this case, both dictionaries must be flushed and the outstanding packets discarded.

The second mode of operation (packet mode) also uses a continuous stream of characters, but the boundaries between packets are maintained and a new dictionary is developed for every packet. The dictionaries need only be synchronized within the packet boundaries. This function already exists in the upper layer of all network protocols (i.e., the transport layer), so a guaranteed link encapsulation mechanism is not required.

The most significant observable performance differences between the continuous mode and packet mode algorithms are packet latency and compression performance. Performance is measured at the compression ratio, that is, the ratio of the size of the compressed output string to the original size of the string. The packet latency is the processing overhead of the algorithm (which affects the algorithm's ability to perform compression). The world's best compression ratio is useless if the algorithm requires an hour of processing time before the packet can be transmitted.

Continuous mode operation typically offers higher compression ratios than packet mode operation; however, somewhat more processing is required to maintain a reliable link connection. Processing overhead affects the speed of compression and decompression and hence packet latency. When latency is great, the benefits of compression may be partially negated.

Packet mode operation creates a fresh dictionary for each network packet. Continuous mode operation continuously examines the incoming data and checks and updates the dictionary, or dictionaries. Dictionary memory requirements for packet mode are thus limited to the amount of memory required to process a single network packet (e.g., 4K bytes). This is because the dictionary is flushed after every network packet is transmitted successfully. In a router with 100 interfaces (to either serial or virtual circuits), packet mode operation requires only a single dictionary (e.g., of 4K bytes). The continuous mode method for the same router requires 100 dictionaries (e.g., 400K bytes of total storage).

Thus, packet mode algorithms are attractive for their low latency and low memory requirements and because the continuous mode can substantially outperform it with higher compression ratios.

THREE METHODS OF DATA COMPRESSION

Compression of LAN traffic for WAN transmission can be achieved by either header compression, link compression, or payload compression. Exhibit 6-2-2 illustrates the differences between the three compression methods.

Header Compression

Header compression as defined by the Van Jacobson algorithm (RFC 1144) lowers, for TCP/IP traffic consisting of small packets (i.e., those with few bytes of data, such as Telnet packets), the high overhead generated by the disproportionately large TCP/IP headers. Transaction-oriented applications (e.g., DEC LAT, Telnet, rlogin, X Windows, and acknowledgment packets) typically can take best advantage of this type of compression.

Because of this technique's processing overhead, header compression is generally used at 64K-bps and not at the higher speeds now critical in LAN and WAN communications. The throughput improvements that header compression achieves across low-speed lines depend on line rate. For example, a 50% throughput improvement can be achieved with Telnet traffic on a 4,000-bps leased line.

In addition, header compression is protocol specific. A different algorithm must be implemented for each protocol. The header compression algorithm has been implemented for TCP/IP (as defined in RFC 1144), for DEC LAT, and for X-Remote. TCP/IP header compression is illustrated in Exhibit 6-2-3.

Option Type	Method	Advantages	Disadvantages
Integrated Compression Solutions	Header Compression	Low latency	Requires a different algorithm for each protocol on the network
		Up to 50% throughput improvement	Uses a lot of CPU capacity to perform compression
		Saves transmission of redundant information in header	No standard except for TCP/IP
		Delivers bandwidth savings for smaller size packets	
		Standardized for TCP/IP	
	Payload Compression	Low latency	Router architecture must be designed to compress and decompress the data before the packet encapsulation
		Maintains full routing information within the header	Uses a lot of CPU capacity to perform compression
		Can use compressed packets over X.25, frame relay, ATM	No standard
		Minimal memory required for dictionary	
	Link Compression	Protocol independent	Only available for point to point links
		Allows a mix of packet sizes and types	Greater latency than other methods
		Provides the best compression ratio	Large memory required for dictionary
External Compression Solutions	External Device	Compresses the full stream of data	Management is separate from router
		Protocol independent	Separate configuration
		Provides the best compression ratio	Greater latency than other methods
			Large memory required for dictionary

Exhibit 6-2-2. Features of Three Data Compression Modes

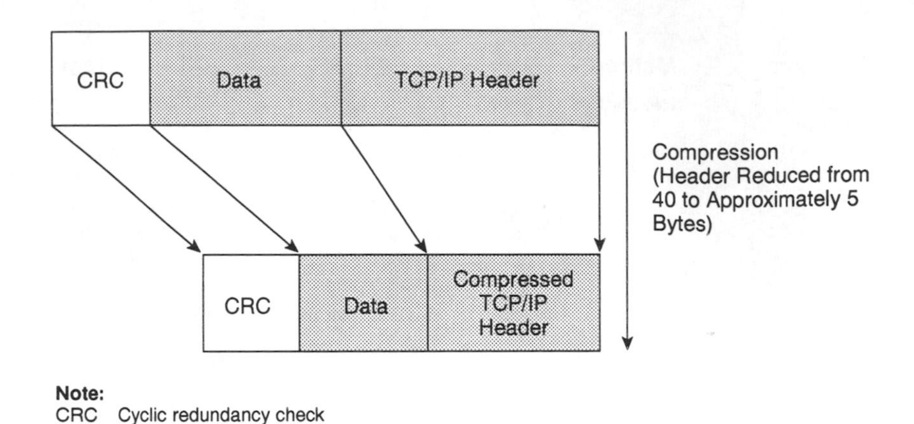

Exhibit 6-2-3. TCP/IP Header Compression

Link Compression

To handle larger packets, support higher data rates, and improve performance across multiple protocols on a LAN, compression can be applied to the entire data stream to be transported across the WAN. With this method, called link compression, the whole WAN link is compressed as if it were one application (illustrated in Exhibit 6-2-4). The link-compression algorithm compresses all the link traffic and then encapsulates the compressed traffic in another link layer—for example, PPP or link access protocol LAP-B—to ensure error correction and packet sequencing. Unlike header compression, link compression is protocol independent.

The very definition of link compression makes it a point-to-point only solution for use in such services as leased line or integrated services digital network (ISDN). Because the complete packet (i.e., the header and data payload) is compressed, the switching information in the header is not available for use by the WAN switching networks. Therefore, link compression cannot be used with such packet-based services as SMDS, ATM, X.25, and frame relay.

The use of link compression imposes another major constraint on network design in multihop topologies. In a network that has a multihop path, the data must be compressed each time it exits a router and decompressed each time it enters a router. The more routers in the path, the more compress/decompress processes. The compress/decompress process adds delay at each router and creates the possibility of significant delay in the network application. The best application for link compression is in environments where the customer knows that the application is a point-to-point, limited-hop connection. Link compression can be applied in router networks using ex-

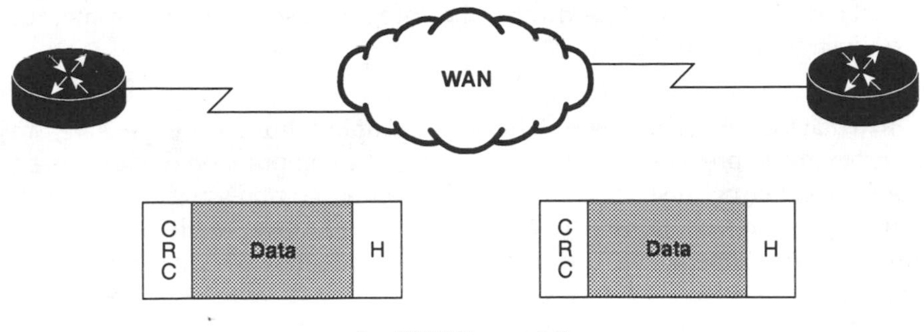

• Any WAN Encapsulation
(e.g., PPP, HDLC, LAP-B)

a. Before

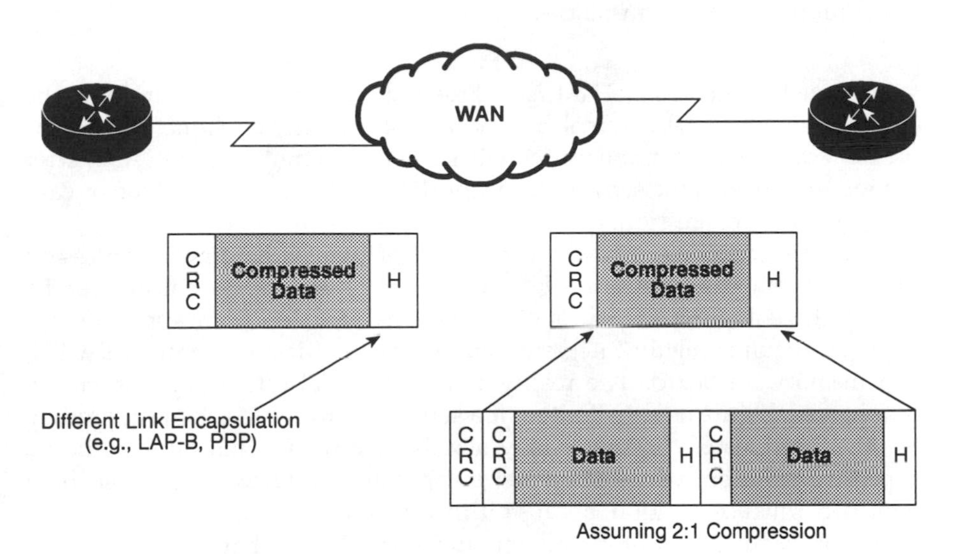

b. After

Notes:

CRC Cyclic redundancy check
H Header
HDLC High-level data link control
LAP-B Link access protocol B
PPP Point-to-point protocol

Exhibit 6-2-4. Point-to-Point Circuits (Leased Line and ISDN) Before and After Compression

ternal devices, integrated hardware, and integrated software (these methods are illustrated in Exhibit 6-2-5).

External Compression Device. This is the simplest, but most expensive, way to provide compression in an internetwork. The compression device acts as a black box, taking in serial data from the router, compressing the data, and sending compressed data to the WAN. The technique is readily available and simple to implement, but this solution is usually not the one preferred by network managers. One reason that most network managers prefer integrated solutions, rather than external devices, is their perception that the integrated technique will be less expensive. This is not always the case. External devices are ill-suited for remote sites that need connectivity to core routers. Usually there is no one at these remote locations to manage and configure the separate components. Management of integrated solutions is easier under these circumstances.

Integrated Compression Hardware. Essentially, this involves putting the components from the external box on a port adapter, applique, or a compression card that fits in the router, with a serial interface to the WAN. After the router performs the serial line encapsulation, the special applique or card performs the compression.

With this technique a router can be retrofitted with some compression capability, but the solution lacks the flexibility of an external device. To deliver the compression power of an external device on an applique, the applique requires highly integrated circuitry, a sufficiently powerful CPU, and memory on board. The reality is that the real estate and computing power available are usually limited. If a separate card is used, some upgrading of other sections of the router may also be necessary. This can also be an expensive solution, when compared to an external device. Yet it can be a high-performance solution, because it involves separate hardware dedicated to compression. The router can function normally, without the additional burden of compression processing.

Integrated Compression Software. This option allows for the compression software to be integrated into the router software while using the same serial interfaces already installed on the router. The encapsulation routine is enhanced to perform link compression after the initial encapsulation for the serial link is done. This method requires that sufficient processing power for compression and memory for the compression dictionary be available in the router. This technique is ideal for access routers, which must support only low-speed lines. The dictionary requirements vary with the algorithm used for compression.

External Compression Device

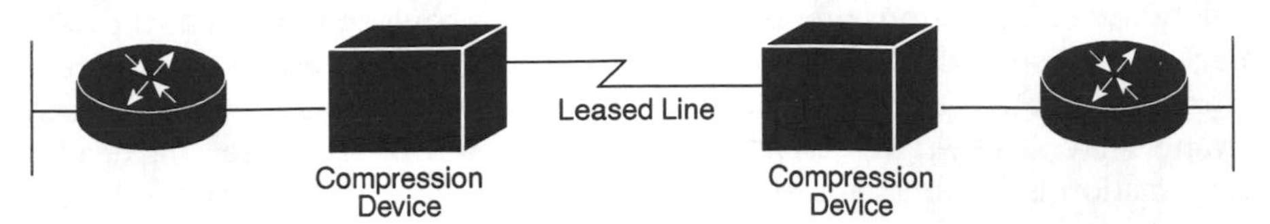

Integrated Compression Hardware

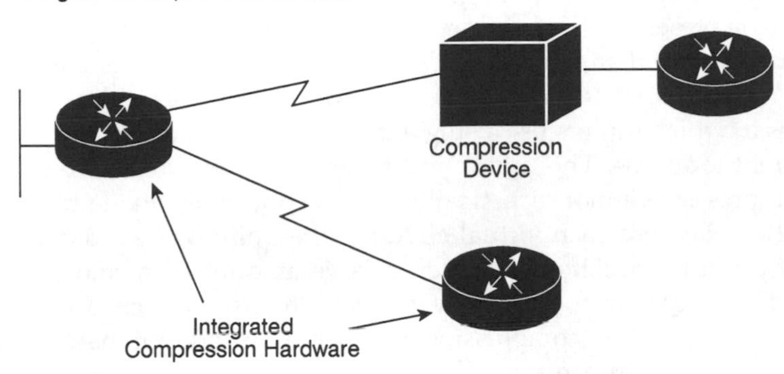

Integrated Compression Software

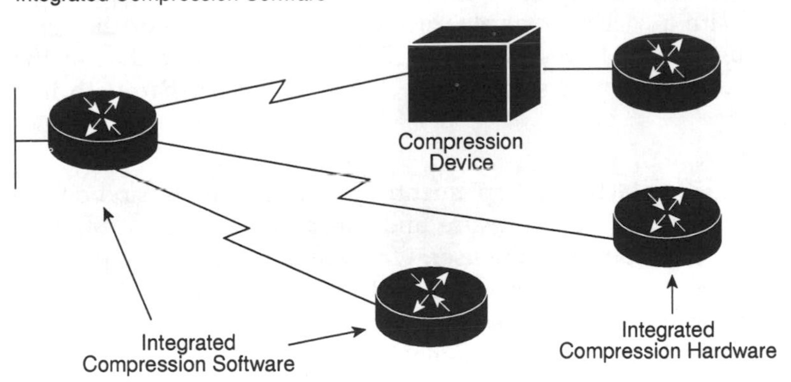

Exhibit 6-2-5. Link Compression Options

Payload Compression

Payload compression compresses only the data payload portion of the data packet and leaves the header intact. This is the required method of compression for operation across virtual network WAN services (e.g., X.25, SMDS, frame relay, and ATM, illustrated in Exhibit 6-2-6). Because the header information is left unchanged, the packet can be switched through a WAN packet network and routed through a router network. Thus compression can be applied while still taking advantage of the typically lower tariffs of virtual network WAN services. In designing an internetwork, if the customer cannot assume that the application will only be going over point-to-point lines, then the header has to be preserved. The compression intelligence has to be placed where it can work on the data just before WAN encapsulation takes place.

Its ability to preserve the header makes payload compression ideal for implementations in which routers use a single interface with multiple virtual circuits to different locations. The use of packet services by routers dictates that payload compression cannot realistically rely on continuous mode compression algorithms, because each virtual circuit will require its own dictionary. With many virtual circuits per router, a large amount of memory is potentially needed to configure the many virtual circuits on each router. Alternatively, a packet mode compression algorithm would not need to maintain large dictionaries to compress the information.

With payload compression, the router still has to compress and decompress the packets at each router hop, thus not improving on the latency in multihop circuits. The best applications for compression still are paths with limited hops. With a packet mode algorithm, the performance of the compression is substantially lower than with link compression, although the latency of the compression is somewhat better (i.e., lower) than with link compression.

Two Implementation Methods. Implementing payload compression on the router involves hardware and software and can be done in two different ways, with different benefits to the user. One way is to apply compression to the data payload after the WAN encapsulation is performed. This process compresses the payload and reduces the overall size of the packet, but the number of WAN packets remains the same. It can be implemented within the existing router software architecture by adding software modules to provide the appropriate compression algorithms. For WAN services (e.g., frame relay or SMDS) that are not sensitive to packet count, this compression method could be very useful. However for the packet networks that charge by packet (e.g., X.25 networks), it may or may not provide significant cost benefits. External compression devices supporting the appropriate WAN interface can also provide this capability and yield higher performance over higher speed links, albeit at higher cost.

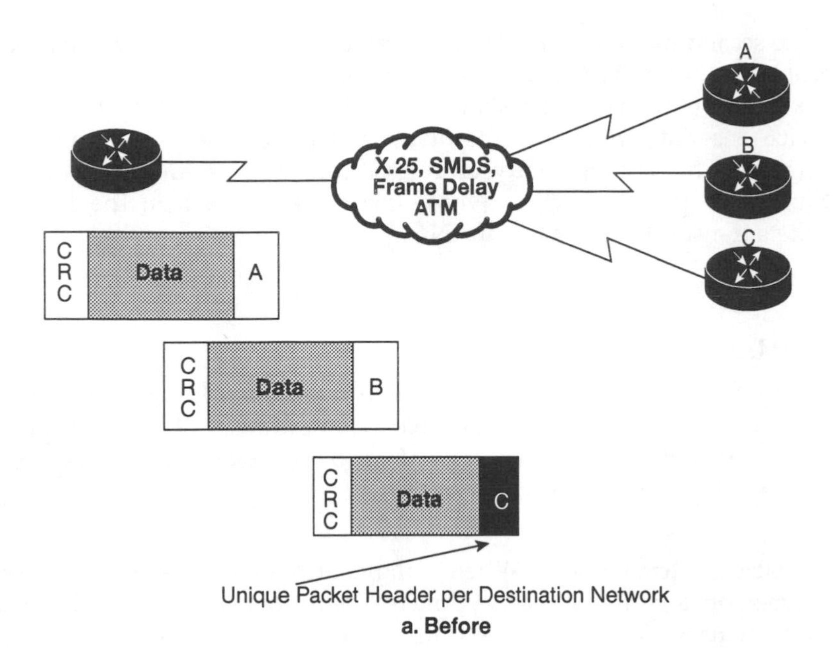

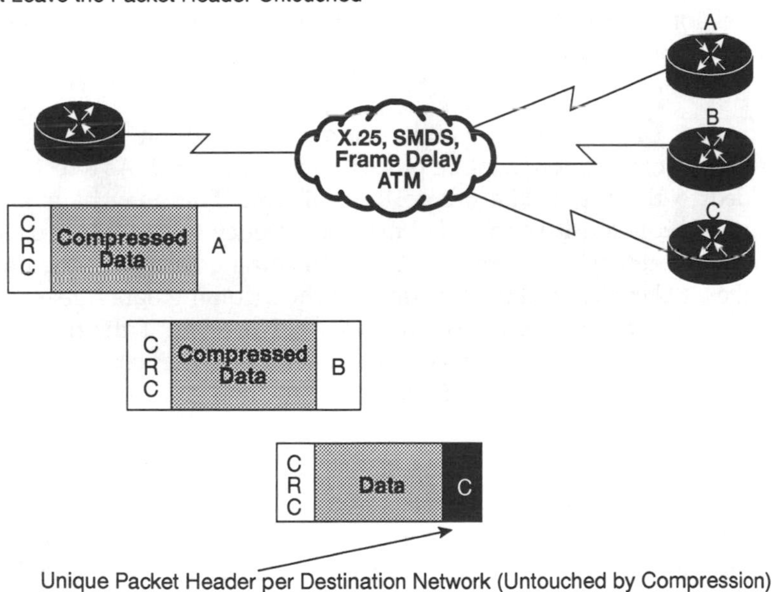

Exhibit 6-2-6. Packet Network Before and After Compression

The second method of payload compression compresses the network protocol traffic before WAN serial encapsulation is performed. The result is that packet size remains at the standard minimum transmission unit for that service or serial interface as the number of packets sent over the network decreases. This is particularly useful for connection to X.25 networks and other WAN packet services, whose charges are based on the number of packets transmitted over the WAN.

SUMMARY

The effective use of compression in an internetwork is very much application dependent. Compression should be considered only if the application is robust enough to function in spite of various characteristics of compression when applied to real-time data traffic.

Number of Remote Sites. When using continuous mode dictionary-based compression algorithms, each point-to-point connection must have dedicated memory for the connection. Hence, the greater the number of connections, the greater the memory required. Usually this is not a problem, but it will add a one-time cost (which is the opposite of why compression is being used) and add latency at the core site, which could affect network performance.

Increased Latency. When compression is applied on the original data stream, the information must be processed and analyzed. As a result, latency is added to the input data before transmission. This may have an impact when protocols sensitive to added network latency (e.g., LAT) are used over the WAN. Latency is determined by the incoming data rate, the dictionary size, the CPU cycles, and the line rate. As the incoming data rate goes up, the router has to process the information faster. However, if the dictionary size is small and there is limited processing power, the router compression will be slow and significant latency will be added to the network.

Congestion Management. The compression process is inconstant, varying with data type and compressibility. An external compression device provides a variable clock signal to the router to manage the difference in the effective data throughput to and from the router and the compressed WAN link. Less compressible data results in a lower clock rate being sent to the router than more compressible data. Clock modulation to the router ensures that input data is not lost during the compression and transmission process. The clock rate to or from the WAN connection is not varied. When compression is performed within router software, normal internal queuing mechanisms buffer the variations in throughput.

Memory Size. All algorithms need some memory to perform compression. The actual amount of memory required varies with the protocol being compressed, the compression algorithm (as noted previously, continuous mode, dictionary-base operation uses more memory per circuit than packet mode operation), and the number of concurrent circuits on the router. If a router has fixed memory and continuous mode compression, then only a fixed number of circuits can have their data successfully compressed.

Speed. Every algorithm performs differently. Some algorithms can process more compression instructions per CPU cycle than other algorithms. This makes them fast and capable of achieving higher line rates. A fast algorithm allows software compression to be implemented on a remote access router while still performing the normal router functions. Another way to view speed is the absence of latency. So the lower the processing overhead, the lower the latency that results from the process.

Compression Ratio. Compression performance is measured as the compression ratio—that is, the ratio of the size of the compressed output string to the original size of the string. This is referred to as the static compression ratio.

Another compression ratio, known as dynamic compression ratio, is defined as the time ratio of transmitting a known amount of data with and without compression. The dynamic ratio is a better representation of actual performance, because it actually measures the savings in use of the WAN link and its value is not affected by the type of encapsulation.

Depending on the type of input, particularly the amount of redundancy in the input, the compression ratio will vary. Because LAN environments are characterized by many types of applications and protocols running concurrently, the compression ratio for such environments will inherently be lower than for those where there is only one application or data type (e.g., plaint ext files or images only).

6-3

Accessing and Using the Internet

Gilbert Held

COMMERCIAL ORGANIZATIONS CONNECT to the Internet to take advantage of the wide range of capabilities it provides both individual employees and business units. This chapter reviews applications either currently on the Internet or expected to be placed on the network, as well as the methods businesses use to gain full or limited access to these Internet applications.

PRIMARY INTERNET APPLICATIONS

Different methods of connecting to the Internet provide different levels of access to Internet applications. As a starting point, an organization must examine each Internet application and determine whether its employees require access to it.

Seven basic Internet applications represent the vast majority of current Internet traffic:

- Electronic mail (E-mail).
- File transfer.
- Remote login.
- File location using Archie.
- Information location using Gopher.
- World Wide Web server access.
- World Wide Web page location using a Web search tool.

Some organizations require access to each of these applications; other organizations may require access to only a few of them.

Electronic Mail

Electronic mail access gives an organization the ability to send and receive E-mail to and from approximately 25 million users currently on the Internet. However, if this is the only requirement an organization has for Internet access and only a few employees actually use electronic mail, it may be more economical to provide this capability through an information utility—a commercial online service such as CompuServe or MCI Mail—that provides a gateway for Internet electronic mail.

For example, an individual subscriber might pay $35 per year for an MCI mailbox plus approximately 50 cents for every E-mail message sent that is under 500 characters. Internet users can reach the subscriber by using the subscriber's MCI Mail address, and the subscriber can send mail to other users on the Internet by placing EMS (for electronic mail service) after the recipients' name when creating a message and then entering their Internet address when prompted by MCI Mail. The economics change, however, when many employees in an organization will be using E-mail. Then, a one-time annual access fee for a dedicated connection to the Internet that supports unlimited usage could cost less than an annual mailbox fee for each employee, especially when their message fees are also calculated.

A similar case to the use of MCI Mail can be made for the use of CompuServe, America Online, Microsoft Network, and Prodigy. Each of these four information utilities can provide an Internet and electronic mail capability for a nominal fee per month. Some utilities, such as America Online and Prodigy, offer unlimited access to the Internet for $19.99 per month, a price that may be difficult to beat on a per-employee cost basis, unless an organization has more than 50 employees and requires a direct Internet connection to enable each employee to send E-mail, as well as perform other Internet-related activities.

Enhancing Business Image with E-Mail. When using the Internet for electronic mail, many organizations choose an E-mail address that reflects the organization's name. An organization named Fastfood can, for example, use that name as its Internet domain name, provided it is not already being used. The domain address for this commercial firm would thus be fastfood.com. In addition, each employee of the organization can use their last name and first initial as a prefix to the domain address to provide a unique electronic mail address (e.g., gheld@ fastfood.com).

File Transfer

For employees in a business, information subject areas available for access through file transfer are virtually unlimited. Organizations commonly access shareware programs and utilities, drafts and finalized standards for computer and communications systems, network management information, and data on currency exchange values.

File transfer to and from servers connected to the Internet allows employees to access literally millions of files representing more than 1,000G bytes of technical, financial, and other information that is valuable for different activities performed by the organization. Through the facilities of Archie search engines, a user can enter one or more keywords and obtain a list of files, as well as their locations on the Internet, including server address and directory locations that meet the search criteria.

If a business user needs information of the consumer price index or producer price index, the user might enter CPI and PPI as search criteria, for example. If an IS employee requires the latest patch for an open data link interface network adapter card, the user might initiate a new search using network, adapter, and open data link as search criteria.

When an E-Mail Interface Is Not Sufficient. File transfer on the Internet requires the use of the file transfer protocol (FTP). Although FTP access usually demands the establishment of an account with an Internet access provider or a direct connection from the corporate LAN to the Internet, access to files can also be obtained through E-mail. In this case, however, the user must know the addresses of E-mail servers that convert E-mail requests to FTP retrievals and then send the retrieved file to the requester as an E-mail message.

Although this technique permits some organizations to satisfy their FTP access requirements by using MCI Mail, or another commercial online service that provides an E-mail interface to the Internet, this method may not be a practical solution to FTP access because of the complications involved with accessing binary files by E-mail. Many electronic mail systems do not support the transfer of binary files. Thus, even if a business can access an FTP server through an E-mail request, it may not be able to receive files returned by the server through an E-mail message.

Remote Login

Remote login lets Internet users access remote computers as if their workstation or personal computer were directly connected to the remote machine. The primary application supporting remote login is Telnet. If, however, employees need to access IBM mainframes connected to the Internet, they will more than likely require the use of a TN3270, a 3270 terminal emulation capability that simulates the type of terminal required to gain access to the remote IBM mainframe.

Telnet and TN3270 access is limited to users that obtain a direct connection to the Internet or use the dial-in services of an Internet service provider.

Information utilities—notably CompuServe, American Online, and Prodigy also provide this capability.

For a business, Telnet and TN3270 access is a mechanism that lets customers obtain price quotations, check delivery schedules, and retrieve other information kept on the corporate computer. One common use of Telnet and TN3270 access for nonorganizational employees concerns just-in-time manufacturing, where suppliers for a company must access the production schedule of the company they support. By providing suppliers with Telnet and TN3270 accounts on the corporate computer, a business can assist its suppliers in delivering equipment to match its corporate production schedule.

Archie

Archie is an Internet application used for searching file names. The user enters one or more search elements to access an Archie server that periodically updates catalogs of local files that are available to the public. Access is accomplished through Telnet; however, an Archie search request can also be sent by E-mail to some archie servers that will respond to such requests with an E-mail message.

Gopher

Invented in 1991 at the University of Minnesota as a campuswide information display, Gopher has evolved into an Internet-wide information display system that allows users to locate and access electronic documents. Information is presented as hierarchical menus to the user. One adjunct to Gopher is Veronica, which lets users search a Gopher server for information using a keyword search.

To access a Gopher server, it is necessary to run a Gopher client on a user's workstation. Thus, Gopher access is restricted: An organization must use an Internet access provider that permits full Transmission Control Protocol and Internet Protocol (TCP/IP) Internet access or establish a direct connection between its corporate LAN and the Internet. If the computer has a World Wide Web (WWW) browser, the user can, as an alternative, access a Gopher server through a WWW server.

World Wide Web

The World Wide Web is a series of Internet-connected servers that integrate data or text with graphics and audio, thus providing a multimedia capability to users accessing the server.

WWW servers present users with hypertext documents created using the hypertext markup language (HTML). To access a WWW server, the client workstation must execute a browser program that can generate hypertext transfer protocol (HTTP) requests as well as display HTML code generated by the server. Examples of popular WWW multimedia browsers include Netscape, Internet Explorer, and Mosaic. Similar to Gopher access, access to a WWW server usually requires the services of an Internet access provider or a direct connection to the Internet.

The number of WWW servers exploded in three years from less than 1,000 to approximately 300,000. Most major corporations have established WWW servers that provide information about the company and its products and job opportunities. The use of WWW servers to sell products has been limited largely because of the security concerns with respect to transmitting credit card information over the Internet.

Security Questions. One company, Netscape Communications Corp., has introduced a WWW server that supports public key encryption as well as a WWW browser with a similar capability. Organizations purchasing a Netscape server software program can receive encrypted credit card information from persons using a Netscape browser.

As more commercial organizations use the Internet, security is a growing problem—but with more commercial users, more effort is also being invested in devising ways to provide security and protect privacy. As security issues are addressed, the ability to sell products and services on the Internet is expected to grow.

World Wide Web Page Location. One of the major problems associated with the use of the World Wide Web is locating relevant information. With approximately 300,000 servers online and each one containing hundreds or thousands of Web pages, the location of relevant information can be a daunting task. Thus the advent of advertiser-supported seach tools, including Alta Vista, Excite, Web Crawler, and Yahoo, among others.

Because many Web browsers have a built-in button labeled "search" or "net search," many browsers can simply click and obtain a list of four to six search tools. What many browsers fail to recognize is that the listed search tools commonly pay a fee to the browser vendor to be listed as a link to the browser "search" or "net search" button. Thus, the user only receives access to a fraction of the available Web search tools. However, by entering "search tools" as a search key or on one of the supported search engines, users can obtain a list of 30 or more search engines.

INTERNET CONNECTION METHODS

Once an organization has identified its primary Internet applications, it must decide on the method for obtaining a connection to this network of networks. Some access methods provide a limited access capability to Internet applications.

There are three basic types of links that connect people and companies to the Internet:

- Dial-up through an information utility (usually one of the five major commercial online services—CompuServe, MCI Mail, America Online, Microsoft Networks, or Prodigy).
- Dial-up through an Internet host computer.
- Direct LAN connection.

An examination of the hardware and software needed for these different types of Internet connections gives a general indication of the cost associated with each access method.

Dial-Up Through an Information Utility

Commercial online services, or information utilities, provide different levels of access to Internet applications. CompuServe provides subscribers access to E-mail, FTP, and Telnet applications, for example. CompuServe subscribers can access just about anything they want from the Internet, including World Wide Web Servers.

Limitations of a Modem Connection. Access to CompuServe is similar to access to other information utilities, such as Microsoft Networks and MCI Mail, in that a dial-up modem connection is used (see Exhibit 6-3-1). As a result, a subscriber must use either a conventional communications program, such as PromcommPlus or Crosstalk, or a communications program developed by the information utility, such as CompuServe's Information Manager.

In fact, because current communications software products from information utilities support the hypertext markup language, most subscribers are able to access the World Wide Web through a commercial service using software programs provided by a particular service.

Costs. The cost of using a commercial online service is generally lower than other methods of Internet access. Most of these information utilities provide a corporate rate as low as $7 to $15 per month per subscriber, plus a nominal hourly usage charge between $3 and $5 per hour (after a fixed number of hours). For residential users, recent reductions in the cost of Internet access result in some vendors providing unlimited access for a flat fee of $19.99 per month.

If an organization has relatively few employees that require access to the Internet and it does not mind being limited to modem-based access, an information utility may satisfy its Internet access requirements.

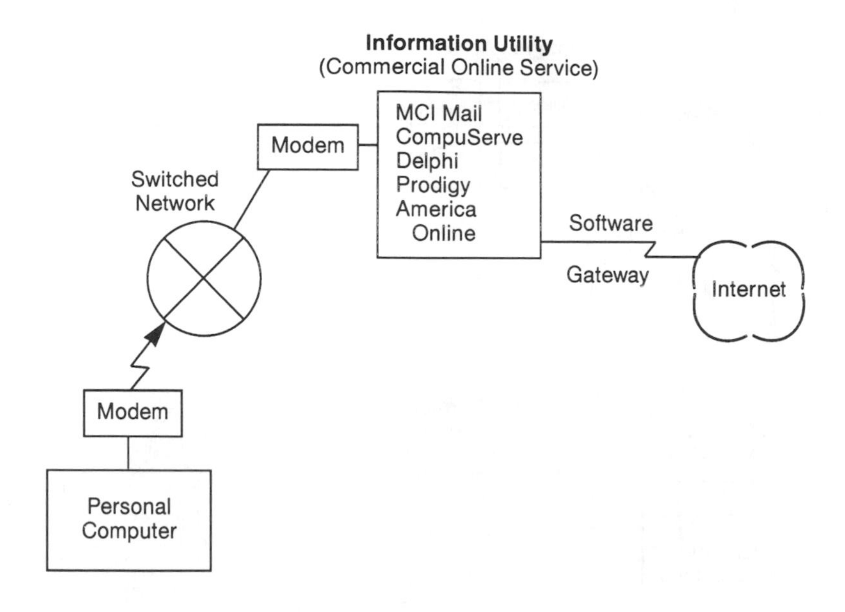

Exhibit 6-3-1. Internet Access Through an Information Utility

Internet Host Computer Access

A second method commonly used to access Internet applications is to directly dial into an Internet host computer. Although this method may appear similar to using an information utility, in actuality there are significant differences between the two methods.

Exhibit 6-3-2 illustrates Internet access using a dial-up connection directly to an Internet host computer. Instead of using a standard communications program, the business must use either a communications program that supports the Serial Line Internet Protocol (SLIP) or the point-to-point (PPP) Protocol. Either of these two protocols supports dial-up access to the Internet through a serial communications connection. The user can access all Internet applications, unless there are specific policies barring host computer access.

Costs. Users must access a computer that is already connected to the Internet, and the most common method of obtaining this type of access is through a contract with an Internet access provider. The typical cost for this type of access is between $20 and $30 per month for either unlimited access or access for core applications, such as E-mail and FTP.

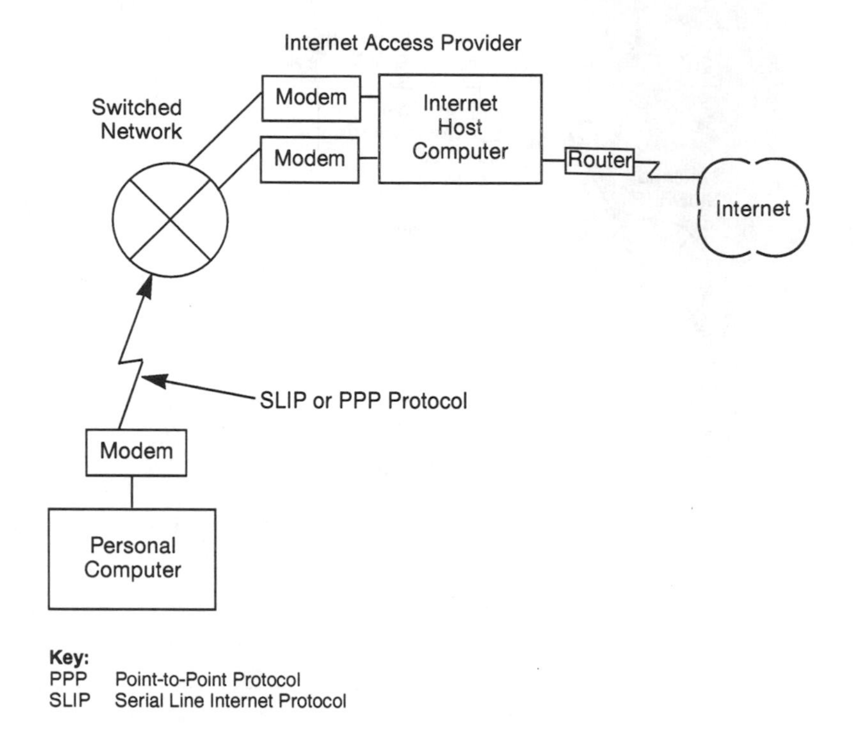

Exhibit 6-3-2. Internet Access Through an Internet Host Computer

Some Internet access providers have a cost schedule resembling a restaurant menu, so careful analysis is needed to predict potential costs. If an organization anticipates that many employees will require access to the Internet and the organization has an existing LAN, it may be more economical to obtain a direct connection from the corporate LAN to be Internet through an Internet access provider.

In addition to the cost associated with using an Internet access provider, an organization must also consider the cost of the communications program that supports SLIP or PPP. Although most Internet access providers now provide such software free, some providers still charge separately for their software. This cost may exceed the cost of a conventional communications program. For example, an SLIP-or PPP-compliant communications program that supports Internet E-mail, FTP, Telnet, TN3270, and Gopher applications can cost between $125 and $495 per program.

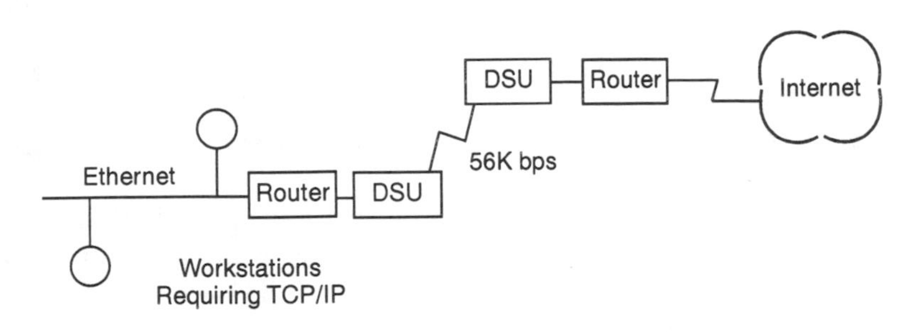

Exhibit 6-3-3. Connecting a LAN to the Internet

Establishing a Direct LAN Connection

For a direct connection of a LAN to the Internet, a variety of hardware and software is required, as well as the installation of a communications circuit between the local area network and an Internet service provider. Exhibit 6-3-3 illustrates the most common method used for connecting the LAN workstations of a commercial organization to the Internet. In this example, a 56K-bps digital circuit is used to connect a LAN to the Internet using the services of an Internet access provider.

Adding Up the Costs. For an annual or semiannual fee, most Internet access providers will furnish a router, a data service unit (DSU) that functions as a digital modem, and the communications circuit connecting their location to a subscriber. The organization is able to have multiple workstations on the network with access to all Internet applications.

Additional Software. For example, BBN Planet (formerly SURANet), an Internet access provider in the southeastern US, provides a 56K-bps connection and router with full access to the Internet for approximately $10,000 per year. If a LAN has 100 workstations needing access to the Internet, the connection fee becomes $100 per user per year, which at first glance appears to be significantly more economical than other methods of access. However, because the Internet requires TCP/IP transmission, each workstation on the

LAN that requires Internet access must obtain a TCP/IP communications program. A TCP/IP communications program by itself does not give LAN users the ability to send and receive E-mail, access FTP servers, and perform Gopher searches, so the organization will probably need to acquire an integrated Internet access application program (e.g., NetManage's Chameleon, Frontier Technologies' TCP/NFS, or a similar product) for each LAN user. On average, the one-time cost for such a program can run from $200 to $400 per LAN user.

If the LAN operates NetWare, the organization must consider the use of a dual-stack software product that allows LAN users to switch between accessing the NetWare server using Novell's IPX and the Internet using TCP/IP. Without such software, workstation users would have to reboot their computer to switch from using IPX to TCP/IP, and vice versa.

Two products that permit a LAN workstation to operate dual communications stacks are ODI and NDIS. ODI (Open Data link Interface) is a standard developed by Apple Computer, Inc. and Novell, whereas NDIS (Network Driver Interface Specification) represents a standard developed by Microsoft Corp. and IBM Corp. Users can obtain a product compliant with either or both of these standards from one of several independent software developers for between $40 and $100 per workstation. Thus, the actual first-year cost to provide 100 LAN users with full access to the Internet using an Internet access provider can range from $35,000 to $60,000. Thereafter, the subsequent annual cost will be approximately $10,000, representing the annual fee of the Internet service provider.

Upgrades. Of course, if many LAN users require access to such graphical applications as the World Wide Web, the use of a 56K-bps circuit will probably be insufficient. Some Internet access providers limit upgrades to a full T1, which can double or triple the organization's annual access bill; other access providers permit line upgrades to fractional T1 circuits that are a more economical approach when additional bandwidth is required. Thus, an organization may wish to consider both the cost required to use an Internet access provider to support the organization's current Internet access requirements as well as the cost required to upgrade access to a higher operating rate when evaluating different vendors.

Exhibit 6-3-4 lists 28 Internet access service providers. The list includes the postal address of each provider as well as the area it currently serves. By examining this list, network managers can locate one or more service providers that support the location or locations they wish to connect to the Internet. They can then contact those providers and compare the services they offer and the cost of each service offered.

Service Provider	Area Served
a2i Communications 1211 Park Ave San Jose CA 95126	408 area code
Advanced Network and Services, Inc. 2901 Hubbard Rd Ann Arbor MI 48105	Continental US
AlterNet 3110 Fairview Park Dr Falls Church VA 22042	Continental US
Bay Area Regional Research Network Pine Hall, Rm 115 Stanford CA 94305	Northern California and San Francisco
CERFnet PO Box 85608 San Diego CA 92186	California
CICNet ITI Bldg 2901 Hubbard Dr Ann Arbor MI 48105	Iowa, Illinois, Indiana, Michigan, Ohio, Minnesota, Wisconsin
Clark Internet Services, Inc. 10600 Rte 108 Ellicott City MD 21042	Washington DC, Northern Virginia, Maryland
Colorado SuperNet CMS Computing Center Colorado School of Mines 1500 Illinois Golden CO 80401	Colorado
CONCERT PO Box 12889 3021 Cornwallis Rd Research Triangle Park NC 27709	North Carolina
CTS Network Services 4444 Convoy St San Diego CA 92111	Southern California
Express Access 6006 Greenbelt Rd Greenbelt MD 20770	202, 310, 410, 703 area codes
Halcyon PO Box 555 Grapeview WA 98546	Seattle WA
HoloNet 46 Shattuck Square Berkeley CA 94704	Berkeley CA

Exhibit 6-3-4. Partial List of Internet Access Service Providers

Infonet Service Corp. 2100 East Grand Ave El Segundo CA 92045	US and International
InterAccess 9400 West Foster Ave Chicago IL 60657	Chicago IL
JvNCnet John von Neumann Center Network G von Neumann Hale Princeton University Princeton NJ 08544	US and International
MichNet 2200 Bonisteel Blvd Ann Arbor MI 48109	Michigan
Minnesota Regional Network 511 11th Ave South Minneapolis MN 55415	Minnesota
MV Communications, Inc. PO Box 4963 Manchester NH 03108	New Hampshire
New England Academic and Research Network BBN Systems and Technologies 10 Moulton St Cambridge MA 02138	New England
NY State Education and Research Network 111 College Place Syracuse NY 13244	New York
Northwestern States Network 2435 233rd Place NE Redmond WA 98503	Northwest US
Performance Systems International 11800 Sunrise Valley Dr Reston VA 22091	US and International
Pennsylvania Research and Economic Partnership Network 305 South Craig St Pittsburgh PA 15213	Pennsylvania
RISCnet InteleCom Data Systems 11 Franklin Rd East Greenwich RI 02818	New England
Southeastern University Research Association Network 1353 Computer Science Center University of Maryland College Park MD 20742	Southeast US

Exhibit 6-3-4. (*continued*)

Texas Higher Education Network University of Texas Office of Telecommunications Services Austin TX 78712	Texas
UUNET Technologies, Inc. 3110 Fairview Park Dr, Suite 570 Falls Church VA 22042	US

Exhibit 6-3-4. ***(continued)***

SUMMARY

The selection of an appropriate method to access Internet applications depends on an analysis of two key factors. First, the organization should determine what Internet applications organizational employees need access to. Next, it should determine the number of employees that require Internet access and whether they represent standalone computer users or are currently using a workstation on a LAN. Using this information as a base, an organization can then analyze the economics and practicality of the two dial-up access methods and the direct LAN connection method of obtaining Internet access and select one or more methods that will cost-effectively satisfy organizational requirements.

6-4 Internet Security Using Firewalls

Vincent C. Jones

AS NETWORKS GROW LARGER—expanding beyond the desktop, even beyond the walls of the organization to support telecommuters and other traveling employees—the reliability and availability of those networks and their attached systems become paramount. In the past, with traditional terminal-based networks, this expansion of accessibility was not a major security headache. Good password discipline and dial-back modems that provided physical security for outside connectivity were sufficient to secure a network, and the emphasis was on quality of service and application usability.

ENTERPRISE NETWORK SECURITY

Business users are no longer satisfied with simple terminal access. The personal computer revolution includes a need for peer-to-peer network access from a user's CPU to the desired information and services. The "anyone-to-anyone" connectivity implied by peer-to-peer enterprise networking has severe security ramifications. Protection can no longer be concentrated at a single point. Instead, all systems on the network must be defended independently because each is an autonomous processor with its own resources needing protection.

This move from point-to-point communications used by terminal networks to shared-media local area networks (LANs) such as Ethernet and Token Ring, where all traffic on the LAN is accessible from any location on the LAN, opens new channels for possible attackers. No special wiretapping tools or skills are required because any legitimate user can see any traffic, making detection of eavesdroppers next to impossible.

INTERNET ACCESS AS A FACT OF BUSINESS LIFE

The increasing popularity of the Internet poses a threat to enterprise network security. Providing connectivity to the resources of the Internet for internal network users can be quite a challenge for the network administrator.

Technically, connecting to the Internet is easy. Nearly all system vendors support the transmission control protocol/Internet protocol (TCP/IP) protocol suite defined and used by the Internet. Many enterprise networks are even based on TCP/IP, as it is the most commonly supported peer-to-peer network architecture. Some enterprise networks are going as far as to use commercial Internet providers as part of their wide area networking connectivity matrix, using the Internet to provide links to other remote offices and on-the-road users.

Even those organizations that have all their connectivity needs covered may find the lure of the Internet irresistible. Internet E-mail addresses are becoming increasingly common on business cards. The information resources freely available to Internet users boggle the imagination and continue to expand exponentially. Although the Internet has a traditional research-oriented, noncommercial use, it also represents a huge listener base for those organizations that want to get their message out, whether selling computer security services or mail-order baby strollers.

Risks of Connecting to the Internet

From a business standpoint, connecting to the Internet is clearly desirable, but it is not without its risks. Connecting to the Internet without adequate protection in place simply opens the enterprise network to the thousands of hackers and vandals who inhabit the Internet along with its millions of honest users.

Although commercial and research users are starting to recognize the importance of network security, security tools have continued to lag behind actual practice, even though it is common knowledge that "business as usual" leaves business wide open to attack. Even well-known weaknesses, such as sending clear text passwords over broadcast networks, continue to be tolerated for convenience.

Most companies concerned with security have limited access to the internal network to trustworthy people. Barriers to entry, such as call-back modems for dial-up access and encryption on external links, may already be in place to protect network traffic and systems from outside attack. Taken in this context, connecting to the Internet is not a new threat; it is simply another avenue for attackers. Concentrating exclusively on Internet connection can therefore be self-defeating unless it is also used as an opportunity to examine all weaknesses. Internal systems are only as secure as the weakest barrier to the outside world.

The Internet is, however, open to virtually every student, researcher, and modem owner in the world. The fundamental paradigm of Internet protocol development has been openness first. Protocols are designed to provide maximum connectivity at minimum cost. The underlying assumption has traditionally been that the only worthy goal is to enable communications, and any feature that might limit those communications (including security firewalls), must be inherently wrong. The result is that many protocols commonly used on the Internet are difficult to control.

ESTABLISHING FIREWALLS

The need to provide connectivity from vulnerable internal network systems to the Internet can be approached in several ways. The correct way, from the viewpoint of the traditional Internet paradigm, is to simply attach the internal network to the Internet using a router and put the responsibility on each individual end system to protect itself. This approach has proved unrealistic because of the lack of security in typical LAN protocols combined with the broadcast nature of LANs, where one device can see all the traffic to all devices on the network.

More common is a firewall approach, where the connection between the internal network and the Internet is filtered through a firewall device to keep out intruders. This method allows those responsible for security to concentrate on a limited number of well-controlled gateway systems rather than having to monitor every user on every system on the internal network.

Router-Based Packet Filtering

Early connections simply programmed packet filters in the routers used for the Internet connection (see Exhibit 6-4-1). However, this method provides a false sense of security because the filters are hard to program, making mistakes likely, and the architecture of popular TCP/IP application protocols makes it impossible to simply filter out dangerous packets, as the potentially bad packets look identical to control packets essential to protocol operation. Moreover, the routers fail to provide any audit or reporting capability, making it impossible to detect if the filters are even being attacked, let alone determine if they are working properly. Users of this approach usually only find out they have a problem when systems start showing signs of corruption.

To get around the limitations of router-based packet filtering, host systems were reprogrammed to serve as intelligent filters between the internal and external networks (see Exhibit 6-4-2). Logically, this configuration is iden-

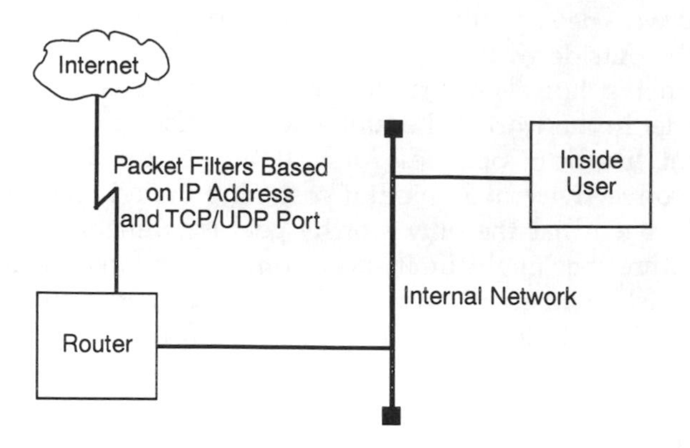

Exhibit 6-4-1. Firewall Based on Routers with Packet Filtering

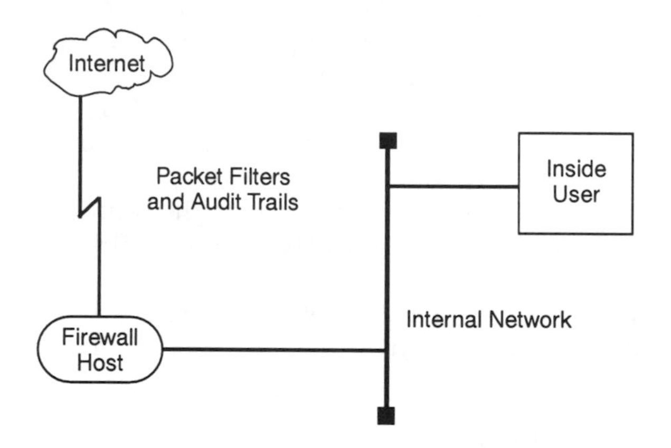

Exhibit 6-4-2. Firewall Based on Intelligent Packet Filtering

tical with packet filtering using routers; the only difference is that now the firewall builder is in control of the source code rather than the router vendor.

Although this solves the problem of missing audit trails and attack alarms, it does not solve the fundamental problem that TCP/IP protocols are inherently hard to secure. It also suffers from the high-defect rate of typical full-powered (and consequently complex) operating systems and network protocol implementations, exposing the internal network to attack through the firewall host operating system.

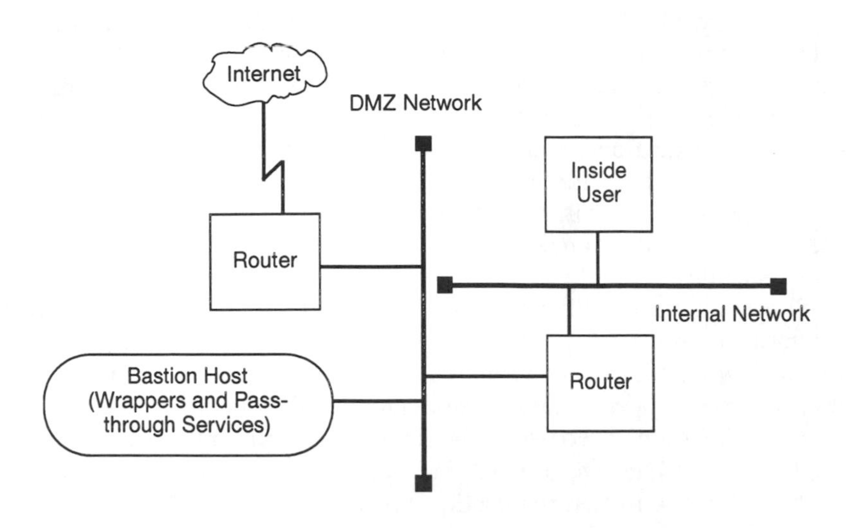

Exhibit 6-4-3. Firewall Based on Bastion Host Application Gateway

Using a Bastion Host

The next step in the development of firewalls was to modify the TCP/IP application protocols to make them "firewall friendly," or at least, less firewall hostile. The firewall host is effectively converted into an application protocol conversion gateway (see Exhibit 6-4-3). By running modified versions of standard services on the internal network, it is possible to defend the internal network from a variety of attacks. The router to the Internet is programmed so that only packets addressed to and from the bastion host are allowed through. Inbound packets for any other internal addresses, including the routers themselves, are discarded.

Similarly, the router between the "demilitarized zone" (DMZ) network and the internal network is configured to only pass packets to and from the bastion host. For added security (just in case the external router is broken into), this filter can be set to filter on the media access control (MAC) address of the bastion host as well as its Internet address and TCP/IP port numbers.

Using Two Addresses. A variation on this scheme is to use two interfaces and two independent Internet addresses on the bastion host; one for connecting to inside hosts and the other for communicating to Internet. Depending on the host platform, this can simplify the programming.

The primary disadvantage of this approach is the need to run special versions of dangerous services, such as file transfer protocol (FTP), on all internal clients. This can be a challenge because there are many different internal platforms and the modified software may not be available. Depending on the modifications, there may also be an impact on transparency. For example, to Telnet to an Internet system may require Telneting to the bastion host and requesting a connection to the ultimate destination. Software is available to make this connection transparent on common platforms. Source code is generally available.

The key to any firewall approach is to keep it simple. Complex software and algorithms are an invitation to intrusion. Generally, the bastion host is a stripped-down UNIX workstation and only those protocols and features essential to firewall operations are implemented. This usually rules out most standard UNIX utilities. For example, many E-mail programs are continually being broken by hackers. Instead, a stripped-down version with no user-friendly features is run on the bastion host.

Using Throttles. Other restrictions can be added to the bastion host to enforce organizational security policies. For example, the DEC SEAL firewall includes throttles on outbound data, limiting the transfer of data to the outside world to an equivalent of 1,200 bps. That way, even when hackers do break in, or a dishonest insider leaks information, the losses are limited by the low-bandwidth channel. Most users will never notice the throttle, because it will not affect the update of screens (inbound data) using Telnet or the ability to download files off the Internet.

The firewall bastion host may also be used in reverse, screening incoming connections to ensure that only legitimate users can access their home systems from other locations using the Internet. This mode of operation requires the use of one-time passwords (or equivalent challenge-response systems, frequently based on credit-card-size encryption calculators) to provide any degree of protection.

Using a Public Access Host

Another approach to controlling Internet access is not to connect at all. Instead, an external public access host on the Internet is made available to internal users using a separate communications channel (see Exhibit 6-4-4), usually asynchronous serial dial-up. This technique tends to be inconvenient because the user's local machine is limited to terminal emulation (with terminal-oriented file transfer such as Kermit or x/y/z modem). It also requires users to learn how to use another operating system, unless the public access host happens to be the same operating system as the internal user's.

The inconvenience of using terminal access to the external system can be

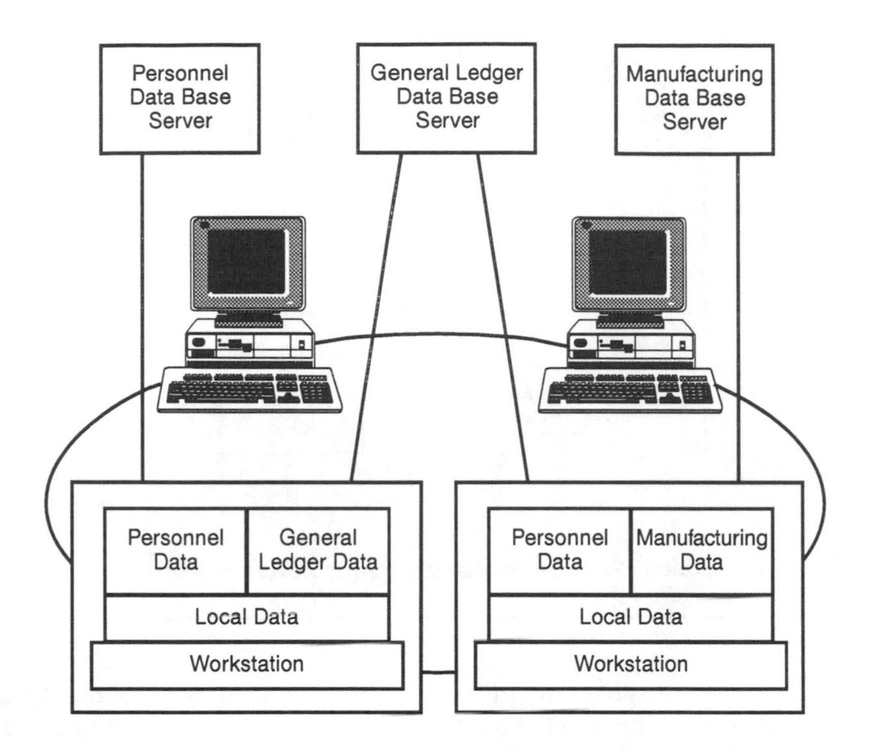

Exhibit 6-4-4. Internet Access Through a Public Access System

alleviated somewhat by using the internal network to access the external system through an outdial terminal server (see Exhibit 6-4-5). That allows all users to take advantage of the highest speed available for local hardwired terminal access to the external host, rather than being limited to dial-up modem speeds. It also eliminates an extra cable hanging out of the user's desktop system and allows some services, such as E-mail to be delivered local without going through a terminal interface. The primary disadvantage of the public access system approach is the limitation of services to those accessible by character mode terminals.

The public access host approach can be very effective at keeping Internet hackers off the internal network, as long as inside users recognize that the public access host is a hostile environment in that it is under the control of outsiders and providers no security. All users should assume that all traffic is monitored and controlled by outsiders and is subject to modification. In other words, unless there is some external mechanism for providing privacy, au-

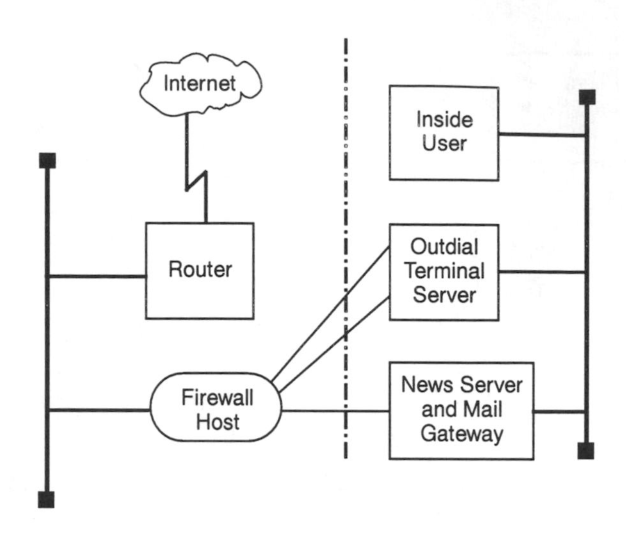

Exhibit 6-4-5. Networking a Public Access System

thentication, and integrity, any information (including account numbers and passwords) passing through the public access system must be assumed to be public knowledge and potentially corrupted. This is usually not a problem, because the Internet is subject to the same limitations (i.e., no assurances of privacy, authentication, or integrity unless provided by higher-level protocols such as privacy-enhanced E-mail).

Unless the public access system is protected at least as well as a direct connection to the Internet, using filtering routers and a well-designed and maintained bastion host, it should still be considered vulnerable. If many users are allowed on the public access host, it becomes very likely that evidence of a skilled attack would remain undetected. In general, once an attack succeeds, it should be assumed that the attackers can cover their tracks. Except for "drop-box safe" style audit trails, the attack is only visible until the intruder succeeds in getting root or equivalent supervisor access, which may be only a matter of minutes after getting any login on the system.

Security Tools

Assuming the Internet connection is used to seek out information, the tools provided through traditional FTP and Telnet are sadly lacking. Better security tools, such as Archie and Gopher servers, are attractive. Unfortunately, the also present challenges to secure implementation. It may make sense to provide them on a protected public access host, rather than trying to secure all systems in the network.

Internet connectivity can also be used to provide public access to press releases, white papers and other information. Many organizations provide anonymous FTP service and some are putting up Gopher servers. Here, too, the level of security depends on the environment. Some form of integrity protection above and beyond that built into an anonymous FTP server may be required. The technology is available to validate documents through the use of message digest algorithms and public key signatures.

SUMMARY

Firewall technology need not be restricted to attachments to the Internet. It may make sense to place firewalls between internal networks to limit the damage from untrustworthy insiders or successful penetration of an exterior firewall. No matter what technology is used for a firewall, it is safe to assume that it will be penetrated. The key is to determine what degree of successful penetration is tolerable and what price is acceptable to legitimate users. This should be part of an overall security policy. The challenge is to put enough roadblocks in front of an attacker to make it likely that any attack will be detected before significant damage can be done.

Firewalls have limitations, however. They will not thwart insider-assisted attacks. Likewise, they do not protect against virus or Trojan attacks through software or data legitimately imported through the firewall. Nonetheless, firewalls can make an effective contribution to an overall security plan. They can provide a tough shell around the relatively unprotected systems common in typical LANs, protecting them from outside attacks. At the same time, they are only one piece of the security solution, and their efficiency and effectiveness depends largely on the particular needs of the organization and network users. Other tools, from the use of one-time passwords to eliminate eavesdropping attacks to encrypting all data on portable computers carried in the field, are equally important.

Although there are tools available to counter every known plan of attack, the problem is that the more effective tools usually are costly, both in terms of purchase price and inconvenience. The challenge to management is to determine the true requirements for security, as well as for usability and connectivity, and select the appropriate level of protection for their needs. Within that context, the variety of firewall approaches described in this chapter becomes just another class of weapon in the arsenal available for selection against the appropriate targeted weaknesses. It is up to management to ensure that other weaknesses are also protected to provide overall strength against attack.

6-5 Managing Coexisting SNA and LAN Internetworks

Anura Guruge

THE IDEA OF WHAT A NETWORK LOOKS LIKE in an IBM-dominated computing environment is in need of a serious update. The all-SNA star network, consisting in the main of 3270 and remote job entry (RJE) terminals multi-dropped off synchronous data link control (SDLC) links to a central S/370 host, is now a relic. New IBM networks consist of microcomputer LANs—both Token Ring and Ethernet—that are appendages of a central, multiprotocol backbone. Systems Network Architecture is just one of the many disparate networking protocols flowing across the backbone. Thanks to the ever-increasing popularity of microcomputers and UNIX workstations in what are still primarily IBM shops, host- and PC-based applications are now routinely front-ended, complemented, or in some instances totally usurped by LAN-oriented, workstation centric, client/server applications.

This increase in LAN-attached, workstation-based applications has not diminished what constitutes, by far, the bulk of SNA applications: host-based, mission-critical applications. Mission- or business-critical applications are in effect the lifeblood of an enterprise. They provide, support, and sustain the vital operations imperative to the enterprise.

USE OF SNA OVER LANs

At least 65% of today's mission-critical applications will be in use for another 10 to 15 years, and this longevity ensures that SNA will be in use for at least that long. Mission-critical applications were invariably developed to be accessed through 3270 terminals. With more and more 3270s being replaced by PCs, 3270 emulation on LAN-attached workstations (using an SNA LAN gateway) is now becoming the main approach for accessing these applications. Access to a host application through LAN-attached workstations currently accounts for most of the SNA-based traffic that flows across LANs.

Access to a host application is one of four distinct scenarios for the use of SNA over LANs. These four scenarios (illustrated in Exhibit 6-5-1) are:

- Access to mission-critical host applications from a LAN-attached workstation or computer system (e.g., AS/400), using SNA logical unit to logical unit (LU-LU) session types 2 and 3 (i.e., 3270 data stream) or LU-LU session type 7 (i.e., 5250 data stream). The host could be an S/3x0 or a minicomputer such as the AS/400 or the S/38. Token Ring is typically the LAN of choice for this type of host access, though it is now possible to access S/3x0 applications using LU-LU session type 2 over other types of networks.
- SNA LU 6.2/APPN-based program-to-program interactions between two LAN-attached systems (e.g., PC to PC, PC to AS/400, AS/400 to AS/400). Many of IBM's contemporary "utility services" for data transfer or remote data access are based on LU 6.2. Hence, the distribution of documents, mail, or files as provided by IBM's office automation packages or remote system access through DSPT can now be realized across LANs.
- Host-to-host SNA paths between IBM 3745 communications controllers, usually over a 16M-bps Token Ring. (For ES/9370s or 43xx hosts with integrated Token Ring adapters, the integrated adapters, rather than 3745s, can be used.) Such LAN connections (subject to their distance limitations) can be cost-effective, high-speed data paths between hosts and attractive alternatives to SDLC link connections or backup paths for channel-to-channel connections between hosts.
- Using an SNA backbone for communications between non-SNA LANs, by encapsulating the LAN traffic within SNA LU 6.2 message units.

A REPERTOIRE OF PROTOCOLS

The popularity of LANs means that enterprises with IBM hosts or minis invariably end up with a repertoire of applications that fall into three main categories: file- or print-server applications, host- and minicomputer-resident traditional applications, and a new generation of client/server, program-to-program applications.

For each of these three applications categories there are native, or preferred, protocols. The protocol set that will typically be used with a given application category can be summarized as follows:

- File/print server. NetBIOS, IPX/SPX, Banyan Vines, TCP/IP, AppleTalk, or XNS.

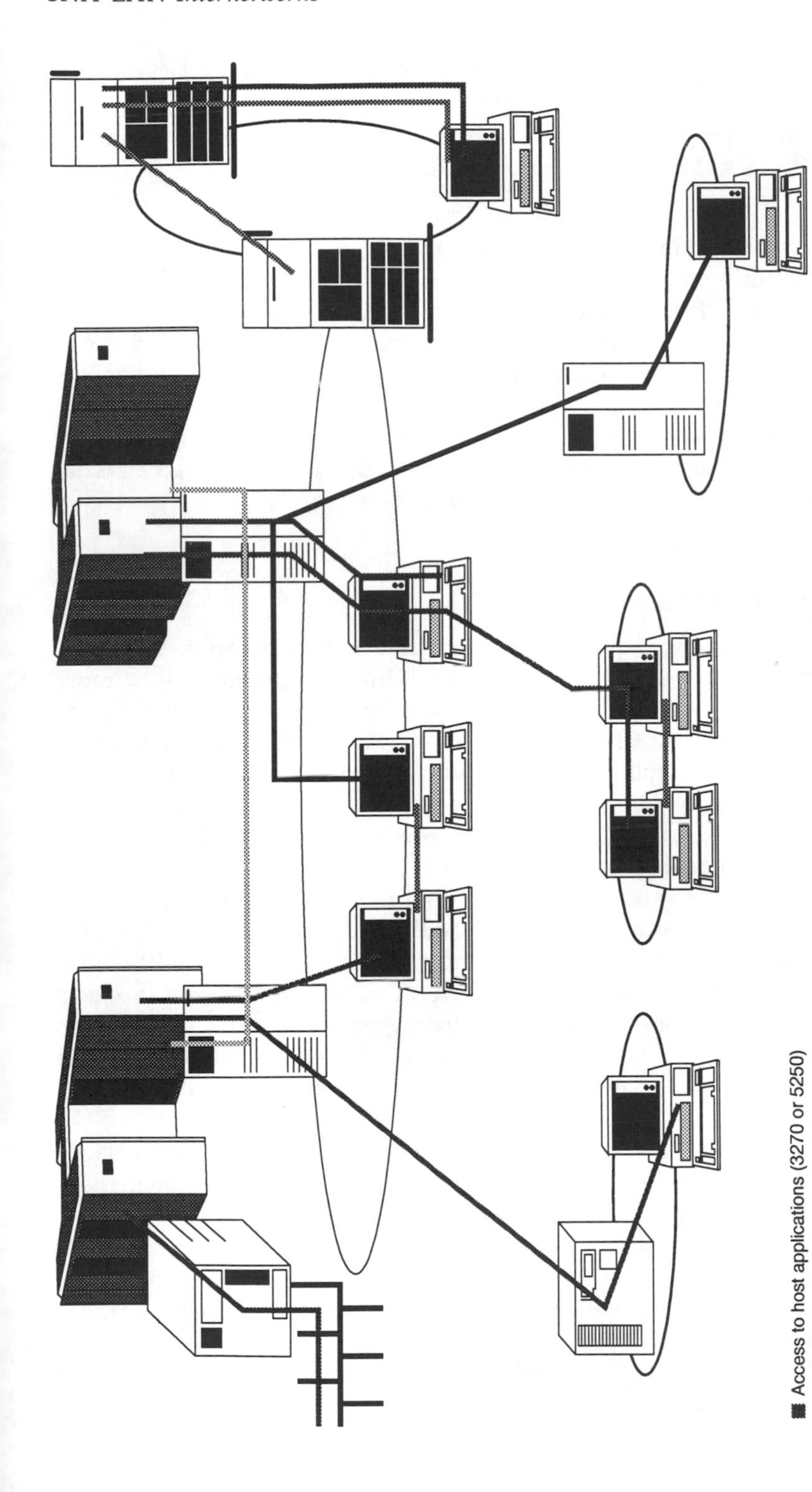

Exhibit 6-5-1. SNA over LAN Applications

- Host/mini application access. SNA LU-LU session types 2 (i.e., 3270 data stream) or 7 (i.e., 5250 data stream).
- Program-to-program communications. SNA LU 6.2 APPN (including SNADS, DIA, DDM, SNA/FS), NetBIOS, TCP/IP, DECnet, IPX/SPX, XNS, or OSI.

The diversity of protocols is not itself a problem because LANs support concurrent multiprotocol data streams. It is not unusual to find SNA, NetBIOS, IPX, and TCP/IP on the same Token Ring LAN.

Networking challenges in the IBM environment are only starting to surface as the scope of LAN-based applications, using the protocol cocktail described, extends beyond LANs that are in close physical proximity to each other to also embrace geographically distant LANs. A workstation on a LAN in Boston requiring access to data on a file server on a LAN in San Francisco or access to an application resident on a host in Chicago is an example.

PARALLEL NETWORKS

Enterprises that now require LAN interconnection between distant locations invariably have an SNA WAN that is hubbed around one or more S/3x0 hosts; this WAN reaches all the remote locations of interest. Companies have sizable investments in these WANs, which also tend to be tried-and-true; therefore, to minimize costs and to standardize on one set of network management requirements and operational procedures, the natural tendency is to strive for a single WAN backbone that supports SNA and the requisite LAN protocol. This is the problem.

For all their pedigree and sophistication (not to mention cost), SNA WANs in actuality can support only SNA traffic, either LAN or link based. Support for LAN protocols, even the more popular ones such as NetBIOS, IPX/SPX, and TCP/IP, is currently available only on an ad hoc basis, using products specifically designed for that purpose. The only potentially general solutions are IBM's frame relay data communications equipment feature for its 3745 communication controller and Computer Communications Inc.'s Eclipse 7020 LAN-over-SNA router.

With these, a user can interconnect LANs across an existing SNA backbone. However, this LAN-over-SNA solution is relatively new to the market, and there is still much doubt as to its feasibility, particularly in terms of throughput and performance. With the exception of these auxiliary, add-on solutions, it is accurate to say that SNA WANs, at present, do not provide a workable cost-effective, high-performance, and general solution for transporting LAN traffic from one location to another.

This frustrating absence of general support for LAN traffic across SNA has forced many enterprises to implement and maintain two parallel WANs: a dedicated 37xx-based SNA WAN for SNA application access and a bridge or bridge/router-based multiprotocol WAN for non-SNA, interLAN traffic, as illustrated in Exhibit 6-5-2.

This dual WAN approach obviously works but requires that users install, maintain, and operate two very diverse networks. In some instances, the duplication of the actual long-distance links can be avoided by deploying a multiplexer (e.g., T1) network.

TOWARD A COMMON MULTIPROTOCOL BACKBONE

Bridge and bridge/router vendors have rushed to the rescue of enterprises dereading the thought—and the expense—of parallel networks. The initial, and intuitive, solution was to bridge the SNA LAN traffic while bridging or routing the other LAN protocols as appropriate. This makes it possible (at least in the case of LAN traffic) to have a single universal backbone WAN that interconnects dispersed locations and permits both SNA LAN and non-SNA LAN traffic to be readily transported between various LANs, as illustrated in Exhibit 6-5-3.

Two techniques for integrating SNA LAN traffic with non-SNA LAN traffic are predominant: source-route bridging and Internet Protocol (IP) encapsulation. The former may be performed by Token Ring bridges or bridge/routers that include explicit support for encapsulating SNA (and invariably also NetBIOS) traffic.

A third option is proprietary but is still relatively popular: CrossComm's protocol independent routing (PIR) scheme. With the increasing popularity of SNA over Ethernet, transparent bridging—the native bridging technique for Ethernet LANs—will become another option for SNA and LAN traffic integration.

Source-Route Bridging

Source-route bridging (SRB) is the original—and in essence the default—method for interconnecting multiple Token Ring LANs. It permits two or more Token Ring LANs to be uniformly and transparently interconnected, such that they appear to form a single, seamless, consolidated LAN.

SRB may be used to interconnect LANs that are either adjacent to each other (e.g., within the same building) or geographically dispersed. The former scenario, referred to as local bridging, does not involve any WAN connection. The latter, which does require WAN connections, is referred to as remote bridging. A bridge that is performing remote bridging is sometimes called a split bridge by IBM, to indicate that the bridging function has been divided and is being performed at either end of a WAN connection.

If the SRB bridge/routers are supporting SNA LAN traffic, these multiple

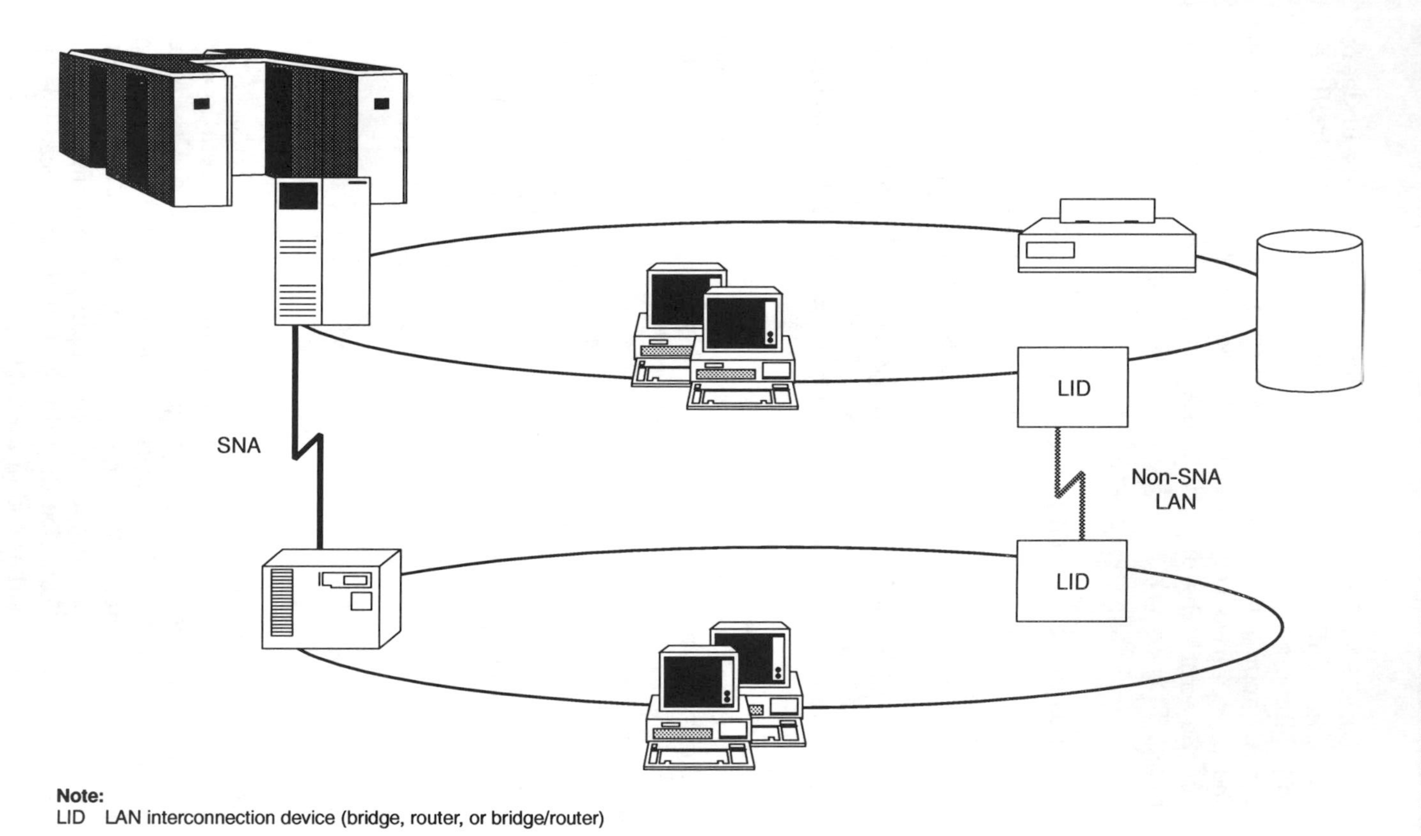

Exhibit 6-5-2. Parallel Backbones for SNA Traffic and LAN-to-LAN Traffic

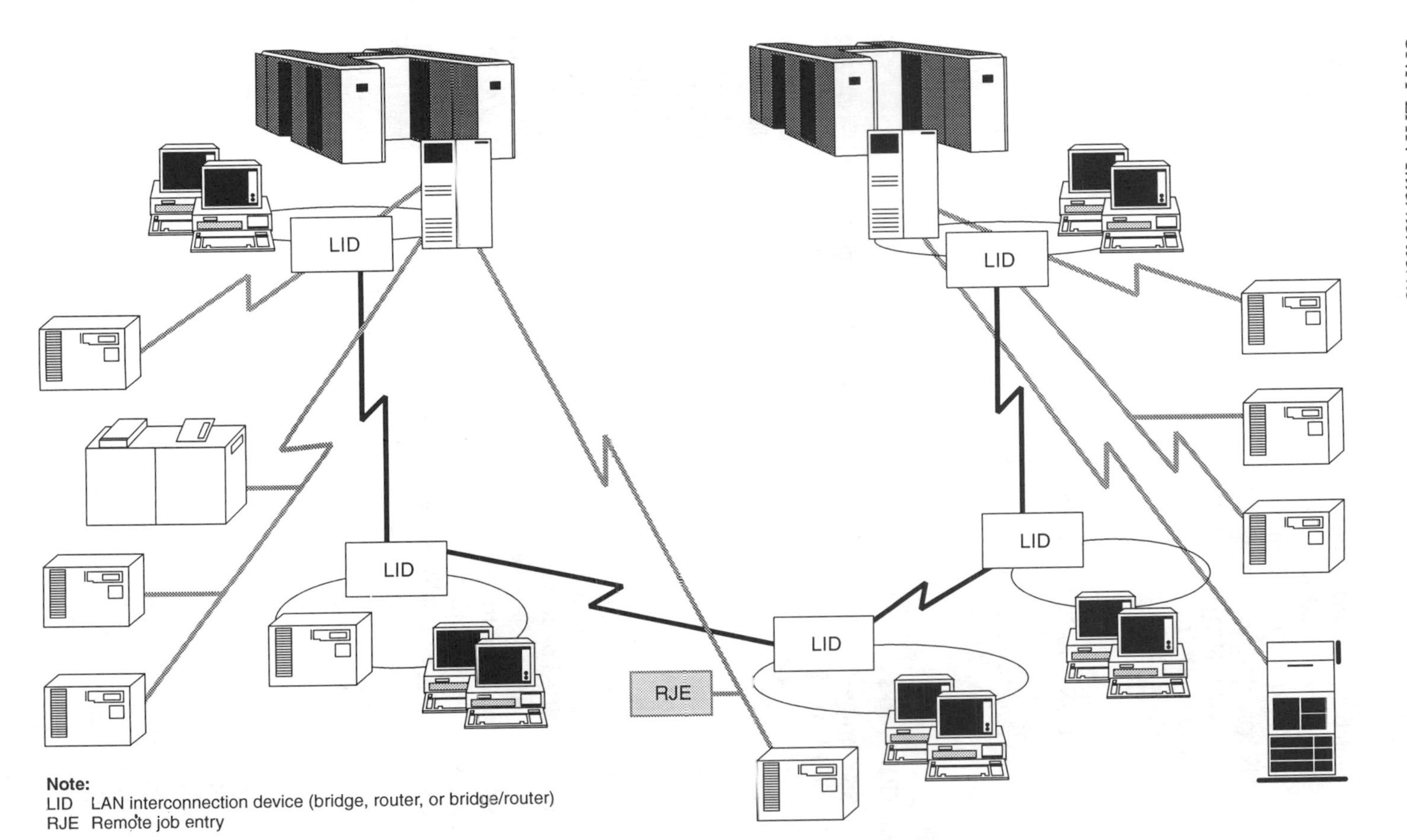

Exhibit 6-5-3. Support of All Traffic on Multiprotocol Backbone Through Bridging & IP-Encapsulating SNA

LANs—which behave logically as a single large LAN—are interfaced to the S/3x0 SNA environment through one or more SNA LAN gateways (e.g., a channel-attached 3745, 3174, or 3172). All the attached devices, including the SNA LAN gateway, behave as if they were attached to the same physical LAN. Thus, host system control software (e.g., ACF/VTAM and ACF/NCP) and consequently the host-resident SNA application programs can view, and treat, the various SNA devices on the dispersed LANs as if they were all on one single LAN.

SRB is a straightforward LAN interconnection technique that can accommodate most SNA LAN traffic consolidation. Like all other bridging techniques, SRB is a data link (OSI layer 2) level process. Thus it is independent and transparent to such higher-level (i.e., layer 3 and above) networking protocols as SNA, APPN, NetBIOS, or Novell's IPX/SPX.

How SRB Works

For SRB bridge/routers to be able to provide interconnection between two devices on two different LANs, a path between the two LANs (a chain of intermediate bridges that the data will traverse) has to be established. A dynamic, broadcast search technique is used to establish this path. The technique used is similar to that employed by SNA to locate undefined resources in multihost environments and what is now available in Advanced Peer-to-Peer Networking (APPN) for dynamically locating remote resources.

The SRB broadcast search technique is initiated by the source device, rather than by one of the bridges. It does so by issuing a Token Ring TEST or XID (exchange identification) command. The destination device being sought is identified by its unique layer-2 6-byte media access control (MAC) address. Because the SRB search process is restricted to layer 2, it occurs without any reference to higher-level addresses or names (e.g., SNA/APPN LU names, SNA network or local addresses, or NetBIOS names).

The source device initially issues the TEXT or XID command without any indication that the command should be broadcast outside the local LAN segment to which it is attached. This initial search is known as an on-segment search. If the source receives no response from the destination before the expiration of a prespecified time-out period, it assumes that the destination is on another LAN.

The source device then immediately resorts to an off-segment search by again issuing a TEST or XID command, but this time with a routing information field (RIF) included by the header prefixing the command, and two flags (represented by four bits) set in the header to indicate the presence of the RIF, as well as the need for a broadcast search to be conducted.

The broadcast search flag denotes that the search is to be conducted in one of two ways:

- All-route broadcast.
- Single-route broadcast.

The method is selected by the software on the source device. In general, SNA software requests all-route searches, whereas NetBIOS opts for single-route searches. When an all-route search is specified in a TEST or XID command, each bridge encountering that command makes a copy of it and forwards that copy to every other Token Ring LAN attached to that bridge. With remote bridging, the copy of the TEST or XID command is sent over the WAN connections to the remote bridges.

Traversing All the Routes

When a TEST or XID command is forwarded to another LAN, the bridge updates the accompanying RIF to reflect the identity of the bridge that is has just crossed. The identification is in the form of a 16-bit segment number that consists of a unique number denoting the Token Ring from which the command was copied and a unique number identifying the bridge. Identifying the bridge, in addition to the Token Ring, permits multiple bridges to be used between a pair of Token Rings.

The current Token Ring approach does not permit a RIF to exceed 18 bytes. Hence a routing information field, which must always begin with a 2-byte routing control field, can at most contain only eight segment numbers. Because the first entry in the RIF has to identify the LAN containing the source device, SRB per se permits LAN interconnections over no more than seven intermediary LAN segments. (The routing control field at the start of the RIF, in addition to denoting the type of broadcast search being conducted, has fields to indicate the number of LAN segments currently present in the RIF, the longest information frame that is supported once a session is established, and a flag to indicate whether the command containing the RIF is flowing from the source to the destination, or vice versa).

With all-route searches, the destination device receives as many copies of the TEST or XID command as there are available routes between the source and destination LANs. Each command received reflects in its RIF the exact path that it has traversed, in terms of intermediate LANs and bridges, in the order in which they were crossed. The destination device returns each command it receives, replete with its RIF, back to the source device. The response traverses the same path as that taken by the original command.

As the source device receives the responses from the destination device, it could, theoretically, determine the optimum route by evaluating the route structure found in the RIF of each response. The source may decide that the route containing the least number of intermediary bridges is the optimum path. It could also record the routes specified in the other RIFs to use as potential alternate routes if the initially chosen route fails during the dialogue. Current Token Ring implementations do neither.

Instead, the route traversed by the first received response is assumed to be the best route, given that this route was obviously the fastest for the round-trip involved in the search process. This, however, may not really be the case. A temporary aberration on what would normally have been the fastest path may have caused a less than optimum path to be selected. The routes specified in the other responses are ignored. The RIF in the first response is then duplicated in the headers of all frames interchanged between the source and destination devices for the duration of that dialogue (e.g., an SNA session). Thus, SRB, just like traditional SNA and even APPN, is an inherently fixed path-routing technique.

The Single-Shot Approach

For a single-route broadcast search, a bridge determines the route over which the TEXT or XID command should be propagated on the basis of a single-route broadcast path maintained by each bridge. This broadcast path is constructed using a spanning tree algorithm. This spanning tree approach, often referred to as transparent spanning tree, is the bridging technique used to interconnect Ethernet LANs. It is also known as transparent bridging.

With single-route searches, the destination device receives only one copy of the TEST or XID command. In marked contrast to the all-route method, the RIF in the command received by the destination contains only a routing control field and does not indicate the route traversed by the command. Just as with an all-route search, the destination device returns the command to the source by toggling the direction flag.

The destination device, however, sets the broadcast flags at the start of the routing control field to denote an all-route search. This causes the response to be returned to the source using all available paths between the destination and source. The route taken in each case is recorded in the RIF as in the case of an all-route search. Thus, in the case of a single-route search, the routing information is collected on the return trip as opposed to the destination-location trip.

The source receives multiple responses and, as with all-route, uses the route taken by the first response received as the optimum path.

SRB's Limitations

The greatest virtue of SRB is that it offers plug-and-play interoperability between devices on different LANs, with minimum predefinition. Usually, just a LAN number and a bridge number are required to set up the bridges.

SRB does, however, suffer from some major limitations. Its fixed-path routing offers no dynamic, alternative rerouting in the event of path failure;

it creates overhead traffic during the broadcast searches; and it has an inherent seven-hop limit on intermediate LANs and bridges between the source and destination. Most of the leading bridge/router vendors have devised methods to circumvent these limitations.

To circumvent the seven-hop limitation, most vendors now offer an extended SRB facility, whereby a bridge/router subnetwork from a given vendor, irrespective of the number of SRB bridges/routers and LANs involved, always appears as a single hop to other internetworking devices in the overall network. The technique is totally transparent to other standard SRB bridges/routers, which see just a single RIF LAN segment entry that happens in reality to correspond to a multisegment subnetwork. Most enterprises have not as yet been inconvenienced by the seven-hop ceiling of standard SRB. But extended SRB sidesteps this problem, and enterprises can consider larger networks for future use without concern that the seven-hop count will be exceeded.

Broadcast Storms. The additional traffic generated during SRB broadcast searches can be a source of major concern, particularly in remote bridging configurations that use relatively low-speed WAN connections. Bursts of SRB broadcast search traffic that interfere with bona fide data traffic are referred to (somewhat dramatically) as broadcast storms. Such storms have been a perennial source of criticism of SRB, especially from Ethernet users who point out that Ethernet's spanning tree bridging technique does not require a broadcast search.

Broadcast storms can be minimized by the use of SRB proxy responder agents. Such agents provide a local cache directory scheme; the addresses of remote destination devices can be saved following an initial SRB search. Before subsequent broadcast searches are performed, an agent checks its cache directory to see whether it already knows of a route to the destination device being sought. If a route is found in the cache directory, the agent inserts that route in the RIF of the TEST or XID command issued by the source device, then returns the command to the source. This avoids an exhaustive broadcast search.

IP-ENCAPSULATED ROUTING

IP encapsulation is now being aggressively promoted by some vendors, including IBM, as an answer to both SRB's fixed-path routing, as well as the inability of bridge/routers to perform SNA routing. (IP encapsulation of SNA traffic is one of the functions performed by the data link switching feature in IBM's 6611 bridge/router.) A LAN layer-2 logical link control (LLC) frame that contains SNA (or NetBIOS) traffic is inserted in its entirety as a piece of data within an IP data-gram packet and routed across a bridge/router network as IP traffic.

IP encapsulation does have some irrefutable attractions. It offers dynamic alternative routing in the event of a path failure, policy-based adaptive routing (e.g., fastest or least-cost path for each packet being transmitted), and avoidance of broadcast searches. On the surface, IP encapsulation would appear to be an ideal solution to some of SRB's weaknesses. Unfortunately, this is not exactly the case.

Because IP encapsulated routing is based on Internet Protocol (IP) addresses, destination IP addresses must be allocated to supplement the already existing MAC addresses and, in the case of SNA, network addresses. To be fair, most vendors do not require that destination IP addresses be allocated on a one-to-one basis for each device. Instead, IP addresses can be assigned on the basis of a destination LAN, LAN group, or destination bridge/router. IBM, with DLSw, dynamically correlates the destination MAC address with the IP address of the 6611 serving the LAN containing that device by first conducting an SRB all-route broadcast search. MAC-to-destination 6611 IP address correlations, thus established, are cached for subsequent use. But for each 6611 that is to participate in DLSw, the IP address must be manually defined.

The drawback to IP encapsulated routing is that it reverses the trend toward plug-and-play networking. Enterprise networking is general and SNA in particular have been definition-intensive activities, but the industry as a whole, including IBM, has been making a genuine, concerted effort to introduce simpler plug-and-play configure-on-the-fly networking techniques. IBM has made contemporary SNA, particularly ACF/VTAM Version 3 Release 4, as dynamic as possible. IBM's APPN is the epitome of a modern, dynamic networking scheme. For all its faults, SRB too is close to a plug-and-play technique. IP encapsulation, even with the DLSw enhancements, requires too much manual predefinition to be accepted as a contemporary networking methodology.

EMERGING OPTIONS

Protocol Independent Routing. PIR is an attempt to offer adaptive as well as alternative routing in addition to broadcast search reduction, without the need for manual IP address allocation. It thus provides the dynamic routing advantages of IP encapsulation with the plug-and-play characteristics of SRB. PIR is a compelling solution for midsize networks (those containing as many as 100 LAN segments) in which the bulk of the traffic is SNA or NetBIOS.

Source-Route Transparent Bridging. A feature that is now offered by many bridge/router vendors is source-route transparent (SRT) bridging. SRT permits a bridge to concurrently perform both Token Ring SRB and Ethernet transparent bridging. SRT thus allows the same bridge to support devices on Token Ring LANs using SRB and devices on the same Token Rings, or on Ethernets connected to other ports, that require transparent bridging.

SRT is a scheme for Token Ring and Ethernet coexistence, as opposed to interoperability. Token Ring to Ethernet interoperability, when needed, is provided by source-route translation bridging (SR/TLB) and is available on IBM's 8209 LAN bridge. SR/TLB is not currently available on the IBM 6611, though it is a feature available on some bridge/routers.

BRING IN THE LINKS

Providing support for LAN-based SNA traffic addresses only a part of the overall problem of integrating SNA and LAN traffic. More than half of the current installed base of SNA devices worldwide are connected to SDLC or 3270 binary synchronous communications (BSC) links. Furthermore, for technical, historic, and financial reasons customers are not likely to upgrade or replace these devices to become LAN compatible.

Link-attached devices in use at SNA sites include IBM 3174s, 3274s, 3770s, 8100s, Series/1s, S/36s, S/38s, 3600s, 4700s, 5520s, and minicomputers from Digital Equipment Corp., Wang, Prime, Data General, and others. Thus, to implement a universal, all-inclusive, multiprotocol WAN that supports all types of traffic, bridges or bridge/routers have to support link-based SNA traffic in addition to LAN-based SNA traffic (see Exhibit 6-5-4). Leading bridge/router vendors and some bridge vendors now offer at least one method for consolidating SNA/SDLC link traffic with LAN traffic.

The three technologically feasible techniques for integrating SNA link traffic with LAN traffic are:

- Straight synchronous passthrough (which may be supplemented with such value-added features as traffic prioritization).
- Remote polling.
- SDLC to Token Ring LLC transformation.

HANDS-OFF HANDS-ON: SYNCHRONOUS PASSTHROUGH

Synchronous passthrough is the easiest, most intuitive, and least risky technique for using bridge/routers to support SNA/SDLC link traffic. It was the first, and for a while the ubiquitous, solution for SNA link integration. For reasons explained in the next section, it is the least desirable technique. Synchronous passthrough cannot generally be used to support 3270 BSC traffic, although it does support other SDLC-like protocols (e.g., High-level Data Link Control, or HDLC).

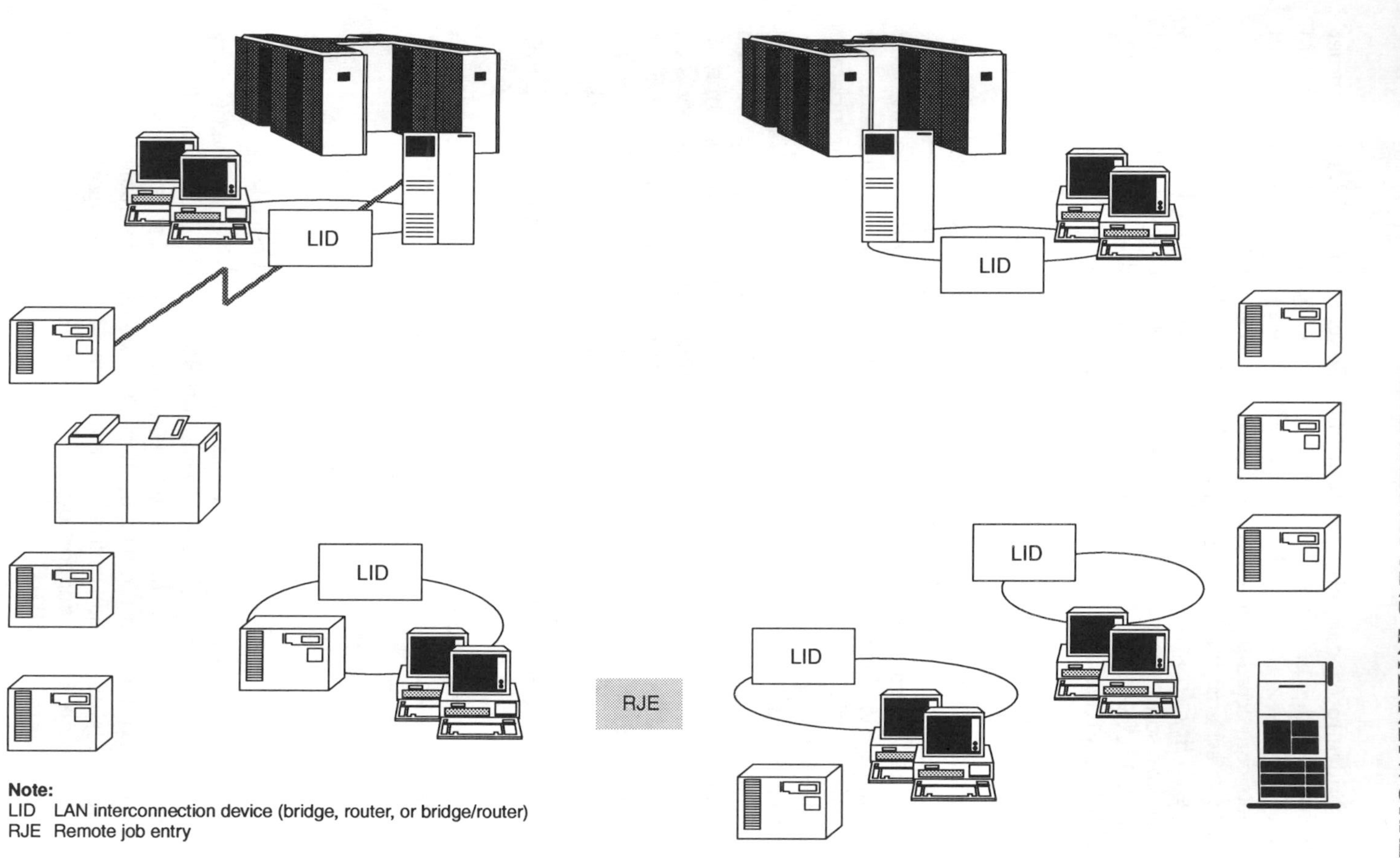

Exhibit 6-5-4. Multiprotocol Bridge/Router-Based WAN for SNA, LAN, and Link-Based Traffic

With synchronous passthrough, each physical SDLC link from a 37xx port to one or more (i.e., in the case of a multidrop link) remote SNA devices is replaced by a clear-channel, port-to-port connection through the multiprotocol WAN. The 37xx port to which the original SDLC link was attached is now connected to a serial port on an adjacent, upstream, local bridge/router, possibly with a modem bypass cable or a short-haul modem configuration. The remote devices are attached to a serial port on a downstream, remote bridge/router. The two serial ports on the bridge/routers are then mapped to each other, over a point-to-point route across the multiprotocol WAN, using the bridge/router's configuration utilities.

SDLC traffic arriving at either of the bridge/router serial ports is cleanly transported to the predefined partner port at the other end. This is achieved by an encapsulating technique, whereby every SDLC frame, from the first bit of the start flag to the last bit of the end flag, is included as data within an IP datagram. The bridge/routers do not read the SDLC frame, let alone modify it in any way. What comes in at one serial port goes out unadulterated at the opposite serial port.

In general, synchronous passthrough is a safe bet that permits SNA links to be cut over to a multiprotocol WAN with a minimum of fuss and certainly without the need for either a new ACF/NCP host gen or a software reconfiguration at the device. Most, if not all, of the glitches that were encountered with the first bridge/router implementations as well as such limitations as 1,500-byte frame size have now been rectified.

Because the technique forwards untouched SDLC frames, which in turn may contain complete SNA message units, neither end sees any changes at the SDLC—let alone the SNA—level, with the exception of a possible degradation in link speed. Thus, both sides continue to work as though they were still connected to each other over a physical link, rather than through a complex multiprotocol WAN. This technique insulates customers from compatibility issues that may arise from host software upgrades or new features added to SNA, because whatever works on a leased line should also, in theory, work with synchronous passthrough, except for functions that are extremely time sensitive.

Synchronous passthrough also provides a high degree of compatibility with SNA-based network management systems like NetView, which continue to have total end-to-end SNA control and visibility. There is no mechanism at the moment, however, for SNA to recognize what is happening within the multiprotocol WAN. Thus, the SDLC links invariably appear to be clean links with few or no retransmissions, because retransmissions on the WAN are invisible to the network management system.

Drawbacks of Synchronous Passthrough

The noninterventionist qualities of synchronous passthrough are also its greatest weakness. By keeping all the SDLC frames intact, synchronous passthrough does not in any way reduce the amount of traffic over a given link. If anything, it increases the traffic, because encapsulating SDLC in IP not only makes the transmitted packets longer but adds additional IP control interactions on top of the SDLC interactions. To compensate for this, many vendors now offer some type of data compression capability; some vendors also offer a mechanism to compress the IP header.

For the new WAN link's response time to be comparable to those experienced with the physical link, the bridge/router network has to ensure that each end-to-end connection is allocated bandwidth comparable to that of the original link. Usually this is not an issue. Most SNA links to peripheral devices, as opposed to those between 37xxs, operate at data rates lower than 19.2K bps. In contrast, the trunks in a bridge/router WAN are unlikely to operate at rate slower than 56K bps. However, from the perspective of each SDLC link, the WAN has to be shared with high-speed, high-volume traffic from LANs as well as with traffic from other SDLC links.

Being able to assign a higher transmission priority to link traffic at the expense of LAN traffic is one way to prevent link traffic from being swamped by LAN traffic. CrossComm Corp. (Marlborough MA) and Cisco Systems, Inc. (Menlo Park CA), the leading vendors, are beginning to offer generic traffic prioritization features. These can be exploited to ensure that SNA traffic—link or LAN—gets precedence over other traffic.

The other unfavorable feature of synchronous passthrough is the amount of spurious, nonproductive polling traffic that must be continually transported across the WAN, consuming valuable bandwidth. Any retransmissions sought by either end also must be made end-to-end, again occupying bandwidth.

REMOTE POLLING

Remote polling overcomes the idle-poll and retransmission problems of synchronous passthrough. It can also be used to effectively and efficiently support 3270 BSC traffic. With this technique, only SDLC frames (or 3270 BSC blocks) containing bona fide data—SDLC I-frames (or BSC text blocks), which in turn contain SNA message units—are transmitted across the WAN. Polling and retransmissions are performed and responded to at the periphery of the WAN by special SDLC (or BSC) link-driver modules.

Primary and Secondary Modules. Two types of these modules are available: primary modules, which issue polls, and secondary modules, which respond to polls. The primary modules are deployed in remote, downstream bridge/routers to which the actual SNA devices are attached, whereas the secondary modules are used in the local, upstream bridge/routers connected to the 37xx ports.

Just as with synchronous passthrough, there will be predesignated port-to-port mapping, associated with each SNA link, between the router port attached to the 37xx and the corresponding port on the remote router to which the SNA devices are attached. However, in the case of remote polling, a link-address-to-link-address mapping is also required on top of the port-to-port mapping.

Once these mappings have been established, the remote primary module polls the devices attached to its port, using a predefined polling table that specifies the link addresses to be polled plus the order and frequency at which to poll the addresses. As each device becomes active, as indicated by a positive response to a poll, the primary notifies its partner secondary module that the device in question is now active and ready for data exchange, using a bridge/router handshake protocol.

As the primary starts its polling cycle, the local secondary activates its port and waits to receive data from the 37xx port. The 37xx reacts to this by issuing polls. If the secondary receives notification that the subject physical device is active, it responds to the 37xx as if it were the real device.

Once this initial activation sequence is complete, one side starts to receive data frames. It accepts them and forwards them to its partner, which in turn ensures that the frames are delivered to the intended destination device. The relaying of such data frames between the modules is performed on a per-address basis. This process of receiving data frames at one end and forwarding them to the other end for delivery to their rightful destination takes place on a routine basis.

Merits of Remote Polling. Remote polling, also referred to as poll spoofing or local acknowledgment, is invariably considered by customers to be more desirable than synchronous passthrough, as it has the unquestionable merit of reducing the amount of link traffic that must be transported across a backbone. Remote polling also reduces the occurrence of time-outs. Remote polling, just like synchronous passthrough, in no way interferes with the SNA protocols and, as an SNA-transparent scheme, is not affected by modifications made at the SNA level. It also provides Net View with full end-to-end SNA access and visibility but now with slightly more distortion of its perception of the underlying link.

Remote polling has an optional capability whereby link configurations are transformed to achieve cost reductions and even improve polling efficacy. Given that remote polling uses a link-address-to-link-address technique, it is possible to transform and map link addresses of devices on multiple links to addresses on a virtual consolidated link at the 37xx side. For example, the devices on four point-to-point links could be presented to the 37xx through the router backbone, as if they were four devices on a single multipoint link.

SDLC-TO-LLC TRANSFORMATION

This technique is in effect a variation of standard remote polling, exploiting remote polling's inherent physical-to-virtual link transformation capability. In standard remote polling, the same data link control protocol (e.g., SDLC or 3270 BSC) is used both at the host and at the remote device. With SDLC-to-LLC transformation, different data link protocols are used at the two ends: SDLC on the link-attached device and Token Ring logical link control type 2 (LLC-2) at the host. Though typically offered today as an SDLC to Token Ring transformation, this same technique could be used between SDLC and Ethernet and between 3270 BSC and either Token Ring or Ethernet.

The primary reason for SDLC-to-LLC transformation is cost reduction. With SDLC-to-LLC transformation, customers can eliminate serial link ports at the host side, both on 37xx communications controllers and on the bridge/routers adjacent to those controllers.

Instead of transporting link traffic to or from the host through serial ports, LLC transformation converts link traffic to Token Ring LAN traffic and conveys it to or from the host over a standard host (e.g., 3745) Token Ring interface. In other words, SDLC-to-LLC transformation converts SDLC link traffic to Token Ring LAN traffic at the host end so that the hosts can treat the link traffic as Token Ring LAN traffic. The link-specific Token Ring traffic is converted back into link format at the remote end so that it can be transmitted over the actual physical links using SDLC. Customers do not need serial ports for the link traffic and a Token Ring LAN interface for LAN traffic. At the host end, a Token Ring interface works with LAN and link traffic.

SDLC-to-LLC transformation permits SNA customers to support remote link-attached SNA devices using non-3745 SNA gateways (e.g., 3174s or 3172s), which, relative to their SNA gateway functions, support only Token Ring-attached SNA nodes. So in some instances, SDLC-to-LLC transformation permits customers to displace 37xxs that were being used primarily to support link traffic in favor of a lower cost 3174 or 3172 (or similar) SNA gateway. These cost-reduction possibilities are making SDLC-to-LLC transformation the preferred option for link-traffic integration. It is, for example, the only technique for SDLC integration offered on the IBM 6611.

Just as with remote link polling, SDLC-to-LLC transformation does not in any way interfere with or modify I-fields (or 3270 BSC text blocks) that contain actual end-user data. Hence, SDLC-to-LLC transformation, like remote link polling or synchronous passthrough, is transparent to all end-to-end SNA interactions. This noninterventionist approach could, however, cause problems later if a customer uses SDLC-to-LLC transformation to support future—as opposed to existing—SNA devices. Recent SNA network management uses unsolicited SNA generic alerts, which contain self-identifying data in the form of subvectors, with some of these subvectors specifying link characteristics. The problem as it applies to SDLC-to-LLC transformation is that the actual link characteristics reported by the physical device relate to SDLC—whereas, because of LLC transformation, the device appears to the host to be LAN attached.

In this respect, it is worth stressing that all SDLC (or 3270 BSC) link support features offered on bridge/routers should be treated as short-term tactical migration aids, rather than as long-term strategic offerings. They are a cost-effective means for integrating existing link-attached SNA devices into multiprotocol WANs. Network managers should not, however, treat this support as justification for continuing to acquire link-attached SNA devices rather than the equivalent LAN-attachable SNA/APPN devices.

SUMMARY

The technology for implementing a single consolidated WAN backbone that supports SNA, LAN, SNA link, and other LAN traffic is now widely available. SNA LAN traffic can be integrated into the common backbone using one of three techniques described in this chapter: bridging, IP encapsulation, and CrossComm's protocol independent routing. SNA link traffic can be incorporated into a multiprotocol backbone using either synchronous passthrough, remote polling, or SDLC to Token Ring LLC transformation. Of these, the latter technique is the one most frequently used.

6-6
SNA Over Frame Relay

Dick Thunen

TODAY MOST TELECOMMUNICATIONS carriers provide frame relay services that allow the IBM Systems Network Architecture (SNA) user to reap a number of benefits, including:

- Investment protection in SNA devices.
- Lower line costs compared to dedicated links.
- Up to 40% increases in network utilization through frame relay's multiprotocol support.
- Sustained integrity and control of the SNA network with NetView and simple network management protocol (SNMP) management.
- Integration of SNA and multiprotocol LANs.
- High-performance access networking for Advanced Peer-to-Peer Networking (APPN) and a migration path to asynchronous transfer mode (ATM) backbones.

Traditional IBM host networks connect users to mainframes via SNA or bisynchronous multidrop lines. These are usually low-speed analog lines that represent a single point-of-failure between user and host. Even though these networks subject network managers to the complexities of dealing with a multitude of leased lines, many organizations continue to maintain their IBM host networks because of the mission-critical applications they support.

IBM Corp. introduced X.25 as a cost-effective alternative to private lines. Many network planners have chosen not to implement it, however, because of higher user-response times from network overhead delays caused by every node in the X.25 network performing error detection/correction, message sequencing, and flow control. Frame relay, however, performs these functions only at the network access points using an end-to-end protocol; thus frame relay uses the network more efficiently.

IBM has developed a set of SNA frame relay products for packet-based, wide area networks (WANs). Frame relay is an integral element of the evolution of SNA networks into the future with full support for APPN and ATM.

FRAME RELAY TECHNOLOGY: AN OVERVIEW

Frame relay offers virtual private-line replacement. As a network interface, it traces its origins to integrated services digital network (ISDN).

When ISDN was being developed, two transport services were envisioned: circuit-mode services for voice and transparent data, and packet (i.e., X.25 and frame relay) mode for data. Frame relay has since evolved into a network interface in its own right, independent of ISDN. It is now specified as a set of American National Standards Institute (ANSI) and International Telecommunications Union (ITU) standards.

The User Perspective

Although services are typically available with transmission rates from 64K-bps to T1/E1 (1.53/2.05M-bps), frame relay is defined as an access interface up to T3 or 45M-bps. By contrast, the typical synchronous data link control (SDLC) multidrop line is a 4.8K or 9.6K-bps analog line. The transmission of a typical two-page text document on a frame relay network takes 1/4 second at 64K-bps and 1/100 second at 1.53M-bps. Transmission of the same two-page text document on an SDLC multidrop line takes 3 1/3 seconds at 4.8K-bps and 1 1/6 seconds at 9.6K-bps.

To the user, a frame relay network appears simple and straightforward. Users connect directly to destinations on the far side of the network. Frame relay provides logically defined links—commonly called data link connection identifiers (DLCIs), permanent virtual circuits (PVCs), or permanent logical links (PLLs)—for a permanent virtual connection.

For example, user A is connected across the frame relay network through separate PVCs to both user B and user C. The PVCs are multiplexed across user A's frame relay interface. Frame relay networks guarantee bandwidth to each PVC, but allow unused bandwidth to be shared by all active users. The guaranteed bandwidth of a PVC is specified as the committed information rate (CIR) of the PVC. A user's traffic can have transmission data rates in excess of the CIR, referred to as the burst rate of the PVC.

User B appears to user A with frame relay address DLCI 100, and user A appears to user B with DLCI 80. A PVC connects user A's frame relay interface through the frame relay network to user B's frame relay interface. Each user's DLCI numbers have local significance only. User A has a second PVC with its own DLCI number connecting to user C. In addition, each user has a local management interface, typically on DLCI 0 (see Exhibit 6-6-1).

The Frame Relay Frame

Each frame relay access station is responsible for transforming the data into frame relay packets for transport (i.e., relay) over the network. Each frame contains the following elements:

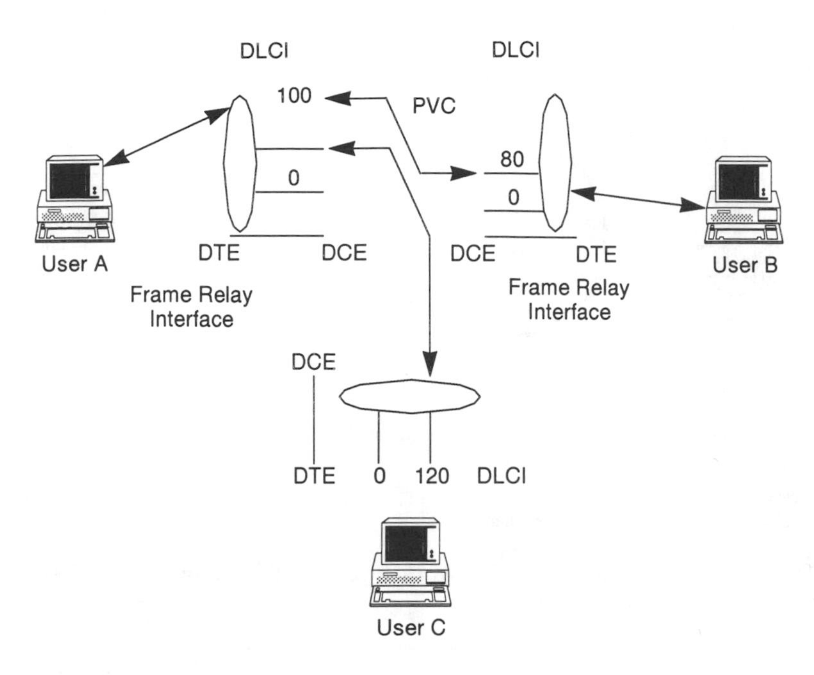

Exhibit 6-6-1. Frame Relay Permanent Virtual Circuit (PVC)

- *Flag.* The flag indicates the start and end of a frame relay packet.
- *Frame relay header.* The header contains the destination of the user data packet and management information.
- *User data.* The user data contains the data to be transported across the frame relay network.
- *Frame check sequence.* The FCS allows the integrity of the data to be validated.

The frame relay network receives, transports, and delivers variable-length frames. The frame relay network consists of a group of interconnected nodes (i.e., switches) that relay the data across the network on the appropriate PVC. A frame relay switch uses only the DLCI information contained in the frame relay header to forward the frame across the network to its destination (see Exhibit 6-6-2).

The path through the network is transparent to the user. The DLCI does not include any description of how the connection transverses the network or the routing topology of the network. A frame relay network operates an

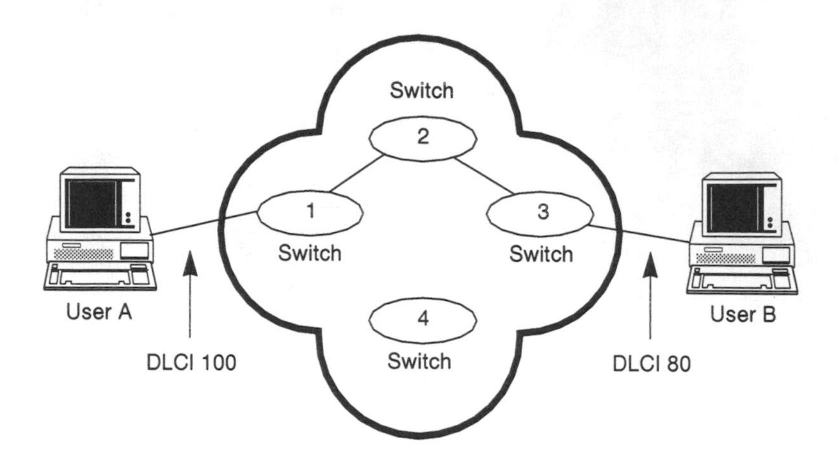

Exhibit 6-6-2. Frame Relay Network Showing PVC Connecting User A and User B

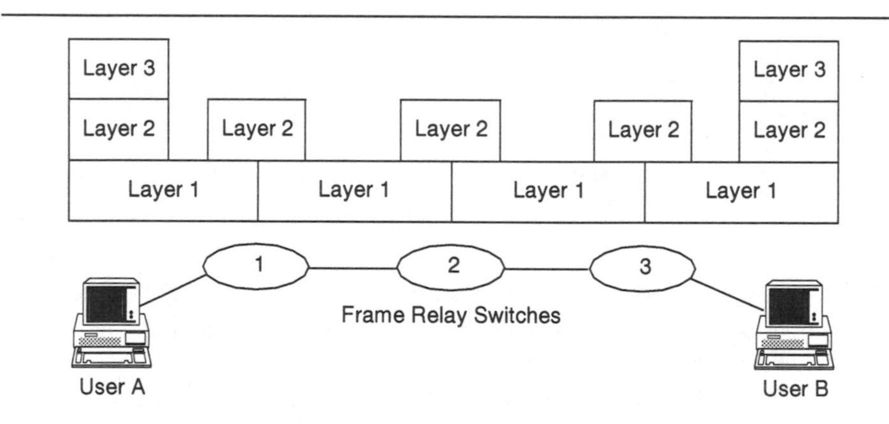

Exhibit 6-6-3. Layer 2 Router Network

Open Systems Interconnection (OSI) layer 2 router network. Each frame relay access node puts the routing information (destination DLCI) in the data link layer (i.e., frame relay header) of the frame. The frame relay network uses only this information to relay the frame across the network. (See Exhibit 6-6-3.) In other words, the frame relay network nodes look only at the frame relay header and the FCS.

The frame relay switch, or node, uses the following two-step review process to forward frames across the network:

- The integrity of the frame is checked using the frame check sequence; if an error is indicated, the frame is discarded.
- The destination DLCI address is validated, and if it is invalid, the frame is discarded. The DLCI destination address is contained in the frame relay header of the frame.

All frames that are not discarded as a result of the FCS or DLCI checks are forwarded. The frame relay node makes no attempt to correct the frame or to request a retransmission of the frame. This results in an efficient network, but requires that the user end-stations assume responsibility for error recovery, message sequencing, and flow control.

Thus, frame relay switches do not look at the user data packets, which makes the network transparent to all protocols operating at levels above OSI level 2.

RFC 1490

Because frame relay networks do not look at the contents of the user data, any format can be used to packetize the data, such as X.25 or high-level data link control (HDLC). IBM uses logical link control type 2 (LLC2) as its frame relay SNA data format.

The IBM format is based on ANSI T1.617a Annex F, which covers encapsulating protocol traffic in frame relay. This process has been approved by the Frame Relay Forum and is included in its Multiprotocol Encapsulation Agreement. IBM's treatment of a frame relay network is based on standards and promotes interoperability with third-party implementations. IBM uses the LLC2 frame format and protocol for transporting SNA across Token Ring and Ethernet LANs.

For SNA data, the RFC 1490 header designates that it is 802.2 (LLC2) data, whether it is SNA subarea, peripheral, or APPN data, and the LLC2 destination and source addresses. This format, illustrated in Exhibit 6-6-4, is also used for NetBIOS data.

Users connected to the network using RFC 1490 frame relay data terminal equipment (DTE) have a logical view of the frame relay network as a virtual LAN. IBM's use of RFC 1490 for its frame relay equipment provides a familiar metaphor to SNA users.

Because the frame relay network does not look at the contents of user data, it allows the multiplexing of multiple protocols across a single frame relay interface. Frame relay network access nodes are responsible for converting the user data into the appropriate RFC 1490 format for SNA and LAN traffic.

In summary, a frame relay WAN:

- Provides packet-mode technology.
- Does not utilize store-and-forward.
- Relies on intelligent endpoints and high-integrity lines.

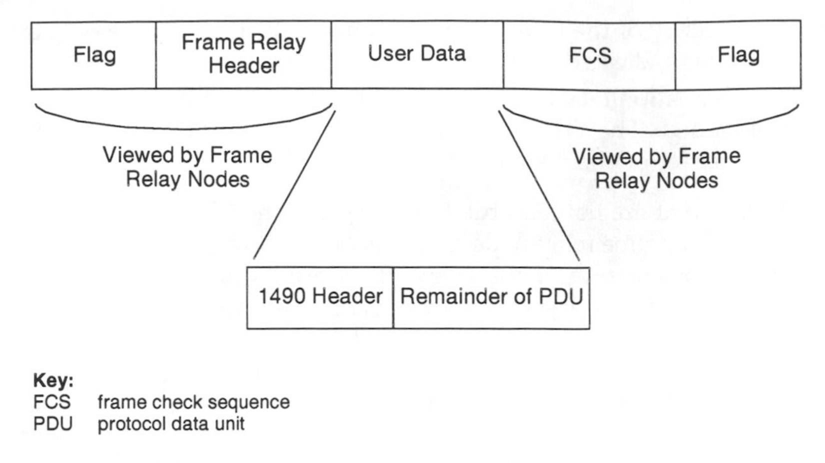

Exhibit 6-6-4. RFC 1490 Frame Relay Frame Format

- Results in low transit delay.
- Is transparent above layer 2.

As a result, frame relay provides a cost-effective alternative to dedicated-line networks.

FRAME RELAY AS A REPLACEMENT FOR SDLC

Frame relay delivers enhanced services compared to alternative SNA WAN techniques such as SDLC. Frame relay:

- Uses the same framing and CRC bits as SDLC. This means that all front-end processor (FEP) SDLC line interface couplers (LICs), modems, and DSUs/CSUs, can be used with frame relay.
- Usually allows for frames up to 2,106 bytes in a frame relay network, but IBM's network control program (NCP) allows for the configuration of up to 8,250-byte frames for use on high-quality, private frame relay networks. Large packets reduce network overhead and improve network performance.
- Allows network access connections from 56/64K-bps to T1/E1 speeds, whereas the typical multidrop connection is 4.8/9.6K-bps. User response times are directly improved by efficient network backbone connectivity.
- Is implemented in software (like SDLC), which means that no hardware changes in either the FEP or remote devices are required to move to frame relay.

- Can be managed by NetView management by NCP for both SDLC and frame relay connections. Therefore, familiar network management tools and practices can be used on the frame relay network.
- Adds multiple protocol transport. All protocols can be transported across the frame relay network; SDLC supports only SNA traffic.
- Provides SNA guaranteed bandwidth through the PVC's committed information rate.
- Requires no host application changes to migrate from SDLC to frame relay.
- Supports point-to-point connections, like SDLC. Frame relay also provides many-to-many connections; SDLC requires a multidrop line to provide one-to-many connections.
- Provides for transparent network routing. SDLC is a single physical path connection.
- Supports burst mode, which lets users exceed their CIR of the link.

SNA FRAME RELAY NETWORKS

IBM provides connections for frame relay networks on all its current networking products, including 3745 communication controller, 3172 interconnect controller, OS/2 RouteXpander/2, 3174 network server, AS/400, and the 6611 network processor. Users can evolve their current SNA networks from an SDLC multidrop backbone to a frame relay WAN. IBM supports all SNA topologies: Intermediate Network Node, Boundary Network Node, SNA Network Interconnect, and APPN across a frame relay network.

IBM's frame relay products are configured as frame relay DTE devices, except the FEP (3745 communication controller), which can also be configured as a frame relay data communications equipment (DCE) device and can act as a frame relay switch.

Intermediate Network Node (INN)

IBM's NCP software provides an INN connection—PU4-to-PU4—between FEPs over a frame relay network. This support was first announced for NCP Version 6, Release 1 (V6R1) in 1992. IBM supports mixed-media, multiple-link transmission groups that can include frame relay, SDLC, and Token Ring links. Thus, frame relay can be incorporated with other data link types in a transmission group to give users flexibility in network design.

Because frame relay is an OSI level 2 routing protocol, it provides fast INN routing, which is an efficient means of interconnecting multiple FEPs. Level 2 frame relay eliminates SNA processing on intermediate FEPs. Furthermore, as each pair of FEPs appears to be directly linked, the intermediate network configuration is transparent to SNA routing algorithms.

SNA Network Interconnect (SNI)

NCP Release 6, Version 1 also introduced SNA over frame relay for interconnecting multiple SNA networks. Two traditional SNA networks can be connected using an SNI link over frame relay so the users of one SNA network can access the resources or applications of another across a frame relay network.

Boundary Network Node (BNN)

NCP Version 7, Release 1 fully expands the role of the NCP to that of providing SNA Boundary Network Node—PU4-to-PU2—connectivity between an NCP and an SNA node (PU2/2.1). The FEP can establish an SNA/BNN connection across a frame relay network with users on a 3174 network processor or users connected through an IBM 6611 network server or RouteXpander/2.

AS/400

IBM's AS/400 supports direct frame relay connectivity to another AS/400, or through a frame relay bridge to a 5494 remote controller or PC workstation. SNA nodes connected to an AS/400 across a frame relay network must be SNA Type 2.1 nodes, such as an IBM 5494 remote controller.

APPN

IBM's APPN Network Node products, 6611 IBM Network Processor, AS/400, and OS/2 Communication Manager (RouteXpander/2) can be configured to establish an APPN network across a frame relay WAN.

APPN end-node applications can thus take advantage of the combined frame relay and APPN network.

IBM Legacy Devices

Many IBM networks include legacy devices that are incapable of supporting frame relay network access, such as 3274 controllers, System 3X computers, and 5394 controllers. A frame relay assembler/disassembler (FRAD) provides connection to a frame relay network for a non-frame relay capable device.

A FRAD translates the SNA controller's SDLC data stream into frame relay frames for transport over the network. FRADs based on RFC 1490 can interoperate across a frame relay network with IBM's frame relay products. Interoperability with IBM requires that the SDLC be converted to LLC2 for encapsulation in frame relay.

In addition to basic framing functions, a FRAD usually concentrates a number of low- or medium-speed SDLC lines into a single, high-speed frame relay link. By combining data from multiple, low-speed controllers onto one or more high-speed lines, FRADs reduce overall network costs.

Private Frame Relay Network

NCP Version 6, Release 2 (V6R2) adds DCE support to the front-end processor. The FEP functions as a frame relay switch (i.e., DCE) for frame relay DTE equipment, such as an OS/2 RouteXpander, so users can create private frame relay networks based on the IBM FEP. Private frame relay networks support both SNA and LAN protocols. In summary:

- All current IBM SNA products provide frame relay network access.
- All SNA topologies are supported across a frame relay network.
- FRADs can be used to provide high-performance connectivity for Legacy IBM SDLC and BSC devices.

IBM MULTIPROTOCOL SUPPORT

IBM's frame relay access products use the RFC 1490 standard, which specifies the frame format and characteristics for multiplexing multiple protocols across a frame relay network on a single, frame relay link.

Treatment of LAN protocols is similar to that described for SNA-over-frame relay. The RFC 1490 header for LAN protocols indicates whether the packet is being bridged or routed. A bridged frame header includes what media it is originating on—802.3, 802.4, 802.5, FDDI, or 802.6—whether it is being source routed or transparently bridged, and its destination medium access control (MAC) address.

Some routed protocols have an assigned Direct Network Layer Protocol Identifier, or NLPID, such as IP. For these protocols the NLPID is used to identify the frame. Otherwise, the Subnetwork Access Protocol (SNAP) header for the frame is used to identify frame contents.

RFC 1490 specifies the transport of both bridged and routed LAN protocols across a common frame relay interface and provides a standard format for the frame relay packets. RFC 1490 specifies for bridged data the protocol being used—source route or transparent—and thus facilitates multivendor networking based on industry standard implementations. For routed data, however, there is currently no means of specifying the routing protocol being used for a given LAN protocol, so interoperability of routed protocols is more complicated.

All of IBM's frame relay products provide for multiprotocol support over frame relay. This support is available over public and private frame relay networks and includes both the bridging and routing of LAN protocols. The

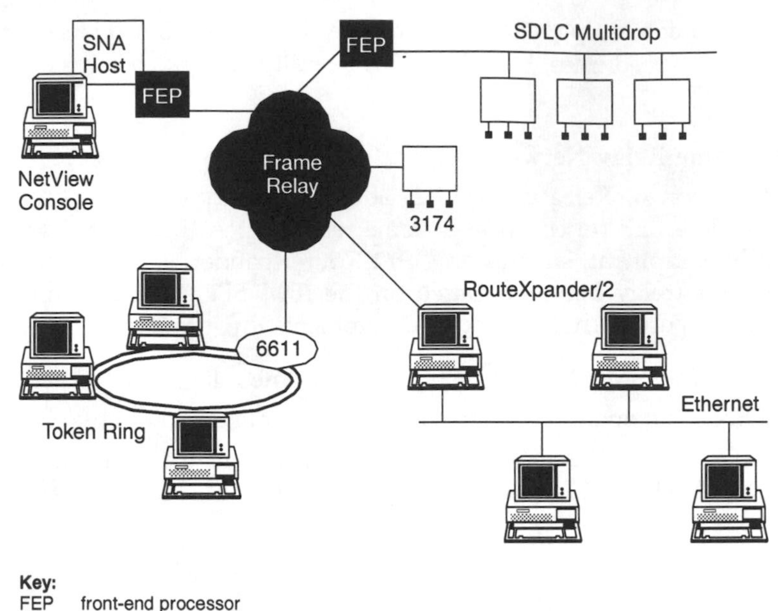

Exhibit 6-6-5. SNA Network Management Topology

IBM 6611 also allows SNA/SDLC traffic to be transported across a frame relay WAN.

NETWORK MANAGEMENT

With the addition of frame relay as a packet-mode WAN supported by IBM's NCP software, IBM incorporates support for frame relay WANs in NetView network management software, including NPM and NTune and NetView/6000, its SNMP manager. IBM provides a complete picture of the SNA and frame relay internetwork including both SNA and non-SNA traffic and DTE devices. Exhibit 6-6-5 shows the SNA network management topology.

NetView Management Services

Although SNMP and other open network management standards continue to evolve, NetView remains the only way to provide comprehensive network management, control, and diagnosis of an SNA network. All SNA network nodes are inherently commandable from NetView and report all network management-related activities directly to NetView for processing by one of its function-specific applications.

IBM's NetView support extends NetView management of the SNA network across the frame relay network to the end user's controller. This support allows complete SNA network visibility and control with no remote-line and physical unit black holes, compatibility with existing NetView tools and applications, and virtually no operator retraining.

Virtual Telecommunications Access Method (VTAM) Network Control Program (NCP)

The VTAM Dynamic Reconfiguration (DR) facility supports the addition of NCP frame relay DLCIs. PVCs can be created or deleted without interrupting the frame relay network or regenerating NCP.

Alternative Routing

NCP provides alternative automatic routing by a private frame relay network if a primary (i.e., public) frame relay network becomes unavailable.

Local Management Interface (LMI)

A reserved link address (local DLCI) is used for communication between the FRAD and the frame relay network. The management interfaces are defined by ANSI T1.617-1991 Annex D and ITT Q.933 Annex A for DLCI 0. Users are able to specify either the ANSI or ITT LMI implementation as part of the configuration. This DLCI is used for communicating network resources (i.e., the list of valid DLCIs), determining the link status of each DLCI, and determining network status.

The LMI DLCI cannot be used for data traffic. A status-inquiry message is used to query the status of the network. The status message is either a keep-alive message or a full-network status report. The status update message reports an unsolicited status change in a network component.

Frame Relay Network Congestion

The frame relay network provides notification of network congestion to end-user devices. Upon encountering congestion a frame relay switch provides forward notification of network congestion along the data route by setting the forward explicit congestion notification (FECN) bit in the frame relay header, as shown in Exhibit 6-6-6.

The network also notifies the sending node of congestion along the pvc by setting the backward explicit congestion notification (BECN) bit of packets going to the sender along the PVC. The bit is changed from 0 to 1 to indicate

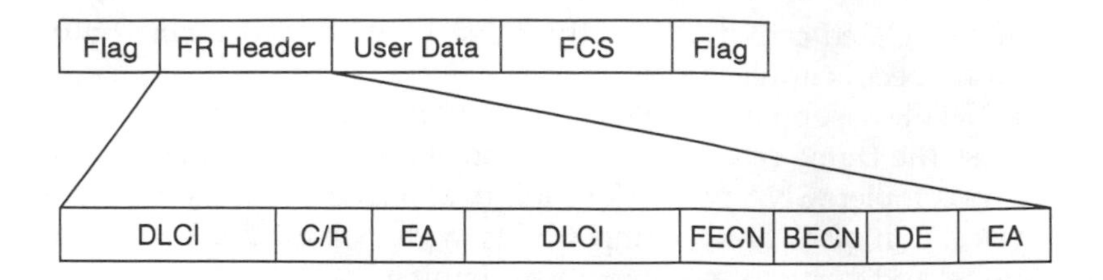

Key:
BECN backward explicit congestion notification
C/R committed information rate
DE discard eligibility
DLCI data link connection identifier
FECN forward explicit congestion notification

Exhibit 6-6-6. Frame Relay Header

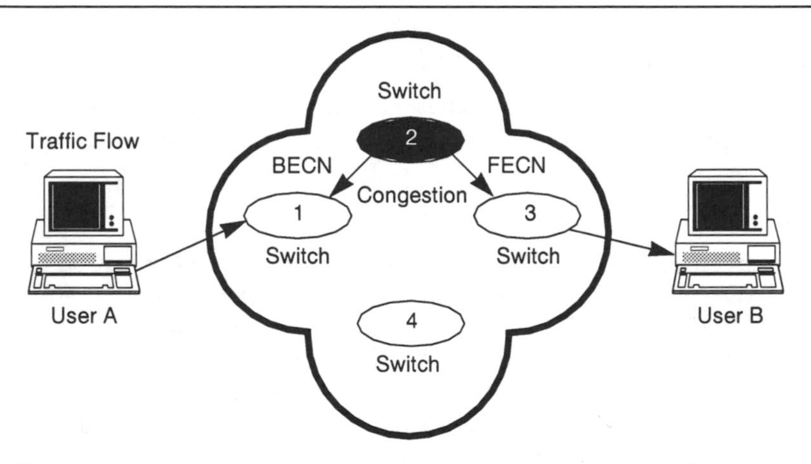

Key:
BECN backward explicit congestion notification
FECN forward explicit congestion notification

Exhibit 6-6-7. Congestion Notification

the presence of congestion. Network congestion is determined by a switch using the switch's queue length or buffer utilization. (See Exhibit 6-6-7.)

It is the function of the frame relay access node, or DTE device, to respond to the FECN and BECN bits. IBM's frame relay devices respond by controlling the transmit window size of devices transmitting on the congested DLCI. When a frame relay DTE receives notification of network congestion, it reduces its transmit window to 1. Once a network has indicated that it is returning to a normal state, the transmit windows are increased a frame at a time until they return to their normal transmit windows.

Consolidate Link Layer Message (CLLM)

If there are no frames returning to the sender, the end node can determine the presence of congestion over a DLCI through the CLLM information on the next query. The network is otherwise prohibited from notifying the sender of congestion on the DLCI.

Discard Eligibility Bit

Frame relay access nodes can mark frames for discard eligibility by the network as a means of reducing congestion during moderate traffic congestion periods. When the discard eligibility (DE) bit in a frame is set to 1, the user has indicated that the frame can be discarded if it encounters network congestion.

The network sets the DE bit to 1 on data that follows on a physical link in excess of the CIR. Thus, the network can be divided into the following three zones:

- *Guaranteed transmission* The data flow is less than the CIR.
- *Transmit if possible* The data flow is above the CIR but less than the maximum rate.
- *Discard excess frames* The data flow is above the maximum rate. The frame relay network does not notify a user of frames being discarded. It is the responsibility of the FRADs to monitor the integrity of the data flow.

SNMP Management

Proliferation of LAN internetworks often leads to a separate management organization seeking a common management platform for multivendor equipment. Often the solution is SNMP. Most of IBM's frame relay products can be configured with an SNMP agent for management by IBM's NetView/6000 SNMP Manager.

Support of concurrent SNMP and NetView enables each functional operations group, SNA and LAN internetwork, to execute their respective network management and control responsibilities through their management platform of choice.

FRAME RELAY COMPARED WITH ROUTER NETWORKS

IBM's products transmit LAN, SNA, and APPN traffic across a frame relay WAN. The following section compares IBM's treatment of a frame relay WAN and the router approach. The major issues include:

- *Backbone.* A frame relay network is compared with a meshed-router backbone network.

- *WAN protocol.* IP encapsulation is compared with native-protocol frame relay transport.
- *SNA support.* Support for all SNA interconnects is compared with data link switching (DLSw).
- *Network management.* Native NetView and SNMP are compared with SNMP.

Backbone

The typical router backbone is a mesh of point-to-point links. In these networks, the router backbone is the network. The router is responsible for routing between end-user clients and application servers. Thus, routers are responsible for the definition and maintenance of network topology and the appropriate routing path for applications. Router networks may be referred to as administratively rich.

In a frame relay backbone, by contrast, the network services are inherent in the frame relay service. Each frame relay access device provides application-transparent communication directly with its corresponding node across the network. This simplifies the configuration and administration of frame relay compared to a router-based network.

WAN Protocol

The router solution to this issue is to encapsulate all traffic in IP packets for transmission over the frame relay (or other) network in conjunction with a proprietary routing protocol. Thus, the router solution is based on adding IP framing overhead to all data prior to adding frame relay framing for transmission over the network. Because most router protocols are proprietary and noninteroperable, a single vendor's product must reside on both sides of the network.

SNA Support

The router solution is to use data link switching (DLSw) to terminate SDLC and LLC2 traffic in the router and encapsulate the SNA data in IP using the DLSw routing protocol over the WAN. This provides a single backbone protocol for SNA and non-SNA traffic over the WAN. Once the SNA data is encapsulated in IP, the WAN treats it as any other IP traffic. The router solution requires a second DLSw-compatible router on the destination side of the frame relay network to remove the SNA data from the IP packet. However, DLSw only covers SNA/BNN PU2 data on SDLC lines and Token Ring LANs and NetBIOS traffic.

IBM uses RFC 1490 for the transmission of SNA data over a frame relay WAN network. RFC 1490 provides for the transport of SNA/BNN PU 2 and

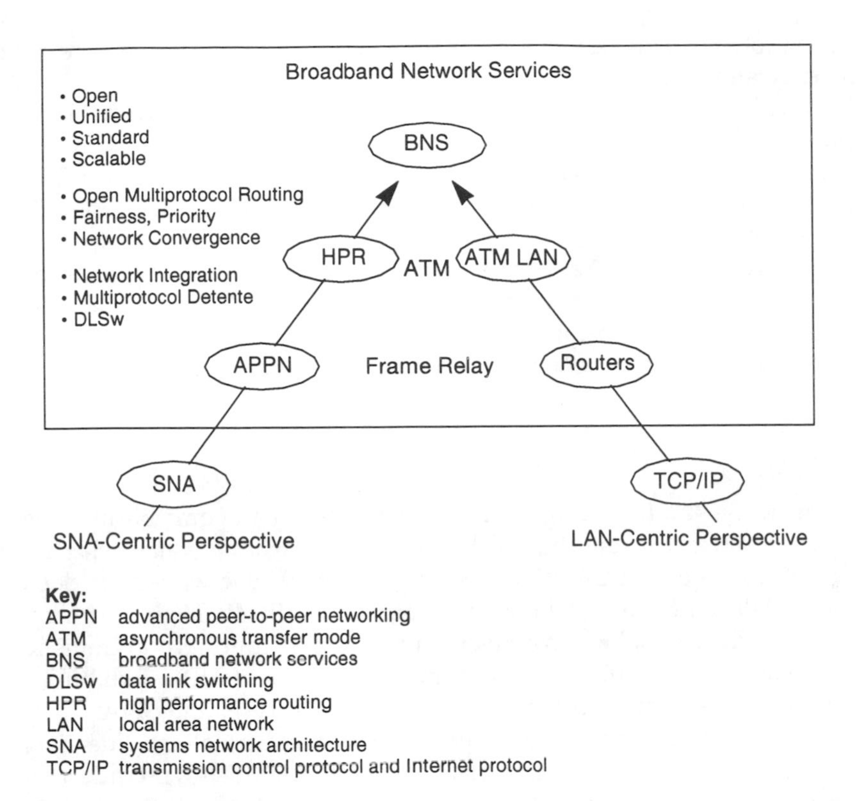

Exhibit 6-6-8. Frame Relay as the Unifying Networks for LAN and SNA Networks

type 2.1, but also SNA Intermediate Network, APPN, and SNA Network Interconnect traffic across a frame relay network. Therefore, RFC 1490 covers all SNA traffic without encapsulation in IP, and DLSw covers only SNA/BNN PU2 traffic and adds the IP overhead.

Network Management

SNMP is the principal network management tool used with routers, whereas NetView is the network management tool of choice for SNA networks. The SNMP management stations are not usually located in corporate data centers, which necessitates a separate set of DLCIs for SNMP management to each remote location.

Such a scheme creates a redundant tier of network overhead that reduces bandwidth availability for data, impedes SNA session responsiveness and reliability, obstructs NetView visibility, and complicates network design and problem solving. This results in poor SNA network performance in terms of efficiency and cost.

When SNA is internetworked using routers that do not provide NetView support, a "black hole" is created in the network, preventing the NetView operator from viewing, managing, or monitoring the frame relay DTE devices. In particular, routers usually do not support SDLC LL2, LPDA-2, or NPM statistic collection.

IBM provides an integral NetView connection in all its frame relay products. NetView connections share the same PVCs as SNA data, thereby eliminating the need for a separate management network for communication with NetView and its component applications.

SUMMARY

Frame relay—multiplexing multiple protocols over a common link—is an efficient solution for unifying LAN and SNA networks. Frame relay is the WAN of choice for organizations moving to APPN. The wide-scale deployment of APPN networks will soon be served by IBM's High Performance Routing (HPR) technology to deliver connectionless routing, and frame relay will be supported by IBM's initial implementations of HPR. Exhibit 6-6-8 illustrates frame relay as the unifying network for LAN and SNA.

Current frame relay network specifications and service-provider implementations are designed for permanent virtual circuits. PVCs provide a direct replacement for leased-line SNA connections, but SNA networks often include switched, dial-up SDLC connections for casual SNA host access. This capability is being added to frame relay. A number of vendors have initiated standardization of switched virtual circuits (SVCs).

Frame relay is also being positioned as the access network for ATM. An RFC that specifies a frame relay interface to ATM networks is currently being worked through the standards process. This interface, referred to as data exchange interface (DXI), covers ATM adaptation layer 1 (AAL 1).

Frame relay provides users with short-term payback and long-term preparedness—immediate economic benefits and a migration path to future, high-performance routing and networking.

6-7
IPv6: The New Internet Protocol

William Stallings

WITH THE CHANGING NATURE of the Internet and of business networks, the current Internet protocol (IP), which is the backbone of Transmission Control Protocol/Internet Protocol (TCP/IP) networking, is rapidly becoming obsolete.

Until recently, the Internet, and most other TCP/IP networks, have primarily provided support for simple distributed applications, such as file transfer, E-mail, and remote access using Telnet. But today, the Internet is increasingly becoming a multimedia, application-rich environment, led by the huge popularity of the World Wide Web. At the same time, corporate networks have branched out from simple E-mail and file transfer applications to complex client/server environments and, most recently, intranets that mimic applications available on the Internet.

All of these developments outstripped the capability of IP-based networks to supply needed functions and services. An internetworked environment needs to support real-time traffic, flexible congestion control schemes, and security features. None of these requirements is easily met with the existing IP.

However, the true driving force behind the development of a new IP is the fact that the world is running out of IP addresses for networked devices. The fixed 32-bit address length of IP is inadequate considering the explosive growth of networks.

IPv6: THE NEXT GENERATION

To meet changing network needs, the Internet Engineering Task Force (IETF) issued a call for proposals for a next-generation IP (IPng) in July 1992. A number of proposals were received and by 1994, the final design for IPng emerged. A major milestone was reached with the publication of Request for Comment (RFC) 1752, "The Recommendation for the IP Next-Generation

Protocol," issued in January 1995. RFC 1752 outlines the requirements for IPng, specifies the header formats, and highlights the IPng approach in the areas of addressing, routing, and security.

A number of other Internet documents defined details of the protocol, now officially called IPv6; these include an overall specification of IPv6 (RFC 1883), an RFC discussing the flow label in the IPv6 header (RFC 1809), and several RFCs dealing with addressing aspects of IPv6 (i.e., RFC 1884, RFC 1886, and RFC 1887).

IPv6 includes the following enhancements over IPv4:

- *Expanded address space.* IPv6 uses 128-bit addresses instead of the 32-bit addresses of IPv4. This is an increase of address space by a factor of 296. Even if addresses are very inefficiently allocated, this address space seems secure.
- *Improved option mechanism.* IPv6 options are placed in separate optional headers that are located between the IPv6 header and the transport-layer header. Most of these optional headers are not examined or processed by any router on the packet's path. This simplifies and speeds up router processing of IPv6 packets compared with IPv4 datagrams. It also makes it easier to add additional options.
- *Address autoconfiguration.* This capability provides dynamic assignment of IPv6 addresses.
- *Increased addressing flexibility.* IPv6 includes the concept of an anycast address, for which a packet is delivered to just one of a set of nodes. The scalability of multicast routing is improved by adding a scope field to multicast addresses.
- *Support for resource allocation.* Instead of the type-of-service field in IPv4, IPv6 enables the labeling of packets that belong to a particular traffic flow for which the sender requests special handling. This aids in the support of specialized traffic such as real-time video.
- *Security capabilities.* IPv6 includes features that support authentication and privacy.

THE IPv6 PACKET

The basic unit of transfer is the IPv6 packet, shown in Exhibit 6-7-1. The packet typically encloses a transmission control protocol (TCP) segment, which in turn consists of a TCP header and TCP user data. To this, IPv6 adds a fixed-length IPv6 header and a number of optional extension headers. The advantage of the multiple-header structure is that a streamlined packet format may be used when optional IPv6 functions are not required.

The IPv6 header and each extension header include a "next header" field.

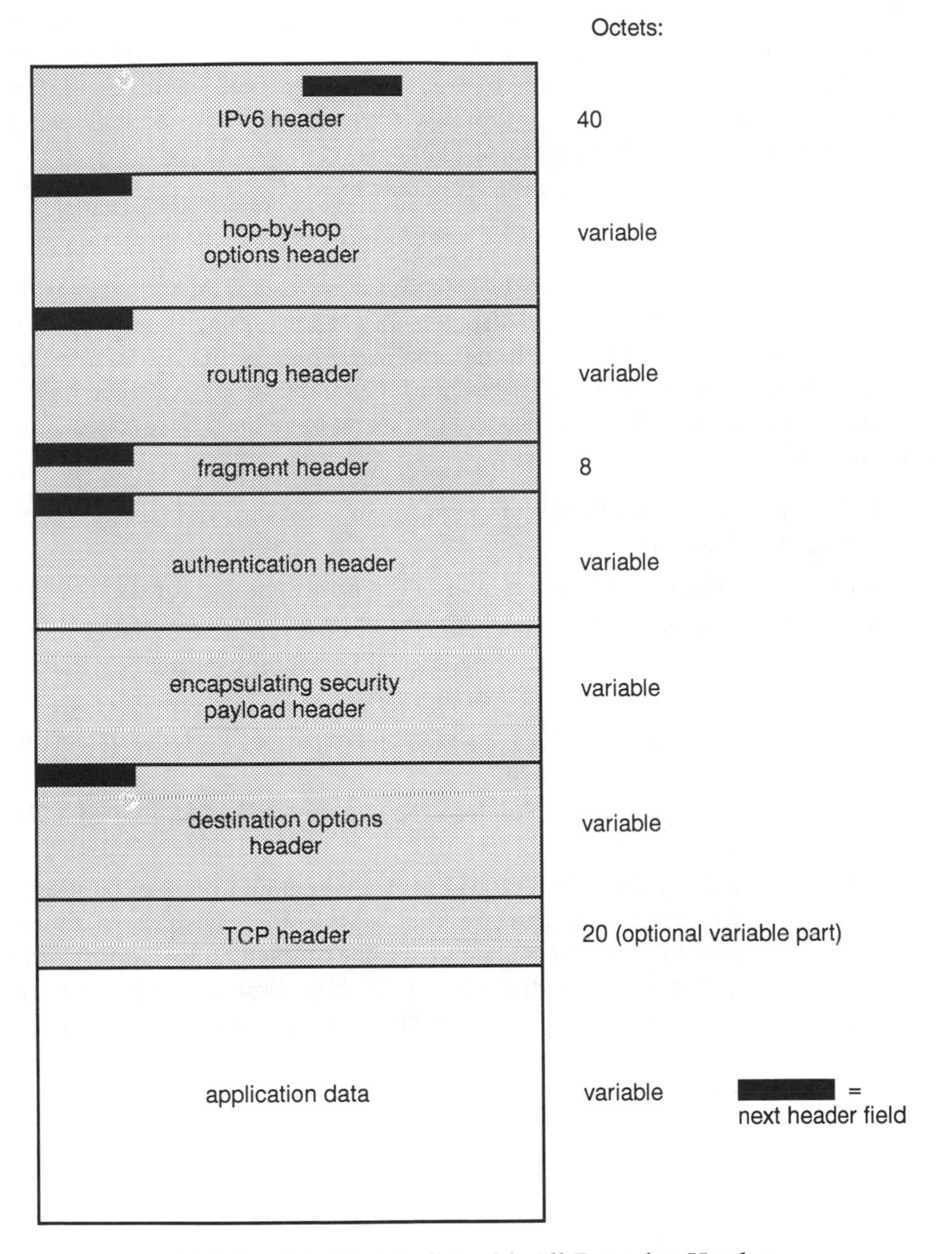

Exhibit 6-7-1. IPv6 Packet with All Extension Headers

This field identifies the type of the immediately-following header. If the next header is an extension header, then this field contains the type identifier of that header. Otherwise, this field contains the protocol identifier of the upper-layer protocol using IPv6 (typically a transport-level protocol), using the same values as the IPv4 protocol field.

The IPv6 Header

Exhibit 6-7-2 shows the IPv6 header and compares it to the current IP header. The new header is a fixed 40 octets long, compared with an IP header of 20 octets plus a variable-length options field. Some of the fields in the IP header, such as those related to fragmentation, are moved to extension headers in IPv6. Others, such as the header checksum, have been abandoned. The header fields are:

- *Version (4 bits).* This field represents the Internet protocol version number; the value is 6.
- *Priority (4 bits).* This field represents the priority value for this packet.
- *Flow label (24 bits).* This field may be used by a host to label those packets for which it is requesting special handling by routers within a network.
- *Payload length (16 bits).* This is the total length of all of the extension headers plus the TCP segment.
- *Next header (8 bits).* This field identifies the type of header immediately following the IPv6 header.
- *Hop limit (8 bits).* This field represents the remaining number of allowable hops for this packet. The hop limit is set to some desired maximum value by the source, and decremented by one by each node that forwards the packet. The packet is discarded if the hop limit is decremented to zero. This is a simplification over the processing required for the time-to-live field of IP. The extra effort in accounting for time intervals in IP added no significant value to the protocol.
- *Source address (128 bits).* This is the address of the originator of the packet.
- *Destination address (128 bits).* This is the address of the intended recipient of the packet.

Although the IPv6 header is longer than the mandatory portion of the IPv4 header (40 octets versus 20 octets), it contains fewer fields (8 versus 12). Thus, routers have less processing to do per header, which should speed up routing.

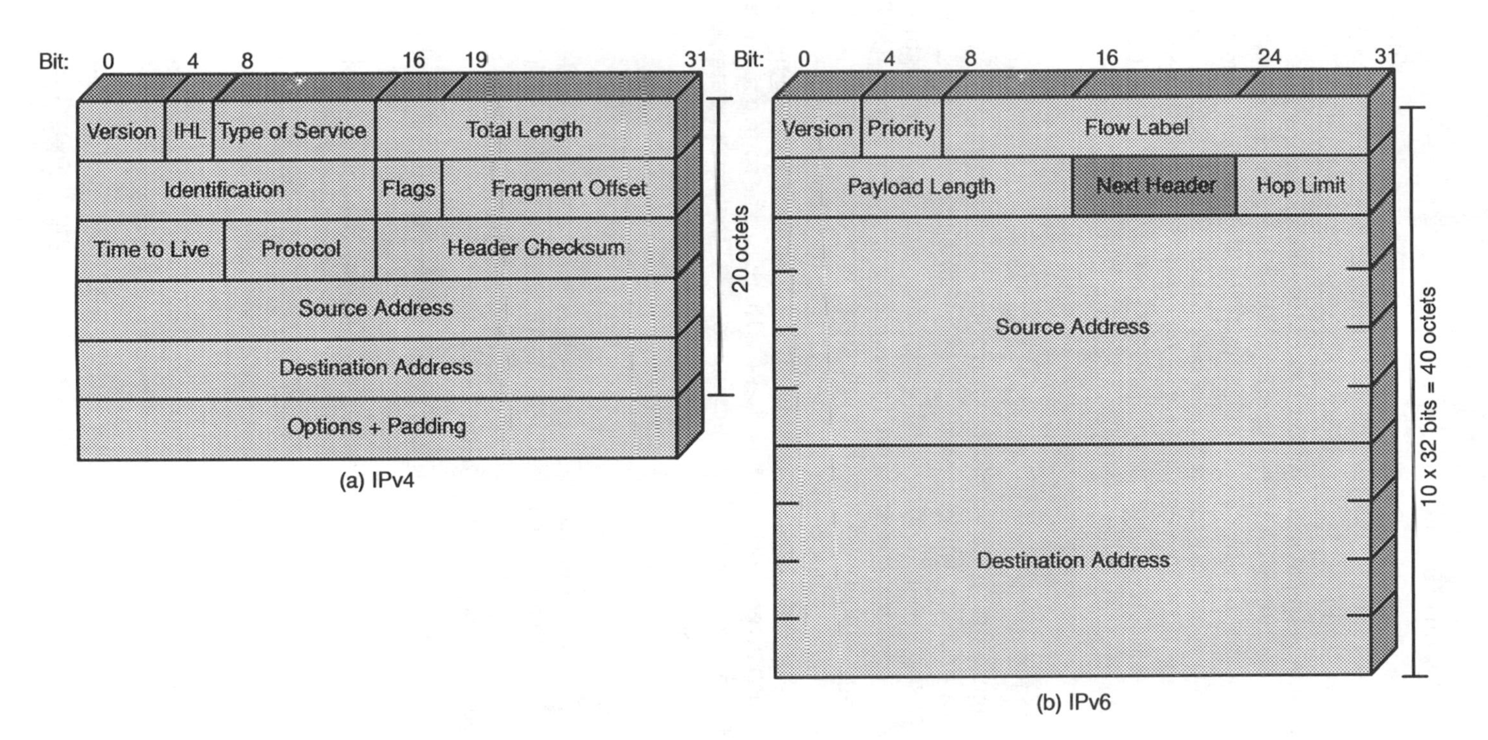

Exhibit 6-7-2. IP Headers

Hop-by-Hop Options Header

The hop-by-hop options header carries optional information that, if present, must be examined by every router along the path. In the IPv6 standard, only one option is so far specified: the jumbo payload option, used to send IPv6 packets with payloads longer than 216=65,535 octets. The option data field of this option is 32 bits long and gives the length of the packet in octets, excluding the IPv6 header. For such packets, the payload length field in the IPv6 header must be set to zero, and there must be no fragment header. With this option, IPv6 supports packet sizes up to 4 billion octets. This allows the transmission of large video packets and helps IPv6 make the best use of available capacity over any transmission medium.

Fragment Header

In IPv6, fragmentation may only be performed by source nodes, not by routers along a packet's delivery path. To take full advantage of the internetworking environment, a node must perform a path discovery algorithm that enables it to learn the smallest maximum transmission unit (MTU) supported by any subnetwork on the path. In other words, the path discovery algorithm enables a node to learn the MTU of the "bottleneck" subnetwork on the path. With this knowledge, the source node will fragment, as required, for each given destination address. Otherwise the source must limit all packets to 576 octets, which is the minimum MTU that must be supported by each subnetwork.

Routing Header

The routing header contains a list of one or more intermediate nodes to be visited on the way to a packet's destination. All routing headers start with a 32-bit block consisting of four 8-bit fields, followed by routing data specific to a given routing type. The four 8-bit fields are:

- *Next header.* This identifies the type of header immediately following the current header.
- *Header extension length.* This is the length of the header in 64-bit units, not including the first 64 bits.
- *Routing type.* This identifies a particular routing header variant. If a router does not recognize the routing type value, it must discard the packet.
- *Segments left.* This is the number of explicitly listed intermediate nodes still to be visited before reaching the final destination.

In addition to this general header definition, RFC 1883 defines the type 0 routing header. The type 0 header includes a 24-bit strict/loose bit map. The bits of the field are numbered from left to right (bit 0 through bit 23), with each bit corresponding to one of the hops. Each bit indicates whether the corresponding next destination address must be a neighbor of the preceding address (i.e., 1 = strict, must be a neighbor; 0 = loose, need not be a neighbor).

When using the type 0 routing header, the source node does not place the ultimate destination address in the IPv6 header. Instead, that address is the last address listed in the routing header, and the IPv6 header contains the destination address of the first desired router on the path. The routing header will not be examined until the packet reaches the node identified in the IPv6 header. At that point, the packet IPv6 and routing header contents are updated and the packet is forwarded. The update consists of placing the next address to be visited in the IPv6 header and decrementing the segments left in the routing header.

IPv6 requires an IPv6 node to reverse routes in a packet it receives containing a routing header to return a packet to the sender.

Destination Options Header

The destination options header carries optional information that, if present, is examined only by the packet's destination node. The format of this header is the same as that of the hop-by-hop options header.

SUMMARY

The Internet protocol (IP) has been the foundation of the Internet and virtually all multivendor private internetworks. This protocol is reaching the end of its useful life and a new protocol, known as IPv6, has been defined to ultimately replace IP.

The motivation for the adoption of a new version of IP was the limitation imposed by the 32-bit address field in IPv4. In addition, IP is a very old protocol, and new requirements in the areas of security, routing flexibility, and traffic support had developed. To meet these needs, IPv6 includes functional and formatting enhancements compared with the current IPv4. In addition, a set of security specifications have been issued that can be used with both IPv4 and IPv6.

With most of the technical details of these enhancements frozen, vendors may begin to move this capability into their product line. As IPv6 gradually deploys, the Internet and corporate networks will be rejuvenated and able to support the applications of the 21st century.

Section 7
Mobile Communications Systems

EVERY COMMUNICATIONS SYSTEMS MANAGER should understand the possibilities of mobile computing and needs to be basically familiar with the technologies used in these networks. The use of mobile communications is growing at a compound rate of 15% to 40%, compared with around 5% for wireline voice and 15% to 20% for data. The US government discovered a gold mine when they decided to auction slices of frequency spectrum to prospective mobile service providers. As a result of sometimes frantic bidding, billions of dollars poured into the treasury. Clearly, these bidders see an opportunity to get into a growth business.

The word "wireless" is often used as a synonym for "mobile." This is acceptable, although technically a misnomer. The communications industry's use of the term "wireless" implies untethered communication, which implies the use of an unbounded transmission media for at least a portion of the connection between users. A much more appropriate term is "mobile communication," which is defined officially by the IEEE as a combination of interrelated devices capable of transmitting intelligence between two or more spatially separated radio stations, one or more of which is mobile.

Wireless communication has been used throughout all of recorded history in the form of heliographs, smoke signals, radio, and television. What many people now think of as mobile communications began with the first deployment of the cellular telephone in the 1980s. Mobile data communications—technology to permit mobile computing—became visible in the 1990s.

Wireless data is a subset of wireless communication. Wireless data networking is defined as mobile radio technology that emphasizes the transport of business and personal data as contrasted with human speech. This data transport is for applications that include messaging, electronic mail, portable computing, image transfer, Internet access, and any other application that can benefit from mobility and location independence. Mobile communication capabilities can be applied in local, metropolitan, and wide area networks.

It is important to distinguish wireless data from wireless voice because of the different technical requirements that are needed to support data and voice. Voice, for example, has a very low tolerance for network latency and no tolerance for delay variance. Data, on the other hand, can tolerate both. Voice has a very high tolerance for error, whereas data has a very low error tolerance. Data transmission tends to be very bursty in nature in contrast with the continuous bit rate of digitized voice. These differences make it hard to optimize a mobile communication system to support both voice and data equally.

One of the impediments to wireless data deployment has been that solving this problem is harder than anyone anticipated. Other impediments to widespread mobile communications are associated with the radio portion of the connection, including technical impairments and signal-to-noise ratio as well as regulatory issues such as spectrum availability.

At the moment, most mobile data systems are dedicated solely to data applications rather than being integrated with voice. Generic data services include packet radio (ARDIS and RAM Mobile are the best known), circuit-switched data over the cellular network, cellular digital packet data (CDPD), specialized mobile radio, and message paging. There are, in addition, many data systems based on microwave technology. These are generally fixed point-to-point rather than mobile. In the future, the low earth orbiting satellite systems such as Iridium and Globalstar will also provide data services.

A mobile data communications system consists of a number of networking elements. Transmission components include the radio frequency link between the mobile station and a base station serving a geographic area. This is typically the only wireless portion of a connection. Transmission components also include the wireline or microwave link between the base station and a mobile switching center and the wireline between that center and the public switched telephone network. Switching elements include those in the public telephone network as well as those dedicated solely to the mobile subscribers. Supporting elements include signaling systems, operational support systems, and data bases storing mobile subscriber profiles, location information, and administrative and billing systems.

Even though growth of mobile data systems is accelerating, it will probably be a number of years before ubiquitous deployment makes these systems as easy to use as the voice network. User surveys indicate that reasons for not going wireless are those associated with any new technology—high costs, confusion over which technology to use, lack of immediate application, insufficient data rates, and lack of corporate commitment.

In the meantime, the communications system manager needs to track developments, be aware of those applications that are potential candidates for wireless service, and include wireless considerations in future plans. The balance of this section is devoted to chapters that provide a broad overview of mobile wireless communication.

Chapter 7-1, "Wireless Communications for Voice and Data," provides an excellent overview of wireless communications. It introduces the various wireless technologies including cellular, wireless local networks, satellite data, and packet radio. Vendor approaches and applications for the future are also covered.

Voice and data have differing requirements, which makes it somewhat difficult to simply overlay data onto the cellular voice network. The difficulties and solutions to the cellular data challenge are discussed in Chapter 7-2, "The Technical Challenge of Cellular Data Transmission."

Managers contemplating wireless systems will gain useful insights from Chapter 7-3, "Developing a Cost-Effective Strategy for Wireless Communications." This chapter explains wireless architectures, access methods, and technology, in addition to offering guidelines for choosing wireless products.

One issue that always arises in any discussion of mobile and wireless communication is the subject of wireless security. For this reason, the handbook includes Chapter 7-4, "Mobile User Security." This chapter reviews the entire spectrum of security issues unique to the mobile user from theft of equipment and virus contamination to unauthorized interception of transmissions.

Telecommuting is often a wireless application. Mobility permits the telecommuting professional to be freed from the bonds that confine the telecommuter to the home office or satellite office park. "Managing a Telecommuting Program," Chapter 7-5, will prove useful to the communications system manager of telecommuters operating in either fixed or mobile environments. This chapter explains the benefits of telecommuting and offers advice on how to prepare an organization for a telecommuting program.

7-1
Wireless Communications for Voice and Data

Andres Llana, Jr.

USE OF WIRELESS TECHNOLOGIES has become one of the fastest-growing communications applications around the world. Recent innovations have greatly increased the availability of the telephone in many parts of the world, yet wireless communications have been around since the early 1900s. Back then, radio served as the principal means of mass communication and, like TV, was the principal means of public entertainment. During World Wars I and II, wireless communications allowed combat forces to communicate. Today, law enforcement agencies, marine agencies, and transportation companies, among many others, use wireless communications to manage deployed resources.

In the 1950s, the Rural Electrification Administration considered wireless radio technology as a means of supplying telephone service to rural populations. This experimentation proceeded through many iterations but was largely abandoned during the mid-1980s as cellular technology emerged.

Today, radio communication is thought of as a new innovation because of its growing ubiquity and its support for personal communications, data, and information collection. Lower costs have made it possible for users to enjoy cellular telephones, personal digital assistants (PDAs), and a host of other devices to simplify the conduct of commerce. Wireless technology has improved intra- and intercorporate communications, enabling more cost-effective control of such business resources as deployed sales forces and technical service personnel.

WIRELESS TECHNIQUES: A STRATEGY FOR WORLDWIDE VOICE COMMUNICATIONS

Cellular Voice

Great strides have been made in the adaptation of cellular radio as a means of supporting local telephone service. In many undeveloped countries there is little or no infrastructure to support telephone services. For this reason, it is not uncommon in some parts of South America, Asia, Russia, and Eastern Europe for a subscriber to wait as long as one year to get local telephone service. Because of this situation, wireless subscriber penetration has grown at about 45% per year. For example, Motorola, Inc. recently reported that it had orders for 150 wireless systems for 21 provinces of China and the three municipalities of Beijing, Shanghai, and Tianjing; an area with a combined population of more than one billion people. As a result, wireless local loop (WiLL) systems are being installed around the world at an accelerated rate to reduce the time to service.

Wireless Radio

Wireless radio is being installed in place of traditional central office systems that require expensive extended copper wire external networks. Service providers are finding that wireless radio central office systems are convenient, fast, and less costly than traditional central office switching systems. Because there are no copper wires to string and no wire plant to maintain, subscribers can enjoy telephone service as soon as the radios are turned on.

Building a traditional central office system with a stationary copper land line network costs between $1,250 and $1,750 per subscriber, depending on terrain and labor. A Motorola WiLL system can be installed for between $800 and $2,000 per subscriber. About 80% of these costs are the construction of cell sites, which can also be used for other forms of wireless communications, such as personal communications systems (PCSs).

Wireless PBX Systems

In companies where operations are widespread, such as chemical and heavy equipment manufacturing, it is often necessary for first-line supervisors and other key employees to cover a lot of terrain in a day. Often these personnel are in high demand and maintaining contact with them is difficult. For these applications, private branch exchange (PBX) manufacturers have developed wireless radio frequency (RF) systems that can be integrated into the architecture of a PBX system.

Lucent Technologies (Basking Ridge NJ), Ericsson Messaging Systems (Woodbury NY), Intercom Computer Systems, Inc. (Woodbridge VA), Northern Telecom, Inc. (Richardson TX), Mitel Corp. (Kanata, Ontario, Canada), and Siemens-ROLM Corp. (Alpharetta GA) offer systems that integrate into their PBX architectures. These systems are integrated through the PBX line cards and support the same line appearances as any hard-wired single line or electronic station set. A base radio operating in the unlicensed frequency range together with a series of antennae spaced around the user's facility comprise the basic network. Low-powered mobile handsets are used with these systems to avoid interference with other frequencies operating in the same area.

Wireless PBX Add-On Systems. Motorola and Spectralink have developed wireless PBX add-on systems similar to those developed by the PBX manufacturers. The Motorola InReach design concept is slightly different because it was developed as an extension to a cellular operator's service offering. InReach handsets can be used either as a cellular terminal or a PBX station set.

For example, when a user enters an InReach-equipped building, the handset can function as an electronic desk telephone. The handset provides access to all of the features on the PBX, including access to the corporate and public network. When the user leaves the building, the handset can then be used to access the cellular network and functions as a mobile handset.

Wireless add-on PBX facilities are expensive because of the addition of a base radio module and antennae infrastructure to the established internal PBX network. A typical midrange (i.e., 75×450 line) PBX system, when configured with a wireless add-on system, can easily double the cost of the basic PBX system. However, as PCS and other handheld terminal-based services proliferate, the costs for PBX wireless systems will continue to decline.

Satellite Voice Services

Satellites are playing an increasing role in establishing still another layer of worldwide voice communications. Two of the most widely heralded services are the Iridium and Teledesic low earth orbital (LEO) systems. These systems offer worldwide telephone service through the use of a small handheld telephone similar to those used for cellular systems. Iridium is owned by a consortium of international companies, one of which is Motorola, Inc. Teledesic is owned by McCaw Communications and Microsoft Corp. (Redmond WA).

Inmarsat now offers voice services through a worldwide consortium of 65 member nations. Special briefcase-size terminals are used to communicate with the satellite. Typical terminal costs range between $18,000 to $22,000 and connect-time costs are approximately $5.00 per minute. A new service that is planned, Inmarsat-P, will compete directly with the LEO systems. Although details of the Inmarsat-P service are still in the making, terminal and initiation costs are expected to be in the range of $1,500, with connect costs of about $1.00 per minute.

These satellite-based voice systems provide the capability to support both voice and data communications in any remote area of the world.

WIRELESS CONSIDERATIONS FOR A DATA COMMUNICATIONS STRATEGY

A variety of services are available to support wireless data communications. Wireless services like cellular digital packet data (CDPD), enhanced special mobile radio (ESMR), Ardis Mobile Data, and RAM Mobile Data Inc. support slightly different needs, although there is some overlap. For this reason, users should not look for a single vendor to supply an all-encompassing wireless service solution. In fact, it is less costly to consider a mix of voice, paging , and data services.

ESMR and CDPD offer competitive data communications services. For example, the Nextel interconnect option on the Motorola Integrated Radio System (MIRS)-based network costs $40 per month for the first 256 minutes plus $0.50 for each additional minute. This assumes that the subscriber also is a dispatch subscriber at about $25 per month for access.

A MIRS Motorola Lingo mobile handset is required to access service on a MIRS system and is priced around $1,000. In comparison, cellular telephones can cost up to $350. Cellular subscribers start out at $14.95 for monthly access plus about $0.45 or more per minute for airtime.

Although there are still many smaller special mobile radio (SMR) operators across the US that will continue to offer dispatch and interconnect services in second-tier markets, major players such as Nextel, Dial Page, and other members of the MIRS-related roaming consortium are likely to maintain their interconnect rates in competition with cellular service providers.

CDPD as a Wireless Option

Implementing cellular digital packet data networks often requires a number of systems applications modifications. For example, a special CDPD modem is required at either end for the transmission of data from one point to a host computer's communications port. This device must be established separately from the other host communications ports and should be installed by the cellular service provider. This task includes assigning the device with an IP address and configuring it for access to the cellular network.

CDPD Costs. Communicating over a wireless network is more costly than using the public network for a number of reasons, not the least of which is the cost for airtime. For example, regular transmission control protocol/ Internet protocol (TCP/IP) applications generate a lot of extraneous traffic that can drive up the cost for transmission on a network that is usage-sensitive. A hardware fix is available to alleviate this type of network condition. For example, products are available to monitor data flow as a means of reducing the number of acknowledgements being sent.

Potential Performance Problems. Another problem that must be taken into account is packet delay. This condition can result in dropped connections or unnecessary retransmissions and is caused by network congestion. Although cellular networks are still relatively lightly loaded, network congestion becomes a problem as greater penetration develops in the wireless market and CDPD networks become crowded. In addition, under some traffic circumstances, it is possible for packets to be dropped, therefore delivery of packets cannot be guaranteed. Noisy lines and poor radio coverage can also present the same types of problems as a congested network.

Under some traffic conditions, duplicate packets can be introduced through retransmission facilities. If the packet acknowledgment is lost, the packet's source will time out and retransmit a second or duplicate packet. Packets can also be thrown out of order when the data path is subjected to delay from rerouting events. These are just a few of the transmission characteristics that must be countered when a CDPD network is used for data transmission. Users should carefully review their applications and develop the measures that may be required to safeguard their data transmissions.

CDPD Test Areas. Cellular digital packet data is being tested by McCaw (in Las Vegas, Dallas, and Seattle), Ameritech Mobile (in Chicago), GTE Mobilnet, and AirTouch Cellular (PacTel Cellular), among others. Bell Atlantic Mobile, Inc. has announced pricing for CDPD services offered in its Baltimore/Washington DC and Pittsburgh test markets. GTE PCS and McGraw Cellular have also initiated trial services in their franchise areas.

Specialized Mobile Radio

Specialized mobile radio (SMR) services began in 1970 when the Federal Communications Commission (FCC) established frequencies in the 800-900MHz range for use in land mobile communications. A typical application of SMR is a radio dispatch for service fleets and taxi cabs. SMR operators are assigned licenses for exclusive use of assigned channels in a given area. SMR operators can also provide interconnection to the public network.

Racotek (Minneapolis MN) is one provider of SMR wireless voice/data service. Racoteck provides a vehicle fleet management service that is based on SMR or trunk radios. A Racoteck communications gateway facility linked to a mobile communications controller (MCC) in a customer's vehicle provides a data communications link between customers (e.g., truck drivers) and their dispatch control centers. A mobile radio collocated in the vehicle with the MCC unit completes the communications link. This system allows the dispatcher to send route information, messages, or other information that cannot be sent over the radio to customers while they are in route to or from a location.

Commercial Mobile Data Communications Services

RAM Mobile Data Inc. RAM Mobile Data Inc. (New York NY) is a joint venture between Bell South and RAM Broadcasting and provides a two-way data communications service that is based on the Mobiltex network architecture. This service is used by many companies for management of their field sales and service operations. RAM Mobile Data provides mobile data communications service in 90% of the urban business areas in the US, covering 6,000 cities and 210 metropolitan trading areas.

Access speeds of up to 9.6K bps can be supported in all areas; in select areas, it is possible to access the network at up to 19.2K bps. Common applications include E-mail and basic information access to the corporate data center for mobile travelers.

Some companies have greatly reduced their cellular telephone use by deploying the lower-cost RAM mobile network to send E-mail and messages to corporate personnel while traveling. A traveler equipped with a radio-enabled laptop or personal digital assistant (PDA) can access the nearest RAM base station. The message is then routed over a leased land line to the corporate data center. Messages can be sent to a traveler over the RAM mobile network where it is routed to the RAM local switch nearest the traveling employee. Conrail uses RAM Mobile Data to transmit train loading information to train crews advising the disposition of freight and empty freight cars. Other user companies, such as TransNet and MasterCard, use RAM Mobile Data to provide access to their central hosts so that merchants in the field can validate credit card purchases.

Ardis Mobile Services. Ardis (Lincolnshire IL) is a joint venture between IBM Corp. and Motorola and is composed of a formerly private corporate network that supported deployed field salesforces and service personnel. Ardis provides data communications services to 4,000 major metropolitan centers and 8,000 cities in the US, Puerto Rico, and the Virgin Islands. The network was originally designed by Motorola to support IBM's 18,000 deployed field service personnel. Access to the network ranges from 4.8 to 19.2K bps and can be reached from within a building or from a moving vehicle. Laptops and PDAs equipped with an Ardis/Modacom modem can be used to access company host computers to retrieve E-mail, enter orders , access diagnostic information, or obtain product information. Salespeople equipped with laptop computers can access product files to provide customers with product specifications as well as check inventories, enter orders , and print on-the-spot order confirmations.

Satellite Data

Satellite systems are composed of a transmission device that is capable of receiving a signal from a ground station. The signal is then amplified and rebroadcast to other earth stations capable of receiving its signal. User signals neither originate or terminate on the satellite, although the satellite does receive and act on signals from the earth that are used to control the satellite once it is in space. A satellite transmission originates at a single earth station and then passes through the satellite and ends up at one or more earth stations.

The satellite itself acts as an active relay much the same as a microwave relay. A satellite communications systems involves three basic elements: the space segment, the signal element, and the ground segment. The space segment comprises the satellite and its launch vehicle. The signal element comprises the frequency spectrum over which the satellite communicates, and the ground segment comprises the earth station, antennae, multiplexer, and access element.

Advantages of Satellite Systems. The advantage of a satellite system can be seen in the transmission costs, which are not distance sensitive, and the costs for broadcasting, which are fixed whether there are one or 100 stations that receive the down signal. Another advantage is the high bandwidth that satellite signals are capable of supporting. Bit errors are random, making it possible to use statistical systems for more efficient error detection and correction. Some satellite service providers are described in the following sections.

American Mobile Satellite Corp. (AMSC). AMSC offers satellite-based mobile data services using its own L-band satellite. AMSC is owned by three major shareholders—McCaw/AT&T, MTEL, and Hughes—although its stock is publicly traded. In the US, AMSC offers service through its Virginia hub. Downlink services may come through the Washington DC international teleport for services sold through the Virginia hub. Pricing is competitive with terrestrial services. For example, a full-time 64K-bps link between Washington DC and Brussels, Belgium would cost $1,350 per month.

OmniTracs. OmniTracs, a service of Qualcomm Inc., uses excess capacity on Ku-band US satellites to provide a data-only mobile tracking service for large trucking companies. The OmniTrac service now has more than 50,000 terminals deployed in trucks in North America. Qualcomm plans to expand its service into Europe, Japan, and South America using excess capacity on existing Ku- and C-band satellites.

Globalstar. Globalstar is the name of a low earth orbit (LEO) system designed for mobile voice services by a joint venture of Loral and Qualcomm. Globalstar has recently extended its ownership to an entirely new set of investors who plan to use excess capacity on available satellites. A series of gateways around the globe will provide an integrated network into the public switched telephone network (PSTN) and the satellite links.

Odyssey. Odyssey, a system proposed by TRW, is composed of four satellites. The TRW system will use fewer satellites for nearly global coverage because the system will be higher in the sky. The Odyssey uses the TRW advanced bus (AB940) L-band dish for mobile-to-satellite links, an S-band dish for satellite-to-mobile links, and two small Ka-band antennae for satellite/ground station links. Each satellite will operate as a bent pipe system, with switching and processing performed at the ground stations using spread spectrum modulation.

Ellipso. Ellipso, proposed by Mobile Communications Holding Inc., is a high elliptical orbiting (HEO) system, consisting of six (although 24 are planned) small satellites deployed in three elliptical orbits. Two of these orbits, called Borealis, will be inclined at 116 degrees. One orbit will be equatorial, which will provide dependable access to users in the northern and southern hemispheres. The Ellipso satellites will be small and use a simple bent pipe design with L-band for uplinks and S-band for downlinks.

Orion Atlantic. Orion is an international partnership of eight companies that operates its own Ku-band satellite composed of 34 transponders. Orion's focus is European business-to-business communications arrangements, as well as transatlantic connectivity. Services include cable distribution, business television, news and network backhauls, feeds, and standard business communications requirements. The service can support a full range of multimedia requirements, including telecommuting and interactive desktop video.

A unique mesh network provides completely independent service for international firms with multiple locations. Uplink/downlink services for 64K-bps access is in the range of about $2,000 per month for an enterprisewide LAN. A dedicated 64K-bps full service point-to-point link can be provisioned for about $1,400 per month for a 36-month contract. This service includes all equipment for rooftop-to-rooftop access, which is configured to support a dynamically allocated bandwidth service supporting both voice and data requirements. Installation for such a service would be about $10,000. Such an international connection is priced below regular internal terrestrial services and completely bypasses all monthly recurring local loop costs. A second system is planned that would cover a large part of Russia, the Middle East, Africa, and South America.

WIRELESS LANs

Wireless LANs are governed by the IEEE Wireless Local Area Networks Standard Working Group Project 802.11. The 802.11 standard establishes the components and interface requirements for a wireless LAN. The basic architecture established by the 802.11 committee organizes wireless LANs into basic service areas (BSAs) and access points (APs). Multiple BSAs can be interconnected at the APs into an extended service area (ESA). The protocols for this model are divided into two groups: the media access control (MAC) specification and physical specifications (PHY). There are different specifications for each radio frequency supported: 915MHz, 2.4GHz, and 5.2GHz.

WIRELESS COMMUNICATION AS AN ALTERNATIVE TO FIXED MEDIA

Traditional fixed-media systems are based on coaxial cable, twisted-pair wiring, fiber optics, or a combination of all three. Over time, the documentation for fixed networks can become lost or rendered inaccurate because of unrecorded equipment moves and changes. As new functions are established or offices rearranged, segments with undocumented cables are often installed to support added network nodes. Some companies that experience a high degree of internal moves and changes find it necessary to abandon at least 30% of their original network media. For these companies, a wireless network strategy superimposed over a base network provides the flexibility to support many permanent and temporary moves. Under this plan, the user is only required to establish a base radio, transmitters for each terminal to be moved, and a series of line-of-sight antennas. Thereafter, relocating network users only requires that the new location has line-of-sight to a network antennae.

The Wireless Cost Advantage

A wireless LAN solution at $750 to $1500 per node may be expensive when compared with a traditional wired solution (approximately $350 to $550). However, when the costs of lost productivity and rewiring are added, a wireless solution may be more cost-effective for organizations that move or change equipment frequently. Wireless solutions find their best fit where there are large unwired manufacturing areas to support, campus buildings that must be interconnected, open office areas without access to wire facilities, or older buildings with concrete partitions and no wire access.

Vendor Support for Wireless Solutions

There are several different vendor approaches for supporting wireless LANs. For example, Motorola's Altair systems use the 18-19GHz frequency range to support a microcellular approach. A series of intelligent antennae is used to establish microcells within the user's building. These microcells are supported with low-powered, high-frequency radios designed to support frequency reuse. This process results in a very efficient network.

Other manufacturers often use two basic components: the radio hub and the transceivers. In some systems, a single hub can support up to 62 transceivers. The transceivers are attached to the terminals and communicate with the hub using a line-of-sight arrangement.

Wireless LAN bridges are used to connect LANs in neighboring buildings. These devices establish a point-to-point connection and may not be a complete system. Examples of wireless bridges can be seen in the Motorola's Altair VistaPoint and the Cylink Airlink.

Infrared and laser technology can also be used to interconnect LANs in different buildings. This technique places information on a beam of light and can support very wide bandwidth over a short distance. In addition, this technology is immune to electric interference and is much more secure than radio transmission. Although infrared and laser techniques do not require an FCC license, users are responsible for any radio interference that develops while they are operating in a densely occupied area. LCI (Lancaster PA) has been developing laser systems for several years and has well over 750 mature systems installed.

OUTLOOK FOR WIRELESS APPLICATIONS

Projections for wireless applications vary depending on the user and the interpretation of the technology. There is no doubt that there will be a tremendous penetration in the basic telephone service market. Wireless local loop access will allow more users in developing nations to enjoy telephone service faster and at an affordable level.

The continued decline in the cost of PCMCIA cards for mobile radio will result in the continued rise in the number of laptops and PDAs used for basic communications functions such as E-mail and information access.

Satellite and radio-based service will continue to support vehicle management and tracking. Services like Qualcomm's OmniTrac provide a cost-efficient method for tracking and establishing a data communications connections with truck assets in the field.

Global positioning systems (GPS) will allow users to track vehicles and provision driver information. Avis rental car agency is testing a system that tracks Avis cars and sends driver information to fleets of specially equipped rental cars.

Hertz, Alamo, and other rental car agencies are using RAM Mobile Data to allow their service personnel to directly process returned vehicles as they are driven onto the company ramps. Using a handheld data entry terminal, the service person is able to enter the vehicle ID code and rental status. This process allows the rental car location to more efficiently manage their available pool of cars.

SUMMARY

Considering that many of the current wireless applications have come into being in only the last few years, new applications are certain to proliferate as users gain confidence in the available services. Mobile workers such as field sales representatives can spend more time with customers. New levels of productivity will emerge as telecommuting employees freed from expensive office space are able to focus more on the delivery of an end product.

AT&T Paradyne's Enhance Throughput Cellular (ETC) can greatly improve the process for sending data over the cellular network. This technology makes the cellular data user transparent to all other cellular traffic. Advancements such as this one will allow wireless users to resolve many of their data transmission requirements that were previously difficult to resolve. There is no question that users are adopting wireless solutions. The important issue to consider is the rate at which this technology is absorbed by mobile workers and the extent to which the penetration of services exceeds the available capacity of the network to support these users' needs.

7-2
The Technical Challenge of Cellular Data Transmission

Steven E. Turner

THE ACCELERATED TIMING OF BUSINESS ACTIVITIES TODAY and the increasing mobility of the work force signal a need for access to information, regardless of a computer user's location. The communications industry has responded by creating radio networks and the cellular telephone service, and the usefulness of these technologies is evident from their tremendous growth.

To date, market considerations and technical constraints have limited mobile radio communications largely to voice. Now these barriers are falling away, and mobile data access is becoming a reality. As this trend continues, the combined capability of wireless voice, data, and facsimile communications will further the growth of the mobile office.

The benefits of the wireless data network are much the same as those of the wireless voice network: portability and mobility. Users can access data from virtually anywhere in the world. As a result, the applications are almost endless; typical users might include sales representatives, real estate agents, insurance adjusters, or delivery services. Emergency services (e.g., fire and police departments) also benefit from the access and exchange of data from a remote location. In fact, anyone who currently uses a modem to transmit and receive data will soon find mobile wireless data access useful, because the laptop computer has made it possible to process and use that data in a portable, and sometimes remote, environment.

CELLULAR ADVANTAGES AND CHALLENGES

There are economic advantages to the wireless data network, most notably the ability to communicate without stringing costly copper wires all the way to the call's source or destination. Because a wireless data call reaches the existing wireline network by means of radio frequencies, the user does not have to extend the wire network or conform to its physical constraints to access it. Rather, the airwaves are the transmission medium. Wireless local area network technologies for intraoffice networks reduce the number of wires that must be run initially and rerun later when an office shuffle occurs. This advantage increases significantly when the radio frequency (RF) network user is in an automobile 10 miles from the office.

Two options in wireless data transmission seem destined for rapid growth. First is the direct-modulation approach, wherein data signals are used to modulate the carrier generated by a mobile FM radio. A number of these direct-modulation radios are then combined to form an RF data network. This is most useful in creating a proprietary, private data network.

The second option is to use the public cellular telephone network to carry data signals rather than voice. This option opens up mobile data access to anyone with a modem and a cellular telephone. To use this network for data, however, it is important to first understand the cellular network—how it works and how the RF channel and cellular telephone equipment affect data transmission.

A broadcast RF signal encounters many perturbations: gain, phase, and frequency attenuation; interference from other RF signal sources; and multipath interference. As a result, it is more difficult for the receiver to accurately demodulate the signal at the far end of the link. Fortunately, the error control and signal enhancement techniques used in wireless modems are also available in cellular modems. These techniques can clean up the RF data signals and make them decipherable and error free.

Other technology challenges are size and weight. Cellular modems must be small enough and light enough to make them truly portable and allow them to fit inside a cellular telephone or laptop computer. Unfortunately, engineering an RF product to meet certain size specifications often interferes with its performance. With care and proper design, this problem is being overcome, however.

CELLULAR SYSTEM BASICS

Much of the land-based cellular data network is the same network originally established to allow cellular voice telephones to communicate with other cellular and wireline telephones. Because the network was designed for voice applications, it has features that optimize it for voice signals. Unfortunately, some of these network properties are harmful to data signals, compounding the challenges in using cellular for data transmission.

The public land/mobile cellular network used to provide mobile communications is illustrated in Exhibit 7-2-1. The procedures for accessing this network are the same for voice and data calls. Wireline users (land units) connect to the cellular network by placing the call over their local subscriber

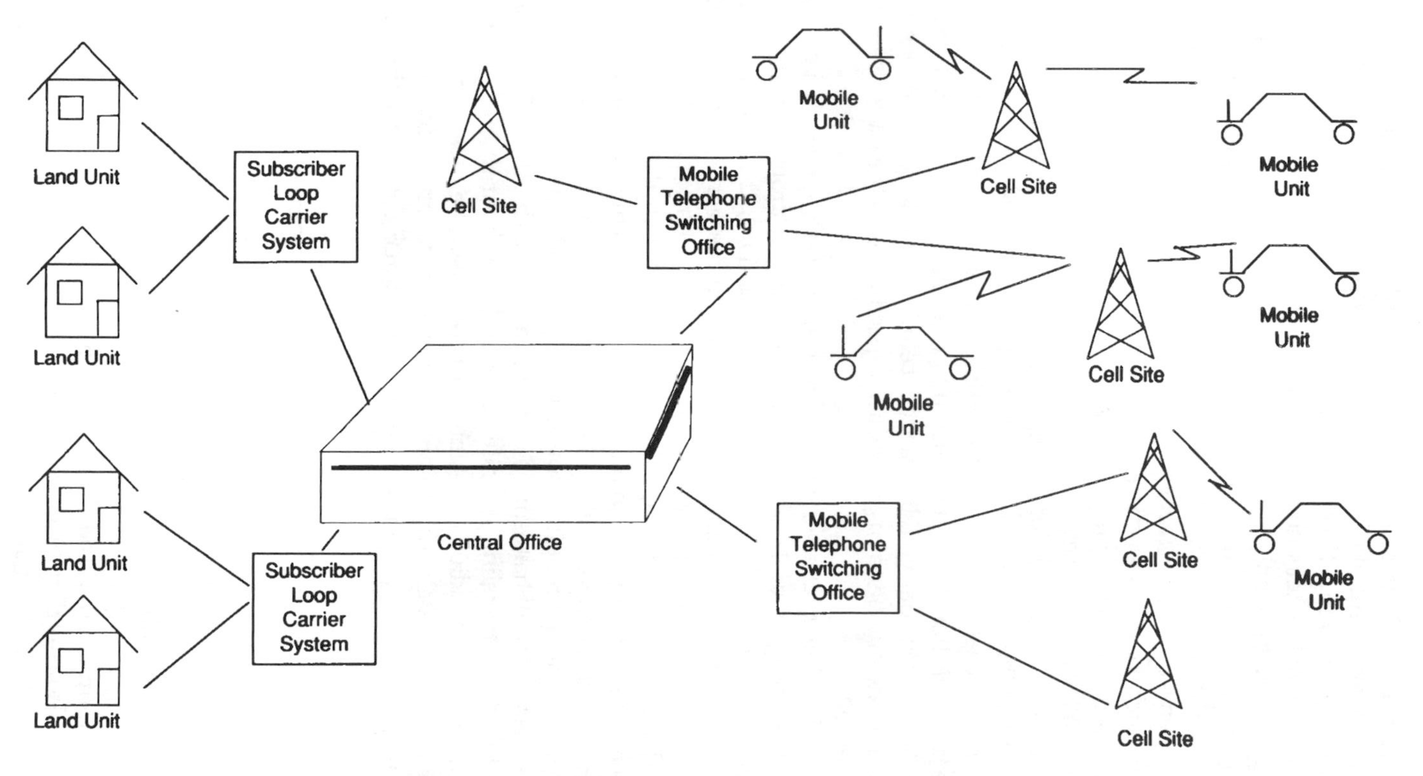

Exhibit 7-2-1. A Land/Mobile Cellular Data Communications Network

loop to their regional subscriber loop carrier (SLC) system and on to the telephone company central office, just as they would do for an all-wireline call. Instead of being routed back to another land unit, however, a cellular call is routed to the mobile telephone switching office (MTSO), which identifies the appropriate cell site antenna for the called cellular unit (i.e., the antenna closest to the mobile telephone at the time the call was placed). The call is then routed through the cell site to the mobile cellular unit. Calls from cellular modems to wireline modems follow the reverse procedure. The wireline side of such a data connection is very similar to a traditional modem call, and standard modem technology works well for that portion of the link.

A single MTSO serves several different cell antenna sites, with a single antenna providing coverage for all calls within its cell. The actual RF coverage differs from the idealized cell boundaries; in an ideal system, hexagonally shaped cells are packed together to provide complete coverage of an entire area. It is from this hexagonal subdivision of the communications system that the term "cellular" is derived. In practice, cellular boundaries are dependent on traffic and the number of users within a specific area—the more users, the smaller the cell size.

As a mobile cellular telephone unit moves about (as happens in a moving car), it crosses geographical cell boundaries and moves from one cell to another. As the mobile unit approaches a cell boundary, the received signal strength from its own cell's antenna weakens, and the signal strength of the neighboring cell's antenna grows. At the point at which the neighbor's signal strength dominates, the cellular call is "handed off" to the adjoining cell. This process is illustrated in Exhibit 7-2-2. By transferring the call from cell site to cell site as the mobile telephone travels, the cellular telephone service provider maintains the required power level to sustain the call.

The frequency of call hand-off is demonstrated in Exhibit 7-2-3, in which a mobile telephone traveling at a rate of 30 miles per hour through a typical cell having a 10-mile radius is assumed. In this case, the mean time to call hand-off is approximately 10.5 minutes. This is an important consideration, as call hand-off is a significant impairment in cellular data transmission.

IMPAIRMENTS TO CELLULAR DATA TRANSMISSION

Unfortunately, the RF environment produces significant problems for data on the cellular side of the connection. For example, FM cellular telephone transmitters use compression techniques on the outbound signal; while this does not adversely affect voice, it can produce significant nonlinearities in data signals.

The waveform limiting and filtering required for signals using the cellular RF network also contribute to trouble for data signals. This unfriendly processing, the audio muting and signaling tones used by the cellular network,

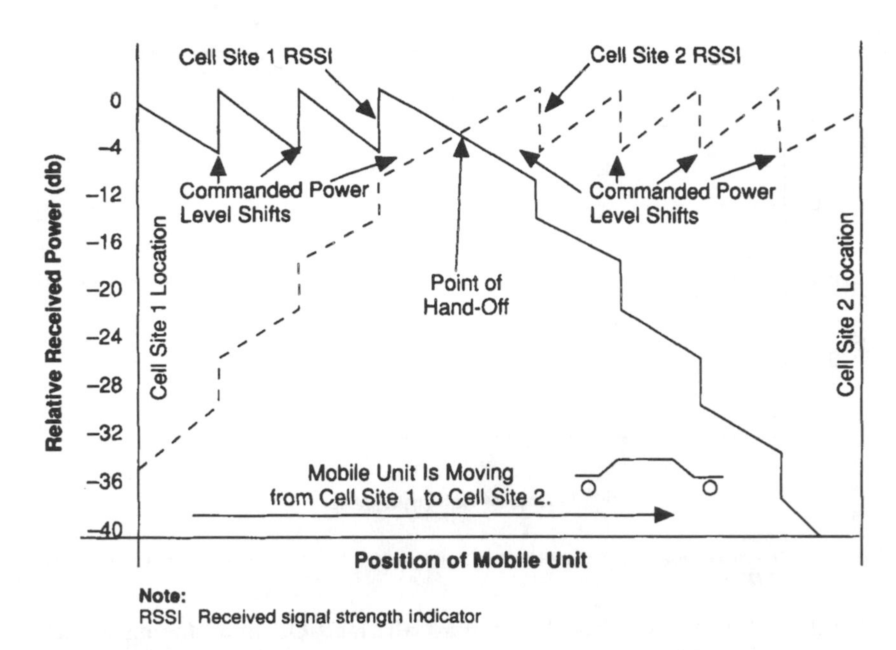

Exhibit 7-2-2. Power Level Control and Hand-Off in a Cellular Network

and the harsh RF environment must be countered in the cellular modem's receiver if the data call is to succeed.

Cellular telephone hand-off is very significant for data calls. During hand-off, dead times of a half-second or more are common; while these dead times go virtually unnoticed in voice calls, they present carrier drops in data signals and can cause a modem to hang up. Cellular modems use hysteresis effects to compensate for this problem, holding over the call through the hand-off from one cell to another.

In addition to the problems induced by the cellular telephone equipment and network, Rayleigh fading and multipath interference in the RF channel can distort the data signal, causing bursts of errors. Rayleigh fading is a result of the multiple paths an RF signal takes as it travels from the transmitter to the receiver. Bouncing off buildings, trees, cars, and other objects, reflections of the signal arrive at the receiver at different times and with different signal phase and strength. The effects of this condition are examined in Exhibit 7-2-4; the signal that the cellular modem receives fades deeply as the unit moves.

With voice signals, the human ear processes the multipath signals, ignoring the ones from unwanted reflections. With data signals, these reflected signals have to be sorted out by the receiver and processed carefully to accurately demodulate the transmitted signal. This is normally accomplished through

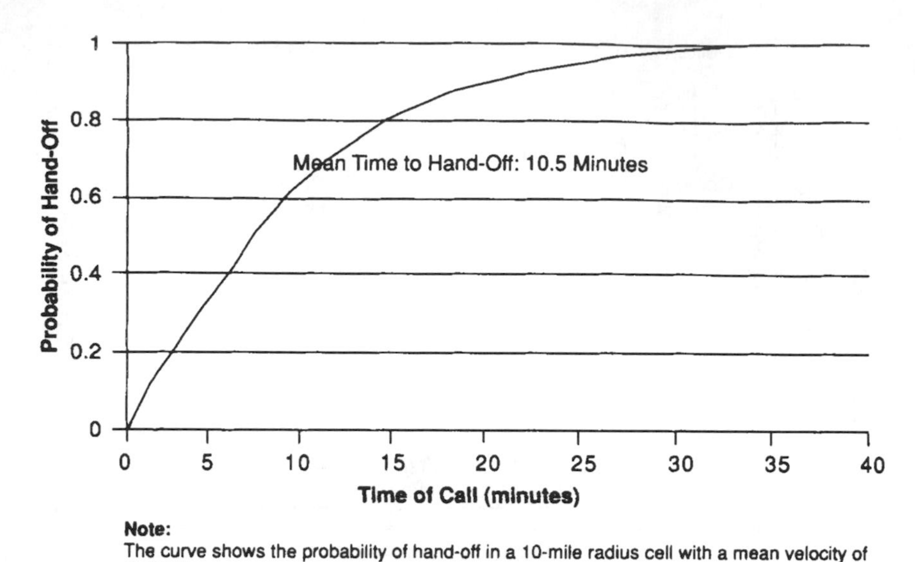

Exhibit 7-2-3. Typical Probability of a Call Hand-Off in a Moving Vehicle

---

advanced channel equalization techniques, which compensate for the channel fades at the receiving modem.

In addition to the problems caused by RF channel disturbances and the function of the cellular network equipment, incompatible modem designs present problems. Inconsistent data formats and protocols, audio muting, and proprietary modulation techniques all restrict the use of cellular modems. The following is a summary of the major transmission problems that cellular modems encounter:

- Low RF power levels can prevent successful connection.
- Incompatible communications protocols can prevent connection.
- Incompatible data formats can impede data transmission.
- Cellular FM radio system can distort data signals appreciably.
- Cellular system call hand-offs can cause data link disconnections.
- Cellular system audio muting can cause data link disconnections.
- Multipath interference can swamp the cellular radio receiver.
- Rayleigh signal fading can diminish cellular data integrity.
- Excessive retransmissions occurring because of errors can reduce data throughput.
- Proprietary modem modulations schemes can inhibit communication.

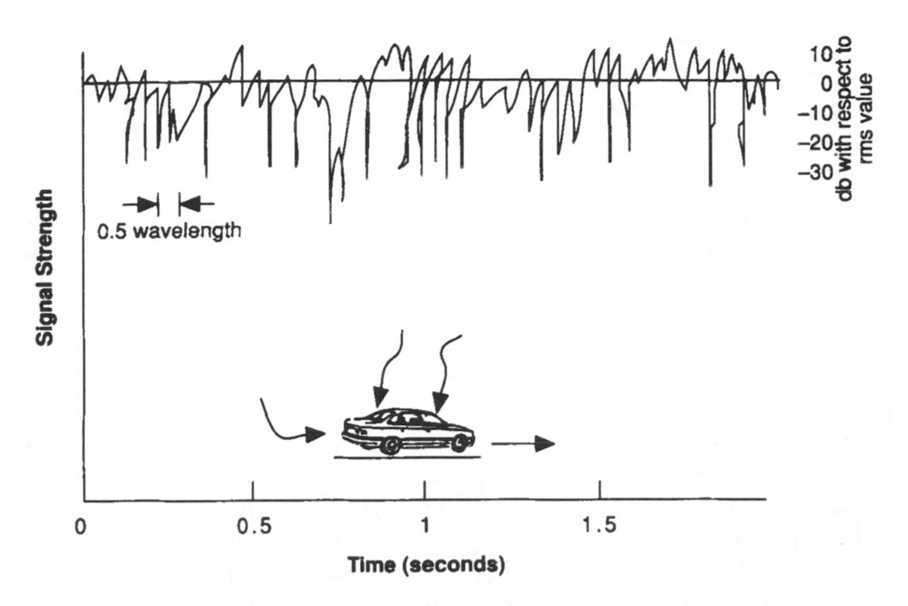

Exhibit 7-2-4. Rayleigh Fading as It Affects a Mobile Cellular Unit

SOLUTIONS TO THE CELLULAR MODEM CHALLENGE

With all the challenges to data transmission over the cellular network, is it feasible to use it for data transmission? It is, in fact, no small challenge, but the problem can be solved several ways. One solution is to build a pair of modems matched to each other and to the cellular telephone network. Such modems contain special, proprietary algorithms to compensate for Rayleigh fades and a special mechanism to adjust the data packet size to adapt to data dropouts and cell hand-offs.

The biggest drawback to this approach is that when a wireless modem connects to a wireline modem, both modems must contain the proprietary cellular protocols for the call to be successful. This severely limits data communication through cellular modems because few wireline modems contain the cellular protocol enhancements. As the number of wireline modems the cellular modem can connect to is limited, so is the value of this approach.

Gateway Service. In response to this issue, a number of cellular network providers offer a gateway service. This service allows cellular data users to establish a connection through a bank of cellular modems operated by the network provider. The call is established by dialing a special prefix before the number. At call establishment, the cellular user's call is routed to a cellular modem. From there, the call is routed over a land-line circuit using a land-line modem to complete the call to the destination. This service typically operates over the analog cellular network, and imposes certain restrictions on the data traffic. For example, at the time of this writing, fax services were not yet available over this service.

V. Series Modems. Another solution to the cellular problem is to enhance existing V. series modems for the cellular side of the data link. In this case, the wireline side of the connection remains the same. Typically, the cellular enhancements include improving channel equalization and filtering of the received signal. Like proprietary cellular modems, these V. standard-compliant modems usually have a mechanism to adjust the data rate up and down in response to the channel conditions encountered during the call. When transmission conditions are poor, the data packet size decreases to allow smaller blocks of data to pass without error; on better channels, the data packet size grows to maximize information throughput. This way, modems that function in the mobile cellular environment can compensate for cellular effects, but at the same time, these modems can be used to connect to any of the installed base of wireline modems already in service. Exhibit 7-2-5 illustrates this approach. The receiver of a standardized (e.g., V.22bis or V.32) modem is modified to deal with cellular impairments, but the modulation, coding, and error-control schemes remain unchanged to ensure full compliance with the International Telecommunications Union (ITU) standards.

STANDARDS

Clearly, a key factor in the success of cellular modems will be their continued compliance with existing modem standards. The ITU Study Group XVII is working to develop a standard for cellular modems that will compensate for cellular channels and equipment while preserving full compatibility with existing modems. At present, the study group is considering several enhancements to existing coding and error-correction schemes. Adoption of a standard would significantly increase the availability and interoperability of cellular modems; however, modems that provide the needed cellular performance and standards compliance are already available from UDS Motorola and other vendors.

Another approach is to develop new transmission standards for the cellular network. A dual-mode cellular network standard (one that supports both analog and digital cellular equipment) is being rapidly expanded throughout much of the world. Coupled with the rapidly emerging personal communication services (PCS) technology, significant advances in wireless data are inevitable over the next few years.

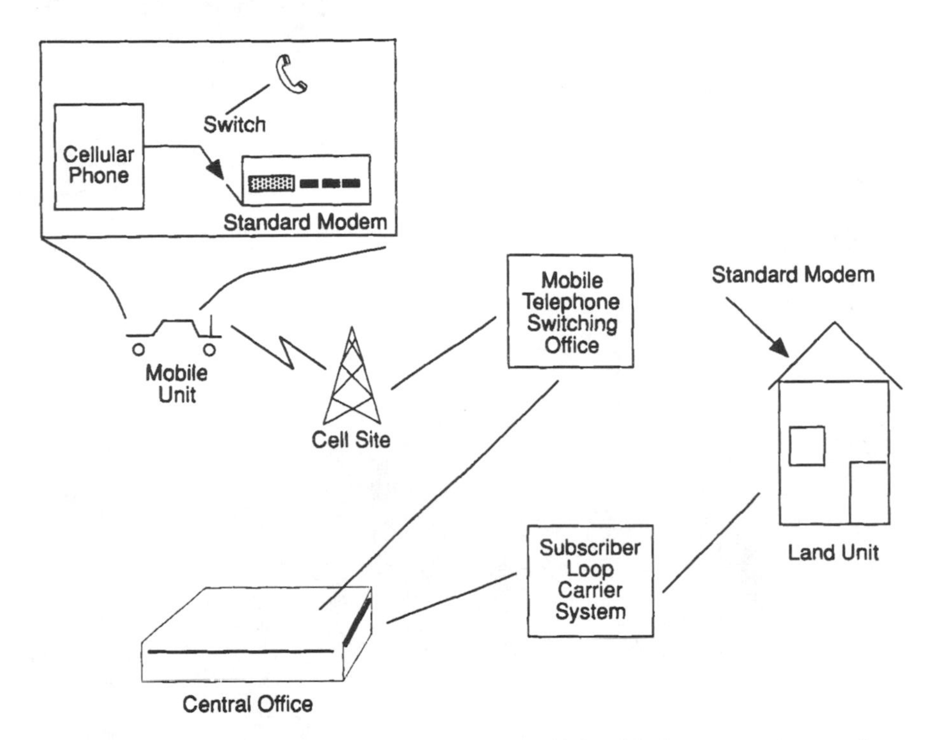

Exhibit 7-2-5. The Optimum Wireline-to-Cellular Modem Data Connection

---

SUMMARY

It comes as no surprise that, just like the wireline telephone network, the wireless networks are being used to send data. Data services are now available over all the major wireless services. The technology behind these services is evolving rapidly, and at present varies widely depending on the type of wireless service.

As business becomes more mobile, the demand for portable data access will continue to grow. This will include not only traditional modem traffic, but mobile facsimile and electronic mail traffic as well. Solutions that include new hardware and new network standards continue to be revised to meet widespread demand for such services.

7-3 Developing a Cost-Effective Strategy for Wireless Communications

Sami Jajeh

MOST ORGANIZATIONS HAVE SOME MOBILE FIELD ACTIVITIES involving sales representatives, field service technicians, telecommuting employees, traveling managers, route-based personnel, or even mobile health care providers. Organizations with significant numbers of mobile field activities require a well-synchronized exchange of information between central information systems and mobile users.

Many organizations are investing in portable computers and software to provide mobile users with the tools they need to accomplish their daily tasks. Some organizations have begun to look at emerging wireless technologies to further enhance communications and streamline information exchange by providing anytime, anywhere access.

Automating business processes through wireless technology offers organizations many benefits, including improved productivity and increased competitive advantage. To achieve these benefits, organizations must thoroughly consider several implementation issues that fall into three broad categories:

- Communications architecture and access methods.
- Application appropriateness.
- Wireless service products.

This chapter aims to help organizations develop a cost-effective wireless communications strategy that meets the needs of mobile and remote workers. Following an overview of wireless network technology and service providers, it discusses each of the three categories of implementation issues and the advantages and disadvantages of the various options within them.

WIRELESS NETWORK TECHNOLOGIES AND SERVICE PROVIDERS

The following sections discuss major network technologies and service providers; they are not intended to provide an exhaustive list of current technologies and players. There are two prevalent technologies for wireless applications:

- Circuit-switched networks.
- Packet data networks

Circuit-Switched Networks

Circuit-switched networks involve establishing a dedicated connection (or circuit) between two points and then transmitting data over the connection, much like a typical telephone conversation. They can be either analog or digital.

Analog Circuit-Switched (Cellular) Networks. Two-way analog circuit-switched cellular (CSC) technology has existed since the advent of cellular phones. To use CSC service, the user requires a cellular phone with a cellular modem. Sending wireless data over a circuit-switched cellular connection offers several advantages, including:

- Wide on-street coverage and availability.
- Suitability for sending and receiving large data files such as long E-mail messages or reports.
- Per-minute (as opposed to per-packet) charges.
- Implementation through standard communications software and a modem attached to a cellular phone.

The disadvantages of using circuit-switched cellular technology include:

- Increased relative cost of sending short messages, because call setup time may become a large percentage of cost.
- Security concerns involving unencrypted files.
- Lack of cellular error-correction or enhancement standards.

Although questionable reliability and poor throughput are often cited as disadvantages of analog circuit-switched cellular connections, the availability of new technology from several vendors, including AT&T, Celeritas Technologies Ltd., Microcom Corp., Motorola, and ZyXEL, is rapidly changing this perception. These mature technologies allow organizations to use analog cellular modem technology to build and deploy enterprisewide dial-up based applications.

Digital Circuit-Switched (Cellular) Networks. Digital communications technology is inherently more reliable for sending data than is analog technology. Examples of digital circuit-switched wireless network implementations in the US are code division multiple access (CDMA) and time division multiple access (TDMA). Because the availability of both CDMA and TDMA is limited, it will be some time before most US organizations will be able to take advantage of digital circuit-switched technologies for wireless data applications.

Packet Data Networks

Packet data networks have been designed for effective and reliable transfer of data rather than voice. They use a method that is comparable to sending a document one page at a time. The document is first broken into pages, and each page (or packet) is sent in its own envelope. The network determines the most appropriate transmission path, and once each page reaches its destination, the document is reassembled (if appropriate).

Packet data networks use radio frequency channels to connect the portable computing device to a network backbone and, ultimately, to the company's host system. The major networks (e.g., Ardis and RAM Mobile Data) use packet radio technology. Packet cellular technology (e.g., cellular digital packet data, or CDPD) is now emerging.

Packet Radio Technology. The two major wireless packet data networks are Ardis and RAM.

Ardis. A nationwide, packet radio network owned by Motorola and IBM, Ardis covers 80% of the US population. Transmitters in the 400 largest metropolitan areas are networked through dedicated land-based lines, although dial-up and radio frequency (RF) connections are also supported. Ardis supports fully automatic roaming. In addition to on-street and in-vehicle coverage, Ardis is said to offer more reliable in-building coverage than do other two-way wireless networks. Pricing depends on the application and is based on both flat-rate and usage charges.

RAM Mobile Data. The RAM Mobile Data network is the result of a business venture between BellSouth Enterprises and RAM Broadcasting Corp. to provide wireless transport for messaging services and products. Commercial service is currently available in more than 6,000 cities and towns. RAM uses the Mobitex architecture for wireless packet data communications originally developed in Sweden and currently in its fourteenth version. RAM's network was designed for message capability with inherent roaming, store-and-forward, and broadcast capabilities.

Packet Cellular Technology. Cellular digital packet data (CDPD) technology is being developed and implemented by a consortium of 10 major cellular carriers, including AT&T Cellular and AirTouch. As a digital overlay of the existing analog cellular network that utilizes unused bandwidth in the cellular voice channel, CDPD is a logical extension of cellular data communications.

Because is based on an open design and supports multiple connectionless network protocols such as the Internet protocol (IP), existing applications require few, if any, modifications to run on CDPD. CDPD claims a bandwidth of 19.2K bps, although typical user rates are closer to 9.6K bps. Approximately 30 markets have access to CDPD technology.

Suitability of Packet Data Networks for Wireless Applications. Packet data networks offer several advantages, including:

- Reliable transmission of data.
- Cost-efficient transmission of short messages.
- Transparent roaming in the locations where the networks exist.
- Fast setup time.

Disadvantages of packet data technology include:

- High costs in certain situations (resulting from a per-packet charge).
- Slow transmission times for large data files (which is less the case for CDPD).
- More-limited coverage and availability than that of cellular technology.
- Limited bandwidth (here, again, CDPD is better than RAM or Ardis).

The suitability of packet data networks for wireless data applications depends largely on the application. The networks provide a solution for applications requiring instantaneous, unconnected delivery of small but valuable pieces of information that can save money or generate revenue. They are therefore used for single-transaction based applications such as remote credit-card authorization or rental car check-in. Use of packet data is more limited in cases of general sales force automation, data base replication, E-mail with attachments, electronic software distribution, and multiple application requirements for mobile users.

CHOOSING AN ARCHITECTURE AND ACCESS METHOD

The first step in implementing wireless technology is to choose an appropriate communications architecture and access methodology. The many wireless and connectivity access methods available generally fall into three categories:

- Continuous extensions of desktop or local area network (LAN) systems.
- E-mail-based systems.
- Agent-based messaging systems.

A solution that fails to address the communications infrastructure of the wireless environment has both financial and systems implications. Although communications costs escalate dramatically with heavy system use and large numbers of users, support and resource costs increase as well.

Continuous-Connection Architectures

A continuous-connection architecture establishes and maintains a wireless connection so that a user can perform work while online to central computing resources, such as a desktop PC or LAN-based PC. This work is accomplished through remote access and file synchronization utilties. Although there are unique variations on how these utilities are implemented, organizations generally use one of two methods: remote node or remote control.

Remote node technology makes mobile users a node on the LAN network and allows them to perform work as if they were locally logged into the LAN, albeit usually more slowly. Remote control technology allows mobile users to connect and see a virtual copy of the remote PC's screen or hard drive so that files can be accessed and applications can be run remotely.

Continuous-connection technologies offer the basic advantage of providing mobile users with access to their central LAN-based PCs and servers; the mobile computing device looks and acts as if it were the user's local desktop PC. Unfortunately, this strategy is inappropriate for the majority of large mobile implementations for several reasons:

- Most field professionals are mobile or remote all of the time. They may not need LAN resources, understand local area networks or logically redirected disk drives, or have a dedicated PC at the central site.
- Even when the complexity of establishing connections is hidden from the remote user, communication time is lengthy and communication costs are high.
- Before performing work online, the user must leave the task at hand to initiate and establish a connection.
- Continuous-connection systems do not provide for communications management or for general systems management; as a result support costs are likely to increase.

E-Mail Based Systems

E-mail based systems use E-mail as both the messaging application and as a general communications transport for other message types or transactions. The basic advantage of using E-mail as the access method for all communications is that it is a prevalent application that users understand.

However, use of E-mail as an access method for other applications is less than optimum because E-mail based systems:

- Lack integral systems management capabilities such as software distribution.
- Involve users with the information delivery process, which is not provided automatically by the application.
- Do not support applications that require queries into data bases.

E-mail is clearly a popular application required by most mobile users. However, it should be considered an additional application that uses the available communications access method, rather than a communications transport or access method in and of itself.

Agent-Based Messaging Systems

Agent-based messaging systems provide a communications architecture built on a client/server platform; a server at the central site acts as an agent on behalf of the mobile users. Software distribution, posting of forms-based data into central data bases, querying of data from central data bases, E-mail delivery, and many other tasks can be automated by agents capable of handling these functions on behalf of mobile users. Wireless or land-line connections can be established automatically and efficiently to synchronize information between the client and the server, with all of the work (e.g., data entry to book an order) being performed offline.

Agent-based messaging systems provide many benefits in extending client/server systems to large field organizations, including:

- Minimized connect times, which yields significant savings in communications costs.
- Minimized user involvement in communications.
- More-efficient applications performance resulting from the tight coupling of applications with the mechanism of information delivery.
- More-efficient management control of system resources and communications.
- High scalability with support for hundreds of remote users per server.
- Capability for more and different types of work (e.g., messaging and transactions) to be accomplished.

In addition to these benefits, agent-based messaging systems are also flexible enough to be used over continuous-connection technologies. For example, a user can establish a continuous connection with a central-site system and then employ the agent-based messaging software to exchange information utilizing that connection. This flexibility is not available with continuous-connection technologies such as remote LAN access.

CHOOSING WIRELESS APPLICATIONS

The next step in implementing wireless technology involves assessing which applications provide mobile workers with the most benefits. Application requirements vary among different classes of users, who may require different products and service providers. Four basic classes of applications are discussed:

- Wireless E-mail and fax systems.
- Remote access and file synchronization utilities.
- Single-transaction based applications.
- Mobile enterprise applications.

Wireless E-Mail and Fax Systems

A survey of telecommunications and IS managers conducted by the Hartford CT based Yankee Group revealed that the two primary drivers behind mobile data networks are customer satisfaction and revenue generation. Similarly, a study by Link Resources Corp. revealed that wireless data solutions were implemented mainly to decrease or control costs and to attain competitive advantage.

The results of these two studies contrast with additional, significant findings from the Yankee Group study, which found that a majority of respondents believed that the greatest potential growth will occur in E-mail and fax applications. Users who believe that E-mail and fax will achieve customer satisfaction and generate revenue generally do so for three reasons:

1. Wireless E-mail systems are being marketed as the next killer application that mobile users must have for the real-time communications necessary to support continuous sales.
2. Personal productivity applications like E-mail and calendaring are believed to be as necessary to mobile users as they are to headquarters-based users.
3. E-mail is thought to be the appropriate transport for routing forms, updating data bases, and performing other critical business functions.

None of these arguments holds true for most field activities. If revenue generation and customer satisfaction is the objective of wireless information exchange, then the most important wireless applications are the line-of-business transactions that generate revenue or improve customer satisfaction. These may include entering and posting sales orders to immediately secure an order, performing a query into a central data base to look up inventory status for a sales manager, or dispatching service requests to mobile field technicians.

Remote Access and File Synchronization Utilities

Personal productivity utilities facilitating wireless remote access and file synchronization functionality are basic utilities that give the mobile user access to local hard drives on a desktop PC or on a LAN drive at corporate headquarters. Generally, the applications perform this function by providing either a wireless remote node connection to a central LAN, a remote control function to a local desktop PC, or a distributed file system that mirrors the remote drives locally and accesses remote files whenever needed. The general idea is to extend the same personal productivity applications and files found in the central office to the mobile user.

Although these utilities provide important functionality for end users temporarily away from their LAN-connected desktop systems, they do little for the requirements of large field organizations involving hundreds of mobile users who rarely, if ever, use a desktop PC. As discussed previously, wireless remote access and file synchronization systems offer solutions that scale poorly and involve high support costs and connection charges. They provide few capabilities for systems management, application management, or connect-time and communications session management—all of which are critical issues for large wireless data implementations.

Single-Transaction Based Applications

Single-transaction based applications use wireless technology to perform one function (and sometimes a few functions) extremely well over a wireless connection. They tend to be oriented toward a large user community.

A single-transaction based application is used, for example, by a rental car employee to enter a returning car's ID number as well as other customer information on a handheld computer that prints a receipt. Another example is a job assignment dispatch application used by an organization with a large field service operation.

To date, these types of systems have produced acceptable rates of return because the applications implemented increase customer satisfaction and generate revenue. However, single-transaction based systems are most appropriate for a small, distinct set of highly repetitive functions. Most mobile users, including salespeople, should not be limited to a single application like order entry. They need a variety of applications to help them perform many functions well. Also, unless additional functionality is custom-built into these systems, single-transaction based application systems do not address application management, update, and maintenance issues.

Mobile Enterprise Applications

Mobile enterprise applications provide solutions to a large mobile user community that needs to exchange information with centrally located systems and users. These applications include transaction-based applications, information distribution applications, and E-mail and messaging-based applications. For example, mobile enterprise applications for a sales force of 500 people may include order entry, inventory status checking, electronic product catalogs, electronic sales report distribution, forecasting, pipeline management, contact reporting, and E-mail. A mobile enterprise application solution could make all of these applications and more available to hundreds of mobile salespeople through a common, easy-to-use interface.

Mobile enterprise application systems provide the most utility and payback of all the wireless solutions for the following reasons:

- They allow an organization to automate within one system many key line-of-business functions that focus on increasing revenue, improving customer satisfaction, and decreasing costs.
- They provide a client/server framework in which to implement a mobile client/server system that is highly scalable because it was designed for hundreds of users.
- They allow for efficient use of land-line, LAN, and wireless networks, so that users can choose the protocol or transport most appropriate to such conditions as the time of the connection or the application type.
- They provide sophisticated application services including posting into central data bases, querying from central data bases, routing and sharing of transactional information, and automatic and efficient updating of messaging-based applications.

Exhibit 7-3-1 summarizes the major application issues and decision criteria that organizations should consider when choosing a wireless application solution.

CHOOSING WIRELESS PRODUCTS

Much of the infrastructure for certain wireless technologies is either immature or under construction, and some wireless service providers require

Wireless Issue and/or Decision Criteria	Wireless E-mail and FAX	Remote Access and File Synchronization Utilities	Single Transaction-based Applications	Tools to Build Mobile Enterprise Applications
Examples of software application or tool	RadioMail Wireless cc:Mail	Airsoft's AirAccess MobileWare	Oracle Mobile Agents In-house developed applications	XcelleNet RemoteWare
Connectivity available	Wireless focus with some landline capability.	Wireless focus with some landline capability.	Wireless focus with some landline capability.	Mixed mode supports wireless, landline and LAN.
System architecture	Peer-to-peer and client/server.	Peer-to-peer focus. Some client/server.	Client/server-based architecture.	Client/server-based architecture, such as client/agent/server.
Application focus	E-mail and fax. Messaging applications.	File transfer and remote access.	Transaction applications, typically vertical in nature.	Messaging-based, transaction-based, and info exchange applications.
Number of applications available	Limited to e-mail and fax	Many	A few applications	Unlimited
End-user profile	Knowledgeable LAN-based PC professional, which occasionally goes mobile.	Knowledgeable LAN-based PC professional, which occasionally goes mobile.	Field professionals who do not have to be computer literate and focus on a single line-of-business task.	Field professionals who do not have to be computer literate and focus on many line-of business tasks.
End-user community	Single user or small department	Single-user or small department	Large field force	Large field force with potentially multiple end-user types.

Systems management	None	Very limited	None	Comprehensive, automatic, efficient
Application services and management	None	Very limited	Very limited	Comprehensive, automatic, efficient
Connect time management	Very limited	Limited	Limited	Comprehensive and flexible
Sharing and workflow capabilities	E-mail only	None	Very limited	Comprehensive, messaging-based
Hardware platform appropriateness	Portable or pen-based computing devices, PDA's.	Portable or pen-based computing devices.	Portable pen-based or handheld devices. PDA's.	Portable or pen-based computing devices.

The software examples provided are used to explain the types of software available. They are not meant to define exactly what a specific vendor's software may or may not do. For example, XcelleNet's RemoteWare is listed in the category of tools to build mobile enterprise applications, but it can be used to build single-transaction-based applications.

Exhibit 7-3-1. Issues and Decision Criteria By Type of Wireless Application

that an application be developed to a nonstandard protocol or application programming interface (API). As a result, organizations should develop a communications and applications strategy that provides the most flexibility regardless of which technologies or services ultimately gain widespread marketplace acceptance. There are two basic ways to do this:

1. Use middleware APIs or developer kits to develop wireless applications.
2. Use a system for communications management that provides an interface based on a high-level graphical user interface (GUI) to set up and maintain multiple wireless technologies.

Using Middleware APIs and Developer Kits

Some vendors offer middleware APIs that shelter organizations from having to learn how to connect over RAM, Ardis, CDPD, or analog cellular networks. By developing to the vendor's API set, organizations can choose different wireless providers or switch from one to another through a simple programming change. Vendors of middleware APIs claim to provide anywhere, any-protocol access.

The basic advantage to using middleware APIs is they allow organizations to skip the details of understanding, testing, and debugging communications. Many APIs also provide communications capabilities for land-line and local area networks.

Middleware APIs also have disadvantages, some of which are:

- Many of them are not based on industry standard messaging APIs, so organizations must develop and maintain applications using a nonstandard API.
- They require organizations to program (i.e., custom build) functionality that is already available in various systems for communications management.
- Many of the companies that offer middleware APIs are new and small; their long-term stability may be less certain than that of existing wireless service providers.

Systems for Communications Management

A system for communications management can provide an organization with support for many wireless technologies. A comprehensive system for communications management can also provide functionality in the area of systems management, software updates, file transfer, E-mail and messaging, and scheduling of tasks to take place over any of the various wireless services.

The major benefits of a system for communications management include:

- Provision of capabilities that organizations would otherwise have to develop for themselves using middleware APIs.
- Powerful functionality in the setup and maintenance of mobile users.
- The benefits of an API set or developer kit, because most systems provide interfaces and APIs.

Systems for communications management also ensure that the messaging layer on which applications are built can support future anticipated wireless services or APIs. In this way, an application can take advantage of future wireless services without additional development.

SUMMARY

Wireless communications offer organizations the opportunity to extend the benefits of automation to hundreds of thousands of remote or mobile workers, across states and continents, who lack access to traditional dedicated networking. The technology available today simplifies the task of synchronizing information flow in the inherently unreliable dial-up and wireless communications environment. Applications that cost-effectively automate remote and mobile business processes can now be built and implemented electronically in days rather than in weeks or months. Wireless and land-line information access can become transparent to the most remote and mobile activities of an organization.

Organizations that thoroughly evaluate the issues involved in choosing a wireless communications architecture and access method, wireless applications, and wireless products have taken the first step toward formulating a cost-effective strategy that generates revenue and increases customer service.

7-4
Mobile User Security

Ralph R. Stahl, Jr.

INFORMATION TECHNOLOGY PROFESSIONALS today are unsure of themselves in a strange new environment. However, end users are telling security practitioners that they can no longer perform optimally and beat the competition if their access to information and processing power is restricted by the mainframes in the corporate data center.

The days when security practitioners arrived at the office to find a stack of computer printouts on their desks are gone. Paper has been replaced by computers on the desktop in the environmentally correct and secure office. In addition, users today expect to be able to connect their notebooks by modem from any location to the server at headquarters. Tomorrow, mobile users may expect connectivity for their notebooks in the air as they fly and on the ground as they drive to their next destination. The business traveler may anticipate that all applications will behave in exactly the same manner on the road as in the office.

Large companies like AT&T are encouraging employees to telecommute for three major reasons:

- State and local governments are requiring companies to take action to reduce air pollution and traffic congestion.
- Office sharing allows companies to reduce their real estate expenses.
- Telecommuting benefits employees by allowing them more flexibility in managing their professional and personal lives.

LINK Resources Corp., a New York City consulting firm, reports that Americans bought for home use a record 5.85 million microcomputers last year. One out of three American households already has a microcomputer. BIS Strategic Decisions estimates that 45 million workers in the United States are considered part of the mobile work force. Other surveys estimate that, in addition to the time spent in the office, the average white collar worker spends six hours a week working at home.

Against this backdrop, the challenge for the application developer is to develop systems that may be used in any environment. The information

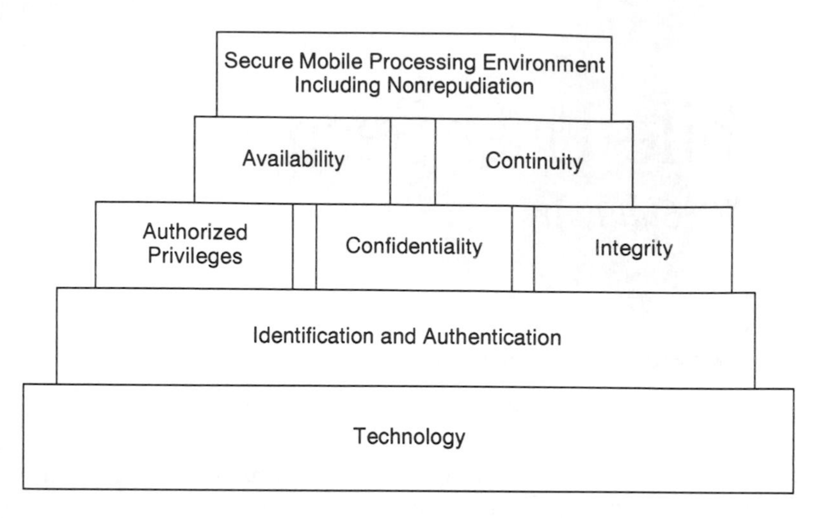

Exhibit 7-4-1. Model of Security Services

architecture for the enterprise must also accommodate many methods of remote connectivity (i.e., dial-up, Integrated Services Digital Network [ISDN], Cellular Digital Packet Data [CDPD], Internet, wireless, video, and image transmission) in addition to the traditional local area network and wide area network connectivity.

This chapter is divided into four major sections: availability and continuity; integrity; confidentiality; and new technology considerations, which briefly reviews the security implications for some of the emerging technologies. The architectural model of the security services in Exhibit 7-4-1 provides a high-level view of the interdependence of identification and authentication, authorized privileges, availability, continuity, integrity, and confidentiality in providing a trustworthy environment that supports nonrepudiation and mobile power user security.

AVAILABILITY

In this chapter, availability is defined as the assurance that an authorized user's access to an organization's resources will not be improperly impaired. Achieving such assurance involves properly categorizing information privilege keys and ensuring that the mobile user's authorized privileges are properly associated with these privilege keys. Availability also involves physical considerations (e.g., theft prevention, device identification, mobile uninterrupted power supply), notebook connectivity (e.g., a power source, telephone communications tools), and miscellaneous toolkit necessities.

Scheduling Considerations

Information availability is an operations scheduling issue, although some organizations believe that all availability needs are covered by their business resumption practices. The security practitioners must be aware of the need to maintain operational schedules. If the backup and batch processing is scheduled to end at a precise time so that the online or remote transaction processing may start, then the credibility of the central staff to meet their commitments to the field are tested every day. Although capacity planning is not a security issue, the complete information protection plan will make sure that the topic is adequately addressed by the appropriate operational staff members.

Physical Considerations

Concerns associated with the desktop microcomputers in the corporate office also apply to notebooks for the mobile user. However, with respect to mobile computing, security practitioners may need to be more creative to achieve the desired results.

Theft-Prevention Devices. Such theft-prevention devices as cabling and bolting plates can be used to minimize the potential of notebook theft by opportunity. The cables are designed so that they may be looped through an opening in a stationary object to tie the laptop down while the user is traveling. Resistance to these devices exists because many users feel that having these devices gives the impression of not trusting coworkers or business associates. However, security administrators who use theft-prevention devices in their companies indicate that they have experienced a significant decrease in loss. Although the products are effective, corporate procedures with strong enforcement practices are usually required before these products are put into use.

Device Identification. Device identification is critical to the ability to identify a misplaced or stolen notebook. In addition to traditional identification methods (e.g., serial number registers, tags, labels, and engraving), microcomputers can be marked by using invisible ink to record the company's name and the notebook's serial number on the inside of the lid just under the monitor display area. The invisibly inked number must match the serial number recorded in the corporation's asset inventory register. This practice can also be used to resolve disputes associated with ownership of the microcomputer.

Mobile Uninterrupted Power Supply. Mobile uninterrupted power supply implies that each mobile user should have a portable surge protector with sufficient electrical outlets for each device that is connected to the microcomputer or notebook. Electricity follows the path of least resistance, and it will reach the microcomputer through any device cable if the power source for the device is not protected. Surge protector plugs are available at most hardware and electronics retailers. It is also recommended that the user carry a fully charged spare battery pack for the notebook. Usually the battery can be purchased from the dealer that sells the notebook.

Notebook Connectivity

To ensure the notebook's power source, the electrical wall connection should be used so that the notebook's battery can be conserved or recharged. For proper grounding, the electrical code requires that all computer male plugs be three-pronged. In some facilities, the female electrical wall receptacles may accept only two-pronged male plugs. In such cases, the problem can be averted by carrying a female/male converter plug that converts three-pronged plugs to two-pronged plugs and has a grounding wire that may be attached to the wall receptacle's holding screw. This type of converter is available in hardware stores. If the user travels internationally, the toolkit must also include an international voltage adapter that eliminates the need to carry different converter plugs.

Telephone Communications Tools

Offices and hotels are updating their telephone PBXs to digital service, but modems and the majority of the PBXs are still analog systems. To ensure connectivity in either environment, a converter should be purchased that can covert the phone line to analog at the modem connection. The complete converter kit should include alligator clips for phones that do not have RJ-11 jacks. The phone line converter requires an AC adapter as the power source; the full-functioning kit will have the capability to use a 9-volt battery when an electrical wall outlet is not available.

Data transmission may be interrupted if a phone system with call-waiting capability is used. The feature can be suspended during the data transmission call by adding *70 (occasionally # 70) at the beginning of the dial string. As a general rule, this is probably not required. However, if a transmission session is interrupted for an unknown reason, call waiting may be responsible.

A miscellaneous toolkit should include the following items:

- A small screwdriver with both a flat and a phillips head.
- An extra-long telephone cord with male RJ-11 connectors on both ends.
- A connector with two female RJ-11 receptacles.

CONTINUITY

Continuity is defined in this section as the processes of preventing, mitigating, and recovering from service disruption. The terms business resumption planning, disaster recovery planning, and contingency planning may also be used in this context; they concentrate on the recovery aspects of continuity that ensure availability of the computing platform and information when needed.

Recovery diskettes will reduce the user's lost time when access to information and remote computing resources is lost due to a major breakdown of the notebook. Recovery diskettes should be customized to the exact configuration of the notebook and should contain the following files:

- DOS system files (COMMAND.COM and the two hidden files), which make the recovery diskettes bootable.
- PARTNSAV.FIL, which contains the hard drive partition table, boot sector, and CMOS information.
- CONFIG.SYS, which contains the appropriate values for files, buffers, stacks, and hard disk drivers.
- REBUILD.COM, which restores the CMOS.
- SYS.COM, which enables the transfer of the operating system to another disk.
- Copies of any special drivers that are necessary to meet the standard operational configuration of the organization's protocol stack.
- Communication programs that support emergency downloading of files and programs.

Creating Mobile Backup

The fastest and easiest backup medium is a cassette tape; however, this represents an additional cost and bulky devices to transport. Most (if not all) mobile users want to travel as lightly as possible. The diskette then becomes the most acceptable medium, but care must be taken to minimize the number of diskettes needed.

The proper organization of the notebook's hard drive will minimize the time and number of diskettes required to create a backup. A directory called *data* can be created with all of the subdirectories necessary to easily organize, track, and access data unique to the user. Application software, data bases, or operating software should not be included in the data directory. These files should be stored in other appropriately established directories. When necessary, these software and data base files may be downloaded from the office server.

The backup diskettes should be kept in a different place than the notebook. If the notebook is lost, the backup diskettes will not be lost.

Loss of Computing Resources

Loss of computing resources as a result of the loss or theft of a notebook or its mechanical failure is the most difficult problem to deal with. The user is away from the office yet may need to repair or replace the notebook immediately, and canceling the next few days' ppointments is not an acceptable option.

The recovery process requires planning and discipline on the part of the notebook's user. The data files on the notebook should be backed up regularly. If any ingredient of the business resumption plan is missing or incomplete, lengthy delays are unavoidable.

Mechanical Failure. When the problem is a mechanical failure, the existence and awareness (by the user) of a national maintenance agreement with a rapid response clause can ensure fast repair. The remote user should be able to easily obtain the location of the nearest repair location. This can be handled by the organization's help desk service, which should provide 24-hour accessibility. After the notebook is repaired, the user must determine if any data or application programs were lost or damaged. Lost or damaged software may be replaced by using the emergency recovery diskettes to download the needed data from the office server. Although this may be a long transmission session, it is preferable to getting on an airplane and flying back to the office. After this is accomplished, the notebook owner can restore individual files from the backup diskettes. This process may consume a full day and may require the active participation of the remote user, but after it is done, the user's machine has been restored with the least amount of lost time.

Lost or Stolen Notebook. When the notebook is lost or stolen, the plan must provide rapid delivery of a new notebook to the user in the field. Spare notebooks with the standard operating software, application software, master files, and current data bases should be available at the data center. After a call is received to ship a backup machine, the only step necessary is to find the quickest method. Airlines and bus depots should be called to determine whether they provide shipment that is faster than that provided by the standard 24-hour service providers. After the shipment arrives, the next step is for the user to restore personal data from the separately stored backup diskettes.

If backup procedures for mobile users are to be effective, they should be tested and adjusted frequently. The recovery process may never be needed, and these procedures may be regarded as taking a lot of time. However, without such procedures, it may take the better part of a week just to obtain a replacement microcomputer. After that, it must be determined what applications and data bases must be loaded. Finally, if backup procedures are not enforced, how will the user's personal information be restored to a new notebook?

INTEGRITY

In this section, integrity is defined as the process of ensuring that the intended meaning of information is maintained. Information integrity is provided by allowing only authorized persons and processes to perform only those tasks that they are authorized to perform. Everything else is prohibited.

Software Considerations

Virus Protection. Remote users are no more or no less susceptible to viruses than their office-based counterparts. Therefore, they are likely to experience a virus infection of their computer within the next few years. Proper procedures can prevent the virus from attaching itself to any of the hard drive's files. A corporate contract should be purchased for one of the leading virus detection and eradication software products. The cost can be suprisingly reasonable, in some cases less than $15.00 per user per year, although some very good products may be even cheaper. The secret of success is to have the detection software active in memory at all times as a terminate and stay resident (TSR) program. Some vendors can relocate the TSR so that base memory is not used at all, while other TSR programs may take as little as 5K bytes of base memory. With the detection software active in memory, the computer's user may remain passive. Many organization's detection programs have limited success because they require the user to execute a scan program to check diskettes or hard drives after disconnecting from a bulletin board. During a normal day's workload, users may often overlook this program. If TSR software is used, however, the microcomputer should lock up when it detects a virus, and it should not allow the user to proceed. The eradication or cleaning program should be run immediately to remove the virus. The remote user should be provided with a system-bootable diskette that contains a virus detection and eradication program. If additional directions are required to remove the virus, the mobile user should contact the organization's help desk for assistance.

Notebook Configuration Integrity. Although the name *personal computer* may have been appropriate at one time, today it is a misnomer because the microcomputer has become such an integral part of information processing for the business community. By personalizing the notebook to the configuration of their choice, users may cause incompatibility with their organizations' requirements.

As part of the processing infrastructure, it is important that the information technologist control the protocol layers to ensure proper connectivity, memory management, and execution of the company's processes. This is not to say that mobile users are not allowed to install some software of their choice. An example of coexistence on the microcomputer would be to control the autoexec.bat by not allowing anyone except the systems administrator to modify the file. However, the last line of the autoexec.bat is a call to an autouser.bat that gives the user the ability to add activity to the booting process. Today, a number of client-based products exist that are designed to establish administrative control over the configuration and accessibility of directories and files on the hard drive of notebooks. These products support multiple-user confidentiality and several levels of administrators.

The content of certain directories on the hard drive should also be under the control of the systems administrator. Again, policies and guidelines should allow a section of the hard drive to be used at the discretion of the mobile user. Procedures and guidelines should clearly state the areas that may not be altered and the latitude the mobile user has to customize the microcomputer.

These same robust access control products provide complete control over the DOS computing environment. Access to printers, ports, disk drives, modems, files, and directories may be constructed so that several users of the same microcomputer may not have the same authorized use of these facilities. In addition, passive DES (or proprietary) encryption and decryption of files or directories may be established. The major vendors provide the ability to configure directories on the hard drive to be encrypted or decrypted on the fly. When a file is written to the designated directory, it is automatically encrypted. Conversely, when a file is read into memory by the authorized owner of the directory, it is automatically decrypted. This prevents unauthorized access to information even though several users may share the same microcomputer.

Nonrepudiation. In this section, nonrepudiation is defined as the process of ensuring that a user—either the originator or the recipient—can be identified as having engaged in a particular transaction. This facility may also be used during the normal course of business activity to identify the originator of information. Nonrepudiation involves the following procedures:

- Electronic identification of the sender is accomplished by digital signaturing. Depending on the nature of an organization's business, a choice of standards can be followed.
- Message hashing ensures that the content of the message is not altered. This is accomplished by cycle redundancy checks, which sum up a total value of the message's bits and stores that hash with the digital signature

in an encrypted envelope for the message. Two major hashing algorithm standards exist: MD5, which is supported by RSA and the Internet (ANSI X930 part 2 & RFC 1321), and secure hash standard, which is supported by NIST (FIPS 180 and 180-1).

- A copy of the message (the hash) with the digital signature of the originator is sent to the message archive. Each message must have an established retention date that will vary according to the message content. The records retention policy should serve as the guide for establishing the date. The message should be automatically removed from the archive data base when this date is reached.
- Proof that the message was delivered requires an electronic acknowledgment containing the date stamp of the activity to be sent to the archive and matched to the message.
- Proof that the message was opened requires an electronic acknowledgment containing the date stamp of the activity to be sent to the archive and matched to the message.
- When the importance of a transaction dictates that nonrepudiation is required, a utility should monitor the message activity to ensure that the message was received and opened. Electronic message status should be returned to the sender for appropriate follow-up as required (e.g., to determine why the message was not received or opened).
- The trustworthy information processing infrastructure must provide assurances that the message and audit details cannot be altered.

Many business uses exist for nonrepudiation in the mobile world (e.g., purchase orders, expense statements, strategic management directives, conflict of interest forms, and other important documents) that can provide technology an opportunity to reduce today's manual administrative efforts. However, nonrepudiation does not address confidentiality of the message; this is accomplished through encryption.

Remote Access Authentication. After the decision is made to allow modem, Internet, CDPD, or ISDN access to the infrastructure, the risk of unauthorized access to the infrastructure increases. Call forwarding and other advances in technology have eliminated the security effectiveness of dial-back modems. The Gartner Group considers dial-back modems to have limited effectiveness.

The most effective way known to authenticate a remote user is through two-phase authentication: with something that the remote user knows and something that the remote user possesses. Another means of authentication is through the use of biometrics, which includes voiceprint, fingerprint, or retinal scan. Currently, the cost and technical problems associated with remote biometrics scans render them impractical for common use.

When selecting one of the myriad products that support two-phase authentication, the following items should be added to the functional requirements list:

- It must provide the capability for centralized administration of access system controls (e.g., personal identification numbers, passwords, alarms, use analysis), while the actual authentication platforms may be decentrally deployed.
- It must function independently from the network infrastructure. The product should be independent of modem type, BAUD rate, or any other characteristics related to transmission of data. When changes are made to the network infrastructure, the product should not require modification.
- It must function independently from the hardware infrastructure. The product must function with all hardware platforms and operating systems. When changes are made to the hardware infrastructure, the product should not require modification.
- The product must function independently from all application software. Changes to the application should not dictate changes to the product. A one-time modification to the application software may be required to request the user's identification to the product.
- It must provide a random-number challenge (algorithm) to the product making the call that is in the possession of the caller. The challenge response (one-time password) must be unique for each authentication session. This ensures that the caller has the product in his or her possession each time a call is placed.
- The product must allow encrypted data to be processed. It is not a requirement for this product to perform the encryption.
- The product must accommodate caller mobility. The caller may need to call different processors or locations that are not part of the infrastructure. In addition, the caller may want to place calls from different devices (e.g., a different microcomputer in a different location); therefore, the authentication process must be capable of being relocated.
- The product must provide magnetic and printed reports of audit trail activity. The following data should be included in the audit log: date and time for all access attempts, the line on which the call entered, entry time, disconnect time, reason for disconnect, caller associated with the call, and system violations or other unusual occurrences.
- If the product in the caller's possession fails, a backup capability should exist that will grant the requester access to the infrastructure. The

backup process must be available at all times. The most economical way would be to place a call to the network help desk; the help desk will then grant one-time access on verbal authentication of the requester. This requires a process that will mitigate social engineering.

- The process device must provide controlled one-time access for some individuals (e.g., vendors or customers) that is granted by a remote authority. An example of this feature would be a one-time password generator that would relay the challenge and response over the phone. The central unit issues a random challenge number, the hand-held password generator calculates a response through an algorithm using the personal identification number of the requestor, and this unique response is compared to the central unit's response for that user. If the two responses agree, then access is granted.
- It must support the establishment of alternate dial authentication hot sites when the primary site goes offline for any reason.

CONFIDENTIALITY

Confidentiality in this section refers to the facilities by which information is protected against unauthorized reading. To facilitate establishing adequate levels of protection for information, data trustees (owners) must provide a classification to all information. This classification is based on the level of damage to the enterprise that may result from allowing individuals to gain access to information that they do not need.

Mobile Employee Information Security Recommendations

Ensuring compliance with appropriate security procedures and practices depends not on the security tools that are provided but on the effectiveness of the security awareness program. Awareness contributes to the success of the effective information protection program. Exhibit 7-4-2 provides a baseline of awareness requirements for the mobile user. Posters in the organization's facilities are very effective and are an important part of the overall awareness program. However, use of E-mail and articles in the organization's internal communication media are more effective than posters with mobile users.

Software Considerations

Version Management. If the application software or data base on the notebook is not the current version, the mobile user may create and transmit incorrect data into the organization's record of reference data bases. If this

1. Be aware of the surroundings.
2. Make portable devices inconspicuous.
3. Shred confidential documents before discarding them.
4. Lock portable devices out of sight when leaving them unattended.
5. Hide physical security tokens; do not carry them in the same case with the notebook.
6. Establish a regular schedule for performing appropriated backup practices.
7. Select nontrivial passwords.
8. Change passwords frequently.
9. Follow the common-sense rule to question whether you are in an appropriate place and time to be working with the company's information assets.
10. Most important of all: treat sensitive company information as if it were the combination to your personal safe.

Exhibit 7-4-2. Safe Practices for Mobile Users

happens, the integrity of the data bases may be damaged and inaccurate information may be given to a business partner.

Each time the mobile user connects into the organization's network servers, a process should be performed to ensure that all of the software (both application and operating system) on the mobile client is current. The same is true for all data base subsets that are mirrored on the notebook. The organizations change control process should notify the synchronization process when production environment changes are made and should have the updates available when a remote connection is made. If the notebook does not have all the current software and data base data, the infrastructure servers must not accept information uploads until the software and data base are synchronized. The mobile user must recreate the information before attempting to upload using the current versions of the processing environment.

When the mobile user connects into the infrastructure and determines that a download of updates is required, the user should have an option to delay the download. The user is not allowed to upload information to the server, but queries may be made. This is important if the user is with a customer and wants to obtain status information; if a potentially lengthy download takes place, the user would waste the customer's time.

Encryption and Decryption. To date, encryption is the most effective security measure to ensure information confidentiality. One type of technology uses a two-part key in which the private key is kept by the owner and the public key is published. The recipient's public key is used to encrypt the data, which can only be decrypted by the recipient's private key. To reduce the computational overhead, encryption is often used to create a digital envelope that holds a DES encryption (symmetric) key and DES-encrypted data. Message nonrepudiation uses document hashing and digital signature as a means of verifying the message sender. This is accomplished by encrypting a message with the sender's private key and letting others decrypt the message with the sender's public key.

The major security concern is maintaining integrity and confidentiality of the keys. Each organization must devise a process to distribute the public keys to everyone who is involved in the encrypted messaging process, including customers and other business partners. The recommendation is to establish a comprehensive public-key data base on a central server that may be accessed by everyone (this means that it is located outside the security firewall) and to have each mobile user keep a subset of public keys on his or her notebook for major business partners.

Another concern is protecting corporate equity. Consideration must be given to the necessity for the corporation to decrypt messages when the owner of the private key is not available. One method may be to include the symmetric DES key (discussed in the previous encryption section) in an extractable format in the message archiving facility (discussed in the nonrepudiation segment). A tightly controlled process to extract the DES key would allow the message to be decrypted without compromising the private key of the originator. Right-to-privacy concerns are outweighed by corporate equity considerations, because company resources were used to create the messages.

NEW TECHNOLOGY CONSIDERATIONS

It is important to have an appreciation for new connectivity technologies so that users may determine their potential threats and vulnerabilities. By looking at what is coming, users should be able to develop the mitigating security measures before deploying the technology. Many security concerns exist, but very few proven answers are associated with emerging technologies. However, the technologist (and to some degree the mobile user) often wants to implement the technology quickly, before the technology itself has reached commercial strength.

PC Card. The unified standard that combines Personal Computer Memory Card International Association (PCMCIA) standard and the Japan Electronic Industry Development Association (JEIDA) standard is called PC Card. The credit-card-sized devices take the form of memory cards, modems, and disk drives that can be plugged into slots in computers. The card's security measures should be the same as those for the hard drive. Experience indicates that many users do not take the card out of the drive when it is not in use; therefore, passive encryption is recommended. For this reason, cards may be ineffective if used as a removable security lock. The card is effectively used in several applications, most notably as a removable modem.

Smart Cards. Although they are the size of an ordinary credit card, smart cards use an embedded processor that gives both the system designer and the system user a powerful authentication tool. Smart cards are a subset of the rapidly growing integrated circuit card industry.

Two types of smart cards exist: contact cards and contactless cards. The contact-type interface uses an eight-position contact located at one corner of the card. A contact card reader also uses a matching set of contact points to transfer information between the card and the reader. The contactless card does not come in direct contact with the card reader, but uses an inductive power coil and transmit and receive capacitor plates to transfer information to the contactless reader. AT&T's contactless card product is essentially an 8-bit computer with a proprietary operating system and either 3K bytes or 8K bytes of user-accessible, nonvolatile memory inside the smart card.

Many applications take advantage of the strong authentication capabilities of smart cards. The most common application is electronic money. Another example of smart-card flexibility is in a building security system protected by card access that otherwise requires a large network of cables connecting the door reader, door controllers, and host computer. By converting systems to a smart-card system and using the onboard data base and encryption capabilities, the miles of network cabling and host computer may be eliminated. However, the need for a special card reader coupled with the mobile user's desire to travel light all but eliminates the smart-card as a practical security mechanism in today's mobile arena.

Cellular Digital Packet Data. Cellular digital packet data (CDPD) technology is rapidly building its infrastructure right alongside the traditional analog cellular infrastructure. In fact, much of the existing analog base stations are being used. Similar to most public communications efforts, a coalition of common carriers have cooperated to ensure interoperability. CDPD was developed by a group of major cellular communications companies. This prestigious group makes it clear that by leveraging existing technologies and infrastructures, they will have a nationwide network available within a very short time.

As the caller moves from one analog cell to the next, the ability to transmit data digitally means faster transmission speeds and a solution to the current problem of lost or repeated transmission. The user will be able to send and receive data while in a moving car.

Technologists believe that because existing common protocols are used, today's applications can use CDPD without modification. The one drawback that may impede CDPD deployment is the development of effective modems that support CDPD.

Three types of CDPD services are provided. CDPD Network allows subscribers to transfer data through their applications as an extension of their internal network. In CDPD Networked Applications Services, the cellular provider provides specific application services like E-mail, directory services, and virtual terminal services to subscribers. CDPD Network Support Services provides network management, use accounting, and network security. The security practitioner should not assume that cellular carrier's security interpretation or objectives are the same as those of the organization. The carrier's direction is intended to protect its investment in the CDPD Network and principally to ensure that only authorized paying subscribers use the network. A secondary concern is providing data privacy. It is not practical for the carrier to comply with each subscriber's security policies.

The user should already be aware of the security risk associated with cellular transmissions. Cellular (analog or digital) transmission is a miniature radio station broadcasting to everyone who has the receiving equipment. Digital transmission will be able to scramble the transmission by channel hopping, which makes interception more difficult but not impossible for the motivated eavesdropper. Therefore, the best solution to maintain data integrity and confidentiality is through digital signaturing and encryption. Because of the administrative overhead associated with key management of symmetric keys (the same passphrase is used to encrypt and decrypt the message), public and private key encryption is recommended.

Wireless Communication. Although cellular communication is a desirable tool for mobile users, wireless communication may be a valuable capability for those roving from place to place within the confines of their own building. Wireless networking employs a number of different methods. One of the most popular of these methods is spread-spectrum radio. Wireless adapters connect into a computer either internally or through the parallel port, and they then communicate to a base-station or what could be called a wireless hub. The communication receiving area of the base station is usually several hundred feet, and the microcomputer and adapter may be placed anywhere within that radius. Other methods use a line-of-sight transmission technology.

Because the transmission takes place within the confines of the company's buildings, the security requirements may be fewer. However, depending on the organization's type of business and the need for confidentiality, consideration might be required to neutralize the motivated eavesdroppers who may be within the company or stationed outside the facility.

26%	Sales
23%	Field Services
18%	Administrative
15%	Field Engineering
10%	Senior Management
8%	Other

1997 Projections by: The Yankee Group

Exhibit 7-4-3. Breakdown of Mobile Work Force

Other technologies use radio frequencies to forward messages to a central station, which in turn (when requested) sends the message to the recipient. This method is known as store and forward, and it is not normally used for interactive messaging. The best answer to ensure confidentiality in all wireless applications is encryption.

The World Wide Web. The World Wide Web (known as the Web or WWW) provides an infrastructure for accessing information. The Web provides a simple means of attaining almost any type of information that is available on a Web server that is attached to the Internet. All information on the Web is stored in pages, using a standardized hypertext language. Many companies use the Web to provide timely information to their customers. They typically provide information about products, upgrades and patches, and feedback areas. The Web is considered a companion to E-mail because it provides information by using an interface. However, the Web is designed to accommodate limited two-way communication.

Security considerations must be the same as with all other Internet connections. The Web server should be placed outside of the firewall. If confidential information is placed on the Web, then encryption should be used. In general, the Web architecture does not provide for the integrity and confidentiality of information. Access to the infrastructure should not be allowed through the Web server.

SUMMARY

Security procedures, guidelines, and practices accepted by end users must enable them to do their jobs. If end users interpret security to be a roadblock, they will often find ways to circumvent security requirements. To ensure that this does not happen, the security practitioner should spend time learning the problems and security concerns of users. The practitioner should consider scheduling one day per month to stay at home and telecommute in addition to dialing in while on business trips. This practice enhances understanding of the remote access conditions, and assists the security practitioner in developing more effective security practices. Exhibit 7-4-3, published by the Yankee Group, provides a breakdown of the professions that make up the mobile community.

Information today is stored not only in the data center but also on desktops, notebooks, and home computers; it is stored wherever mobile users have taken the data. By understanding the implications of this fact of business life, practices can be better established to secure assets while supporting employees' requirements to perform optimally and competitively through having access to the most current information and computing power.

7-5
Managing a Telecommuting Program

Sheila M. Jacobs
Mary Van Sell

TELECOMMUTING enables office employees to work effectively in nontraditional settings. Telecommuters work in remote locations, such as their homes or neighborhood satellite offices, one or more days a week. Using personal computers, these employees can link to their companies' computer systems via telecommunications lines.

The IS manager may be involved with telecommuting in the organization in three major ways:

- Helping functional area managers establish and manage telecommuting in their departments.
- Deciding which products or technology the organization should use for telecommuting arrangements.
- Establishing and managing telecommuting in the IS department.

SOCIETAL BENEFITS OF TELECOMMUTING

For millions of people in America and other high-technology nations, the work day is bracketed by stressful rush-hour commutes over clogged and often decaying highways or railways to congested urban centers where the costs of parking and office space rise continually. Mass transportation systems, where they are available, are overcrowded and often unpleasant.

The efforts of many nations to ease traffic congestion and air pollution, or to curb spending on public roadways, have included the promotion of telecommuting programs. For example, four southern California counties began requiring companies with more than 100 employees at one location to develop plans for cutting commuter traffic. Similar laws exist in other states, including Arizona, Hawaii, Texas, and Washington.

If telecommuting could replace 10 to 20% of US road trips, it could save as much as $23 billion per year in energy, transportation, and environmental costs. A telecommuting program can also help a company lower real estate costs in urban areas by allowing employees to share a smaller office space by spending different days in the main office.

Telecommuting also facilitates employment for workers who have physical difficulty getting to an office, including employees who are disabled, recovering from illness or injury, or are on maternity leave.

Why Employees Want to Telecommute

Most telecommuters are successful professionals who want more from life than traditional corporate mobility. Employees who request telecommuting often do so because telecommuting enables them to combine work with another valued goal. This goal may be related to personal enrichment, such as travel or enrollment in a graduate degree program, or balancing work and family commitments more effectively. Telecommuters value the flexibility to attend their children's special needs (e.g., medical appointments, school visits) during normal work hours. Telecommuting gives employees the freedom to spend more time with their families.

The Benefits of Autonomy. Most telecommuters feel that working at home, rather than in the office, enhances their ability to concentrate on their work because there are fewer distractions. Other motivations for telecommuters are autonomy and control over their time. For example, telecommuters in IBM's nine-month telecommuting pilot program said that telecommuting had given them time to experiment, as well as the ability to increase job turnaround time, by scheduling their work at faster off-shift times.

Many telecommuters report that their primary motivation for working at home is that they can get more work done. The reason for this is not apparent. It may be that successful telecommuters are individuals with "night owl" circadian rhythms who work after normal office hours when they are most energetic.

ORGANIZATIONAL BENEFITS OF TELECOMMUTING

Organizations, whether or not they allow telecommuting, want to minimize their labor and operating costs. At the same time, they want to increase levels of productivity. Having a telecommuting program makes it easier for an organization to attract top-quality employees who may not wish to move or commute. Telecommuting also helps reduce labor costs by lowering the rates of absenteeism and employee turnover.

Increased Productivity

Telecommuters are not only more productive at home than they are in the office; they are also more productive than their in-house counterparts—by approximately 30%. Telecommuters are generally more effective because they:

- Work at times of day when they are most productive.
- Are more likely to finish projects ahead of schedule.
- Work for longer periods of time without interruptions.
- Experience improved communication with the work group.
- Are more available for consultation with clients and supervisors at home by phone than when in the office.
- Are more creative because they concentrate better at home.

Reduced Costs

Although 30% of cost savings attributed to telecommuting are due to decreased labor costs, 70% of cost savings result from reduced overhead expenses. Organizations save between $1,500 and $6,000 per telecommuting employee on reduced office space and related overhead expenses. Telecommuting also lowers the costs of using and maintaining equipment, such as mainframe computers that can be used by telecommuters at off-peak hours.

Telecommuter Job Satisfaction

Several studies have found that the majority of telecommuters are satisfied with their arrangement and prefer telecommuting to working full time in the office. Satisfied telecommuters also report:

- Higher self-rated productivity.
- Satisfaction with the performance appraisal system for telecommuters.
- Technical and emotional support from managers.
- A lack of family disruptions.
- Greater loyalty to the organization as a result of being trusted by managers.

Predictions that telecommuting employees would become isolated from the organization and overlooked for promotions have not been supported. Telecommuters are usually more visible in their companies because they are almost always more productive than office workers. Some telecommuting employees report a feeling of missing daily social and professional interaction. The establishing of a neighborhood shared workspace, referred to as a "telecenter" or a satellite work center, can alleviate this problem.

PREPARING THE ORGANIZATION FOR A TELECOMMUTING PROGRAM

Not all companies are good candidates for telecommuting programs. The structure of the organization must be considered. Companies with little work autonomy that use time-based methods of work supervision, and whose decision-making processes are centralized and hierarchical, would find it virtually impossible to supervise the work of telecommuters without changing all these aspects of organizational structure.

If the company is flexible and results-oriented, then the organizational structure of the company is suitable for telecommuting. However, although most companies have more volunteers for telecommuting than they can use in a pilot program, there may still be resistance to telecommuting from managers who fear losing control of their employees or who do not trust employees. Telecommuting presupposes supervisors managing employees by results, communicating expected outcomes clearly, and controlling output and quality rather than time spent and processes used.

Jobs Suitable for Telecommuting

Jobs tasks that are suitable for telecommuting have been classified as output tasks. Output tasks typically produce discrete pieces of work, such as project reports produced by one person working alone. Jobs that consist of information processing, that result in measurable output, and that do not involve physical contact are candidates for telecommuting programs. Telecommuting enhances productivity in jobs requiring creativity and analysis, where the need to interact with others is not critical.

Many categories of jobs are suitable for telecommuting programs. Several of these jobs are in the information systems field, such as:

- Computer programmer.
- Software engineer.
- Computer systems analyst.
- Technical writer.
- Data-entry clerk.
- Consultant.
- Technical supporter.
- Word processor.

Outside the IS department, the range of jobs that lend themselves to telecommuting includes:

- Translator.
- Sales representative.

- News reporter.
- Public relations professional.
- Stockbroker.
- Lawyer.
- Accountant.
- Engineer.
- Architect.
- Real estate agent.
- Travel agent.
- Writer.
- Insurance agent.
- Purchasing agent.
- Claims processor.
- Marketing manager.
- Customer service representative.

Successful telecommuters tend to be technically skilled and usually have substantial professional experience. Telecommuters often have to perform tasks that would be done for them by others in the office (e.g., quality testing, job completion time estimation), so they need a variety and depth of skills. Because telecommuters will have limited face-to-face contact with their supervisors and coworkers, good communication and organization skills are important. Telecommuters should also be adept at scheduling, preparing, and documenting their work.

Technology Requirements

An organization does not need a large capital investment in equipment to initiate and support a telecommuting program. An employee can telecommute with a personal computer, a modem, and the appropriate software. Many prospective telecommuters already own these products.

Computer Equipment. Personal computers are the foundation of telecommuting. Laptop computers, portable computers that can send faxes, and client/server computing equipment are critical. Modems and printers are important, and fax/modems are desirable for additional flexibility.

LANs. Remote access to the company's LAN is important. Some telecommuters need remote access to a LAN only for short time periods (e.g., for E-mail); others need their home computers to behave as nodes on the network. Remote software packages facilitate this arrangement. Telecommunications technology and services make it possible to network home computers to office computers.

Sophisticated Phone Systems. A customer or a coworker should be able to reach a telecommuter at home or at the office by dialing the same phone number. Today's phone systems can make the physical location of the telecommuter irrelevant. The phone technology in the telecommuter's home (or satellite office) should include the same features found at the office, such as speed dialing, redialing, return dialing, caller ID, priority call, and select forward. Interactive voice services, which provide telephone recordings such as "If calling for billing information, press 1," may also be important.

Voice Mail. Voice mail, like E-mail, enables a telecommuter to remain informed about office activities and to communicate with the office effectively. Voice mail messages can be recorded and retrieved at any time of day, so telecommuters on flexible schedules are not disadvantaged.

Videoconferencing. Videoconferencing allows people at geographically separate locations to have face-to-face meetings. As the cost of videoconferencing drops, remote conferences will increase. A telecommuter may be able to participate in a videoconference with the company office by going to a telecenter or satellite office that has video equipment or by going to a videoconferencing facility at another location, such as a phone company office that provides this service. Desktop videophones, with small screens, are also available and may be placed in the homes of telecommuters.

Other Products. Other products, such as fax machines and cellular phones, can facilitate telecommuting programs. An electronic imaging systems is another useful product. With this equipment, an employee at the office can scan a document onto the main computer. The telecommuter can then access this document using a home computer and a modem.

Future Needs. Some of the future needs for products and services to support telecommuting include:

- Equipment that is light, compact, and easily portable, such as portable fax machines and printers weighing less than five pounds.
- Better paging and remote access technologies.
- Client/server operating environments that use applications programming interfaces to support remote access technologies.
- Remote access technologies with built-in security controls.

IMPLEMENTING A TELECOMMUTING PROGRAM

A company's first telecommuting program should be a pilot program, lasting six to eighteen months. The pilot program will help the company determine needed training and equipment, program costs and benefits, and program management. At the beginning and end of the pilot program, levels of productivity and other outcomes, such as morale and overhead expenses, should be formally measured.

When the actual program is launched, mandatory core times should be established. These are times when the telecommuter is to be available by phone or E-mail to customers or supervisors. Usually, telecommuting employees have scheduled days in the office to keep them in touch with their co-workers, managers, and the day-to-day affairs of the company, and to prevent feelings of isolation.

Telecenters

There is a growing trend to establish neighborhood telecenters as alternatives to employees working at home. A telecenter is somewhat like a branch office, except that its location is convenient for employees rather than for customers. Employees from a variety of departments, or even from a variety of companies, may work together in one telecenter. The telecenter may be set up by the company or by an independent organization that rents space to several companies.

Telecenters can give a telecommuting employee a structured work environment while still saving on transportation, real estate, and overhead costs. A telecommuting employee can work at a convenient location without feeling isolated and without missing social interaction. Employees who do not have room at home to set up office equipment can still telecommute. A telecenter is advantageous economically because telecommuting employees share equipment. The telecenter may also serve as a pilot program before the company fully invests in telecommuting.

Managing Telecommuters

Although careful preparation for a telecommuting program is essential, the ultimate success of the program depends on the way it is carried out by the telecommuting employees and their managers. Managers and telecommuting employees should agree on what work is expected, how it should look, and when it should be completed. A written agreement, signed by both parties, may be helpful. The most common reason for failure of a telecommuting program is inadequate communication between managers and employees.

Managers of telecommuting employees need training in techniques of results-based management, job analysis, work specification, and performance appraisal. They may also need training in communications skills.

IS managers must know how to prepare organizations for the changes induced by telecommuting. They should help functional managers establish telecommuting in their departments, and determine the technology requirements for telecommuting. Finally, IS managers should know how to implement telecommuting programs, meeting the information and communications needs of the telecommuters while monitoring and controlling telecommunications costs for the company.

SUMMARY

Tips for implementing a successful telecommuting program include:

- Start with a pilot program.
- Decide in advance what the desired outcomes are (e.g., increased productivity, reduced turnover) and how to formally measure them.
- Select employees who are motivated and qualified to work independently.
- Consider both the personality of the employee and the needs of the company when deciding how many days per week the employee can telecommute. Decide each case on an individual basis.
- Make sure the telecommuters can be reached by those working in the main office.
- Schedule regular office visits for the telecommuters.
- Be sure the supervisors of telecommuters have (or learn) management skills that focus on results and communication.
- Give telecommuters regular feedback on quality control.
- Try to avoid impromptu office meetings; telecommuters may feel excluded.
- Do not assign telecommuters tasks that could cause delays or backlogs for in-office workers.

ACKNOWLEDGEMENT

This research was partially supported by a 1994 School of Business Administration Spring/Summer Research Grant.

Section 8
Implementation and Case Studies

THE COMMUNICATIONS SYSTEMS MANAGER'S APPROACH to technology procurement and implementation often depends heavily on corporate culture. There are three major organizational cultures that influence the manager's approach to expansion of an existing system or deployment of a new one. First, there are the innovative, leading-edge, early adopters—institutions where performance is much more important than price. Universities, medical centers, the military, and governments are examples of this group. A subset of this group includes community-centered applications such as public safety, utilities, and libraries. The technology being considered allows the organization to do something that could not be done previously or provides a significant improvement in performance or efficiency. The new technology might permit a rapid emergency remote diagnosis that can save lives, provide a tactical or strategic advantage on the battlefield, or provide a better way to collect taxes. Developing and experimental technologies are often sponsored by universities.

The second organizational culture may be called conservative risk takers. These organizations will deploy a new technology only after it has demonstrated an out-of-the-box operational capability. These buyers do not expect to have swarms of engineers and programmers on their premises trying to support new technology products. This segment of the market is much more price sensitive than the institutional segment. These tend to be high-volume users. New technology generally penetrates this segment first with high-technology companies that want to be on the leading edge. Next will come manufacturing, financial, and service industry users. Finally, the technology will begin to penetrate the small business and retail markets.

The third cultural segment of the market is the consumer market. Here, price is all important. The technology must be the basis for a product that does something useful, entertaining, and uncomplicated. These are nonsophisticated users who do not care about the elegance of the technology. They care only about price and function. The applications are in support of personal communication, entertainment, or service access.

All successful technologies seem to pass through these various cultures as they move from introduction as a unique new widget to recognition as a common everyday gadget. Which stage a product is in impacts the implementing manager's approach. The manager, for example, may be interested in the availability and widespread deployment of the product based on the new technology. In the early, leading-edge phase, deployment will be spotty and perhaps unavailable when or where needed. If, however, the technology is being used by more mainstream enterprises, the manager is likely to find products easily available in most business locations. If consumers are using the technology, deployment will be virtually universal.

Other implementation issues are also affected by the current target market cultures of a product. The early adopters, for example, recognize that there will be open technical issues, that things may not always work correctly, that solutions may be proprietary, that costs will be high, and that product stability will be low. Enterprise users expect that technical difficulties will be minor, that approaches will be industry standard, that cost will be moderate and justifiable, and that reliability will be acceptable. The most demanding market is the consumer. In this culture, the technology should be completely invisible to the user, the approach should be universally standard, the price for perceived value equation must be favorable, and product reliability must be unquestionable.

Chapter 8-1, "Acquiring Systems for Multivendor Environments," explores the critical issues surrounding the operation in a heterogeneous equipment and software environment. Many factors are involved in a decision to move away from a single vendor position. These factors are discussed along with the possible tradeoffs that may be involved. The chapter also includes useful tips and checklists on evaluating and selecting vendors.

Before final selection of a vendor, managers should review Chapter 8-2, "Evaluating Vendor Support Policies." This chapter is a very useful guide for asking potential vendors pertinent questions about their after-sale support. A sample schedule is included with a recommendation that it be part of any request for a quotation.

Chapter 8-3, "WAN Network Integration: A Case Study," addresses a situation that is becoming more common—the integration of two distinct wide area networks made necessary because one organization is merged into another. This chapter contains useful lessons for any manager faced with such a challenge. The reason for integration as well as the benefits and difficulties are detailed. Readers will follow the process step-by-step from inception to a successful conclusion.

The reason for the success of frame relay as a wide area networking solution becomes evident in the case study offered as Chapter 8-4, "Frame Relay in an IBM Environment." This chapter explains why leased-line costs were an impetus to a new approach, discusses why frame relay was chosen, and details how vendors were selected and the project managed to a successful conclusion.

An important aspect of the communications manager's job is monitoring and controlling. "Network Controls," Chapter 8-5, highlights those local and wide area network functions that have the most impact on meeting system requirements. These functions are the most likely candidates for careful monitoring. The chapter offers techniques to control physical, environmental, software, administrative, and operations aspects of the network.

This section ends with a case study on implementing remote access to a campus network. The specific problem addressed is how to accomplish the task without heavy expenditures for new equipment and personnel at a new remote location. How this was accomplished is the story told in Chapter 8-6, "Remote LAN/WAN Connections: A Case Study."

8-1 Acquiring Systems for Multivendor Environments

Thomas Fleishman

IS AND COMMUNICATIONS SYSTEMS MANAGERS may not always initiate the decision to purchase hardware or software packages, but they are often responsible for selecting, implementing, and operating the associated peripherals and such supporting equipment as communications systems. Selection is often complicated, however, by third-party vendors whose products are not compatible with the CPU manufacturer's hardware.

Many factors, particularly lower cost, influence the decision to establish a multivendor systems environment; the price of equipment from third-party vendors is generally competitive and equipment performance is at least equivalent to, if not better than, that of the hardware available from the CPU manufacturer.

Managers must recognize that the decision to install equipment from multiple vendors involves additional administrative responsibilities (e.g., special policies and procedures). This chapter addresses the critical issues in multivendor installations.

SELECTION FACTORS

To determine whether the acquisition of equipment from a vendor other than the mainframe vendor is appropriate, managers should weigh the following technical and financial factors:

- *Availability.* If a CPU supplier does not have a particular component required for a specific application, equipment from another vendor may be the only alternative.
- *Performance.* Only one vendor's product may be capable of meeting specific performance criteria for an installation.
- *Reliability.* With the proliferation of both equipment and vendors, man-

agers have many alternatives. A specified degree of reliability (i.e., a defined mean time between component failures) may dictate equipment purchased from a different vendor.

- *Cost.* Equipment from one vendor may cost less than the equivalent components offered by another. Cost is the most difficult issue to address because quality may be sacrificed for savings.
- *Equipment use and location.* Managers must consider the environment in which the components will be installed. Certain equipment is manufactured to meet more stringent environmental conditions; if such conditions exist, the manager must acquire products that can operate in those conditions.
- *Ergonomics.* Ergonomics is particularly important when terminals and microcomputers are widely used. The size, screen, keyboard, and other physical attributes of the equipment should all be weighed in the acquisition decision.

TRADE-OFFS

Managers must consider whether equipment from a third-party vendor meets the performance levels established for equipment currently in the department and assess the risks of installing equipment that does not meet this level. For example, replacing disk drives in an online environment entails a significantly higher risk than exchanging tape or cartridge units. Managers may not be willing—or even able—to accept the risk of a new vendor's equipment not performing at required levels. The cost savings must be enough to offset the risks and the potential expense of remedial backup measures.

Fitting the Current Configuration

Equipment supplied by a new vendor must function in the installation's current configuration. For example, in one department that acquired a new line printer, preinstallation testing showed that the printer could not handle all the special and multipart forms required by users. The printer worked well on some forms but not on others. In other instances, the performance of disk drives was outstanding in installations that had minimal pack-mounting requirements, but fell below acceptable levels in operations that required a great deal of pack mounting and demounting.

Quality of Service

Managers must also investigate a vendor's ability to provide and sustain acceptable levels of service. Before making such an assessment, the manager should establish internal tracking criteria based on the requirements of the installation. These criteria might include:

- *Response time.* What is the average time that a vendor requires to respond to a service call? Is the vendor willing to commit to such response through a contract?
- *Repair time.* What is the average repair time for a given component?
- *Resource availability.* Does the vendor provide resources (e.g., technical specialists) in a timely manner?
- *Incidence recurrence.* If a specific component develops problems, is it fixed on the first service call or is it necessary to recall the customer engineer?
- *Dispatch location.* Where is the service location from which the repair resource is dispatched? Is the parts depot in the same location? If not, where is it located?
- *Management attention.* In cases of serious or recurring problems, does the vendor's management get involved?
- *Escalation procedures.* Are established escalation procedures in place in case of extended problem resolution?

Customer Engineering

In a multivendor installation, the level and quality of support from the manufacturer largely determines the success or failure of the installation in terms of equipment performance. Managers should initially establish whether the manufacturer can provide:

- *On-site support.* Would a vendor's customer engineers be on site or dispatched from a central service depot?
- *Continuity.* Would the same customer engineers provide support on every occasion or would customer engineers be randomly assigned from a branch or regional pool?
- *Experience.* Are the supporting customer engineers experienced with the type of installation and configuration in the manager's department?
- *Customer engineer support organization.* Are the customer engineers employed by the equipment manufacturer or employed by a third-party maintenance organization contracted to provide service support?
- *Parts.* Are parts stored locally in case of hardware failures that require part exchanges? Or could there be major shipping delays in procuring the necessary items from distant warehouses?

Managers must address these issues before acquiring third-party equipment. Clearly stated expectations, policies, provisions, and if possible, contractual guarantees should be established with the vendor to minimize problems or to allow the organization to withdraw from the acquisition agreement without financial penalties.

For example, one vendor acknowledged that the usual repair support provided for purchased equipment was minimal and that timely response could not and would not be guaranteed. The vendor did, however, offer an alternative maintenance service contract at additional cost, guaranteeing the arrival of customer engineers at an installation within four hours. In this case, the manager was sufficiently impressed by the vendor's equipment to buy it and pay the higher maintenance charges to obtain guaranteed service.

In another case, the manager found that a third-party service organization had been contracted to support a piece of hardware. Although the vendor agreed to certain service requirements, the service organization actually performing the maintenance service was not committed to providing the necessary support. Only after lengthy negotiations and considerable expense was the manager able to obtain the appropriate service agreements.

Vendor's Financial Conditions

Managers must evaluate the financial condition of prospective vendors, deciding whether a vendor has the resources to provide support and service for the life of its equipment. If the acquisition is sufficiently large in terms of price or the equipment being considered is sufficiently strategic, the organization should consider reviewing a Dun & Bradstreet analysis of the vendor before entering any contractual commitments.

Contractual Criteria

The contracting organization might want to specify certain performance and service criteria in the acquisition contract (e.g., service response time and availability schedule). Financial penalties or component replacement may be defined for failure to meet contractual requirements. All contractual demands should be negotiated by legal representatives for the vendor and the organization. The degree of cooperation shown by the vendor during negotiations usually indicates what the manager can expect after equipment installation.

AN EYE ON THE VENDOR

Managers must establish objectives for equipment performance and service support before evaluating a vendor's products. Establishing benchmarks and tracking component performance are two generally accepted methods of evaluation.

Benchmarks

Before equipment is acquired, the systems department staff should become familiar with the vendor's hardware and its specifications to establish performance benchmarks. Benchmarks can help determine whether the equipment meets its advertised capabilities and whether the vendor complies with the requirements in the installation agreement.

Performance Tracking

An organizational unit must be established to track the performance of hardware components. This unit can establish a comparative rating system to help the manager determine which vendors provide satisfactory service. Tracking can be performed manually (e.g., for operations personnel file incident or trouble reports, which are transmitted to an employee who compiles the data and issues periodic reports) or through an automated tracking and reporting system. Because manual tracking is cumbersome, inefficient, and expensive, especially in large departments, automated tracking and reporting systems are preferred. These packages can be developed or acquired, depending on the capabilities of the staff and the requirements of a particular installation.

Studies conducted in large installations indicate that component tracking can deliver a significant payoff. Besides being a straightforward way to rank vendors, component tracking is a proven operations aid. Because a faulty disk, tape drive, or channel may cause failures in other components or systems degradations, the ability to quickly detect a faulty device is a valuable aid in maintaining appropriate service levels.

The following scenario illustrates the usefulness of component tracking. An automated tracking system provides a daily report of all failures (i.e., hardware and software) of each component in an IS department. The vendor of the tracking package provides a monthly comprehensive data base compiled from data collected from all installations that subscribe to the service. The data base indicates an installation's performance relative to the other organizations using the same configuration and brand of equipment. For example, if an organization has a CPU A with disk drive B installed, and 12 other organizations use the same CPU and disk configuration, the monthly report shows a performance ranking for all 13 subscribers. For managers with multivendor environments, such systems are almost mandatory, especially in large, multiple CPU installations.

Vendors' Mutual Interest

Problems that involve more than one vendor often occur because vendors compete in markets that discourage communications between vendors. To maintain smooth operations in a multivendor environment, managers or members of their operations staff must coordinate communication between vendors. This communication can be encouraged through frequent meetings that include representatives from all vendors. Even when there are no specific problems, these meetings help open channels of communication between the vendors whose products are installed in a particular department.

CASE STUDY: RECOGNIZING RESPONSIBILITY

Under pressure to reduce equipment costs, a manager decided to reduce costs by replacing 24 disk spindles acquired from the manufacturer of the facility's CPU. The manager contacted several disk drive vendors and leasing companies, and after a period of evaluation, chose to replace the 24 spindles, saving, in theory, approximately $73,000. The spindles were acquired on a three-year lease. Shortly after the replacement, it became evident that the acquisition of the new devices was a mistake.

Although the disk manufacturer had an impressive facility and was supplying thousands of units worldwide, the product was relatively new in large-scale, 24-hour-a-day, online business environments. Because the support organization was poorly trained, with no prior exposure to the CPU manufacturer's equipment, diagnosing problems was extremely time-consuming. In addition, CPU-device interfaces were not well understood; therefore, communication between the disk and CPU vendors was nearly impossible. No repair personnel were permanently assigned to the data center's account and trouble calls became exasperating as clusters of inexperienced and poorly trained staff tried to resolve problems, while major online systems remained unavailable to users, resulting in losses of thousands of dollars for every minute the systems were down.

After several months, the equipment appeared to stabilize, but the department began to experience serious channel interface problems. Much of the manager's time was spent mediating between the CPU and disk vendors, neither of whom was willing to accept responsibility for the problem.

As the interface problems were resolved, the incidence of head crashes increased dramatically. It took approximately four months to trace the problem to the foam-rubber-seal stripping insulating the spindle door. The frequent mounting and demounting of packs hastened the deterioration and disintegration of the stripping, eventually contaminating the packs and causing head crashes. After analyzing this situation, the manager decided to absorb the penalty in the lease contract and replaced the disks at a cost of $27,000. (The penalty had been agreed to during the contract negotiations as a contingency escape arrangement.)

The disk manufacturer was only partly at fault for marketing a device that was not fully tested and for not planning and staffing a support organization of trained personnel for servicing the product. The manager was negligent in failing to perform an in-depth evaluation of a new product that was being introduced into a critical business environment. With a project plan outlining specific, defined performance objectives, the manager might have deferred or completely avoided the product in question. After contacting other installations, the IS manager learned that other users had experienced similar problems.

SOFTWARE VENDORS

IS and communications systems management is complicated by the increasing number of commercial software packages. A turnkey package may be acquired as an integrated system that requires specific equipment that may not be compatible with the existing mainframe configuration; or the software may be acquired to process a specific application (e.g., payroll, general ledger, accounts payable) on the mainframe.

As they do for third-party hardware purchases, managers must implement formal procedures and tools to collect performance data on the vendor's products and institute communications processes that allow the multivendor environment to be managed effectively.

This is becoming especially true as smaller minicomputer- and microcomputer-based systems spread throughout user areas as well as in the IS department. Further complexities are being introduced by the disappearing delineation between information and communications technologies. These factors, along with the development of complex technologies and products, essentially preclude the ability to maintain a single-vendor environment. IS and communications systems managers must confront the realities of today's information technology industry: the proliferation of software, hardware, and service vendors is likely to increase.

SUMMARY

Managers should approach the multivendor installation by considering the risk/benefit trade-offs in the context of their IS or communications environments. Establishing a project plan that details equipment and software performance and support should be mandatory. If possible, contractual contingency arrangements should be specified and agreed to by all parties before the installation or implementation of a vendor's product.

To carry out a multivendor installation plan, IS and communications systems managers should:

- Identify objectives to be achieved through a third-party vendor, emphasizing:

  - — Performance (e.g., the vendor may be the only source to meet requirements).
  - — Cost (i.e., costs are reduced or benefits per unit cost are enhanced through third-party purchase).
  - — Environment (e.g., strict operational constraints that must be accommodated).
- Review the performance history of the vendor, emphasizing:
  - — Reference checks of current and past customers.
  - — The performance history of the product being considered.
- Review the vendor's financial condition, emphasizing:
  - — Financial performance history and trends.
  - — Installation base trends.
  - — Market penetration in comparison with competitors.
  - — Whether it is privately owned or publicly financed.
  - — Annual revenues and profits.
  - — Debt-to-equity position.
  - — Ability to withstand short-term financial setbacks.
- Review the vendor's installed base, emphasizing:
  - — The number of customers serviced.
  - — The location of the customer base.
  - — The rate of new customer acquisition.
  - — The rate of existing customer defection.
- Perform a risk analysis that considers:
  - — The effects on the procuring organization if the vendor fails to perform as expected.
  - — Contingency plans for vendor replacement.
  - — Adequate and appropriate contractual protection in case of bankruptcy.
  - — Unchallenged use of the product.
  - — A statement of fair value purchase price for bankruptcy trustee.
- Review vendor management, emphasizing:
  - — Whether the organization is managed by the owner or professional management.
  - — Tenure of management.
  - — Management turnover history.
  - — Management remuneration plan (e.g., is a percentage of management's annual compensation incentive based?).

8-2
Evaluating Vendor Support Policies

James A. Papola

THE PRIMARY CRITERIA FOR SELECTING communications equipment are usually current and future user requirements and the performance characteristics of the hardware and software. However, vendor support services (both before and after installation), hardware maintenance contracts, software maintenance practices, and training programs may be just as critical to the system's success. Buying a system on the basis of hardware and software performance alone, without considering the support services the vendor offers, can be precarious. Vendors often market products and services separately, so asking vendors to submit a quotation schedule with their bids will help to clarify the actual support services they offer.

DEVISING A QUOTATION SCHEDULE

For managers to evaluate vendor policies effectively, they need information that is current, objective, and in a quantified form. The sample quotation schedule in Exhibit 8-2-1 can be used to structure the information-gathering process. The communications systems manager should first review the quotation schedule to become familiar with the specific information used to compare vendor policies and then choose those areas most pertinent to the desired project or application. Potential vendors should then be asked to respond to each item in writing. Questions can be added to the quotation schedule if the acquisition has specific requirements, but the quotation schedule should remain clear and to the point. The communications systems manager must ensure that the items are worded to elicit brief, quantifiable responses from the vendor.

The vendor must address all the items on the quotation schedule. The vendor may use specifications and other printed material as long as exact references are noted in the quotation schedules. If one vendor asks for a

Instructions

All bidders must supply a completed quotation schedule with submitted bids. All items must be completed where appropriate. The quotation schedule from the successful bidder will become part of the contract with that vendor. All submissions must be valid for at least 60 days.

Hardware Pricing

1. Purchase price of recommended system: ______
2. Lease or rent plans (operating lease):
   - Month to month ____/mo.
   - 12-month term ____/mo.
   - 24-month term ____/mo.
   - 36-month term ____/mo.
   - Other ____/mo.
3. Lease plans (finance and full payment lease):
   - 12-month term ____/mo.
   - 24-month term ____/mo.
   - 36-month term ____/mo.
   - Other ____/mo.
4. Discounts:
   - Are quantity discounts available? If yes, explain (include quantities).
   ____
   ____
   - Are there discounts on dollar volume? If yes, explain (include thresholds).
   ____
   ____
5. Installation costs: ______
6. Installation time: ______
7. Shipping costs: ______
8. Delivery time: ______

Hardware Installation

Attach a copy of installation specifications, including environmental and power requirements. Be sure that specifications include clearance dimensions.

9. Will shippers deliver to dock or installation site?
   ____
   ____
10. Is installation by user possible? If yes, is there any impact on warranty?
   ____
   ____
11. Is factory acceptance testing available?
   ____
   ____
12. Is site acceptance testing available?
   ____
   ____
13. Duration of performance test period:
   ____
   ____

Exhibit 8-2-1. Sample Quotation Schedule

Hardware Warranty

14. Duration of warranty period: ______
15. Start of warranty period: ______
16. Is warranty affected by user installation? If yes, explain.

______

______

17. Working hours for warranty service: ______
18. Extra charges for warranty service:
    - Outside regular business hours ______
    - Travel expenses ______
    - No defect found ______
19. Guaranteed response time for warranty service: ______
20. Is response time upgradable for extra fee? If yes, what is the cost?

______

______

21. If system is purchased, who owns replaced parts or components?

______

______

Hardware Maintenance

22. Types of maintenance service available:
    - On call 8 hours ____/mo.
    - On call 16 hours ____/mo.
    - On call 24 hours ____/mo.
    - Per call or incident ____/hr.
23. Extra charge for:
    - Weekend service ______
    - Outside regular business hours ______
    - Travel ______
    - Overtime work that began during regular business hours ______
24. Guaranteed response time: ______
25. Cost of improved response time: ______
26. Cost of improved system uptime: ______
27. Is equipment refurbishment included in maintenance?

______

______

Software Licensing

Please provide the following information for each proposed software item, both system and application.

28. Owner of software: ______
29. Types of license available:
    - Renewable Cost/yr ______
    - Perpetual or fully paid Cost ______
    - Other Cost ______
30. Software is licensed to:
    - CPU ______
    - System ______
    - Site ______

Exhibit 8-2-1. (*continued*)

- User ______
- Other (specify) ______

31. License extends to:
    - Object code ______
    - Source code ______
    - Quantity discount ______
    - Secondary license ______
    - Upgrades ______
    - Backup copy for disaster recovery ______
32. Distribution method (e.g., tape, disk): ______

Software Installation

33. Installation cost: ______
34. Performed by:
    - Salesperson ______
    - Customer engineer ______
    - Systems engineer ______
    - User ______
    - Other (specify) ______
35. Warranty:
    - Duration of warranty period ______
    - Start of warranty ______
    - Acceptance testing (vendor or user)

Software Maintenance

36. Maintenance cost: ______
37. Maintenance includes:
    - Debug ______
    - Fixes ______
    - Enhancements ______
38. Average response time to trouble reports: ______
39. Method of reporting trouble:
    - Phone ______
    - Mail ______
40. Method of forwarding repair information:
    - Phone ______
    - Mail ______
    - New copy ______
    - Online ______
41. Repairs performed by:
    - Vendor representative ______
    - User ______
42. Discounts for new or enhanced release: ______
43. New releases installed by:
    - Vendor representative ______
    - User ______
44. Discounts for second license: ______

Exhibit 8-2-1. (*continued*)

Availability of Source Code Escrow Agreement

45. Under what conditions is source released from escrow?

____

____

46. Ownership of user-development enhancements:
    - User ______
    - Owner ______
    - Joint ______
47. Number of copies of software documentation distributed with license: ______
48. Cost of additional copies: ______
49. Is software license transferable if hardware is sold?

____

____

Exhibit 8-2-1. *(continued)*

clarification of a point, the manager must provide that clarification to all bidders.

All potential vendors must be advised that the quotation schedule for their system will be included in the contract with that vendor. By taking this step, the manager can eliminate much vendor sales hype.

Another dilemma that may face the communications systems manager is the fact that equipment, software, and services can be acquired from a wide variety of sources. These sources include: direct from the manufacturer, a third-party distributor, electronic specialty stores, and office supply retailers. Depending on what is being acquired, any one of these sources is acceptable, but the level of support that may be required should always be considered.

Hardware and Software Comparison Categories

Each category in the quotation schedule is briefly described in the following sections. The categories cover most of the necessary evaluation criteria. Not all categories may be needed, because an application or user may require only certain types of support. The manager may also need to add questions to evaluate a support area in more detail, or more specific equipment, such as LANs or multimedia networks.

Hardware Pricing. Policies regarding a manufacturers pricing structure are one of the primary considerations to be made before purchasing any computer system. This section asks vendors to supply information pertaining to sales posture, available discounts, trade-in allowances, original equipment manufacturer equipment resale, available lease programs, sales, shipping charges, and delivery schedules. Although this is an important category, pricing and sales policies should never be the only criteria when a purchase is contemplated.

Hardware Installation. After a hardware configuration is chosen, attention must be paid to the concomitant installation planning and the amount of support the vendor can actually provide. Installation specifications for facility and site preparation, physical planning, and system checkout are all extremely important. A competent, cooperative, and thorough vendor can usually make all the difference between a relatively trouble-free installation and an installation beset with delays and aggravation.

Hardware Warranty. Another important consideration is the extent and breadth of the hardware warranty. The warranty's duration, services included, charges for nonwarranted items, and field service and repair personnel response times are all extremely important determinants of the quality and usefulness of the warranty.

Hardware Maintenance. After the warranty period expires, contract and per-call maintenance must be carefully considered to plan for most contingencies. The communications systems manager should understand the standard and expedited services available as well as the services that affect different price structures. The availability of maintenance of foreign peripherals and the costs of maintenance training may also need to be evaluated.

Another critical element in maintenance is the size of the IS shop. In some cases, it might be advisable to have only one or two technicians specifically responsible for a particular account.

Software Licensing. Software support is important in terms of the cost of license fees and the extent of the license. The form of the software, the updates, and the availability and cost of new versions are critical in choosing a vendor, unless independent software will be developed or purchased. The communications manager must remember that he or she is purchasing a nonexclusive "license to use" the software, nothing more. The same software is also being used by many others, including possibly a competitor.

Initial Software Installation. The vendor's professional services are extremely critical to the smooth start-up of a system. Especially important is the length of the warranty period, the status of the software delivered, and the personnel responsible for the installation.

Software Maintenance. After the system is installed, the types and levels of maintenance must be ascertained, including the time frame of maintenance and software fixes, the updates, the cost, and other important software maintenance factors.

Software Documentation. Software documentation is extremely important to ensure the effective use of the hardware installation. The availability and cost of the vendor's offerings in software documentation should be carefully evaluated.

Internet Access Providers. Internet access providers (IAP) are popping up all over the world. Most are acceptable, but care must be taken in selection for specific types of use. For example, an IAP whose busiest times are between 8:00 a.m. and 4:00 p.m. may not be able to provide a suitable response time to some business applications.

SUMMARY

The quotation schedule provided in this chapter is only a guideline; a more detailed schedule can be developed as a proposed system becomes more specific. Including the quotation schedule in the final contract will eliminate interpretation hassles after the sale. Furthermore, every point covered in the quotation schedule, and ultimately in the contract, can be negotiated.

The single most important element in evaluating vendors is planning. Allow enough time for vendors to respond to the quotation schedule and to quantify and evaluate their responses. Taking the time to completely understand the vendor's position on key points will give the manager the necessary leverage in negotiations.

8-3

WAN Network Integration: A Case Study

Charles Breakfield

MERGING TWO WIDE AREA NETWORKS (WANs) is a process comprising several steps that, when carefully planned, can ensure a successful implementation. In the case of the Resolution Trust Corporation (RTC) and the Federal Deposit Insurance Corporation (FDIC), considerations included network operating systems compatibility, enhancing E-mail and directory services, applications support, server and workstation memory and processor upgrades, applications support, and routing software upgrades. This case study details the methodologies that brought together two WANs with more than 17,000 end users.

COMPANY BACKGROUND

The original project was created to merge the Banyan WANs at the FDIC and the RTC. The RTC's corporate charter expired on December 31, 1995, and its assets were integrated into the FDIC's operations.

The RTC's origins—and its reason for being—date back almost 10 years. When banks and savings and loans (S&Ls) became insolvent in the late 1980s, they called on the FDIC and the Federal Savings and Loan Insurance Corporation to prop them up. The task was so overwhelming that Congress was petitioned to assist. As a result, the RTC was created, designed to handle the disposition of the assets and liabilities of the failed banks and S&Ls until 1995. When the RTC's corporate charter expired, all activities and assets rolled over to the FDIC. Its systems were developed independently of the FDIC's but with this inevitability in mind.

The FDIC had no plans to keep RTC employees after December 31, so FDIC personnel will assume all existing RTC computer network activity. This would be practical only if both organizations were similar in design and layout. Therefore, the project, which began in August 1993 and took 10 months to complete, was implemented to make the RTC's WAN compatible with the FDIC's.

THE CHALLENGES AND BENEFITS OF IMPLEMENTATION

The RTC's Banyan WAN encompassed the 7,000 employees and contractors located in California, Colorado, Texas, Kansas, Georgia, New York, Virginia, and Washington DC. The WAN needed to be connected to the FDIC's WAN, which encompassed 10,000 employees who are similarly located. All the RTC Banyan servers targeted for upgrade were production servers being used 8 to 12 hours a day, six days a week. Server upgrades had to be coordinated so the end users were not disrupted during the day and so there would be a fall-back position if a server upgrade failed.

Several other areas for upgrades were also targeted. An operating system upgrade was required as part of the integration proposal. The FDIC's 337 Banyan servers, from Banyan Systems Inc., are running on one network operating system, Vines 5.52(5); the 329 RTC Banyan servers were running on Vines 4.11(5), two revisions old. This network operating system upgrade was required to bring the RTC up to the same level the FDIC was running before merging the two organizations. Also, Banyan no longer wanted to provide fixes (i.e., software patches that solved "bugs" or problems) or enhancements for an older version of Vines, so the RTC had to upgrade its network to continue to receive operating system support.

As part of the operating system upgrade, a server hardware upgrade also was necessary, to support the enhanced version of software. Additionally, the Cisco routers in the RTC network needed to be upgraded.

The Business Opportunity

As part of the implementation, new functionality offered by the newest revision of Vines version 5.52 would allow the corporation to use the new features to operate the business better.

The enhanced functionality of the new E-mail under Banyan Vines 5.52(5) was one of the main reasons for the upgrade, because the RTC's most important application was the E-mail system, and the most heavily used applications were the E-mail and the directory services, Streettalk Directory Assistance (STDA). The full listing of the combined organizations (a total of 17,000 users) under STDA would provide access to all end users by all end users.

Additional network management tools and diagnostic tools were included as part of the new network. Better performance tuning and use of resources would also be easier to accomplish under Vines 5.52(5).

Technical Benefits

The technical benefits of the Vines 5.52(5) operating system were:

- The new file format, S10, under Vines 5.52(5) allowed up to 4G bytes of inodes per file system, up from the old file format, S5, under Vines 4.11(5), which allowed 64K bytes of inodes. Each file or directory was represented by one inode. More files or directories per volume were required by the RTC.
- The Access Rights List (ARL) functionality was greatly enhanced under 5.52(5). Access control could then be given all the way down to the file level within directories or within subdirectories to a user or a group of users. Security issues could be tailored to the file level, which was not possible before.
- Greater network management was offered through the network management statistics on memory usage, server processes, hard drive activity, and CPU usage. The improved tools promised better server monitoring and network management.
- Valuable to the RTC were the tape backup and restore processes built into the operating system. Under Vines 5.52(5), the operator can back up and restore not just file services, but also specific directories and files with much greater ease.
- New printer functionality was also gained by migrating to Vines 5.52(5). Multiple print queues were serviced by one printer, and print queues were set up to print to other printers if a printer was overloaded. Print jobs were redirected by queue operators or done automatically.
- A less obvious benefit, but still an important feature, was the new routing matrix algorithm that was implemented into Streettalk. The overhaul to the routing matrix in Banyan Streettalk all but eliminated the sporadic occurrence of network "broadcast storms," which were an inherent problem.
- New functionality was offered under Banyan Vines STDA. User IDs could be defined to include additional attributes on each user put into the system. The security could be set so no one could look at the associated attributes in the Attribute View Definition (AVD) file that holds them. The information contained in the AVD file could be masked and display only the users under STDA that the personnel department and the end user wanted displayed for E-mail purposes.

PRIORITIES

The majority of the mainframe applications used by the RTC and FDIC were located at the Virginia location in Rosslyn VA and required 3270 emulation for host access. The RTC network was independent from the FDIC network, to begin with, but there were certain priorities to consider to bring them together.

As these are government institutions dealing with assets of failed banks, network security was a priority. The original decision to go with Banyan was based in part on its WAN capabilities and the tight security design of the user log-in process. Each time a user logged in, his or her unique network ID was time and date stamped to prevent unauthorized access by a computer hacker. The flexibility to log in anywhere in the network and still maintain a high level of security was key to the way the RTC did business.

A requirement of the new version of the network operating system was that it support all existing off-the-shelf software. All standard applications, word processing, spreadsheets, data bases, and E-mail were provided on the local Banyan servers. The current suite of standard software packages that the RTC supported included WordPerfect 5.1, Lotus 2.4, and Paradox 3.5. All workstations ran either DOS 5.0, which was preferrable, or DOS 3.3 with 386 Max version 5.0 as the memory manager. Saber, a software package, was used for the menuing front-end for all users to access all applications. All the applications were to be compatible with the Vines 5.52(5) upgrade.

ELEMENTS TARGETED FOR CHANGE

All applications in use at the RTC when the project began needed testing in a lab environment before delivery into the production environment. Preliminary testing showed complete compatibility across all applications with the Vines 5.52(5) upgrade.

The workstations in the field were configured with only 2M bytes of random access memory (RAM). RAM requirements of the new operating system shell, as well as the applications to be run for daily applications, had to fit in the 2M bytes of memory. There were no plans or budget to upgrade RAM in field workstations.

Each of the two styles of Banyan server (the RTCs and the FDICs) required identical configuration in naming conventions of groups, file services, print services, 3270 services, E-mail naming standards, amount of disk space, memory configurations, server names, network interface card (NIC) settings, and cabling specifications for connectivity.

Several Banyan servers also required hardware upgrades before the new operating system could be installed. Banyan server hardware upgrades included faster processors (i.e., Intel 486 chips), larger server hard drives, and faster NIC cards, when feasible.

The Cisco routers used to connect each office to the WAN also required an upgrade, as they supported only Vines 4.11(5) IP protocol stack. The revision of the routing software in the Cisco routers was 9.x, and the RTC needed a 10.x revision.

UPGRADE COSTS

An upgrade server key option had to be purchased for each RTC server—a $5,000 option for each server. Because RTC servers needing the upgrade numbered 329, the cost totaled $1,645,000.

Each Corporate Network Server (CNS) required one 1.3G byte hard drive per server before the upgrade could take place. Each CNS server also required a drive 0 larger than the existing 80M byte drive. The upgrade process required that each CNS have a 330M byte drive in slot 0, a 330M byte in slot 1, and the 1.3G byte drive in slot 3. The larger drive 0 would accommodate the new operating system requirements. The actual number of CNS Banyan servers was not known but was estimated at 40% of the total 329 Banyan servers, for a total of 132 CNS servers to be upgraded. Each 1.3G byte upgrade drive was valued at $5,400, for a total of $712,800 to upgrade all CNS servers. The Compaq SystemPro accounted for the balance of the RTC 329 servers, or about 197 servers. Of the 197 SystemPros, roughly 50% needed hard drive upgrades, roughly 30% needed CPU upgrades, and approximately 40% needed to upgrade to 32-bit NIC cards (either Token Ring or Ethernet). Hard drive upgrades (and controllers) cost approximately $4,500 each, for a total of $445,500. CPU upgrades and system PROMS cost approximately $1,900 for each of 59 servers, or $47,400.

The hardware upgrades for the servers were required only for drive 0 of the CNS. The hardware upgrades for the other servers were done in conjunction with the Vines 5.52(5) upgrade for two reasons. First, the hardware upgrades had already been planned for and budgeted before the Vines upgrade. The existing server hardware was two to four years old at the time of the project and required upgrading because hardware failures were occurring more frequently. Second, because the servers needed to be rebuilt from scratch, it made economic sense to do the hardware upgrades just before laying down the operating system and files. This would save overtime costs that would be incurred if the servers were refurbished after upgrading the operating system.

Router Upgrades. The last items to be upgraded in the network were the Cisco AGS and AGS+ routers, which were scheduled for upgrade after the servers were completed. The Ciscos were running the 9.4.0 software version, which did not fully support Vines 5.52(5). Full functionality for Vines 5.52(5) was promised if RTC upgraded the hardware and software. Additional memory was required to run the software update 10.0.4. The total number of

Banyan Server Upgrades	$1,645,000
CNS Server Upgrades	$712,800
SystemPro CPU Upgrades	$112,100
SystemPro Hard Drive Upgrades	$445,500
SystemPro NIC Card Upgrades	$47,400
Cisco Hardware Upgrades	$45,000
Cisco Software Upgrades	$90,000
Upgrade Labor Costs (OT only)	$55,000
Total (Estimated)	$3,153,730

Exhibit 8-3-1. Total Estimated RTC Upgrade Costs, Per Component

Cisco routers that needed an upgrade was 30. (Some of the 3000 series did not need the hardware upgrades but did need the software.) The hardware upgrades were approximately $1,500 each, and the software upgrades were $3,000 each, for a total of $45,000 and $90,000, respectively.

Man Hours. Each production server upgraded from Vines 4.11(5) to Vines 5.52(5) needed two RTC network engineers, each working four to six hours (five hours, average) of overtime per server if the following methodology were used. If a figure of $17 per hour for labor costs is used, each server would cost approximately $170 to upgrade. If the 329 servers were upgraded at that price, the labor cost to upgrade all the servers was $55,930. The total costs are summarized in Exhibit 8-3-1.

UPGRADE MODELING AND REFINEMENT

All the first-line network engineers were trained on the Vines 5.52(5) operating system first and then were involved in testing specific aspects of the operating system at each site. The results were communicated via Lotus notes, videoconferences, and face-to-face roundtable meetings.

The upgrade methodology was tested, refined, and then taught to the staff members who would actually be doing the upgrades. The modeling and refinement activity was particularly valuable in that it gave the staff time to practice before operating on live servers and corporate data. These practice sessions were the greatest insurance policy available in this process. In fact, the real success for the project began in these early planning sessions.

Each type of production server, Compaq SystemPro and Banyan CNS, needed upgrading to Vines 5.52(5). Every server required the internal hard drives be reformatted to take advantage of the S10 file format before having the data restored from tape. Within the restore-from-tape process, all the files needed the ARLs converted to the new 5.52(5) format.

The overall upgrade approach to both servers was the same: to build a like server, either a SystemPro or CNS, offline to replace the production server. By starting on Monday morning, an identical Banyan server could be built, tested, and ready for live data to be restored beginning at 5:00 PM on Friday. Typically, a server was fully tested and ready for the final data files by Wednesday night or Thursday morning. The offline servers were built during regular business hours, so overtime was needed only on Friday night, for the last stage of the conversion process.

The SystemPros were upgraded with Intel 486 processor boards, new ROM chips, and new 510M byte hard drives array pairs and SCSI II controller cards. The upgrade process was greatly hindered by these extra variables, because of a 50% failure rate with the Seagate 510 drive array pairs. Not all SystemPro hard drives were upgraded, and the existing Conner 210 drive array pairs and SCSI I controller cards were reused. In each configuration, the Compaq SystemPro was left to run the comprehensive drive array and to run controller card diagnostic tests for 24 to 48 hours before being formatted with the S10 file format under Banyan Vines 5.52(5). Although this was a time-consuming action, the servers that made it into production were very reliable.

Additionally, any Token Ring or Ethernet NIC that were not 32-bit Enhanced Industry Standard Architecture (EISA) were replaced with 32-bit EISA boards for improved performance.

The Banyan CNS servers, for the most part, had a like configuration. Each was a 386 class machine with an 80M byte drive as drive 0, a 330M byte drive for drive 1, and a 330M byte drive for drive 2. The new 5.52(5) operating system would just fit on drive 0, but that left no room for file swap space, for growth in the internal routing table, or for E-mail service. The upgrade position was to remove the 80M byte drive and replace it with the 300M byte drive in slot 3. A 1.3G byte was purchased for each CNS for installation as the third drive. All applications, all noncore Banyan services, and all data are placed on this drive. All core Banyan services were installed by default onto the first drive, and the 330M bytes provided ample room. For the most part, the second drive (a 330M-byte hard drive) was left for growth but was largely unused. No CNS servers were scheduled for upgrades with 486 processors.

The Process

The beginning process was the same for each style of server. Diagnostics were run on each; each drive was formatted multiple times (time permitting); and the Banyan Vines operating system was loaded. At this point, the Banyan server key on the production server was carefully removed and placed on the offline server. The offline server was brought up and given the same name as the production server. The identity of the new 5.52(5) server was critical and required the server key to build its internal routing tables. It was essential that the two servers not be connected to the network at the same time.

Once the new server had been brought up and named properly, it needed to be patched with a total of 10 Banyan patches (i.e., fixes, to be installed for full functionality) before anything could be done. The order of patch installation was important, and several versions were tried and discarded before the list was stable. However, some patches were more trouble than the problems they fixed—even having to be replaced only hours after they had been applied. Patches were installed after the data was loaded and converted as well as before. Both methodologies worked.

The offline Vines 5.52(5) server targeted to replace the production server then had all the same services created using the same naming convention for file restoration and conversion. It is important to note that if a file service was not within the naming standards, the name was converted.

Print and File Services. All print services were rebuilt, as opposed to restored. The new print services at the FDIC were different enough to warrant setting up the RTC services from scratch. The Banyan operating system also would not permit moving and converting the print services. Print services were created on the target server during the week.

All file services were created on the target server to match the production server, and each service containing data was moved individually and converted. The applications file service was restored from the standard application tape that was created at the beginning of the upgrade process. Because this was a static file system, meaning that usually no changes to the applications occurred, this service could also be restored. After the first server was converted, a Vines 5.52(5) tape backup of the applications was created and then used for restoring target servers for upgrades. This process saved time converting ARLs. The ARLs, however, still needed to be reworked to be server specific—so the restore was not as trouble free as it sounds.

Any other service on the production server, such as asynchronous dial-in, 3270 SNA service, server to server, and WAN links, were also created and set up on the target server before Friday evening.

Any lists created on the production server needed to be printed out and rebuilt by hand before the target server was brought online. The list itself was added to the server before the data files were restored and converted. Names could be added to the list only after the server was brought online.

The Group Moves. The only activities that could not be done before Friday afternoon were the group moves, and the transfer of the E-mail, data files and updates to the Adminlist for the target server.

Promptly at 5:00 PM on Friday, the groups were moved from the production server to tape and then restored to the target server. The group moves were done before the file services, because in the process of the file conversion, the ARLs were updated to the new Vines 5.52(5). The file services (i.e., the actual data) were backed up to tape and then restored to the target server. ARLs and the group and user IDs were matched. If the groups with their users had not been there before moving the files, it would have been necessary to go back in and edit the ARLs on the target server.

Moving E-Mail and STDA. The E-mail service was moved from the production server to the target server via tape backup. This process was straightforward and reliable.

The STDA was recreated on the target server, once the target server was moved to production status and cabled to the WAN and the old server was removed. Again, the proper sequence was removed and the new target server moved into place. The STDA service could not be moved with reliable results, so Banyan recommended that it be recreated; however, it could not be created before the server was placed on the WAN. The target server needed the connection to the WAN so it could rebuild the internal STDA data base from upstream neighbors. The creation and forced rebuild on Friday night gave three days for the rebuild process to operate before Monday morning. Sometimes, though, three days is not enough time for STDA to rebuild properly, and the data base had to be killed and recreated before it would work properly.

Moving the ARLs. The ARLs were one of the last major areas left to update once the target server became the production server. The administrators were usually part of another group that were kept on a different server. Sometimes the Adminlist could not be updated with the names of administrators until the new production server could "see" the target server. At those times, two different methods to force the update process on the new production server were used. A dummy group was created on the new server, which forced a Detail broadcast to the WAN that updated all the surrounding servers, and, as a result, the entire WAN began to "see" the new server. When the server could "see" and "be seen," the dummy group was deleted.

The other method was to use a Banyan utility that would "goose" Streettalk and force the surrounding servers to exchange routing information and update their respective routing tables. The STSYNC utility is effective in getting servers to synchronize with each other.

The last upgrade detail was updating the boot disks for the users. This process typically required that someone physically visit each workstation that was physically attached to the upgraded server and run the NEWREV PROGRAM, which copied down the new boot files to the workstation. This process was quickly automated by the staff members so when a user of a group that resided on a newly upgrades server logged in, a batch file ran that copied the necessary files to their workstation and then required them to reboot. This process saved an enormous amount of time.

Because entire groups with the associated users were being moved, no password changes for the end users were required. This helped make the conversion more transparent to the end user.

Some Glitches

Final issues for the conversion were to double-check the ARL of the root in each file service, as they did not convert reliably and had to be modified by hand. Oddly enough, the remaining ARLs in the rest of the directory tree were usually correct and did not need modifying. Some tailoring of the ARLs did occur after the fact, so as to comply with the new RTC standards that were published before the upgrade was started.

One ARL problem encountered was with lists on the server that were used in the ARLs of files and directories. If the ARLs were being changed or updated with the Banyan Vines utility Netpro, the workstation would receive an out of memory error and terminate. The reason for this was that Netpro could not update ARLs with a list that was not there, so it buffered the transaction and went on to the next ARL update, and so on, until the workstation ran out of memory. The lesson learned here was to make sure that all lists were at least placed on the server, empty or not, so the ARL conversion would operate properly.

In the particular upgrade scenario for the site, two and sometimes three server conversions a week could proceed with the equipment on hand. As servers were replaced with new production servers running Vines 5.52(5), the 4.11(5) servers coming offline became the new target machines in the upgrade process for the following week. The assorted 20 Banyan servers took 10 weeks to convert from Vines 4.11(5) to Vines 5.52(5). All staff members worked in teams that rotated each week, so no network administrator was needed for consecutive weekend work, and no mental "burn out" that would have resulted from such a demanding schedule occurred.

As in all well-thought-out plans, there is always an overlooked X variable that comes back to foil complete upgrade success. As all sites reached the final stages of the server upgrade process, a problem surfaced that no one had anticipated: Paradox 3.5 would not support over 25 concurrent users on a Banyan Vines 5.52(5) platform.

Lab applications testing did not explore the Paradox application to this extent. Both Banyan and Borland were contacted about fixes or workarounds. Banyan declined to assist RTC in this problem, saying it was an application problem. After many phone conversations with Borland, their technicians did concede that it was a problem with Paradox 3.5. But because

Borland was no longer supporting that product version, the technicians suggested that RTC upgrade to Paradox 4.0 to solve the problem.

There was no money budgeted, however, for upgrading approximately 329 servers, with only 2M bytes of RAM, to the 4M bytes of RAM that Paradox 4.0 required to operate. Therefore, Borland's solution was cost prohibitive.

Ultimately, the solution was somewhat unsatisfactory but effective. Each site left one Banyan Vines server at 4.11(5) that contained the Paradox NET and SOM files. These files allowed Paradox 3.5 to run normally for all the end users that were attached to Vines 5.52(5) servers. Fortunately, Banyan supports mixed versions of its Vines operating systems, so long as they are within one version of each other (e.g., 4.X and 5.X systems can be mixed and still talk to each other).

THE FINAL INTEGRATION

The last stage was opening the main Cisco router, which was connected to the RTC and FDIC networks, to full Banyan Vines traffic from both sides. As part of the initial procedure, certain action items were completed before the wave of router table updates went surging through the respective networks.

First, all RTC Vines servers had their communication buffers set to at least 400,000 before the cut over. All CNS servers had to have a primary drive that was larger than 80M bytes (300M bytes were preferred). All network servers were scanned to ensure that no duplicate server names (or server key numbers) were in the two systems. All end users were prohibited from logging on and using the network beginning at 5:00 PM that Friday. The STDA directory services on each server were stopped on Friday morning and were not allowed to rebuild again until Sunday night. Finally, no production servers were trying to get their Streettalk updates across a WAN link that was slower than 56K bytes per second. The STDA was also stopped to prevent any unnecessary WAN traffic while Streettalk was trying to update the internal routing tables of all the production servers.

Full server backups were completed and verified before the process began, to provide an absolute fall-back position in the unlikely event that each organization had to return to a preintegration status. The actual process of the integration was that the servers in each network needed to have their internal routing tables updated to know about the additional 300-plus servers on the other network. Because Vines does this dynamically, the staff's primary job was to monitor the health of each server and, in fact, the whole RTC network while the update process unfolded. Once the main Cisco router was allowed to pass Streettalk traffic and updates both ways, it became much like an information tidal wave slamming into each site. At 9:00 PM Friday night, the information updates were rolling toward the location, and server memory utilization stats were watched to see if any server started swapping information from memory to disk, a sure sign that the server was in danger of being swamped.

By 10:00 PM, the Streettalk routing updates were in full swing, and the available server memory was dropping to levels too low for comfort. Staff stopped all unnecessary file server processes in an effort to free up and make available more server memory for the routing table updates. By 11:00 PM, the servers in the network had stabilized, and available server memory began to climb, a sign that the critical time had passed and that the network was stabilizing.

At 9:00 AM Saturday morning, the staff returned to restart all the services that had been stopped the night before and began testing the services to which the end users would need access on Monday morning, with the exception of STDA. Once all services were tested, and no lingering problems existed, the staff began to deal with the STDA issue.

It was decided to permit the regularly scheduled rebuilds to go as planned, but the FDIC would not be included in the rebuilds for another 10 days. This would allow Streettalk time to "settle" before subjecting it to another data "blitz" from an STDA that would have added another 10,000 names to the existing 7,000-name data base. The final integration of the two STDAs proved to be modest and concluded the entire data integration process rather quietly.

SUMMARY

In retrospect, this network integration might have been overplanned, considering how smoothly it went; however, the Vines 5.52(5) server upgrades that blindsided the team on the Paradox issue were underplanned. Data center operations managers looking at integrating similar or even dissimilar networks save time and aggravation if they include field personnel who will be participating in the upgrade/integration process. Bringing in the senior network personnel, training them, and including them in the integration process made all the difference in the merger of the FDIC and RTC Banyan WANs. People who are actually scheduled to do the work who "buy in" to the project help breed enthusiasm and interest in the success of the project and provide unexpected benefits.

For example, field personnel identified many incidental problems that no one at corporate headquarters had even considered. The success of the overall project was ensured by the corporate people involving the field network engineers.

Every effort should be made to get the hardware and software vendors involved at the beginning of the project to assist and comment during the planning stage. Banyan Systems placed one of its certified Banyan engineers on site at headquarters to help with the integration process. Additionally,

Compaq sent its engineers out to RTC sites to help resolve the SystemPro hardware problems that were first encountered when the new drives were installed before the operating system upgrades.

Each vendor that was contacted and included in the upgrade planning before implementation improved the chances for success with either technical advice or on-site involvement. RTC suppliers also contributed in the upgrade process when involved.

Finally, communication among all the players in an upgrade of this size is critical. Information is useless unless it is in the hands of the right people at the right time. Using Banyan E-Mail, Lotus Notes, and teleconferencing, headquarters staff and field engineers were able to communicate critical pieces of information to the proper people in a timely manner. As a result, the upgrade and integration success was accomplished with existing personnel who were also responsible for the day-to-day operations of the existing user community.

8-4
Frame Relay in an IBM Environment

Glenn R. Brown

THE SPOKE AND HUB IS THE TYPICAL WAN DESIGN for most IBM computing environments. Usually, a mainframe is located at a central point through which all data from every remote site must pass. This topology is illustrated in Exhibit 8-4-1. If a site in northern California must communicate with a site in southern California, the data is sent from northern California to the mainframe at corporate headquarters on the East Coast and then to southern California. Both time and money are wasted, as the sites are bearing the cost of sending the data on the long trip through the hub. If the spoke and hub network connects IBM Token Rings LANs, the seven-hop count limit must be considered; it may be exceeded if widely distributed LANs have to communicate with each other through the central point.

A meshed or webbed network, which is an alternate to the spoke and hub, provides multiple paths to sites but can also be needlessly expensive because of the costs of the many lines. Transparent source route bridges can link nodes on a meshed network so that the seven-hop Token Ring limit is avoided, but the higher line costs remain. If the nonroutable NetBIOS protocol is used, the overhead involved can mean an 8% to 25% loss of available bandwidth.

The central point in a spoke and hub network is a potential operational problem. As a single source of failure, a relatively minor accident (e.g., a cut cable) can bring down the network. For the company described in this case study, lost business may cost $250,000 to $500,000 per hour, and a cable cut can take four or more hours to repair.

CURRENT NETWORK COSTS

A large manufacturing company with headquarters in the Southeast has an agency distribution system of 36 warehouses and five central distribution

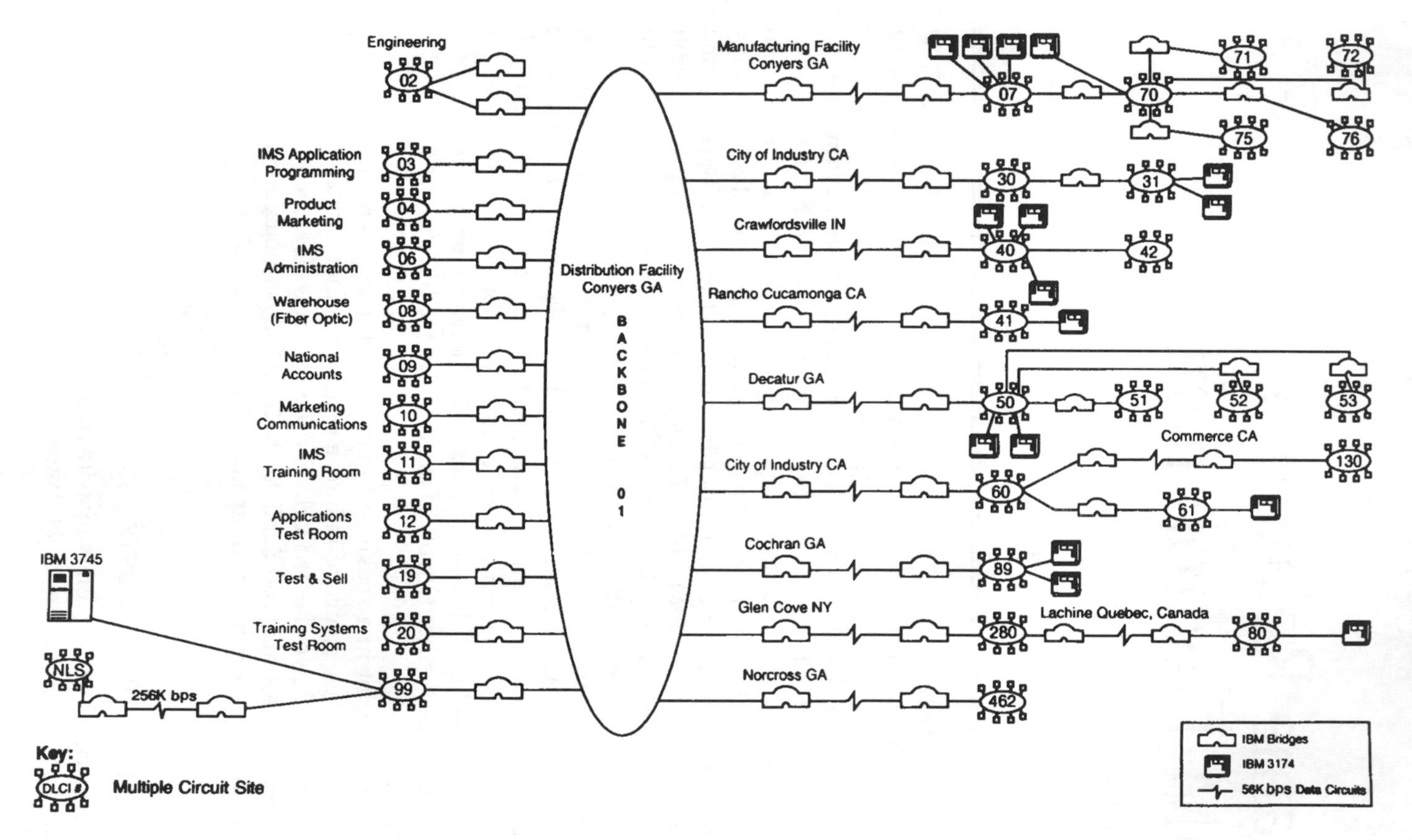

Exhibit 8-4-1. Spoke and Hub WAN-LAN

centers in the US and Canada. The company also has 11 manufacturing facilities in the US and Canada, with yearly revenues of approximately $700 million.

The company's networking environment consists of IBM's OS/2 LAN server, with MS-DOS, Windows, and OS/2 on the requesting stations. The mainframe is an IBM 3090/400 connected to 3,000 or more combined mainframe and LAN requesters. All remote sites must communicate to the central site as well as with each other for manufacturing schedules, distribution requests, inventory tracking, mainframe reports, E-mail, word processing, spreadsheet information, CAD drawings, and data base information. The central site is headquarters for general ledger, payroll, and human resources data processing.

The network is made up of 12 56K-bps leased digital circuits, which cost, altogether, $286,000 per year. Two additional voicegrade phone circuits with 14.4K-bps modems cost $16,800 per year. To web or mesh into the network three additional sites, with each site having one or two permanent circuits to other sites, would cost an additional $248,014 per year. Management refused to approve this approach to expanding the network.

Once the decision had been made to redesign the network, the following requirements were established:

- At the minimum, the same level of service must be provided, even if no cost reduction is possible.
- Faster service for site-to-site communications must be provided.
- Modification of the existing IBM 3174 controllers and the IBM 3745 front-end processor must be avoided.
- Redundancy and backup circuits must be provided.

THE NETWORK REDESIGN

First, the bridges in the spoke and hub network were changed for bridge routers, in order to route NetBIOS and SNA traffic by encapsulation in Internet Protocol (IP) packets.

The spoke and hub solution was replaced with a frame relay network. Data link connection identifier (DLCI) is a frame relay value that identifies a logical connection. Each site has a DLCI, allowing the definition of multiple virtual circuits with specified committed information rates over a single actual local exchange loop circuit. This eliminates the need to obtain (and pay for) multiple circuits. Using the DLCI, multiple permanent virtual circuits (PVCs) were defined within a single circuit, as shown in Exhibit 8-4-2.

It was necessary to ensure that the frame relay service carried the current NetBIOS and SNA traffic, that it eliminate the central site as a single point

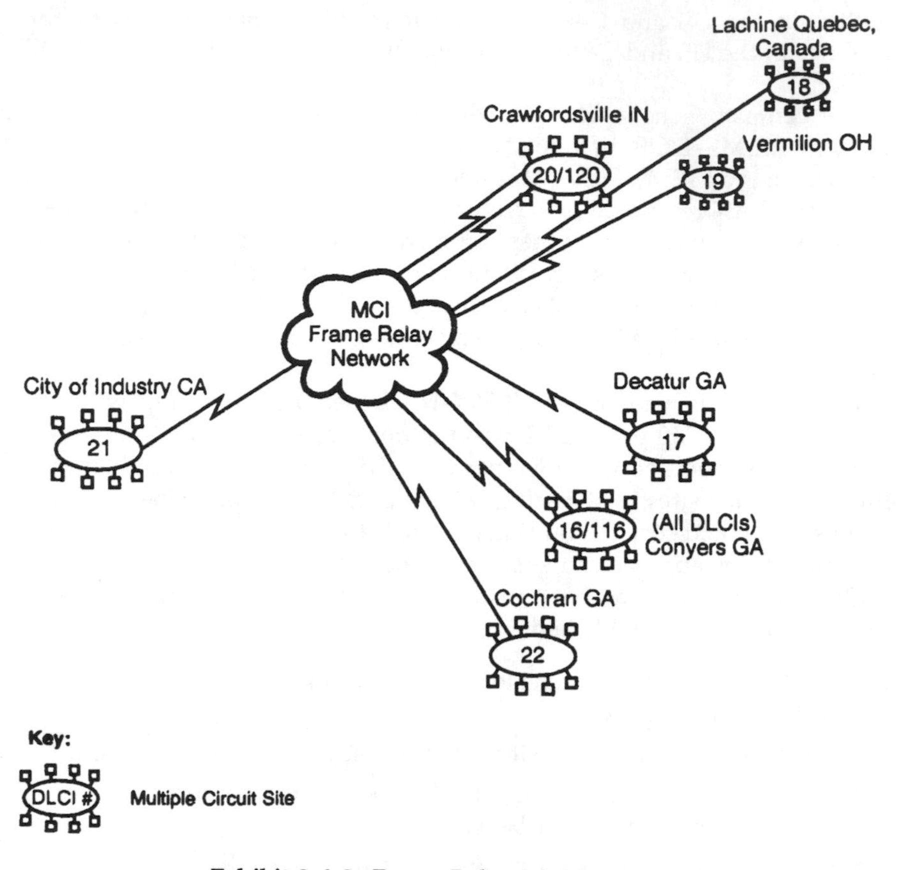

Exhibit 8-4-2. Frame Relay DLCI Network

of failure, and that the bridge routers operate in a NetBIOS and SNA environment. For SNA traffic, the central site remains a potential single point of failure even with frame relay, but LAN-to-LAN traffic is not disrupted if the central site is lost.

Local acknowledgment or spoofing was a requirement to help reduce the traffic load on all circuits. Mainframe traffic acknowledgments would be kept local and microcomputer traffic would be routed intelligently through IP.

Challenges for Equipment Providers

Customers premise equipment for Bay Networks (Billerica MA), CrossComm Corp. (Marlboro MA), and Cisco Systems, Inc. (Menlo Park CA) was tested. The frame relay providers tested were Worldcom (Tulsa OK), AT&T (Bridgewater NJ), and MCI (Dallas TX).

The largest challenge for the vendors was routing the nonroutable NetBIOS and SNA protocols while maintaining communications to the mainframe from the Token Ring interface coupler (TIC)-attached controllers. Another requirement was the attachment of an SDLC controller to the WAN through an LLC2 (logical link control type 2.A) conversion, which converts the SDLC to a Token Ring or NetBIOS transmission.

Of the equipment vendors, only Cisco was able to do this. One vendor's equipment would not operate with a TIC-attached IBM 3174 controller, and the other vendor's equipment could not keep the IBM 3174 controllers in session for more than a few hours. IBM has since introduced the 6611 multiprotocol router, but this was not available at the time of this evaluation.

Services from all three frame relay carriers worked with few problems. Cost became the deciding factor, and MCI was selected.

Test Phases

A frame relay connection between two nearby sites was tested to determine the effects of frame relay on remote NetBIOS and SNA traffic, as illustrated in Exhibit 8-4-3. Crucial performance characteristics that required testing were speed degradation, reliability of TIC-attached IBM 3174 controllers, and reliability of local acknowledgment.

Next in the testing phase was to bring up additional sites and determine the gains in transfer time on a fully meshed frame network. Three sites—Georgia, Indiana, and California—were upgraded. California can communicate directly to Indiana and Georgia; Indiana can communicate to Georgia and California; and Georgia can communicate directly to Indiana and California. The file transfer time savings should average around 30% .

Reliability in the Canadian circuits had to be tested. All US carriers must be compatible with Canadian Bell, but response times and problem resolution were unknowns. Carriers are currently establishing alliances with Canadian Bell, so US equipment can be placed in the Canadian Bell regions, thereby reducing circuit mileage costs.

Next in the test phase was to check the conversion of SDLC to LLC2 for non-TIC-attached IBM 3174 controllers. The test results showed that Pin 11 of the cable on the RS-232 connection to the SDLC IBM 3174 controller should be removed. Pin 11 is needed only for IBM testing when connected to a digital sharing device and is not needed in a Cisco router configuration.

Each controller required its own RS-232 data circuit-terminating equipment (DCE) serial port, which meant that all sites having non-TIC-attached controllers required four port serial cards instead of two. Users reported

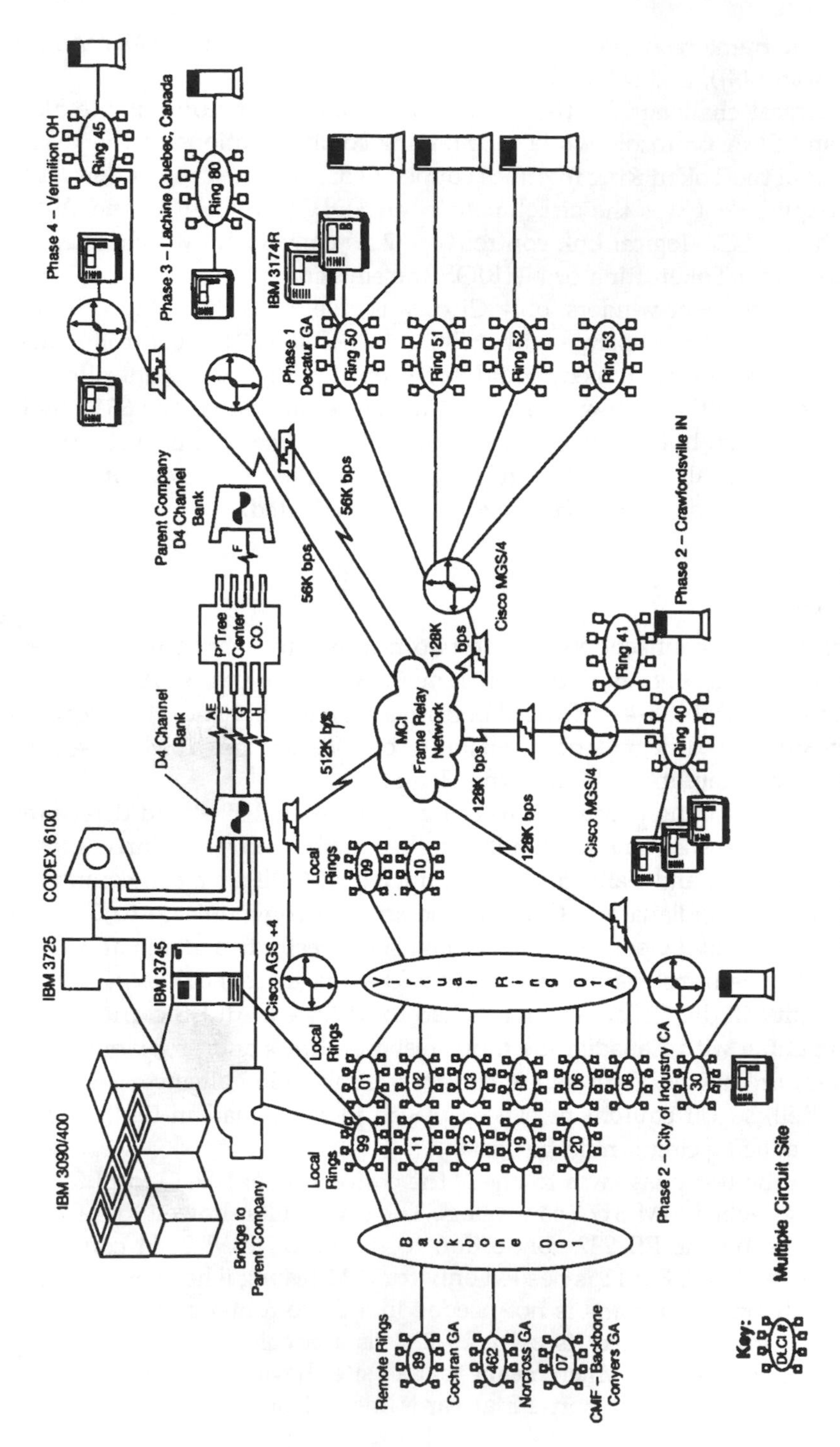

Exhibit 8-4-3. Frame Relay Test Phases

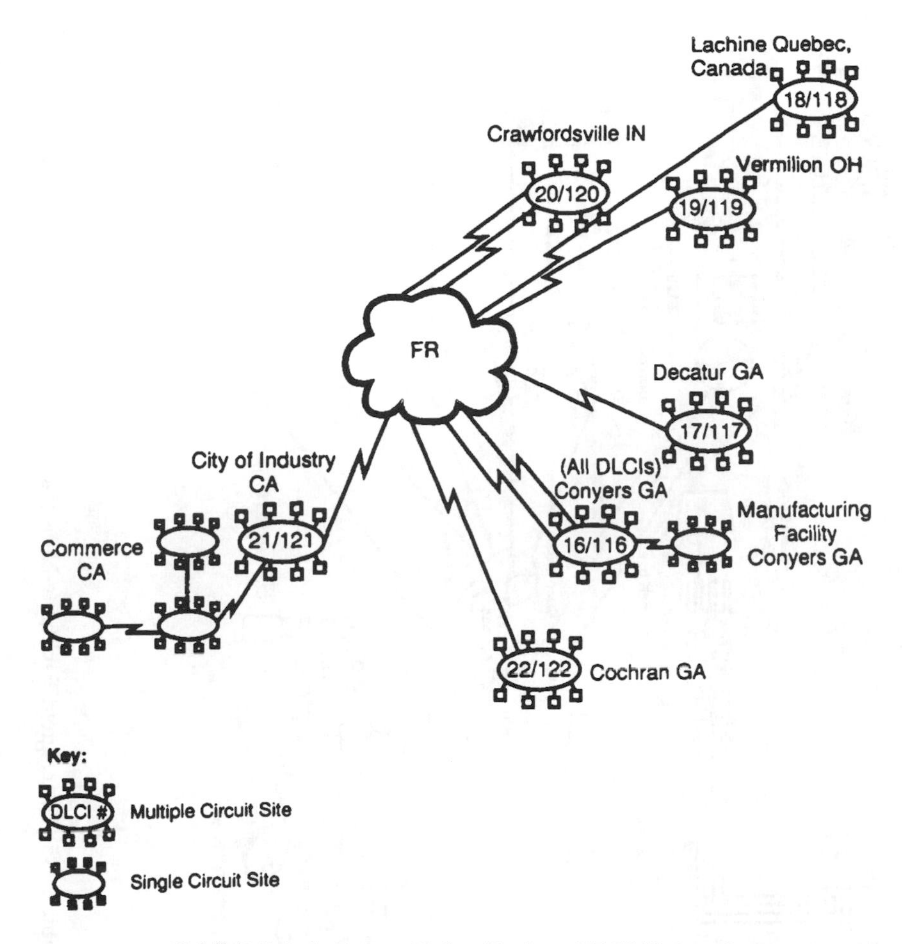

Exhibit 8-4-4. Frame Relay Backup DLCI Network

---

faster response time on their terminal sessions using frame relay than on the shared 19.2K-bps dedicated circuit previously used.

Redundancy

A carrier's frame relay network is self-healing. If a carrier switch goes down, the carrier can temporarily reroute the traffic from the disabled switch to another switch to keep a network functional.

Backup of the connection to the frame relay point-of-presence is usually done through a dial-up device, but this did not provide adequate bandwidth. The solution was to obtain two local circuits, routed over different paths, and

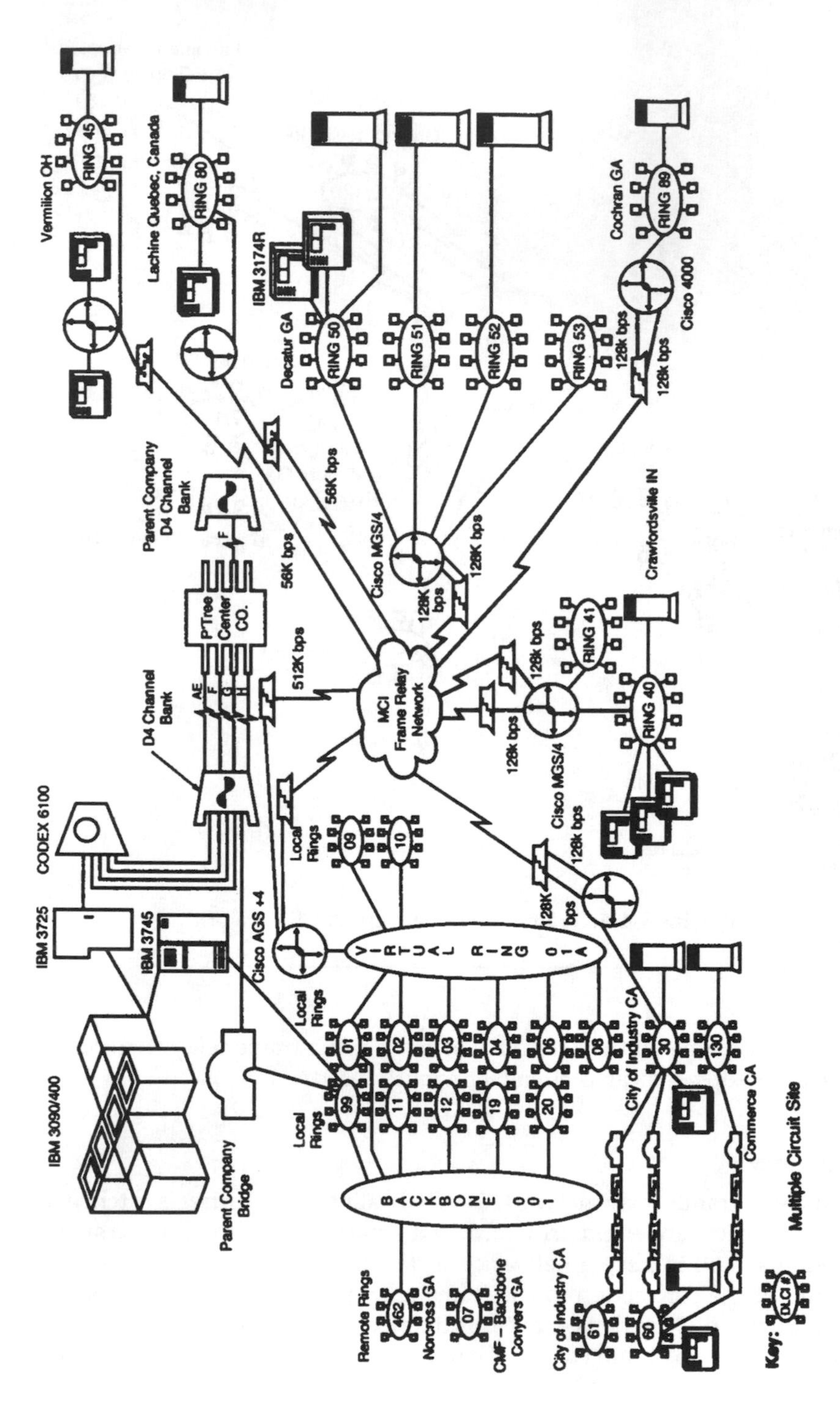

Exhibit 8-4-5. Redesigned Backbone Network

divide the load between them. If one circuit goes down, the entire load is shifted to the remaining path. A DLCI network does not automatically provide this type of backup, so it must be addressed by the network's designers. This is illustrated in Exhibit 8-4-4.

BENEFITS OF FRAME RELAY

For five sites, voice and data communications were combined on a T1 link to the point of presence. The elimination of the separate local data circuits saved approximately $12,000 per month. Voice and data communications were combined using a drop and insert channel service unit (CSU), which provides 20 voice circuits for four sites and 16 voice circuits for the host site. Data and voice are separated by the local provider with the data then being forwarded to the MCI frame relay point presence. The drop and insert CSU provides the alternating mark inversion signaling for the Northern Telecom voice switches.

Frame relay provides the ability to overcommit the circuit. The committed information rate can be significantly higher than the actual available bandwidth, allowing communication to several sites simultaneously through one circuit.

Costs for the original SNA spoke and hub network exceeded $300,000 per year. Costs to fully mesh the network would have run over $500,000. Costs for the frame relay network, however, are slightly more than $200,000 per year. Frame relay offers low-cost connectivity to additional sites (at $75 to $150 for each additional permanent virtual circuit) and entails less overhead than an X.25 network. The structure of the new frame relay network is illustrated in Exhibit 8-4-5.

SUMMARY

The frame relay network described in this chapter markedly reduced delay time as perceived by network users, reduced network costs, improved network reliability, and lessened the cost and difficulty of network expansion.

Costs and user's networking requirements are driving the industry toward the implementation of high-speed frame relay, or hyper-frame relay, which offers committed information rates of either 4M- or 16M-bps for LAN interconnection.

8-5
Network Controls

Frederick Gallegos
Steven R. Powell

THE GROWTH IN USER ACCESS TO INFORMATION processing has increased the amount of data, software, and other information being shared across the local area network (LAN). The term "intranet" is widely used to describe the application of Internet technologies on internal corporate LANs.

Although a LAN can consist of a variety of independent devices, such as intelligent terminals, printers, and workstations, this chapter emphasizes the networking of microcomputers. Any references in this chapter to LANs and personal computer networks refer to the networking of microcomputers. Exhibit 8-5-1 illustrates the five most common topologies for networked distributed processing systems. Most distributed or wide area networks (WANs) use a hierarchical or tree topology. Today, about half of distributed or wide area networks use hierarchical or tree topology and the rest are mesh topology.

MAJOR ISSUES IN NETWORK INSTALLATION

The control issues that this chapter focuses on go hand in hand with the planning, implementation, and operation of networks. The following list provides a quick reference identifying the installation issues most relevant to audit and control:

- *Wiring and cabling.* Approximately 80% of network problems are due to improperly installed or poor-quality cabling. When quality is sacrificed for cost, operations may be detrimentally affected once the network is put into service, resulting in retrofit costs.
- *Throughput or traffic.* This is an area in which effective planning and visits to organizations that have implemented a similar network can pay off. Major decisions include specifying the type of file servers needed to support the desired application and determining connectivity with other

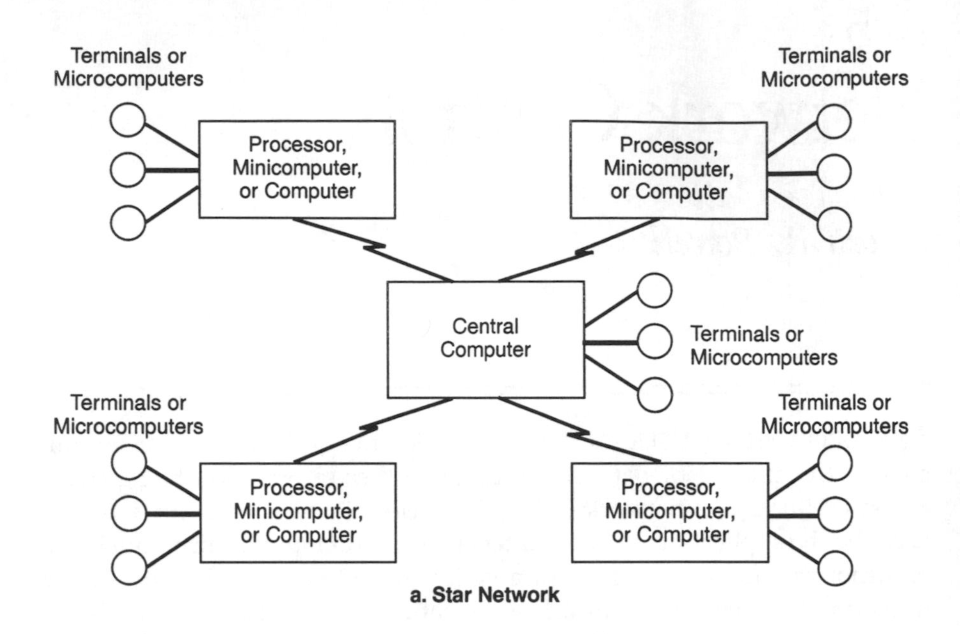

a. Star Network

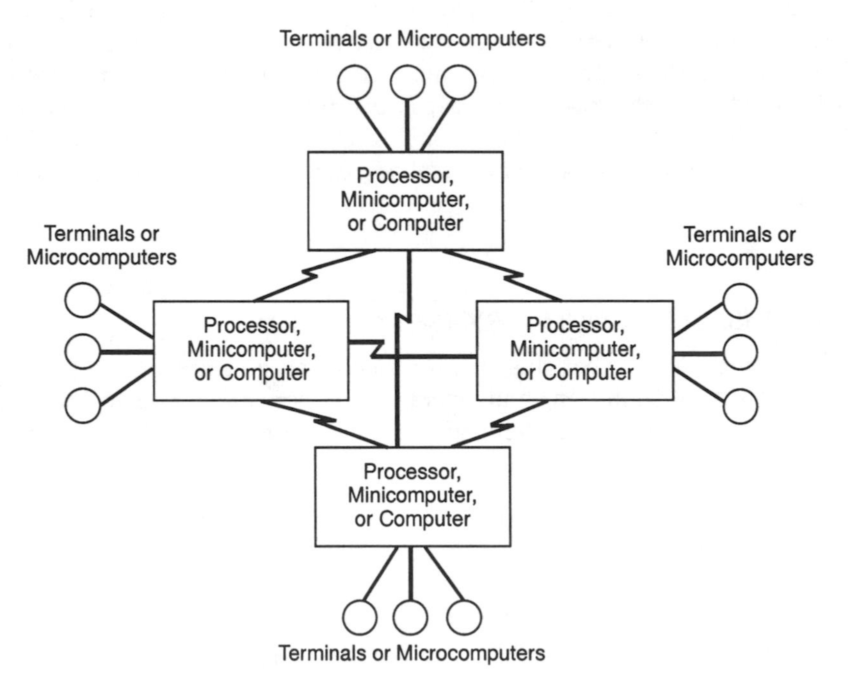

b. Fully Connected Mesh Networks

Exhibit 8-5-1. Network Topologies

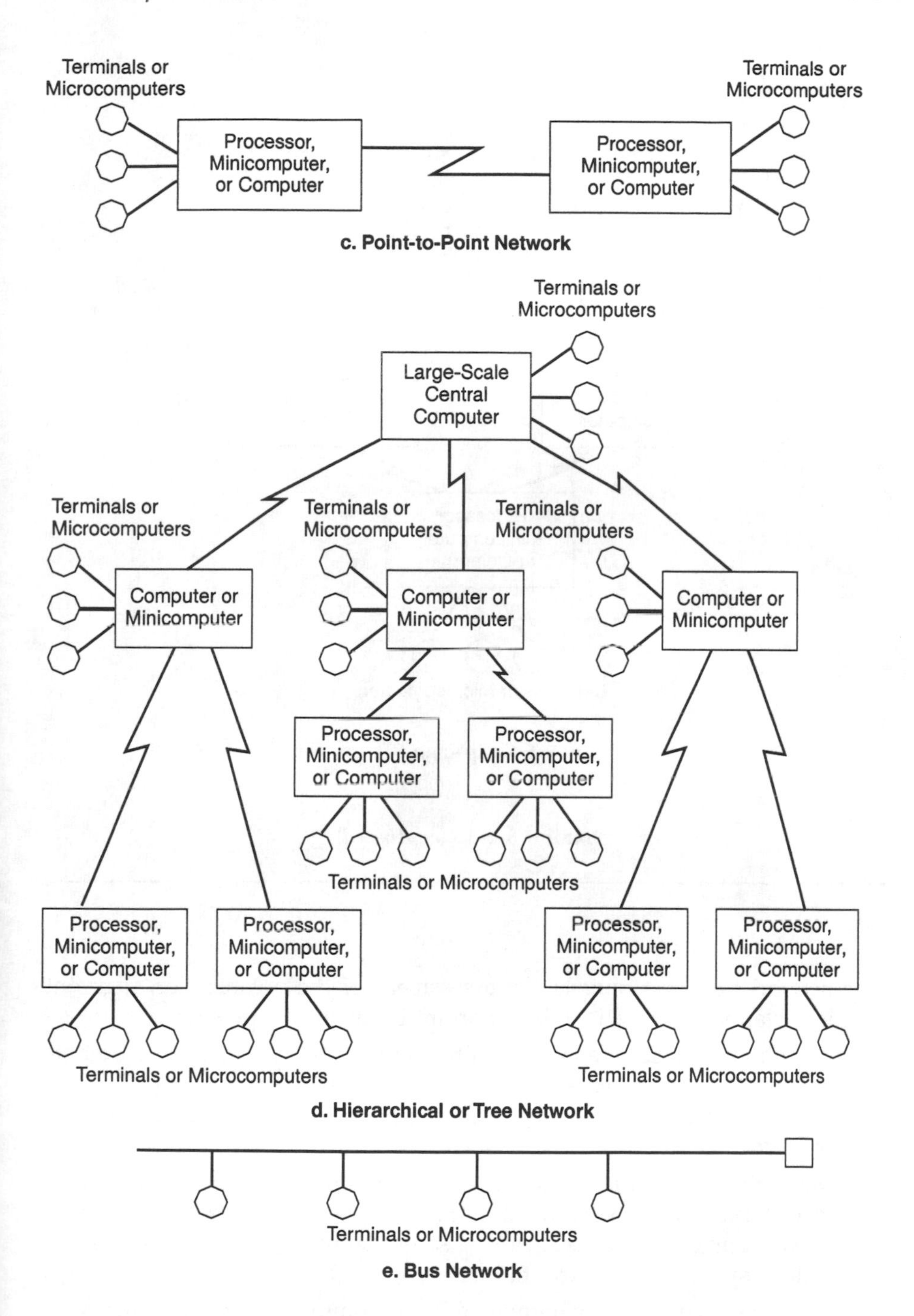

Exhibit 8-5-1. ***(continued)***

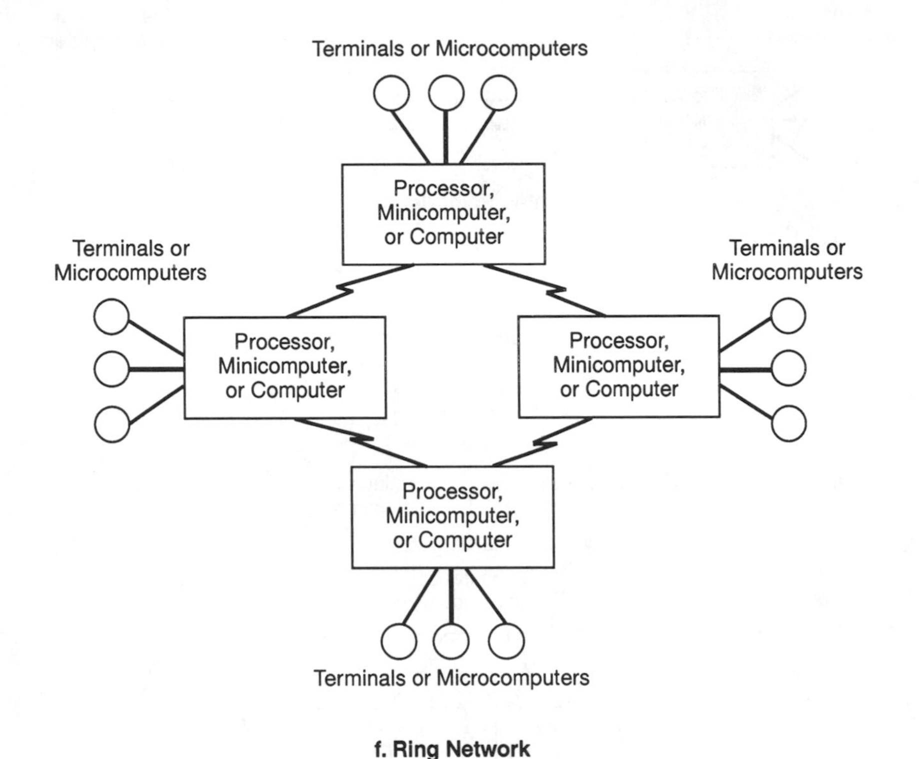

f. Ring Network

Exhibit 8-5-1. *(continued)*

networks and peripherals. An assessment of the organization's current and planned applications is important input to this process.

- *Layout.* To facilitate enhancements and the ability to adapt to organizational change, layout should be part of the planning process. This can save an organization money in the long run and help identify control concerns.
- *Measuring performance.* Tools that help monitor performance and analyze change can be very valuable to the network manager. In addition, network analysis products can assist in diagnosing data bottlenecks as well as system and control failures.
- *Intranet assessment.* Techniques to help monitor performance and analyze change are also invaluable to the network manager.

Because a network represents a substantial investment in PC equipment, network equipment, network operating software, shared software, individual user software, data, personnel, administration, and training, the network or the information contained in the network files may represent an organization's major assets. Therefore, an organization must evaluate the controls to be used in the LAN. Controls can be classified as physical security and access controls, environmental controls, software and data security controls, and administrative security controls. Each of these is discussed in the following sections.

PHYSICAL SECURITY AND ACCESS CONTROLS

The objective of physical security and access controls is to prevent or deter theft, damage, and unauthorized access and to control movement of network-related equipment and attached devices. Some physical controls also prevent unauthorized access to data and software.

General physical controls that can be used to protect office equipment and personal computer networks include personnel badges, which help employees identify authorized personnel, and alarms and guards, which deter theft of network equipment. In addition, placement of the network equipment and office design will further secure the network.

For example, network equipment should be placed in areas in which office traffic is light. If possible, the microcomputers, printers, and other equipment should be placed behind locked office doors. Data center staff may want to use combination locks to prevent duplication of keys; another alternative is to use a locking device that operates on magnetic strips or plastic cards—a convenient device when employees regularly carry picture ID badges.

Network equipment should be attached to heavy immovable office equipment, permanent office fixtures, special enclosures, or special microcomputer workstations. The attachment can be achieved with lock-down devices, which consist of a base attached to permanent fixtures and a second interlocking base attached to the microcomputer equipment. The bases lock together, and a key, combination, or extreme force is required to remove the equipment. All network equipment must be locked down to prevent unauthorized movement, installation, or attachment.

Many PCs and other equipment attached to the network may contain expensive hardware and security-sensitive devices. The removal of these devices not only incurs replacement costs but could cause software to fail and may be a means of circumventing security or allowing for unauthorized disclosure of such company-sensitive information as customer lists, trade secrets, payroll data, or proprietary software. Internal equipment can be protected by lock-down devices, as previously discussed, and special locks that replace one or more screws that secure the top of the equipment. These special locks are called CPU locks because they prevent access to the CPU area.

Cabling enables the various users and peripheral equipment to communicate. Cabling is also a source of exposure to accidental or intentional damage or loss. Damage and loss can occur from the weather, or by cutting, detaching or attaching to and from equipment, and other incidents. In many networks, if the cable is severed or damaged, the entire system will be impaired.

Cabling should not be accessible to either the environment or individuals. The communications manager may want to route and enclose cabling in an electrical conduit. If possible and if the exposure warrants the cost, cabling can also be encased in concrete tubing. When the cable is encased, unauthorized access through attachment is lessened. In addition, unauthorized movement of the cabling will not occur easily and will enable the network manager to more efficiently monitor and control the network and access to it.

To alleviate potential downtime, cable may be laid in pairs. In this arrangement, if one set is damaged, the alternate set can be readily attached. The second pair is usually protected in the same manner as the original but is not encased in the same tubing, thus preventing the same type of accident from damaging both cables.

Notebook computers should also receive care and attention. They are even more vulnerable than other equipment because they can be taken offsite, where they are susceptible to theft or sabotage, such as viruses.

ENVIRONMENTAL CONTROLS

All network equipment operates under daily office conditions (e.g., humidity, temperature, smoke, and electrical flow). However, a specific office environment may not be suited to a microcomputer because of geographical location, industrial facilities, or employee habits. A primary problem is the sensitivity of microcomputer equipment to dust, water, food, and other contaminants. Water and other substances not only can damage the keyboard, CPU, disk drive, and diskettes but may cause electrocution or a fire. To prevent such occurrences, the network manager should adhere to a policy of prohibiting food, liquids, and the like at or near the microcomputer.

Although most offices are air-conditioned and temperatures and humidity are usually controlled, these conditions must nonetheless be evaluated by the network manager. If for any reason the environment is not controlled, the network manager must take periodic readings of the temperature and humidity. If the temperature or humidity is excessively high or low, the microcomputer equipment and the network should be shut down to prevent loss of equipment, software, and data. When microcomputer equipment is transported, either within the building or especially outdoors to a new location, the equipment should be left idle at its new location for a short time to allow it to adjust to the new environmental conditions.

Airborne contaminants can enter the equipment and damage the circuitry. Hard disks are susceptible to damage by dust, pollen, air sprays, and gas fumes. Excessive dust between the read/write head and the disk platter can damage the platter or head or damage the data or programs. If there is excessive smoke or dust, the microcomputers should be moved to another location. Small desktop air filters can be placed near smokers' desks to reduce smoke, or the responsible manager can limit smoking to specific locations, away from microcomputer equipment.

Static electricity is another air contaminant. Static electricity can be reduced by using antistatic carpeting and pads placed around the microcomputer area, antistatic chair pads and keyboard pads, and special sprays that can be applied to the bottoms of shoes. Machines can also be used to control static electricity in an entire room or building.

Major causes of damage to network equipment are power surges, blackouts, and brownouts. Power surges, or spikes, are sudden fluctuations in voltage or frequency in the electrical supply that originate in the public utility. They are more frequent when the data center is located near an electrical generating plant or power substation. The sudden surge or drop in power supply can damage the electronic boards and chips as well as cause a loss of data or software. If power supply problems occur frequently, special electrical cords and devices can be attached to prevent damage. These devices are commonly referred to as power surge protectors. Users who take laptops offsite should be issued the appropriate support peripherals such as surge protectors and electric connectors.

Blackouts are caused by a total loss of electrical power and can last seconds, hours, or days. Brownouts occur when the electrical supply is diminished to below-normal levels for several hours or days. Although brownouts and blackouts occur infrequently, they are disruptive to continuing operations. If microcomputer use is essential and the organization's normal backup power is limited to necessary functions, special uninterruptible power supply (UPS) equipment can be purchased specifically for the microcomputer equipment. UPS equipment can be either battery packs or gas-powered generators. Battery packs are typically used for short-term tasks only (e.g., completing a job in progress or supporting operations during a transition to generator power). Gas-powered generators provide long-term power and conceivably could be used indefinitely.

SOFTWARE AND DATA SECURITY CONTROLS

Data and software security and access controls are the key controls over a microcomputer network. (Software and data security controls are referred to as data security throughout this chapter, because the same controls provide security for both data and software.) A PC network has two levels of data security. The first is access and use of local standalone microcomputer capabilities. The second is access and use of the network system and its capabilities. These two levels can be integrated through the installation of certain security software and hardware. However, the organization must be aware that these two levels exist and are necessary to provide the security for all the network's functions, as required.

The objective of data security is to prevent access by unauthorized users and to restrict authorized users to needed data and functions. Authorized users should be restricted to the use and access of specific data, screens, software, utilities, transactions, and files. Exhibit 8-5-2 provides an overview of the security and administration capabilities of the major LANs.

Password and Data Access

The key to network security is user authentication. Although badges and personal IDs are common authentication tools, they can fail at all levels of security. Other methods of authentication can be obtained by the computer system itself. Special software and hardware products exist that allows authentication of users through the entering of an ID number and a password.

A PC network should have several levels of password and user ID security requirements. One level protects local microcomputer access and use, and the second protects network access and use. The user ID and password used for the local microcomputer will be a double safeguard to network access. To access the network, the user must have a valid user ID and password for local use, and that password can be set up to restrict access to the network user ID and password screen. In addition, this double security can be used to restrict users to specific applications and to limit specific users to the network. Restricting a user to a specific workstation could be accomplished by matching the user's ID and password against embedded firmware; alternatively, a system call-back facility could be used to reauthenticate the user. Another layer of protection is at the intranet level, which requires validation of user IP addresses.

User IDs and passwords are practical and economically feasible in most situations. If the risk of loss is high and the cost can be justified, card IDs or a voice authentication system can be added to the user ID and password system.

In addition to the use of passwords and user IDs, a security system must list the user IDs and the locations of all microcomputer security violations. The actual violation (e.g., an attempt to use an incorrect password or to access

	NetWare Version 2.2	NetWare Version 4.0	LAN Manager Version 2.1	Banyan Vines Version 5.5	NCR Star Group Version 3.6	Appletalk Version 7.0
Security						
Encrypted passwords can be sent over the network	Yes	Yes	Yes	Yes	Yes	Yes
Access control is provided by:						
—Time and date	Yes	Yes	Yes	Yes	Yes	Yes
—Group	Yes	Yes	Yes	Yes	Yes	Yes
Historical error and log status can be checked	Yes	Yes	Yes	Yes	Yes	Yes
Names of users can be displayed	Yes	Yes	Yes	Yes	Yes	Yes
Administration						
Open files can be monitored	Yes	Yes	Yes	Yes	Yes	Yes
Fault tolerance is checked through disk mirroring	Yes	Yes	Yes	Yes	Yes	No
Can report:						
—Network errors	Yes	Yes	Yes	Yes	Yes	Yes
—Faulty packets	Yes	Yes	Yes	Yes	Yes	Yes
—Current server load as a percentage of total	Yes	Yes	Yes	No	Yes	No

Exhibit 8-5-2. Overview of Major LANs: Security and Administration Capabilities

unauthorized systems) must be logged, and the security administrator must investigate and resolve the problem.

Duties must be segregated in microcomputer networks as they are with local microcomputer use, because the traditional separation of duties seen in mainframe systems is circumvented. Computer operations, systems development, systems programming, and application programming in a microcomputer environment are usually at the hands of the user, who can perform various programming, operations, and application functions by simply reading a book and applying the knowledge.

To segregate duties on the network, security software must be in place. The software not only limits access to specific programs and data but limits the user's capabilities to DOS commands and programming tools. In addition, the software must monitor what a particular command or program is doing—for example, formatting disks or making global copies or deletes. The security software will prevent accidental errors and possible theft of sensitive data. This is especially true with an intranet, where potential intruders or viruses can be contained at the firewall.

Encryption

Encryption is a technique of creating unintelligible information (i.e., ciphertext) from intelligible (i.e., cleartext) information. The same algorithm is used to create the ciphertext from cleartext and convert the ciphertext back to cleartext. The algorithm uses a key to tell the mathematical equation where to start producing the ciphertext. The key must be given to the program that will decipher the text. For encryption to be effective, the passage of the key and the algorithm must be kept secret. Programs must not be hard-coded with the key, nor should the key be written down. The key and the algorithm are often encrypted or written in machine language to prevent the casual user from intercepting code messages.

Encryption is useful in protecting sensitive information (e.g., password, user ID, and payroll files). Any sensitive message or transmissions must be encrypted to prevent interception either by someone on the network or by someone using highly sophisticated electronic equipment to capture emitted radiation. (The tapping of such emissions is more of a problem with twisted-wire cable than with the microcomputer itself.) Use of fiber-optic cabling in place of twisted-wire cabling can help protect transmissions, because fiber-optic cable resists tapping. However, the use of sniffers (i.e., traffic analyzers) to intercept transmissions continues to pose a serious threat to the security of transmissions.

A combination of encryption and fiber optics ensures transmission security. Encryption can be provided by either hardware or software. It can be as uncomplicated as a simple algorithm or as complex as the National Institute of Standards and Technology Data Encryption Standard (DES). The choice depends on the level of security required or desired by the organization. If encryption is widely used within the organization, the public key system should be used. (Public key encryption is also a good way to transmit a DES key.) The public key system requires that the sender create the ciphertext on the basis of the receiver's public key. The receiver deciphers the information using his or her private key. The public key system eliminates the passing of the key, which can threaten its secrecy.

Backup, Recovery, and Data Storage

Data and software on local microcomputers and on the network must be copied and archived periodically. Backup copies can quickly and accurately restore misplaced diskettes and erased files and diskettes. The backup method must allow the restoration of individual files, directories, partitions, and disks. Tape backups that allow restoration only of the entire hard disk can cause data integrity errors on data for files and directories that do not require restoration. If one file requires restoration on the basis of a previous day's backup, any files updated since the backup will not reflect those updates after the restoration.

Users should perform the backups for shared data. The data backup should be in the same format as that on the original disk. Encrypted files must be backed up as encrypted files, and files that require a password to be accessed must be stored in that manner. This prevents circumvention of security by accessing the backup and copying its contents. Moreover, if the backup format is not the same as the original, a backup tape that is used to restore information there will weaken security until the files are reprotected.

Software and Other Controls

Software that accesses shared data on the network must be designed specifically for a multiuser environment. Single-user software may cause data integrity programs, data access problems, or network crashes. For example, if user A accesses a file and begins changing information, the applications software does not actually change the data on the disk but changes the data in the memory of user A's microcomputer. Further assume that while user A is editing or inputting, user B edits the same file. When users A and B then save their copies of the same file, the only changes that were made to the file are user B's—all of user A's changes were overwritten when user B saved the file.

One method of avoiding the previously mentioned access problems is to design access locks. These locks prevent the user from updating specific data while another user is already updating that data. Locks can be provided by either applications or network operating system software. These locks can be user requested or automatically implemented. They limit access to data fields, records, file transaction types (i.e., read or write), directories, or software systems.

Access locks must be used for all files accessed by applications that are used on the network. Applications must lock data that is necessary to complete a particular task in order to prevent deadlock, which occurs when application A has locked access to a file or piece of data and then requires a second item. Meanwhile, application B has locked application A's desired second item and now requires A's locked data. Neither application will release the data until the other releases its data; as a result, both applications are left waiting for the other to release its data. Applications software and the network operating system software must provide a method of releasing inaccessible files or data without stopping the application.

Data and software used in the PC network can be controlled by using diskless workstations. The major loss of data, security, controls, and privacy occurs through use of diskettes. Users can copy data or software either up or down onto the network. If applications already exist on diskettes, the diskettes could be given to the network administrator to review their contents and load the software. Thus, the network would be controlled from loss of privacy of sensitive data. The diskless workstation can provide greater control over local and network access but does require greater administrative and organizational controls.

ADMINISTRATIVE SECURITY CONTROLS

The administration of the PC network is similar to the management of any information systems facility. Audit, security, and network management's main objective is to prevent, detect, and correct unauthorized access to the network's hardware, software, and data to ensure the network's sound operation and all security surrounding local equipment and the PC network processing.

Before any security controls can be implemented, the responsible IS auditor must perform a risk assessment of the network. Risk assessment is the task of identifying the assets, the threats to the assets, and the probability of loss of the assets as a result of a particular threat. The loss is determined by quantifying the dollar value of the assets and multiplying that value by the probability of the loss occurring.

After implementation of the necessary controls, daily management of the network is required. Daily management ensures that security controls are maintained, though changes occur in the software, applications, and personnel. Daily management can be classified into various categories, which are discussed in the following sections.

Physical and Environmental Controls Management. All such controls in active use must be tested periodically. Such testing includes the evaluation of the effectiveness of current controls and the implementation of additional controls as determined to be necessary. The results of testing of physical and environmental controls should be reported to senior management.

Data Access Management. The security or network administrator must assign and maintain user IDs and passwords and associated file and data access schemes as well as receive computer-generated reports of attempted unauthorized accesses. Reports on data access and traffic analysis should also be reviewed. These reports will allow the administrator to manage network growth and help foresee future security needs.

Policy and Procedures Documentation Review. The objectives here are to provide standards for preparing documentation and ensuring its maintenance. The network manager must set documentation standards so that when employees change jobs, become ill, or leave the organization, replacement personnel can perform the tasks of that employee. The manager must also periodically test the documentation for clarity, completeness, appropriateness, and accuracy.

Intranet Management Review. A major problem associated with an intranet is security. First, an updated, continuously maintained security policy needs to be established and actively used by personnel. The policy should answer the questions: What data needs securing? Should data be secured? What type of password security is to be used?

Once the security policy is established, the firewall is the next decision point. If a LAN is connected to the intranet, it is vulnerable to penetration and its contents can be compromised from anywhere on the Internet. One of the major solutions is the dedication of a firewall or computer to protecting the LAN. Some of the major checks should be: ease of administration, security services offered, security of the firewall, and availability of the firewall to support new services.

Data and Software Backup Management. Backup media must be labeled, controlled, and stored in an appropriate manner. The network manager must maintain control logs of all backups as well as provide documentation on how to recover files, data, directories, and disks.

Other Management Controls. The internal audit department, external auditors, contingency or disaster recovery planning, personnel background checks, and user training are included in this category. The auditor can aid in establishing proper testing requirements and in reviewing, testing, and recommending the proper controls to establish the necessary safeguards. Contingency planning or disaster recovery is essential to ensure the availability of the network. The contingency plan establishes the steps to recover from the destruction of hardware, software, or data.

Personnel background checks must be performed on all data center employees who have access to key organizational information. The background check should involve a review of credit history, financial health, personal problems, and other areas that may identify potential risks. This information can help avoid potential integrity breaches before they occur.

User training must be established for all network functions. Users must be trained in microcomputer use, general computer knowledge, security, policies and procedures, consequences of noncompliance, and general network use. In addition, users should undergo more specific training for the different software on the network as required. Such basic training can prevent many problems from occurring.

NETWORK OPERATIONS CONTROL

The major objective of network operations is to provide a high-performance network that meets user needs. To meet this objective, the following functions must be performed:

- Monitoring network performance statistics.
- Network testing.
- Logging of network activity, including access attempts.
- Real-time restoring of the system in the event of a system failure.
- Monitoring and maintaining the firewall.

Currently available network management systems provide the tools needed by network administrators to manage network performance. Such network management software offers facilities for online surveillance, modeling, collection of current or historical network statistics, traffic analysis, and the recording and reporting of information that can assist in the prompt detection and correction of system failures or security violations. Thresholds can be established for reporting monitored information.

In addition to integrated network management software, individual products are available that perform single, specialized tasks, such as monitoring, diagnostics, and testing. For example, vendors offer products for monitoring and testing network interfaces, terminals, multiplexers, modems, transmission lines, and network software. Some products can be used to create color-coded network maps showing each router, bridge, host, terminal, and connecting link in real time.

The following sections describe each of the key functions that must be performed to effectively manage a secure network.

Monitoring Network Performance. A high-performance network is a reliable network with minimum downtime. The performance of a network can be measured by regularly monitoring such statistics as long mean time between failures and short mean time to repair.

In addition, the network administrator can analyze traffic statistics (e.g., by reviewing operations logs) to determine ways to improve network efficiency. For example, a network status report might indicate that the network slows down when certain frequently used programs are run; the administrator might respond by rescheduling or limiting the use of these programs or by adding an extra file server to handle the load. Other events that affect network reliability and availability and therefore should be monitored include transaction priority, circuit errors, and hardware and software component failures.

The LAN protocol analyzer is a key tool for monitoring real-time network status. The protocol analyzer captures and analyzes packets of information transiting the LAN. For example, Network General Corp.'s Sniffer captures and reports such transaction-related information as file length, user name, source, destination, last access time, and last update date and time. It should be recognized, however, that protocol analyzers can pose a security threat because these products can be used to read IDs and passwords being transmitted to a host or server if such information is not encrypted.

As networks become more complex, any network failure may have serious consequences for the organization. Effective network monitoring and testing is essential to minimize this threat.

Network Testing. Network testing products are designed to reveal the following common network failures:

- Faulty cable connections.
- Malfunctioning Ethernet controllers.
- Overloaded servers and node devices.
- Defective software.
- Improperly configured networks and network software.
- Improperly installed or malfunctioning transceivers.
- Excessive network traffic.
- Overlapping network addresses.
- Firewall malfunction.

The major operational tests can be grouped into the following categories:

- Functional testing to make sure that the system functions as expected with actual data.
- Integrated testing to ensure that software components function together as a complete system.
- Stress testing to ensure that the system provides optimal performance with expected traffic loads.

If a problem is identified in testing, the key step is to isolate the problem; the source of a failure may be hidden among various programs or devices. It should also be recognized that most interruptions result from relatively minor system failures and human error and not from major software or hardware problems. To reduce the potential for human error, it is important that manuals of operations and correction procedures be available for employee reference and training; these manuals should be updated whenever changes are implemented.

Many small organizations are unable to afford expensive test equipment and in-house experts. Such enterprises can use carrier-provider test services, which can be accessed from the users' local sites by means of dial-up devices. To provide additional monitoring capabilities, many companies use a secondary, low-speed data channel (independent of the production data channel) to perform network surveillance without disrupting normal system processes. New technologies are also being developed for network testing and monitoring; these include such devices as LAN/WAN integration test sets, fiber distributed data interface (FDDI) testers, frame relay testers, and integrated services digital network (ISDN) testers.

Network Logging. A network system should have a built-in, automated program for recording all significant network activity. Network system logs (also referred to as audit trails) are a critical resource for security administrators, network administrators, and auditors. The purpose of a network system log is to provide information that can be useful for investigating illegal log-on attempts and for system auditing. The data collected typically includes normal traffic flow statistics, operator errors, unauthorized and authorized access attempts, and authorized accesses at various security levels.

Critical problems identified by network system logs should be promptly investigated by the network or security administrator. These logs should be properly saved and archived.

Network Backup and Recovery. Potential network failures can result from such causes as cabling problems, power failures, fires, natural disasters, and malicious attacks by network intruders. To ensure information availability and reliability, data must be protected from the damaging effects of system failure. For example, daily backup of critical data is mandatory.

In addition to data backup and recovery, consideration must be given to recovery of facilities. Most large organizations have established backup facilities to provide ongoing processing capabilities in the event of a disaster. For example, a major New York newspaper with 700 installed workstations on its LAN established a second main processor and twin gateways, bridges, servers, and network lines to prevent an interruption of service during an election night or other critical period.

Prevention is the key to reducing damage experienced in a disaster. Therefore, an adequate contingency and recovery plan must be designed and implemented. The major concerns in developing a contingency plan include:

- Ensuring that backup facilities provide all necessary equipment, including processors, printers, software, power supplies, modems, and alternative communications lines, gateways, and bridges.
- Forming and training an emergency response team.
- Designating priority levels that list resources in their order of importance to the organization. For example, such major applications as decision support systems may require faster recovery than would an E-mail system.
- Maintaining spare equipment for major system components at the backup site and checking their working condition periodically. An inventory of equipment should be prepared and maintained.
- Arranging a temporary site for essential services, an external service center, or equipment rentals that might be required in the event of a disaster.
- Evaluating and, where appropriate, implementing fault-tolerant support systems to provide backup. Fault-tolerant systems use disk mirroring, optical disk backup, and other technologies to assist in system recovery.

Monitoring and Maintaining the Firewall. Firewalls are a combination of hardware and software tools that are designed to help manage network access and aid in security policy implementation. All firewall reviews should focus on the following areas:

- Ease of administration. The firewall is easy to administer and maintain.
- Security services. Firewalls provide security through filtering methods, authentication schemas, and decryption.
- Support for new services. Firewalls can migrate to and add new support services.

The firewall has the ability to implement full-tolerant support systems to provide backup. Fault-tolerant systems use disk monitoring, optical disk backup, and other technologies to assist in system recovery. Potential network failures can result from causes such as malicious attacks by network intruders or the passage of viruses or other destructive elements. To ensure information availability and reliability through the intranet, the firewall must continually be monitored and evaluated, and maintained to ensure that corporate data and files must be protected from the damaging effects of system firewall failure.

SUMMARY

Controls must be established to protect the information stored and processed on PC-based networks. The controls to be implemented include:

- Physical security and access controls.
- Environmental controls.
- Software and data security controls.
- Administrative controls.
- Network operations controls.

Passwords provide one of the key controls for ensuring network security. Passwords establish data and software access schemes that provide for segregation of duties. Password controls must be established and maintained by the network administrator in conjunction with the information security manager. The network administrator must ensure that all controls operate as intended and that they provide an appropriate level of security in accordance with organizational objectives and policies.

8-6 Remote LAN/WAN Connections: A Case Study

Charles Breakfield
Roxanne Burkey

THE ORGANIZATION DISCUSSED IN THIS CHAPTER is a national corporation headquartered in Virginia. Currently, the company's corporate locations interconnect through a Banyan wide area network (WAN) for access to business software, national data bases, and E-mail. Each location contains a network service area for user support and connects to other locations, including the Virginia headquarter offices.

Space reconfiguration at a corporate location in Texas prompted a move of approximately 125 support staff to the fourth and sixth floors of an adjacent building, approximately 100 feet away from the primary building. The Banyan WAN resided on the seventh floor of the primary building, and the main support areas for data access were to remain there.

The relocating staff, however, required network data access from two of the existing servers on the seventh floor to perform their work, but providing network servers and the associated support staff to the new location was not an option. Exhibit 8-6-1 illustrates the wiring layout of the primary building.

An additional consideration was that the organization was consolidating business activities. Broader plans for the year were to consolidate network operations to enable an overall reduction in support staff. Therefore, any system changes made for this relocation of the 125 support staff had to:

- Be limited in cost.
- Use existing materials.
- Require no additional staffing.
- Adhere to the existing network standards.
- Contain limited leasing and licensing agreements.

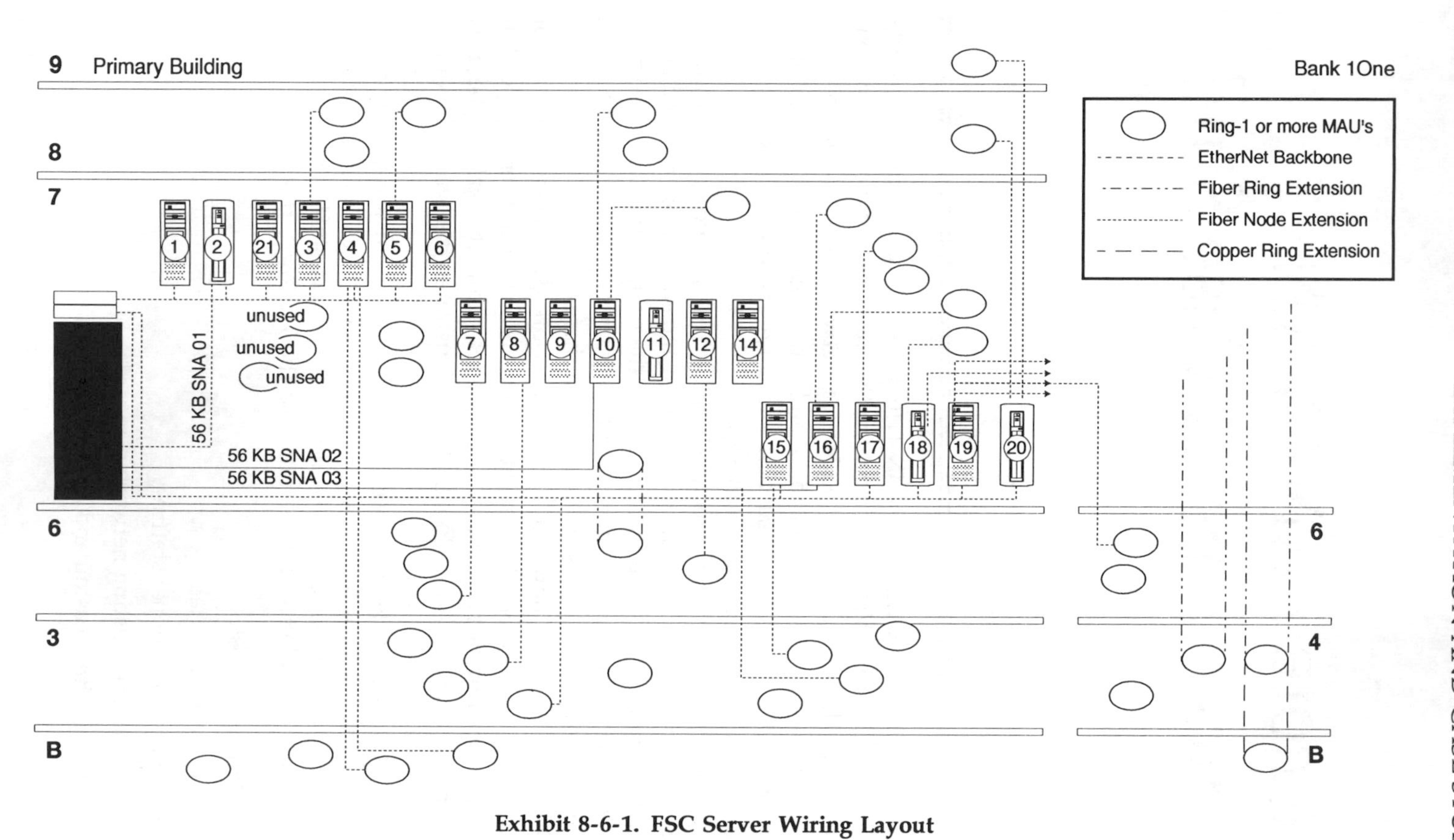

Exhibit 8-6-1. FSC Server Wiring Layout

A feasibility study determined the technology needed for data connection from the fourth and sixth floors of the adjacent building to the seventh-floor server room of the primary building.

This article describes the three connectivities (i.e., fiber optics, infrared, and microwave) considered in the feasibility study, the advantages and disadvantages of each, the estimated time to complete installation, and the draft estimated costs. The article also puts all these factors in the context of the company's business operations.

Finally, the article also supplies a proposed plan for the solution's implementation and installation. The recommended course of action is based on the technology feasibility, time frames, costs, and flexibility for known future organizational changes.

ESTABLISHING A BASELINE

A feasibility study certified the data connection method among the existing services and the remote new services and outlined the most viable option for connectivity within the confines determined by upper management. The study began with an itemization of existing resources and criteria for deciding on technology options.

Existing Resources

Staff members familiar with the existing Banyan network conducted the study over a nine-week period. Detailed reviews of the existing and planned systems included a review of the wiring configuration for the new space, data specifications, building requirements, and existing AT&T services to determine the project requirements.

The staff investigated the possibility that appropriate technology might already be in use elsewhere in the company, as well. They conducted a nationwide, corporatewide investigation of the options available for data connectivity requirements. A step was also completed to locate in-house support staff with experience with similar installations, problems that might come up with a specific technology, and available materials.

DEFINING THE OPTIONS

The criteria for evaluating the options were determined based on the initial request and follow-up discussions with staff. The criteria, in order of importance, included the operational time frames, connection reliability, technical support provisions, and cost.

The physical hardwire and wireless connection possibilities were reviewed based on the criteria, and three viable options were selected to undergo in-depth analysis. After the technology investigation was complete, a review with building management was conducted to ensure acceptance of the recommendation.

DEFINING THE EXISTING RESOURCES

The existing Banyan WAN servers were located on the seventh floor of the primary building. Staff was in place at the primary location to support the requirements of Banyan WAN users at the primary building and of users selected for the move to the adjacent building.

The adjacent building's floor configuration included free-standing cubicle work areas and walled offices with free-standing desks. Each workstation required power to support the electrical equipment, an RJ-45 data jack for the phone connection, and an IBM data connection to support the existing Token Ring topology.

Wiring closets were installed on each floor at the remote location; the closets had three wire racks as well as punch blocks. The wire racks contained the data cables terminated into patch panels and mounted in the top third of the wire racks. The power drops for the walled offices included voice, data, and power connections. Power poles fed voice, data, and power connections for each cubicle. Each workstation contained Type 2 cable, Type 1 data cable for the Token Ring connection, and 4-pair wire for the voice connection.

DEFINING THE BUSINESS OBJECTIVE

Banyan support service for the remote location was possible by means of one of the following:

- Hiring additional staff for Banyan server support.
- Moving existing support staff to the new location.
- Providing remote support by the existing staff running back and forth between the buildings.
- Establishing a remote connection into the existing server configuration while maintaining current staffing levels.

The option that best suited the company's long-term needs was establishing a new connection into the existing configuration at the primary site, without adding support staff. The project then was to provide a data interface between the two buildings.

The requirements for data access included a network connection from the three wire racks in the fourth-floor wire closet and three wire racks in the sixth-floor wire closet of the adjacent building to the wire racks located in the seventh-floor server room of the primary building. The final link between the selected technology and the target servers would require connection to a Token Ring Type 1 drop cable. The cost considerations had to include any adapters or converters needed for linking the existing system to the recommended technology.

There were, in this case, possible limitations on the use of physical hardwiring networking items; the building management might determine certain equipment to be dangerous to the property, building staff, or other tenants. Because physical wiring above ground was not allowed, the access had to take place below ground under the parking garage or via equipment to be located on the roof.

CONNECTIVITY OPTIONS

The method selected had to provide reliable connectivity for data while remaining cost-effective. To meet this requirement, both physical and wireless connectivity options were explored. Several potential methods were reviewed; three of these were selected for final consideration.

Physical Connections

Physical wiring of a WAN is the single largest cost factor of materials and installation labor.

Twisted-Pair Cable. Cable standards include either twisted-pair or coaxial cable. Hardware connection performance for a Token Ring network environment is measured by the rate at which it can move data. Twisted-pair cable can reliably move data at the rate of 10M bps and coaxial cable can reliably move data at 16M bits.

Fiber Optic Connectivity. Fiber optic connection is considered by many today as the premiere connection for voice and data. It provides a reliable connection and allows for an increase in system throughput.

Wireless Connections

Today, wireless network connection is concentrated in three areas, including spread spectrum UHF, infrared, and microwave radio technologies.

Option 1: Fiber Optic	Option 2: Microwave	Option 3: Infrared
Impervious to electrical noise and interference typically present in all office environments	Low signal attenuation	Freedom from government regulation, no licensing or usage fees
Systems using fiber are immune to RFI	Portability of equipment, owned by the organization	Portability of equipment, owned by the organization
Fiber optic poses no risk of carrying lightening charges to computer equipment	Cost-effective for line-of-sight connection and when frequency bank congestion is low, this is cost-effective	Ease of interconnecting additional locations
Fiber is conducive to a Token Ring environment	Conducive to a Token Ring environment	Immunity to radio interference
Rapid installation	Reasonable installation time	Reasonable installation time
High data security	Signal scrambling is available through some vendors	Security of the data
Upwardly scalable		

Exhibit 8-6-2. Advantages of Fiber Optic, Microwave, and Infrared Technologies

---

Spread Spectrum UHF. This is designed to appear as background noise in most radio frequency transmitters and receivers. Consequently, data is very secure; only an authorized user is able to access the data. Data is not susceptible to interference from other signals or electronic devices; however, transmission speed is limited and best suited for a small LAN environment.

Infrared. Infrared is the type of signal used over most fiber optic links but without the fiber media. These devices can achieve speeds of 16M bits. Products available include those specifically designed for a Token Ring environment with multiple-access units. This technology is effective in environments with an unobstructed line of site.

Microwave. Microwave technology offers speed for transmission to 6.7M bits at a range of 130 feet and supports Ethernet. This technology has the benefit of providing ownership of materials, allowing lower costs on future relocations.

Eliminating Inappropriate Connectivity Technologies

Two of the technologies described in the previous sections were eliminated from any in-depth consideration. Twisted-pair cables were dismissed as an

Option 1: Fiber Optic	Option 2: Microwave	Option 3: Infrared
Additional repeaters are needed to boost the signals for long distances	The capacity of 6.7M bits is less than the 16M-bit needed at its peak times	Atmospheric conditions affect reliable data transmission
Fiber cable runs cannot be subjected to sharp turns	Outside installation support required and no existing training available	High up-front equipment investment and installation support
Cable terminations and splices must be specially prepared	Signal or transmission unreliability during excessive rains	Potential safety issue for retina damage caused by looking directly into the beam
The materials belong to the property	Potential delay of licensing from the FCC, possible installation delay	

Exhibit 8-6-3. Disadvantages of Fiber Optic, Microwave, and Infrared Technologies

option because the preliminary review indicated a lack of reliability at this project's distance—100 feet. Twisted-pair cables also have limitations in handling the anticipated data traffic.

Spread spectrum UHF was also eliminated. Even though spread spectrum UHF is a secure connectivity option, it is limited in speed and best suited for small LANs. Consideration of this application to future, smaller-scale projects was recommended.

The three options given serious consideration are discussed in the following sections. Exhibits 8-6-2 and 8-6-3 summarize the advantages and disadvantages of each technology.

THE FIRST OPTION: FIBER OPTICS

Fiber optics is a communications media linking two electronic circuits by a strand of glass. It is lightweight and small, making it more attractive for projects in which space is at a premium. A graded index fiber performs best, because of its ability to carry multiple signals, with the least amount of signal loss, due to dispersion. Wide bandwidth, low signal loss, and electromagnetic immunity are the three most outstanding features of fiber optics.

Fiber Optic Data Transmission

Fiber optic systems transmit data as a series of light pulses, generated either by light-emitting diodes (LED) or lasers. The bit error rates (BERs) for fiber optic cabling are as much as 10,000 times lower than standard electrical media. The connections are further simplified by the absence of ground loops, crosstalk, ringing, and echoing.

Light propagation through fiber depends principally on three factors, including the composition, size, and light injected into the fiber. Transmitting sources commonly use LED in place of laser. With a longer lifetime than a laser light, LED is easier to use and maintain. A LED source has a higher and broader output pattern but is not capable of single-mode compatibility. The transmitter output power is coupled with the diameter of fiber, so that power increases with core diameter.

Detectors perform the opposite function from the source by converting optical energy to electrical energy. The photo diode produces current in response to incident light. Detectors are typically packaged in the same receptacles as sources. Receiver sensitivity specifies the weakest optical signal it will receive. This is affected by the amount of noise, or signal clarity, during signal receipt as measured by bit error rate (BER) or the signal-to-noise ratio (SNR).

Fiber Optic Cabling

Fiber optic cable is available in either single mode or multimode. Single mode has an aperture of about nine microns, has a low attenuation rate, and is ideally suited for long-distance networks. Multimode is available at apertures from 50 to 100 microns and has a higher attenuation rate, because signals enter at an angle and bounce off the fiber walls as they travel. This allows for use of multiple paths, making it best suited for short-distance applications. Fiber cable comes in simplex (containing one fiber), duplex (containing a sending and receiving fiber), or hybrid, in which duplex is combined with twisted pair.

A network requires only two strands of fiber; however, a multiple strand cable is often used for backup reliability for transmission. The fiber diameter is used for this type of connection is 62.5 μm and has a standard cladding diameter of 125 μm. This combination offers high speed, low attenuation (3.75 dB/km), and a high bandwidth of 1,000 MHz/km at 1,300 nm.

Fiber optic cabling is versatile, with the ability to serve as a backbone, front-end, and back end of LAN networks. Both Ethernet and Token Ring network configurations are adaptable to fiber optic cabling in place of standard copper wiring. The cost of this media typically makes it best suited to campus, building, and data center network environments. Exhibit 8-6-4 displays the potential connection configuration considered for this project.

Advantages

Fiber optic connections offer a wide variety of advantages over other hard-wire and wireless options. Fiber:

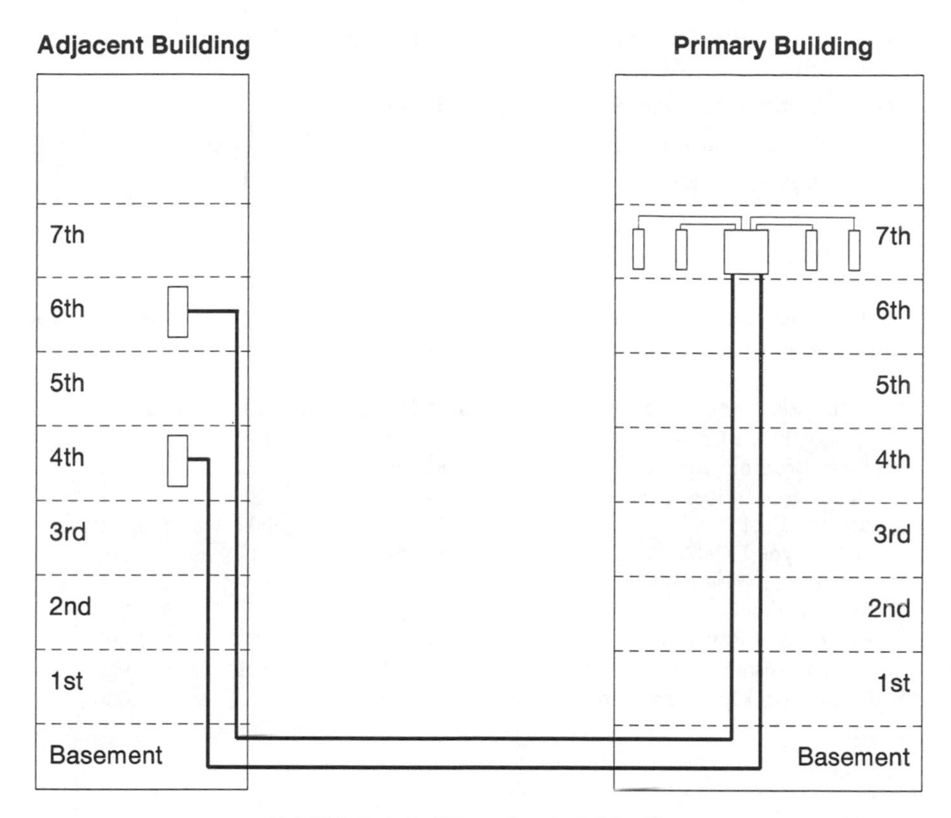

Exhibit 8-6-4. Fiber Optic Cable Run

- Transmits data as a series of light pulses, making it impervious to electrical noise and interference typically present in all office environments.
- Is immune to radio frequency interference (RFI).
- Does not conduct electricity, thus protecting computer equipment from lightning charges.
- Is a highly secure data transmission medium, because it does not radiate energy, and tapping data from it is extremely difficult.
- Is conducive to an Ethernet or Token Ring environment, as with most hardware installations.
- Is upwardly scalable.

In addition, the installation of fiber is rapid.

Disadvantages

The disadvantages of this technology include:

- Additional repeaters are needed to boost the signals over long distances, which incurs costs.
- Fiber cable runs cannot be subjected to sharp turns.
- Cable terminations and splices must be specially prepared.
- Materials cannot be removed once they are installed.

Costs

The basic cost to set up the fiber optic option for the remote location would be as follows:

Materials Description	Unit Price	Quantity	Total Price
15-foot Patch Cables	$ 45	10	$ 450
Fiber Breakout Box	$ 150	4	$ 600
Eight-strand Fiber Cable 62.5/125 μm per Foot	$ 2	1,500	$ 3,000
DB-9 Type 1 Cable	$ 50	4	$ 200
Additional Wire Racks for the Seventh Floor			$ 250
Fiber Optic Repeaters	$1,400	8	$11,200
Outside Installation	$ 30	32	$ 960
Total Fiber Optic Installation			$16,660

THE SECOND OPTION: MICROWAVE

A microwave radio system can transmit voice, data, and video and uses direction radio broadcast transmission methods, operating in the 2 GHz to 40 GHz frequency bands, as displayed in Exhibit 8-6-5.

A line-of-sight relationship is required between transmitting and receiving antennas, which must resist winds reaching 70 miles per hour. The transmission is primarily stable, though certain conditions, such as excessively hard or prolonged rain, can reduce reliability of data transmission.

There are two types of microwave systems available, the short haul and long haul. The short haul system is for transmissions up to and including 250 miles. This method is typically used by universities, businesses with multiple locations, hotel chains, and hospitals. The long haul system is designed for transmissions greater than 250 miles. This method is typically used by common carriers (i.e., AT&T, GTE, and Sprint), utility companies, oil companies, broadcast companies, and paid television.

The increased amount of satellites available for transmitting signals is reducing the overall ongoing costs of microwave systems. For international businesses, satellite transmission is an alternative to poor voice and data

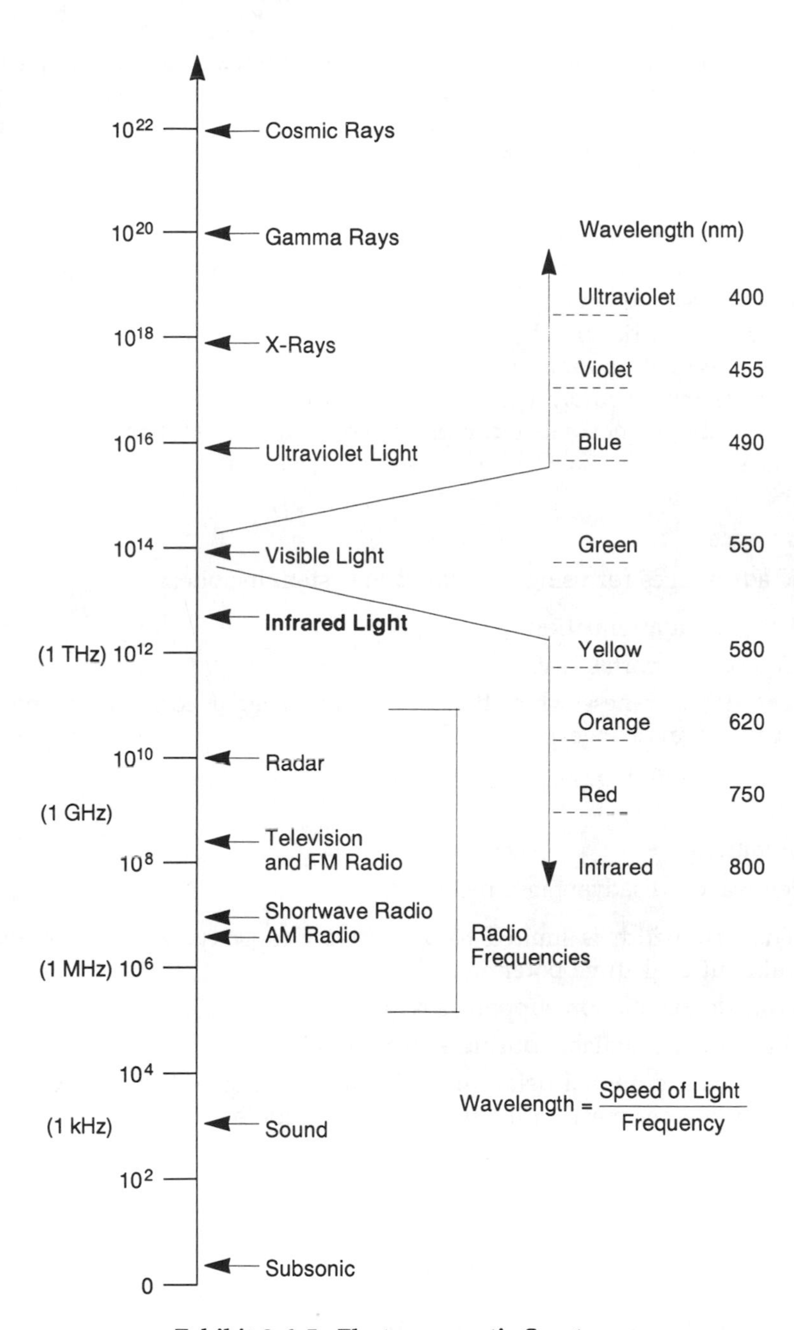

Exhibit 8-6-5. Electromagnetic Spectrum

communications in some countries' public communications systems, particularly those classified as third world. The wireless aspect of the system provides portability for companies that change locations frequently or those planning future location changes.

Licensing Requirements

Microwave technology requires licensing, based on the frequency bandwidth. Licenses are issued by the FCC and take from two weeks to several months to obtain, depending on the request load. Licensing is granted for one year, with annual renewal options.

Antennas must comply with FCC standards for acceptable performance. The connections must also comply with Underwriter Laboratories (UL) standards.

Advantages

The advantages for using a microwave system include:

- Low signal attenuation.
- Equipment portability.
- Cost-effectiveness when there is a line-of-sight connection and frequency bank congestions are low.
- Signal scrambling for increased data security.

Disadvantages

Microwave's disadvantages include:

- The bandwidth is limited to a 6.7M-bit capacity, with an estimated 16M-bit peak-time potential.
- Outside installation support is required.
- Signals are unreliable during excessive rains.
- There is a potential delay in receiving licensing from the FCC, which could prevent meeting the installation schedule.

Costs

The basic cost to set up the microwave radio system for the corporate remote location would be as follows:

Materials Description	Unit Price	Quantity	Total Price
Microwave Radio System	$10,000	1	$10,000
Microwave Interface Unit for Networks	$ 3,000	2	$ 6,000

Materials Description	Unit Price	Quantity	Total Price
Antennas	$2,100	2	$ 4,200
Subcontracted Installation	$2,200	1	$ 2,200
Licensing Fee One Year (Requires Annual Renewal)	$ 520	1	$ 520
Total Microwave Radio Installation			$20,950

There was an optional system maintenance contract available that included immediate replacement of faulty equipment, to minimize downtime, for $250 per month. This annualized to $3,000. Staff training would also be required to provide system maintenance. Backup equipment would increase overall costs by 50%.

THE THIRD OPTION: INFRARED

A Free-Space Infrared Local Area Network (FIRLAN) system is based on infrared (IR) technology and can be used to build or replace a traditional hardwire network. The FIRLAN provides point-to-point or point-to-multipoints transmission of Ethernet signals between segments or stations, as well as T1 line signal transmission, as shown in Exhibit 8-6-6.

IR technology replaces cables with wireless optical links using line-of-sight IR transmission. Some systems use lasers as a basis for optical transmission. This is superior to standard IR transmission devices, especially under high ambient lighting and poor weather conditions.

It is recommended with IR systems that the installer have an IR viewer for alignment accuracy. The units are mounted on the corner, at the windows, or on the roof of the building to which they are attached and preferably are mounted to masonry construction. The supporting structure for the viewers cannot be wood or sheet metal.

IR systems offer industry-standard LAN and WAN interfaces, enabling the use of standard network interface cards and network software. FIRLAN provides seamless Ethernet integration, setting the required bit rate at 10M bits, and defines the physical interfaces and operating characteristics for the hardware, as shown in Exhibit 8-6-7.

Conditions Affecting Effectiveness. The three most significant atmospheric conditions that affect laser transmission include absorption, scattering, and shimmering. All three conditions can reduce the amount of light energy received by the receiver. The phenomena affect the laser transmission to varying degrees.

Absorption along the transmission path is caused mainly by the water vapor and carbon dioxide content in the air, which in turn depend on the

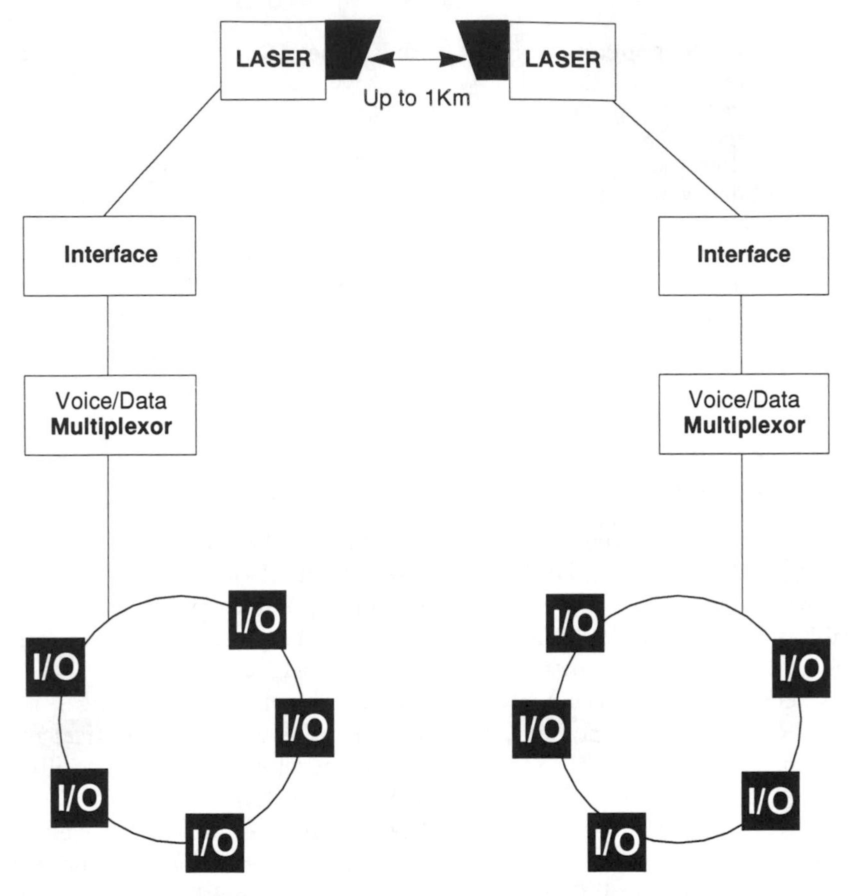

Exhibit 8-6-6. The Typical T1 Connection Setup

humidity and altitude. The gases that form in the atmosphere have many resonant bands (i.e., transmission "windows") that allow specific frequencies of light to pass. These windows occur at various wavelengths including the visible range. If a system uses a near IR wavelength of light (820 wavelength, or 820 nm) for laser transmission, however, absorption is not a great concern.

Scattering has a greater effect than absorption, because a smaller percentage of the transmission beam reaches the receiver. The atmospheric scattering of light is a function of light's wavelength and the number and size of scattering particles in the path.

Specific conditions that cause scattering problems include the following:

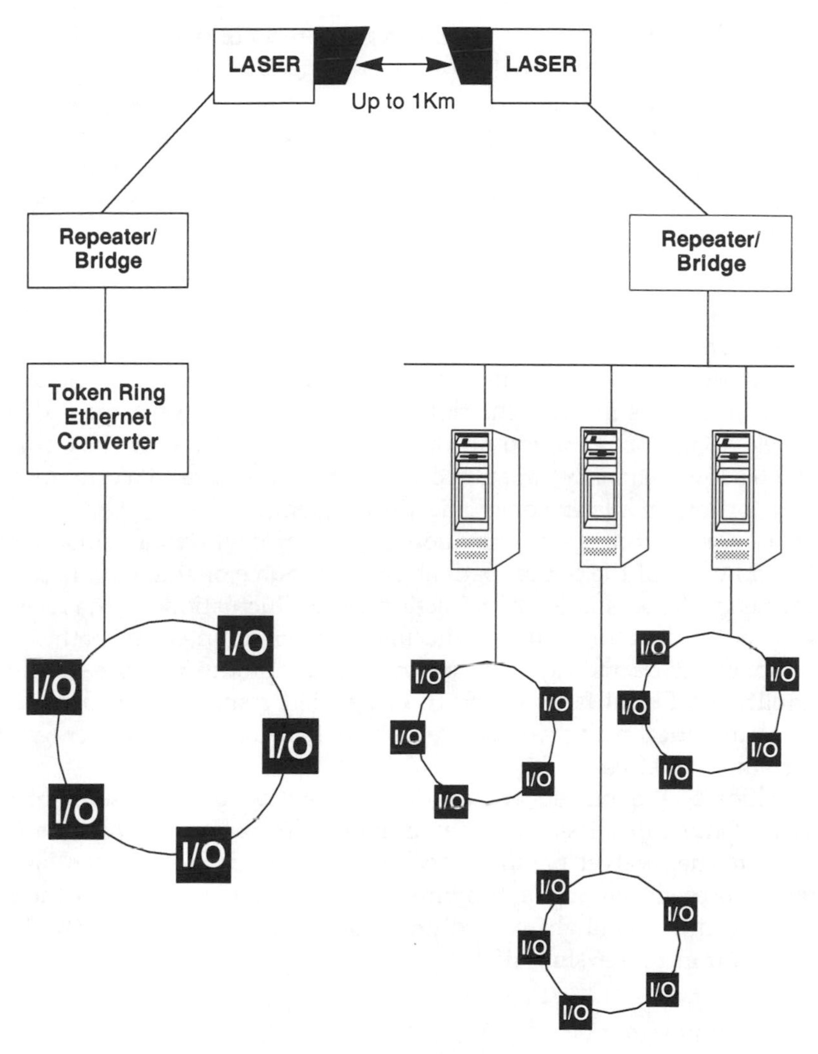

Exhibit 8-6-7. The Typical Ethernet Setup

---

1. *Fog.* The worst condition, fog, produces a scattering effect in all directions. This is attributed to water drops with a radius less than a few microns. Smog has a similar effect.
2. *Snow.* Its effect varies depending on its water content. A very wet snow is closer to rain, which has less of an effect on visibility than fog. An extremely dry snow, on the other hand, is closer to fog and has a similar

effect. The severity of the problem is in direct correlation to the radius of the particles; therefore, snow's scattering effect varies.

3. *Rain.* Rain-related attenuation is one hundred times less than that of fog. Although the liquid water content of a typical heavy shower is ten times that of a typical dense fog, the raindrop radius is about one thousand times larger than that of a fog droplet. This causes less scattering effect.

There is a way to minimize the effects of scattering, too. A product designed with a fade margin of 17 Db, for a 1,000-meter link would activate an automatic turn off when visibility drops below 800 meters.

Shimmer, also known as mirage or atmospheric turbulence, is the visual distortion of images in heat situations and imposes a low-frequency (normally below 200 Hz) variation on the amount of light detected by the receiver. This low frequency variation can result in excessive data error rates or video distortion on a laser communication system.

The shimmer effect is a combination of time of day, terrain, cloud cover, wind, and height of the optical path above the source of shimmer. Localized differences in the air's index of refraction cause fluctuations in the received signal level by directing some of the light out of its intended path. Beam fluctuations may degrade system performance by producing short-term signal amplitudes. Signal fades below the threshold result in error bursts. Selection of an optical path several meters above heat sources, however, greatly reduces shimmer effects.

In addition to the atmospheric conditions mentioned, direct sunlight into the front of the transmission or receiving unit affects performance. Sunlight can saturate the receiver photo diode, resulting in outages that can last for several minutes, depending on the time of the year and the angle of the sun. In addition, direct sunlight can saturate the feedback diode in the laser, resulting in transmitter shut off.

Licensing Requirements

There are no licensing requirements. However, the Center for Devices and Radiological Health (CDRH) of the US Food and Drug Administration is the agency responsible for reviewing IR technology. The technology complies with the federal regulations covered by 21 CFR 1010 and 21 CFR 1040 (HHS Publication FDA 88-8035).

Advantages

FIRLAN advantages include:

- Freedom from government regulation through licensing.
- Equipment ownership when a company changes locations.

- Connectivity ease when locations need to be added.
- Immunity to radio interference.
- Data security.

Disadvantages

Specific FIRLAN disadvantages include:

- The effect of atmospheric conditions on reliable data transmission, which may prohibit its use in some locals.
- High up-front equipment investment and installation support.
- Potential safety issue of retina damage caused by looking directly into the beam.

Cost

The basic cost to set up the system is as follows:

Materials Description	Unit Price	Quantity	Total Price
Ethernet Link IEEE 802.3M bits to 10M bits. Includes Two Laser Sets with AUI Connection	$15,000	1	$15,000
Mounting Cost	$ 600	2	$ 1,200
Bridge Ethernet to Token Ring	$ 2,000	2	$ 4,000
Converter	$ 2,000	1	$ 2,000
Total for Basic System Setup			$22,200

Options for system reliability include:

Optional Items	Cost	Description
Spare system	$9,250	Backup Set
Replace Service Contract	$1,300	Annual Contract for Unit Replacement Within 24 Hours
Onsite Survey and Installation Charge	$4,000	To Ensure All Required Equipment Is Available for Installation
Installation and Maintenance Training Cost	$1,200	Training for Actual Installation

Suggested options to purchase are:

Suggested Options	Cost	Description
Onsite Survey and Installation	$4,000	To Ensure the Installation Is Well Planned
Installation and Maintenance Training	$2,400	Sending Two People for Installation and Maintenance Training
Total Costs:	$6,400	

THE FINAL ANALYSIS

Several new factors became apparent during the research process. These became part of the decision-making process: the availability of fiber repeaters with no other designated use; discovery of an existing infrared installation at a company location on the East Coast; and building management, fearing harm to its employees, resisting the use of infrared lasers.

To complete the analysis, these items were included in the final recommendation. The analysis of the technology benefits is provided to gain insight from the available technology and present the applicable situation scenarios for each option. (Exhibits 8-6-2 and 8-6-3 are grid comparisons of these technologies.)

Weighing the Factors

The comparison of the technologies alleviates concerns for data security, as all three methods provide for secure data. Both of the wireless technologies, however, offer a lower data reliability factor on a daily basis. The weather in the region of the remote corporate location is subject to dramatic shifts, which are detrimental to the wireless technologies. Obviously, electrical outages could affect the users in the same way for physical wire options; these outages would also effect any work they might conduct.

All three technologies are available to install within the time frames required, with the possible exception of the FCC licensing required for the microwave option. Certainly, the licensing for the microwave option could have a serious impact on the ability of the staff to work on the scheduled date. This is a high consideration from a work-due-date standpoint and the organization's potential liability for missed due dates.

It would be advantageous to own the equipment as offered by either the microwave or infrared options for net costs, given the organization's business goals to downsize in the near future. The high up-front costs for the technologies were also noted. This investment in materials would have to be to the long-term advantage of the organization.

Cost Comparisons

Costs are certainly a consideration, but they are not always the prime consideration. The following cost comparisons are based on the pure costs of the three systems for network data connectivity. Items to be considered in the final cost estimates include estimates for equipment required for connectivity, purchased equipment depreciation, and use of any existing equipment.

Total Cost of fiber optics	$19,600
Total Cost of microwave	$20,950
Total Cost of infrared	$28,600

The depreciation on the purchased equipment averages approximately 50% based on the schedule of depreciation that was current at the time of decision making. This would reduce the cost of the microwave by $10,475 and infrared options by $14,300. A significant savings in using fiber optics would come through permission to use the existing company fiber optic repeaters for the remote location installation. This would reduce those costs by $11,200. These considerations factored in yield the following numbers:

Total Cost of fiber optics	$ 8,400
Total Cost of microwave	$10,475
Total Cost of infrared	$14,300

From a cost comparison aspect, fiber optics would be the most cost-effective method of installation.

RECOMMENDATIONS

To recap, the considerations for the recommended solution must address the following issues in order of importance. The solution must:

1. Be fully functional in time to coincide with the contractor's relocation to the company's new space in the adjacent building.
2. Provide reliable connectivity and an acceptable response time similar to that which is currently available.
3. Be supportable by the existing technical staff.
4. Come in line with the costs of the other solutions.

Each of the technologies was reviewed based on the above criteria. The recommendation discussed in this article was based on this specific situation and may not meet criteria for other situations. The technical notes following the recommendation provides highlights for future technology reviews.

The Decision on Microwave Technology

Given the required installation time for this project, the microwave solution was eliminated, because it requires a licensing lead time before equipment is operational. Bandwidth throughput of the microwave solution was also a point of concern, because the cabling installation at the remote location workstations called for 16M bit Token Ring, and total peak throughput of the microwave solution is only 6.7M bits. Also, existing data traffic at peak usage would reach bandwidth saturation points. The microwave technology solution, therefore, fails the first two points of the selection criteria.

The Decision on Infrared Technology

Several factors in the final review of the infrared laser link requirements eliminated this solution. Network uptime would be compromised by the lasers being mounted on the roof from east to west because of the building management's requirement for worker safety. In spite of staff in other locations being available to lend support, the local staff lacking experience in the necessary skill set to maintain or repair the laser units was a concern. In addition, the need to purchase a backup unit to have an on-site spare would make the solution cost prohibitive. The infrared solution failed to satisfy the last three points of the selection criteria.

The Decision on Fiber Optics

A fiber optic installation could be completed within the stated time frame, and the use of Token Ring fiber optic repeaters would maintain high levels of data throughput to reduce the threat of bandwidth saturation. The existing staff had some expertise in fiber optic repeaters and fiber optic cable installation, which would reduce maintenance response times. Further, if any new technology that provides greater bandwidth were to appear on the horizon, it would be best served through the fiber optic installation. Lastly, the cost was not prohibitive. Therefore, this technology provided the response best suited to the selection criteria and is the most viable solution.

As a final note, the microwave and infrared solutions required total backup units to be quickly recoverable. The cost for backup units would increase the overall cost of these options by approximately 30%. The fiber optic recommendation, on the other hand, contained a backup system with the extra strands included in the cable runs. Short of someone cutting the entire cable, these strands would be available if a failure were to occur. The switch to the available strands could be completed quickly, increasing the overall reliability of this connectivity solution.

IMPLEMENTATION AND INSTALLATION PLANNING

As with most projects, the final installation phase was the most critical. The situation under discussion required precise timing for user access to the Banyan network on a specific day. The implementation and installation phase for this project centered around connectivity on December 15, the scheduled date for the staff's move to the adjacent building. The build-out schedule for the space showed that the equipment installation was finished on December 12. To complete the connection for network access between the two buildings, the following process was recommended. It allowed for potential delays and provides some recovery time. The process involved:

1. Issuing a final request for proposal by November 22, on the selected option with a maximum two-day response time. At least three vendors were to be considered.
2. Issuing a purchase order by November 28 to cover the materials cost and installation fees.
3. Receiving and inventorying materials by December 9.
4. Installing materials on December 12 and 13. Testing connections and workstation access as each floor was completed.
5. Testing all connections on December 14.
6. Going fully functional on the morning of December 15. The staff was prepared to troubleshoot all problems encountered as the user community attached to the Banyan network.

SUMMARY

Network expansion, in any form, requires a baseline knowledge of the systems used, the organization and business direction, the liabilities a business may face with each major process change, and the effects of any downtime. Information systems are the lifeline of most businesses today, and these systems must exhibit enough flexibility to change as quickly as business directions change.

Technology is changing so dramatically that many system needs are available if the time and effort are applied to research what is available and to understand how to make it work for a given application. This effort is best served with staff experienced with the business, systems, and available technology.

Each of the technologies reviewed in this chapter is applicable to certain situations; therefore, the criteria requirements for each situation must be carefully weighed against the direction of the business organization. The proper knowledge base determines the correct direction for this type of connection or for any system expansion plans. Full analysis and good planning on the front end increase the project success rate.

Section 9
Network Operations and Management

THE NEW MODEL OF TELECOMMUNICATIONS is characterized by rapid changes in both technology and requirements. The communications systems manager must be very nimble, open to new ideas, and technically adept. In a less dynamic environment, using mature, off-the-shelf solutions, the manager could concentrate on monitoring and controlling the network systems. In an environment where the technology is changing every six to eighteen months, however, the communications manager must also cope with a large measure of uncertainty and doubt.

Were the right choices of technology made? Are there hidden development implications? Is the network architecture robust enough to support the anticipated applications? Have the network elements been deployed in the most efficient manner? Are the desired services available where the users need them? What changes are pending that will affect performance in six months or a year? What regulatory decisions are going to impact our deployment? These, and a myriad of other similar questions, do not have simple answers—they are the reasons that network management is classified as a high-stress job.

There has never been a manager who did not have doubts about the choices made—whether frame relay instead of cell relay, switching instead of routing, or vendor A instead of vendor B. Technology choices of this kind depend on what the objective or requirement is and on when the objective must be achieved. Some solutions may be available now. Others will be available a little later. A totally new solution may be available in eighteen months. Some current solutions may endure, and others will fail. It makes no rational sense to second guess choices. If a decision was made based on a solid business and technical case, the choice was the correct one.

The deployment of new technology always has development implications. The best approach is to anticipate implications and then deal with them as they arise. New, faster, technology, for example, may not yield the expected increase in performance. Maybe the increased transfer rates exposed previously hidden bottlenecks in the workstation-to-network links, in data rates of installed ports, or in communications software. It may be discovered that new protocols are required to take advantage of new technology features, speeds, or error rates. Adjusting and fine tuning will always be necessary as new systems are deployed.

High-speed technologies require high-speed architectural structures. Most managers are familiar with the layered approach to communications architecture. Although this is an excellent model of what needs to be done, it is a poor model of how it should be done.

Layering, when taken too literally, is much too slow for high-speed communications. Managers should think of putting multiple layers of communications functions in silicon. This does not change the architecture, only the implementation of the architecture. The lower in the architecture a function can be accomplished, the faster it will be. The result of this is the movement of switching and routing functions, formerly restricted to the network layer, down to the link or even physical layer. The lesson for the communications manager is to be flexible and less dogmatic about architectural issues.

There are many aspects of deployment and availability to worry about. Deployment decisions are made on the basis of the configuration at a specific point in time. If the configuration changes, and it most certainly will, the manager will need to deploy the necessary resources to meet the new requirement. That does not mean that the initial deployment was wrong. It is simply the price of rapid change.

Regulatory issues will also have to be dealt with as they arise. "Regulatory" implies government, and government can be an enigma in a high-technology communications environment. The government can be a source of research funding to help advance technology. It is also a large customer for advanced technology solutions and can, therefore, influence which technologies will succeed in the marketplace. The government is a major supporter of national and international standards. Finally, in its role as a regulator, the government can inhibit or encourage innovation and competition. The best that the communications system manager can do is to stay aware of what is going on, and, if the organization has a regulatory liaison department, maintain contact with those individuals in a position to know about or even influence the outcome.

The network operations and management task is certainly not a walk in the park. It is a demanding job. The manager must evaluate user needs, decide on a leading edge or conservative philosophy, stay with standard choices when they are available, build the business case and, if the direction is clear, introduce new technology solutions. The manager does this knowing full well that at about the time network operations settle into a routine, the cycle will begin again.

The chapters that comprise this section zero in on more specific aspects of the network management task. Chapter 9-1, "Managing Distributed Computing," for example, describes a model for delivering services to users in a distributed environment. This chapter offers five specific pieces of advice that can help managers evaluate their own management styles.

Management tools are discussed in "An Overview of Network Management Systems," Chapter 9-2. These combinations of hardware and software serve the manager by providing status information on the state of network elements and facilities. The chapter provides feature checklists for both wide area and local network systems.

Chapter 9-3, "Introduction to LAN Efficiency," offers some useful insights into networking. The chapter emphasizes the importance of using the right tools, obtaining the proper training, and paying attention to detail to accomplish efficient management of the network.

Security is a prime concern in the operation of a network. "Protecting Against Hacker Attacks," Chapter 9-4, is an excellent primer on the subject of people breaking into communications and computer systems. This chapter provides background on the subject, discusses hacker profiles and techniques, and suggests defensive measures.

Chapter 9-5, "Documenting a Communications Recovery Plan," addresses the absolute necessity of having disaster recovery plans documented and available. Too often such plans are kept in the desk drawer of some network administrator. The problem is that neither the administrator nor the desk may be available when disaster strikes at 3 a.m. on a Sunday morning.

A related subject is explored in Chapter 9-6, "Points-of-Failure Planning." As the distribution of interconnected network elements extends into remote locations, it becomes harder to grasp the entire scope of the network, and especially where points of failure are located that can impact overall network performance. Tools for recognizing and mapping these points is provided in the final chapter of this section.

9-1
Managing Distributed Computing

Richard Ross

MANY OF THE TOP CONCERNS of managers in IS departments relate directly to the issues of distributing information technology to end users. The explosive rate at which information technology has found its way into the front office, combined with the lack of control by the IS organization (ostensibly the group chartered with managing the corporation's IT investment), has left many IS managers at a loss as to how they should best respond. The following issues are of special concern:

- Where should increasingly scarce people and monetary resources be invested?
- What skills will be required to implement and support the new environment?
- How fast should the transition from a centralized computing environment to a distributed computing environment occur?
- What will be the long-term impact of actions taken today to meet short-term needs?
- What will be the overall ability of the central IS group to deliver new standards of service created by changing user expectations in a distributed computing environment?

In large companies during the past decade, the rule of thumb for technology investment has been that the opportunity cost to the business unit of not being able to respond to market needs will always outweigh the savings accruing from constraining technology deployment. This has resulted in a plethora of diverse and incompatible systems, often supported by independent IS organizations.

In turn, these developments have brought to light another, even greater risk—that the opportunity cost to the corporation of not being able to act as a single entity will always outweigh the benefit of local flexibility.

For example, a global retailer faces conflicts if it has sales and marketing organizations in many countries. To meet local market needs, each country has its own management structure with independent manufacturing, distribution, and systems organizations. The result might be that the company's supply chain becomes clogged—raw materials sit in warehouses in one country while factories in another go idle; finished goods pile up in one country while stores shelves are empty in others; costs rise as the number of basic patterns proliferate. Perhaps most important, the incompatibility of the systems may prevent management from gaining an understanding of the problem and from being able to pull it all together at the points of maximum leverage while leaving the marketing and sales functions a degree of freedom.

Another example comes from a financial service firm. The rush to place technology into the hands of traders has resulted in a total inability to effectively manage risk across the firm or to perform single-point client service or multiproduct portfolio management.

WANTED—A NEW FRAMEWORK FOR MANAGING

A distributed computing environment cannot be managed according to the lessons learned during the last 20 years of centralized computing. The distributed computing environment is largely a result of the loss of control by the central IS group because of its inability to deliver appropriate levels of service to the business units. Arguments about the ever-declining cost of desktop technology are well and good, but the fact of the matter is that managing and digesting technology is not the job function of users. If central IS could have met their needs, it is possible users would have been more inclined to forego managing their own systems.

It is not just the technology that is at fault. Centralized computing skills themselves are not fully applicable to a distributed computing environment. The underlying factors governing risk, cost, and quality of service have changed. IS departments need a new framework, one that helps them to balance the opportunity cost to the business unit against that to the company while optimizing overall service delivery.

DEFINING THE PROBLEM: A MODEL FOR DCE SERVICE DELIVERY

To help IS managers get a grip on the problem, this chapter proposes a model of service delivery for the distributed computing environment (DCE). This model focuses on three factors that have the most important influence on service as well as on the needs of the business units versus the corporation—risk, cost, and quality (see Exhibit 9-1-1). Each factor is analyzed to understand its cause and then to determine how best to reduce it (in the case of risk and cost) or increase it (as in quality).

Risk

Risk in any systems architecture is due primarily to the number of independent elements in the architecture (see Exhibit 9-1-2). Each element carries its own risk, say for failure, and this is compounded by the risk associated with the interface between each element.

This is the reason that a distributed computing environment will have a greater operational risk than a centralized one—there are more independent elements in a DCE. However, because each element tends to be smaller and simpler to construct, a DCE tends to have a much lower project risk than a centralized environment.

Thus, one point to consider in rightsizing should be how soon a system is needed. For example, a Wall Street system that is needed right away and has a useful competitive life of only a few years would be best built in a distributed computing environment to ensure that it gets online quickly. Conversely, a manufacturing system that is not needed right away but will remain in service for years is probably better suited for centralization.

One other difference between a distributed environment and a centralized environment is the impact of a particular risk. Even though a DCE is much more likely to have a system component failure, each component controls such a small portion of the overall system that the potential impact of any one failure is greatly reduced. This is important to take into account when performing disaster planning for the new environment.

Cost

Cost is largely a function of staff levels (see Exhibit 9-1-3). As the need for service increases, the number of staff members invariably increases as well. People are flexible and can provide a level of service far beyond that of automation. Particularly in a dynamic environment, in which the needs for response are ill-defined and can change from moment to moment, people are the only solution.

Unfortunately, staff is usually viewed as a variable cost, to be cut when the need for budget reductions arises. This results in a decrease in service delivered that is often disproportionately larger than the savings incurred through staff reductions.

Perceived Quality

Quality is a subjective judgment, impossible to quantify but the factor most directly related to the user's perception of service where information

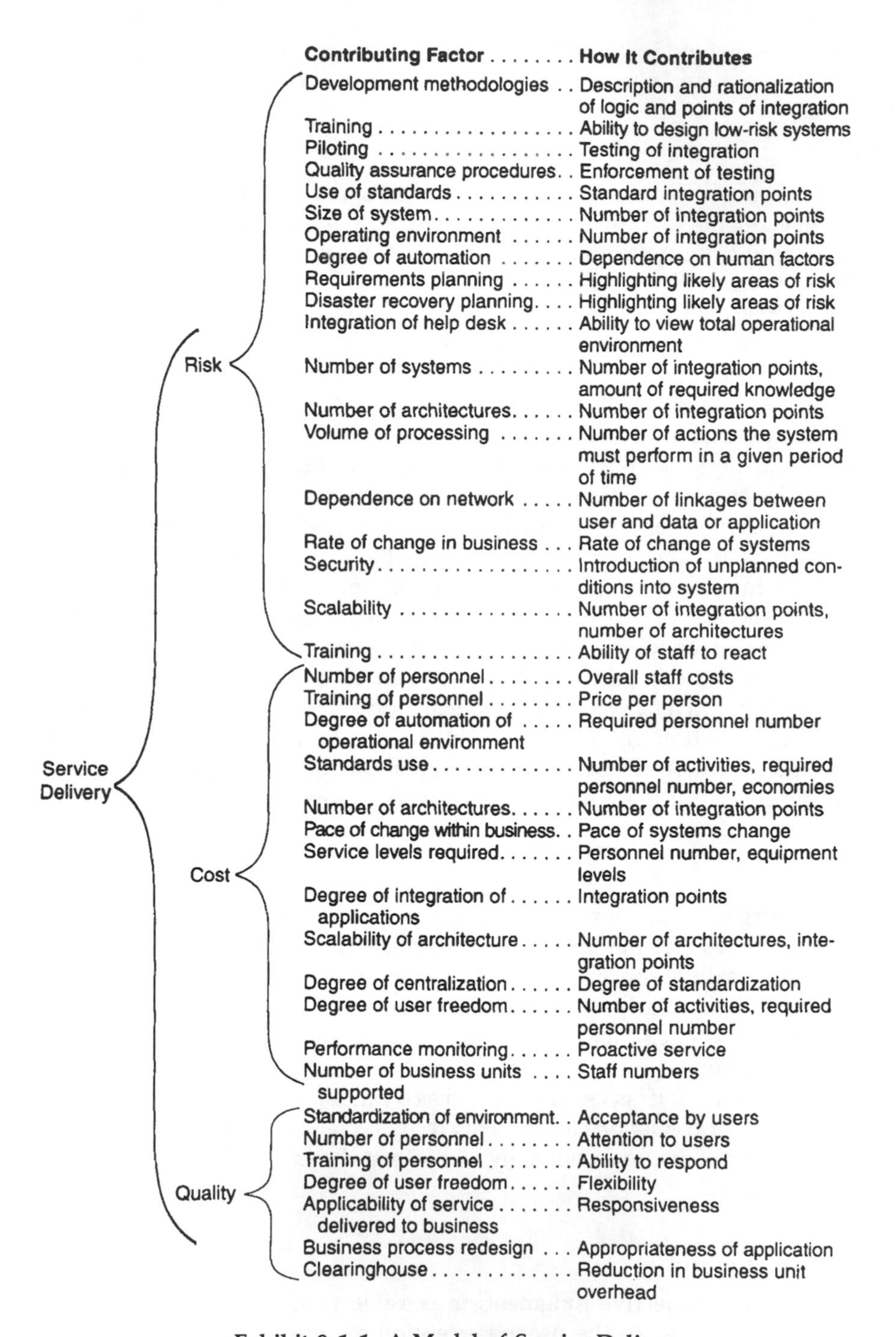

		Contributing Factor	How It Contributes
Service Delivery	Risk	Development methodologies	Description and rationalization of logic and points of integration
		Training	Ability to design low-risk systems
		Piloting	Testing of integration
		Quality assurance procedures	Enforcement of testing
		Use of standards	Standard integration points
		Size of system	Number of integration points
		Operating environment	Number of integration points
		Degree of automation	Dependence on human factors
		Requirements planning	Highlighting likely areas of risk
		Disaster recovery planning	Highlighting likely areas of risk
		Integration of help desk	Ability to view total operational environment
		Number of systems	Number of integration points, amount of required knowledge
		Number of architectures	Number of integration points
		Volume of processing	Number of actions the system must perform in a given period of time
		Dependence on network	Number of linkages between user and data or application
		Rate of change in business	Rate of change of systems
		Security	Introduction of unplanned conditions into system
		Scalability	Number of integration points, number of architectures
		Training	Ability of staff to react
	Cost	Number of personnel	Overall staff costs
		Training of personnel	Price per person
		Degree of automation of operational environment	Required personnel number
		Standards use	Number of activities, required personnel number, economies
		Number of architectures	Number of integration points
		Pace of change within business	Pace of systems change
		Service levels required	Personnel number, equipment levels
		Degree of integration of applications	Integration points
		Scalability of architecture	Number of architectures, integration points
		Degree of centralization	Degree of standardization
		Degree of user freedom	Number of activities, required personnel number
		Performance monitoring	Proactive service
		Number of business units supported	Staff numbers
	Quality	Standardization of environment	Acceptance by users
		Number of personnel	Attention to users
		Training of personnel	Ability to respond
		Degree of user freedom	Flexibility
		Applicability of service delivered to business	Responsiveness
		Business process redesign	Appropriateness of application
		Clearinghouse	Reduction in business unit overhead

Exhibit 9-1-1. A Model of Service Delivery

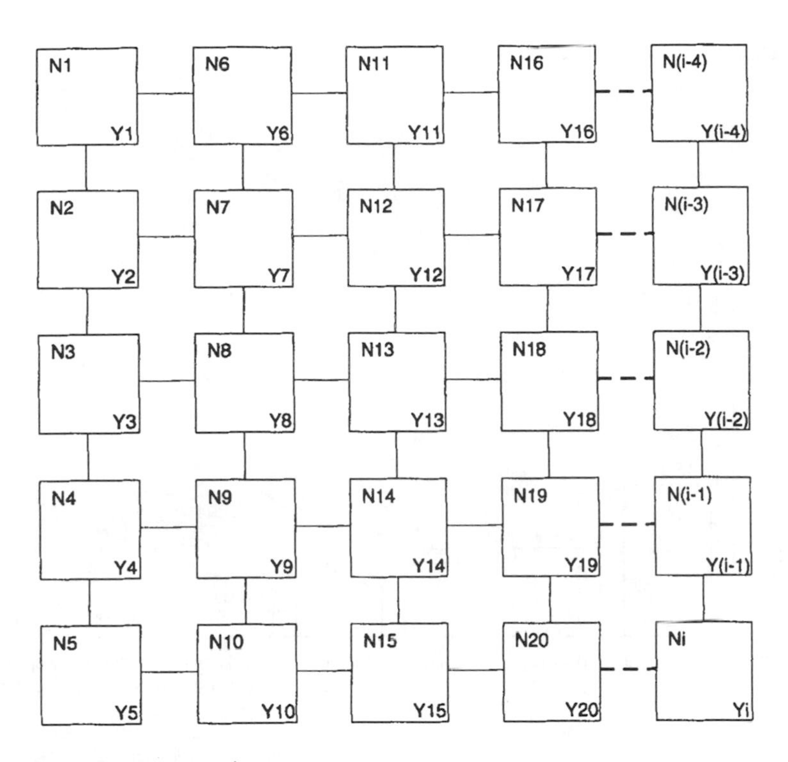

N = Component number

Y = Component risk

Given fully independent components, total network risk is equivalent to the sum of the individual component risks, 1 to i. Thus, the way to minimize risk is either to minimize i (i.e., to have a centralized computing environment) or to minimize Y for each component by standardizing on components with minimum risk profiles.

Exhibit 9-1-2. Optimization of Risk in a Network

technology is concerned. In essence, the perception of quality is proportional to the user's response to three questions:

- Can I accomplish this task?
- Am I able to try new things to get the job done?
- Am I being paid the attention I deserve?

Except for the sheer ability to get the job done, the perception of quality is not necessarily a factor of how much technology a user is provided with. Instead, quality is largely a function of the degree of freedom users have to try new things and whether they are being listened to. This may mean that

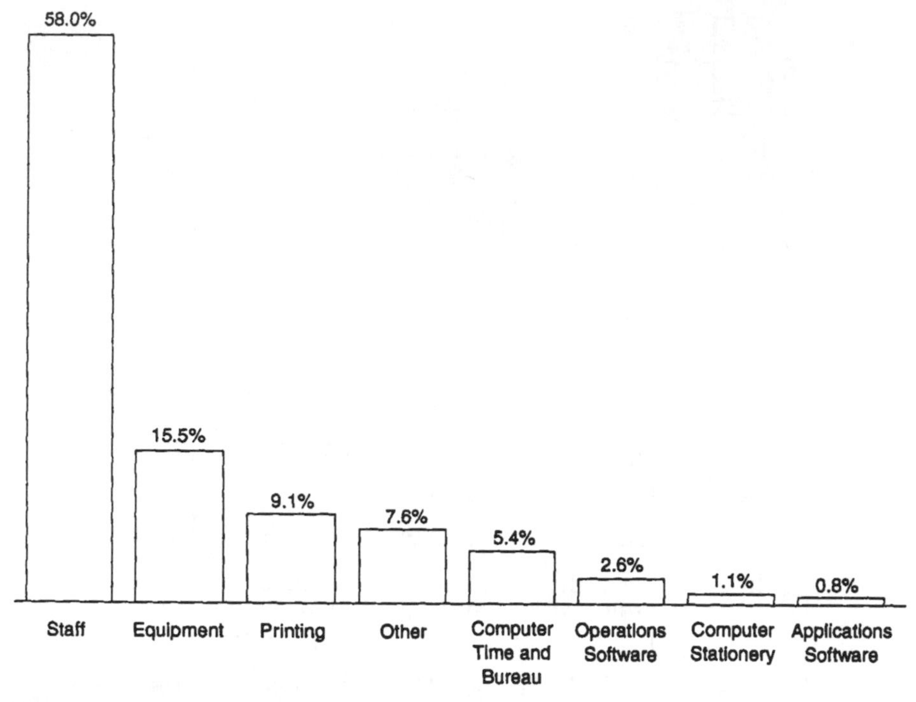

Exhibit 9-1-3. Cost Profile of the IS Function

for many organizations, a simpler environment—one in which the user has fewer technological options but has ready access to human beings—will be more satisfying.

This view is in direct contrast with the approach many companies have taken of delivering more capabilities to the users in an effort to increase their perception of service delivered. One of the most important factors in the perceived quality of service delivery is the ability of the support technology to work unnoticed. Because of the similarities between this need and the way in which the US telephone network operates (a customer picks up the phone and the service is invariably there), the term dialtone is used to describe such a background level of operation.

"Dialtone" IS Service. One problem with highly functional IS environments is that users must think about them to use them. This is not the case with the telephone system, which operates so dependably that people and businesses have integrated it into their routine working practices and use it without much conscious effort. The phone companies maintain this level of usefulness by clearly separating additional features from basic service and letting the customer add each new feature as the customer desires.

Contrast this with the typical business system that represents an attempt to deliver a package of functions on day one and to continually increase its delivered functionality. The impact on users is that they are forced to continually adapt to changes, are not allowed to merge the use of the system into the background, and must continually stop delivering on their jobs just to cope with the technology. This coping might be as simple as looking something up in a manual or changing a printer cartridge, or it may mean not working at all while the system is rebooted.

COMPLEXITY: THE BARRIER TO SERVICE DELIVERY

In general, the basic driver to each of the three service factors is complexity. Complexity increases risk by increasing the number of interfaces between system elements as well as the number of elements themselves. It increases cost by increasing the need for staff as the only way to deal with ill-defined environments. Finally, it affects quality by making it harder to provide those services that users base their perception of quality on (i.e., dialtone and personal attention), in response to which even more staff are added.

This is the paradoxical environment in which IS operates. To improve the quality of service, they find themselves increasing the risk and cost of the operation. Improved application delivery cycles result in more systems to manage. End-user development tools and business unit-led development increases the number of architectures and data formats. Increasing access to corporate data through networks increases the number of interfaces.

Conversely, trying to improve the risk and cost aspects, typically through standardization of the environment, usually results in decreased levels of service delivered because of the constraints placed on the user freedom. This paradox did not exist in the good old days of centralized computing, when the IS organization dictated the service level.

FIVE PIECES OF ADVICE FOR MANAGING DISTRIBUTED COMPUTING

The measure of success in a distributed computing environment is therefore the ability to deliver service through optimizing for the factors of risk, cost, and quality while meeting the needs of both the business units and the corporation. It sounds like a tall order but it is not impossible. There are five key practices involved in corporate information processing:

- Manage tightly, but control loosely.
- Organize to provide service on three levels.
- Choose one standard—even in a single bad one is better than none or many good ones.
- Integrate data at the front end—do not homogenize on the back end.
- Minimize the use of predetermined architectures.

Manage Tightly, Control Loosely

A distributed computing environment should be flexible and allow users to be independent, but at the same time should be based on a solid foundation of rules and training and backed up with timely and appropriate levels of support.

In the distributed computing environment, users must react with the best of their abilities to events moment by moment. IS can support users in a way that allows them to react appropriately, or can make them stop and call for a different type of service while the customers get more and more frustrated.

BPR and Metrics. Two tools that are key to enabling distributed management are business process redesign (BPR) and metrics. BPR gets the business system working first, highlights the critical areas requiring support, builds consensus between the users and the IS organization as to the required level of support, and reduces the sheer number of variables that must be managed at any one time. In essence, applying BPR first allows a company to step back and get used to the new environment.

Without a good set of metrics, there is no way to tell how effective IS management has been or where effort needs to be applied moment to moment. The metrics required to manage a distributed computing environment are different from those IS is used to. With central computing, IS basically accepted that it would be unable to determine the actual support delivered to any one business. Because centralized computing environments are so large and take so long to implement, their costs and performance are spread over many functions. For this reason, indirect measurements were adopted when speaking of central systems, measures such as availability and throughput.

But these indirect measurements do not tell the real story of how much benefit a business might derive from its investment in a system. With distributed computing, it is possible to allocate expenses and effort not only to a given business unit but to an individual business function as well. IS must take advantage of this capability by moving away from the old measurements of computing performance and refocusing on business metrics, such as return on investment.

In essence, managing a distributed computing environment means balancing the need of the business units to operate independently while protecting corporate synergy. Metrics can help do this by providing a clear basis for discussion. Appropriately applied standards can help, too. For example, studies show that in a large system, as much as 90% of the code can be reused. This means that a business unit could reduce its coding effort by some theoretical amount up to that 90%—a forceful incentive to comply with such standards as code libraries and object-oriented programming.

Pricing. For those users that remain recalcitrant in the face of productivity gains, there remains the use of pricing to influence behavior. But using pricing to justify the existence of an internal supplier is useless. Instead, pricing should be used as a tool to encourage users to indulge in behavior that supports the strategic direction of the company.

For example, an organization used to allow any word processing package that the users desired. It then reduced the number of packages it would support to two, but still allowed the use of any package. This resulted in an incurred cost to the IS organization due to help desk calls and training problems. The organization eventually settled on one package as a standard, gave it free to all users, and eliminated support for any other package. The acceptance of this standard package by users was high, reducing help calls and the need for human intervention. Moreover, the company was able to negotiate an 80% discount over the street price from the vendor, further reducing the cost.

This was clearly a benefit to everyone concerned, even if it did not add up from a transfer pricing point of view. That is, individual business units may have suffered from a constrained choice of word processor or may have had to pay more, even with the discount, but overall the corporation did better than it otherwise would have. In addition to achieving a significant cost savings, the company was able to drastically reduce the complexity of its office automation environment, thus allowing it to deliver better levels of service.

Organize to Provide Service on Three Levels

The historical IS shop exists as a single organization to provide service to all users. Very large or progressive companies have developed a two-dimensional delivery system: part of the organization delivers business-focused service (particularly applications development), and the rest acts as a generic utility.

Distributed computing environments require a three-dimensional service delivery organization. These services are:

- Dialtone, or overseeing the technology infrastructure.
- Business-focused or value-added services, or ensuring that the available technology resources are delivered and used in a way that maximizes the benefit to the business unit.
- Services that maximize leverage between each business unit and the corporation.

Dialtone IS services lend themselves to automation and outsourcing. They are complex, to a degree that cannot be well managed or maintained by human activity alone. They must be stable, as this is the need of users of these services. In addition, they are nonstrategic to the business and lend themselves to economies of scale and hence are susceptible to outsourcing (see Exhibits 9-1-4 and Exhibit 9-1-5).

Value-added services should occur at the operations as well as at the development level. For example, business unit managers are responsible for overseeing the development of applications and really understanding the business. This concept should be extended to operational areas, such as training, maintenance, and the help desk. When these resources are placed in the business unit, they will be better positioned to work with the users to support their business instead of making the users take time out to deal with the technology.

People do not learn in batch. They learn incrementally and retain information pertinent to the job at hand. This is why users need someone to show them what they need to accomplish the job at hand. Help desks should be distributed to the business units so that the staff can interact with the users and solve problems proactively, before performance is compromised, and accumulate better information to feed back to the central operation.

The third level of service—providing maximum leverage between the business unit and the corporation—is perhaps the most difficult to maintain and represents the greatest change in the way IS does business today. Currently, the staff members in charge of the activities that leverage across all business units are the most removed from those businesses. Functions such as strategic planning, test beds, low-level coding, and code library development tend to be staffed by technically excellent people with little or no business knowledge. IS managers must turn this situation around and recruit senior staff with knowledge of the business functions, business process redesign, and corporate planning. These skills are needed to take the best of each business unit, combine it into a central core, and deliver it back to the businesses.

Common Operational Problem	Responsiveness to Automation
Equipment hangs	High
Network contention	Medium
Software upgrades	High
Equipment upgrades	Low
Disaster recovery	High
Backups	High
Quality assurance of new applications	High
Equipment faults (e.g., print cartridge replacement, disk crash)	Low
Operator error (e.g., forgotten password, kick out plug)	Low
Operator error (e.g., not understanding how to work application)	High

Responsiveness
High ●
Medium ◑
Low ○

SOURCE: Interviews and Decision Strategies Group analysis

Exhibit 9-1-4. Responsiveness of Operations to Automation

Choose One Standard

If the key to managing a distributed computing environment is to reduce complexity, then implementing a standard is the thing to do. Moreover, the benefits to be achieved from even a bad standard, if it helps to reduce complexity, will outweigh the risks incurred from possibly picking the wrong standard.

The message is clear: there is more to be gained from taking inappropriate action now than from waiting to take perfect action later.

It should be clear that information technology is moving more and more toward commodity status. The differences between one platform and another will disappear over time. Even if IS picks a truly bad standard, it will

Dialtone Function	Applicability
Equipment maintenance	●
Trouble calls	●
Help desk	●
Installations	●
Moves and changes	●
Billing	◑
Accounting	◑
Service level contracting	○
Procurement	○
Management	○

Applicability
High ●
Medium ◑
Low ○

SOURCE: Interviews and Decision Strategies Group analysis

Exhibit 9-1-5. Applicability of Outsourcing to Dialtone

likely merge with the winner in the next few years, with little loss of investment. More important, the users are able to get on with their work. In addition, it is easier to move from one standard to the eventual winner than from many.

Even if a company picks the winner, there is no guarantee that it will not suffer a discontinuity. IBM Corp. made its customers migrate from the 360 to 370 architecture. Microsoft Corp. is moving from DOS to Windows to Windows NT. UNIX is still trying to decide which version it wants to be. The only thing certain about information technology is the pace of change, so there is little use in waiting for things to quiet down before making a move.

Integrate Data at the Front End

At the very core of a company's survival is the ability to access data as needed. Companies have been trying for decades to find some way to create a single data model that standardizes the way it stores data and thus allows for access by any system.

The truth of the matter is that for any sufficiently large company (i.e., one with more than one product in one market), data standardization is unrealistic. Different market centers track the same data in different ways. Different systems require different data formats. New technologies require data to be stated in new ways. To try to standardize the storage of data means ignoring these facts of life to an unreasonable extent.

To try to produce a single model of all corporate data is impossible and meaningless. The models invariably grow to be so large that they cannot be implemented. They impose a degree of integration on the data that current systems technology cannot support, rendering the largest relational data base inoperative. And they are static, becoming obsolete shortly after implementation, necessitating constant reworking. Moreover, monolithic data models represent unacceptable project risk.

The standardization approach also ignores the fact that businesses have 20 to 30 years' worth of data already. Are they to go back and recreate all this to satisfy future needs? Probably not. Such a project would immobilize the business and the creation of future systems for years to come.

Systems designed to integrate and reconcile data from multiple sources, presenting a single image to the front end, intrinsically support the client/server model of distributed computing and build flexibility into future applications. They allow data to be stored in many forms, each optimized for the application at hand. More important, they allow a company to access its data on an as-needed basis. These integration systems are an important component to successfully managing future growth.

Less Architecture Is More

To overdesign a systems architecture is to overly constrain the organization. Most architecture arises as a function of rightsizing of applications on the basis of where the data must be stored and used. Understanding this helps the IS manager size the network and associated support infrastructure.

The management of risk and impact also drives architecture by forcing redundancy of systems and, in some cases, mandating the placement of data repositories regardless of user preferences. Assessing project versus operational risk helps to determine whether a system is built for central or distributed use.

Economics. This view is one in which the business needs drive the shape of the architecture. It results in a dynamic interconnection of systems that respond flexibly to business needs. Under a centralized computing environment, it was impractical to employ such an approach. It took so long and cost

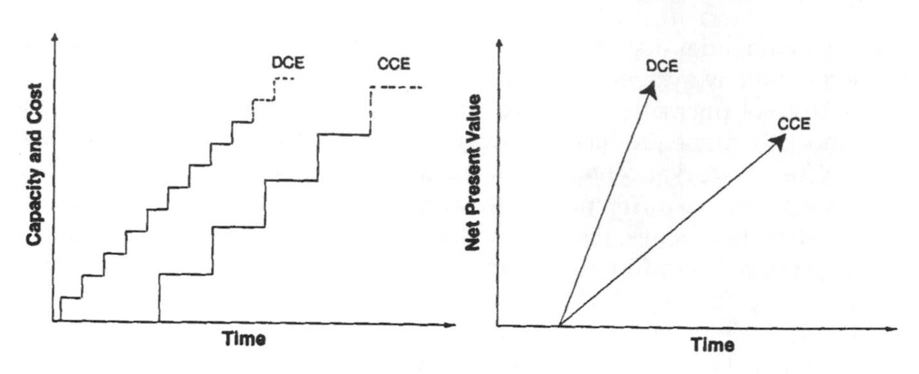

Distributed computing environment (DCE) has a higher net present value because its capacity can be used sooner relative to its marginal costs when compared with the centralized computing environment (CCE).

SOURCE: Decision Strategies Group

Exhibit 9-1-6. Net Present Value of Distributed Versus Centralized Computing

so much to implement a system that investment had to come before business need. This necessitated preplanning of an architecture as an investment guide.

The economics of distributed computing are different. Systems cost much less and can be quickly implemented. This means that their use can be responsive to business needs instead of anticipative. It also results in a greater net present value, for even though their operational costs might be higher, distributed computing environments are more immediately useful for a given level of investment (see Exhibit 9-1-6).

SUMMARY

This framework for managing the new computing environment relates directly to the business. If managers of IS functions are able to master it, they can enhance their opportunities to become members of the corporate business management team instead of simply suppliers of computing services.

Success in a distributed computing environment requires a serious culture shift for IS managers. They must loosen up their management styles, learning to decentralize daily control of operations. They must provide direction to staff members so that they can recognize synergies among business units. Some jobs that were viewed as low-level support activities (e.g., value-added services such as help desk and printer maintenance) must be recognized as key to user productivity and distributed. Others, viewed as senior technical positions (e.g., dialtone functions such as network management and installations), might best be outsourced, freeing scarce IS resources.

Most important, IS must understand the shift in power away from themselves and toward users. The IS organization is no longer the main provider of services; it now must find a role for itself as a manager of synergy, becoming a facilitator to the business units as they learn to manage their own newfound capabilities.

9-2 An Overview of Network Management Systems

Gilbert Held

NETWORK MANAGEMENT SYSTEMS monitor the status of network components and line facilities. In general, they are owned and operated by the end user, though commercial carriers typically operate large-scale network management systems to monitor and control line facilities used by their customers.

Although most network management systems are based on the use of a microcomputer, some systems are minicomputer based, and a few systems (e.g., IBM Corp.'s NetView) operate on a host computer system. Regardless of the type of computer system used, each network management system performs a core set of functions.

OPERATIONAL GOALS

Network management systems can help users meet several objectives. These include:

- A reduction in technical staff requirements.
- Rapid isolation of network failures.
- Reduced or eliminated downtime.
- Access to a data base of information automatically generated by the system.

Each of these operational objectives is discussed in the following sections.

Reduction in Technical Staff Requirements

Because network management systems can be configured to perform many functions automatically, the number of technical staff needed can be substantially reduced if such a system is implemented. By automating functions that previously required technicians to perform manual measurements on a periodic basis, large organizations can either reduce their technical staff requirements or make more effective and valuable use of their staff members' time.

Rapid Isolation of Network Failures

Network management systems provide the ability to isolate network failures rapidly through many of the same system functions that permit a reduction in technical staff requirements—that is, the ability to constantly monitor network components, set thresholds, and generate alarms. For example, thresholds can be set to generate alarms when a network component or line facility is still operational but has deteriorated to a point at which communications could be affected in the near future. Once an alarm is received, the data center staff member can use the network management system s built-in testing capability or standalone diagnostic testing equipment to isolate the potential or actual network failure and arrange for corrective action.

Reduced or Eliminated Downtime

Network management systems can reduce or eliminate the downtime that usually results when corrective action must be taken. Some systems permit a network operator to place a spare device into operation, thus bypassing the failing or failed device. Systems that do not have this capability can still be used to reduce downtime. In such a case, the capability of the network management system to note the degradation in performance of a network component enables the data center technician to make arrangements for its replacement or repair before it becomes completely inoperative.

Access to the System's Data Base

Network management systems enable users to gain access to the system's data base. Data base information generated by most network management systems include files of historical performance records, configuration files, and trouble report files. This type of information can be used to provide equipment vendors and communications carriers with a record of the performance of their equipment or facilities over a period of time, which may assist them in providing more efficient service to the organization.

OPERATIONAL USE

The operation of network management systems can be categorized into three areas: enterprisewide, wide area network (WAN), and local area network (LAN).

An enterprisewide network management system provides monitoring and resource control on an end-to-end basis encompassing both local and wide area network devices and media. Although many vendors now offer enterprisewide network management systems based on devices compliant with the Simple Network Management Protocol (SNMP), such systems are not truly enterprisewide. Many commonly used network devices, such as T1 multiplexers, channel service units (CSUs), data service units (DSUs), and modems, are not SNMP compliant. Therefore, many vendor products designed as SNMP systems require separate vendor-specific network management systems to provide a full end-to-end monitoring and resource control capability. This chapter focuses on the operational characteristics of WAN and LAN network management systems that can be used to obtain enterprisewide network management capability.

WAN-BASED SYSTEMS

Unfortunately for the network manager, most WAN-based network management systems are currently restricted in regard to the devices they support. Some support modems; others support multiplexers. Even systems that support multiplexers may be limited to working with either statistical multiplexers or T1 multiplexers.

Another limitation associated with network management systems is a lack of true heterogenous component management, whereby one vendor's network management system works with, and can control, another vendor's network component. Fortunately, several vendors of network management systems and network components have agreed to develop standards under the aegis of an open systems interconnection (OSI) forum. In addition, the growth in the use of SNMP agents within routers and bridges has been extended to a few vendor modem and DSU and CSU products. Unfortunately, because it will probably take until the late 1990s for a sufficient base of standardized products to be manufactured, most technical control centers will continue to be more like battle control centers, in which the ability to monitor multiplexers, digital service units, T1 multiplexers, and other network devices requires the purchase and operation of several systems, each containing one or more consoles.

Types of Control

Two methods are used by a WAN-based network management system for controlling devices: out-of-band and in-band signaling. Each is discussed in the following sections.

Out-of-Band Signaling. This is designed only for the control of modems and uses a form of frequency division multiplexing, in which a region from

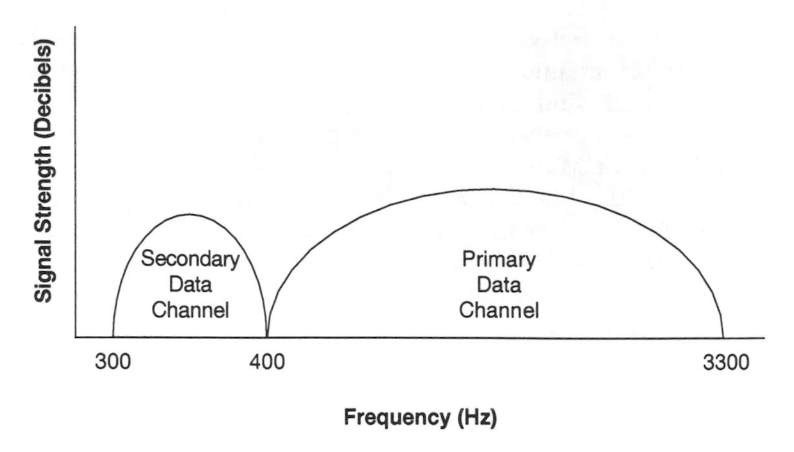

Exhibit 9-2-1. Out-of-Band Signaling

---

approximately 300 Hz to 400 Hz is used to support a slow-speed asynchronous transmission of 75 bps to 150 bps. This slow-speed transmission carries the control signals from the network management system to remote modems as well as the modems' responses to system queries. The bandwidth from 400 Hz to 3,300 Hz, which is the remaining usable portion of a voice channel, is used to carry the actual data that flows between modems. Exhibit 9-2-1 illustrates out-of-band signaling.

Because the control signals and responses to queries flow on a secondary channel, data transmission between modems can continue while the network management system polls each device to determine its status. However, the use of frequency division multiplexing to derive a secondary channel restricts this signaling technique to the control of modems.

In-Band Signaling. In-band signaling requires each network component supported by the WAN-based network management system to interpret data and respond to control codes that may be intermixed with the data flow between devices. This type of signal control is generally used to govern the operation of multiplexers and digital service units as well as modems. In fact, network management system vendors have designed wraparound units that are controlled by in-band signaling. The wraparound unit is a microprocessor-controlled device contained in an elongated U-shaped housing that wraps around a modem. Exhibit 9-2-2 illustrates the use of a pair of wraparound units that enables modems produced by one vendor to be supported by a different vendor's network management system.

The wraparound unit placed around the modem at the top left of Exhibit 9-2-2 is connected to both a port on the mainframe as well as to the network

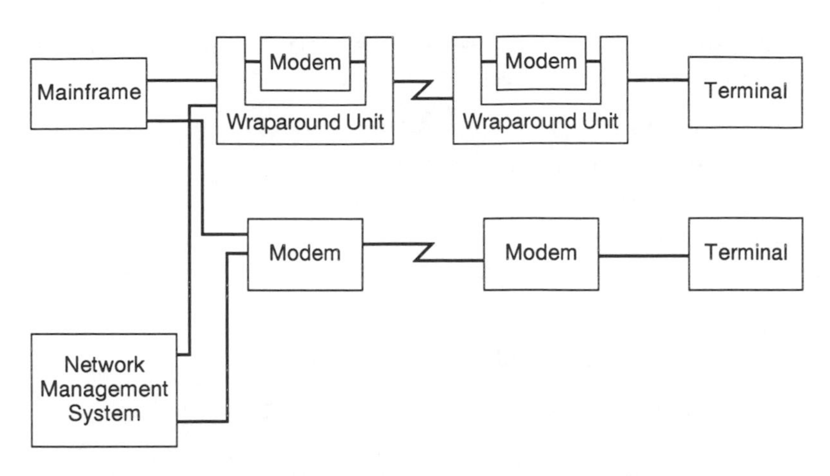

Exhibit 9-2-2. Sample Wraparound Unit Configuration

management system. Control signals sent from the network management system are interpreted by the first wraparound unit. If the control signal is interpreted as a request for the remote modem to perform a function or return its status, the local wraparound unit passes the control signals to the modem for modulation. If the control signal requests the local modem to perform a function or return its status, the local wraparound unit will respond directly to the WAN-based network management system.

Control signals passed to the local modem for transmission to the remote site are demodulated by the remote modem and interpreted and acted on by the remote wraparound unit. The wraparound unit spans both sides of a modem to permit the device to report information concerning both the circuits to which the modem is attached and the physical interface between the modem and the computer port or terminal.

In the lower portion of Exhibit 9-2-2, the pair of modems installed on the second circuits are supported by the network management system used. Because of this, the network system can be directly cabled to the local modem on the second circuit. In turn, control signals transmitted to the remote modem on the second circuit are first examined by the local modem. Because these signals affect the remote modem, the local modem modulates them as it would data. The remote modem demodulates the data. When it interprets that data, it recognizes the control signals and then responds to them.

System Features

Although vendors market a variety of WAN-based network management systems that include a wealth of unique features, most systems provide users with a core set of features. The seven WAN-based network management system features common to most systems are:

- Password access.
- Color-coded graphics or text display.
- Status reporting.
- Performance monitoring.
- Configuration management.
- Threshold setting and alarm generation.
- Data base maintenance and report generation.

Each of these features is discussed in the following sections

Password Access

Most WAN-based network management systems are password protected to prevent unauthorized personnel from changing equipment configuration, placing equipment out of service, or similar activities that can have serious consequences. Some network management systems provide several layers of access control, which are usually broken down into operator and administrator categories. The data center manager generally controls all operator access and privileges, and if the network management system has multiple consoles, the manager controls who can use which ones.

Color-Coded Display

To facilitate the visual identification of problems, most WAN-based network management systems use a color code to identify the importance of text or graphics. This display enables a technician at a distant location within a technical control center to easily determine whether the contents of a display should be read.

Most network management systems use a red-yellow-green color-coding scheme. Red identifies alarm conditions that require the immediate attention of a technician. Yellow indicates a serious condition that, if unchecked, could result in an alarm condition. Green indicates no unusual activity (e.g., a message reporting that line parameters fall within acceptable limits).

Status Reporting

One of the key functions of a WAN-based network management system is to reduce the burden of the technician testing and observing network component status. The status-reporting feature of most network management systems is actually the response of network components to predefined status requests that the system transmits to each device at regular intervals. By displaying the status of network components using the previously discussed color-coded scheme, a manager can tell at a glance whether a component's operation requires further investigation. In addition to displaying the status of network components, most systems log the information to appropriate data base files, providing a historical record of network components' performance.

Performance Monitoring

The performance monitoring feature of a WAN-based network management system can be included in its status reporting capability or may be provided as a separate feature. In general, status reporting refers to the operational status of network components, and performance monitoring refers to the capability of the communications facility connected to a network component to permit data transmission. Status reporting can therefore include such information as the state of the control circuits at the physical interface; performance monitoring includes such measurements as signal strength, impulse hits that occurred over a period of time, and similar parameters.

Configuration Management

Configuration management provides the user of a WAN-based network management system with the ability to vary the operational state of devices that can be placed into one of two or more modes of operation. Devices that can have their configuration changes may include multiport modems, multispeed modems, and multiplexers.

One of the most common uses for configuration management is to set up different device modes of operation to correspond to the differences in transmission requirements between shifts. Typically, a day or prime shift will favor interactive transmission. Second and third shifts generally favor batch transmission because many batch jobs are deferred for execution until most interactive users have left for the day.

For example, the capability to alter the configuration of network devices helps establish a daytime and nighttime network configuration. The daytime configuration allocates resources to include bandwidth or port speeds to favor interactive users. Similarly, the nighttime configuration is set up to allocate resources to favor batch transmission devices (e.g., remote job entry workstations) that have high-speed printers to which system output can be directed. Once daytime and nighttime configurations are established, a change to the appropriate configuration can be initiated as part of the organization's standard operating procedure. Because each configuration change is designed to favor the data transmission requirements associated with a particular shift, the network is more responsive to end-user requirements than a network whose configuration remains static.

Some WAN-based network management systems have increased the automation of configuration management to the point at which a network operator can assign a start and stop time and date to each configuration. Once this is accomplished, routine changes in a network s configuration occur automatically, with only a message generated on the network management system's command console to indicate that the change has occurred. With this capability, technical control center personnel can focus their attention on resolving problems rather than on performing routine functions.

Threshold Setting and Alarm Generation

Through the ability to set thresholds and associate them with alarms, the WAN-based network management system can forewarn of situations that could result in a communications failure if left unchecked. One example is the assignment of a signal level to a threshold. If a signal level below what a modem normally should receive but above its sensitivity level is selected as a threshold, the network management system indicates when a level indicative of an abnormal condition has been reached; if such a condition is not corrected, the modem is unable to receive a transmitted signal. By associating the threshold with an alarm, the network management system may sound a buzzer, display a message on the console in red, print a message, or perform a combination of these when the threshold is reached.

Data Base Maintenance and Report Generation

Most WAN-based network management systems collect a file with a variety of data. The systems provide an information retrieval system that can rapidly and accurately retrieve information concerning historical operational status or the current level of performance for specific devices or lines.

Another key feature incorporated into many WAN-based network management systems is the ability to generate, update, file, and retrieve trouble tickets. For large networks, the ability to generate and update trouble tickets provides data center managers with a record of each problem as well as the corrective action taken by their personnel to resolve the problem. Because trouble tickets can also be files and can be retrieved, the data base facilities of the network management system can be used to retrieve problems associated with a specific network component or line facility. Information concerning repetitive problems can then be provided to equipment vendors or communications carriers to assist them in resolving problems associated with their equipment or line facilities.

A WAN-based management system can be a valuable tool to assist in the operation, maintenance, and management of a network. In evaluating the

Feature or Operational Reguirement	Organization's Requirement	Vendor A	Vendor B
Type of System			
Support Modems			
Support DSUs			
Support Statistical Multiplexors			
Support T1 Multiplexers			
Type of Control			
In-Band Signaling			
Out-of-Band Signaling			
Support Wraparound Unit			
Features			
Password Access			
Color-Coded Graphics and Text			
Status Reporting			
Performance Monitoring			
Configuration Management			
Threshold Setting			
Alarm Generation			
Data Base Maintenance			
Report Generation			

Exhibit 9-2-3. Network Management System: Features and Operational Requirements

operational capabilities of network management systems, the individual organization's requirements should be compared with the features and operational capabilities of different vendor products. Exhibit 9-2-3 is a checklist of features and capabilities of WAN-based network management systems that should be evaluated with respect to the organization s requirements.

LAN-BASED SYSTEMS

Because LANs postdate the development of wide area networking, it is not surprising that LAN-based network management systems borrowed many features from WAN-based systems. Many LAN-based network management systems include most, if not all, of the seven features previously described. The key difference between LAN- and WAN-based systems is that two distinct areas must be monitored on a LAN: transmission media and file servers. Unfortunately, the present state of technology is such that separate LAN management systems are typically required to monitor each area.

Media Monitoring

Media or traffic monitoring provides the data center manager with an indication of the activity of each workstation on a LAN as well as a composite of LAN traffic. This requires a hardware probe connected to the network or the execution a software program on an existing workstation connected to the network.

Exhibit 9-2-4 illustrates the screen display seen when using EtherVision, an Ethernet network management product developed by Triticom, Inc. The program was set up to monitor the source address of packets flowing through the network to determine if one or several workstations were performing an excessive amount of file-transfer activity that was responsible for some users complaining of slow network response. Although only 3 seconds elapsed when the monitoring screen was captured, one source address transmitted 441 frames, whereas the second-highest address accounted for 187 frames. If a significant period of monitoring resulted in a few users accounting for a majority of transmissions, the EtherVision screen provides the source addresses of workstations that should be checked.

Even though Exhibit 9-2-4 shows individual workstation network use statistics, often information concerning composite network use is of primary interest. Exhibit 9-2-5 illustrates the EtherVision Skyline display of network use as a percentage of media bandwidth capacity. Many network and data center managers prefer to set an alarm based on a percentage of network use. Then, if the alarm occurs, the network management product is used to focus on the activities of individual users to determine who specifically was responsible for the majority of network use. This indicates that one or a few individuals should consider performing long file transfers or network print jobs at a different time of day. The alarm can also determine whether the situation was caused by a large number of users accessing network resources; if this is the case, it indicates that network segmentation was required to reduce network delays.

On both Ethernet and Token Ring networks, different types of frame errors commonly affect network performance. Some errors on an Ethernet are caused by collisions resulting from the carrier sense multiple access collision detection (CSMA/CD) protocol used by Ethernet. Other frame errors result from improper cabling, routing of wiring near electric machinery, and other impairments. Thus, most LAN network management systems provide users

Monitoring SOURCE Address; Started Mon June 19, 1995 at 12:19:55 12:19:58

02608C00000C	127	08002B000023	9
02608C00000E	441	08002B000021	28
02608C000007	61	08002B00001E	27
0000CA000019	29	08002B00001D	25
02608C000009	63		
0000CA000011	187		
02608C00000A	81		
02608C000008	58		
0000CA00001A	31		
02608C000002	28		
02608C00000D	183		
0000CA000014	70		
02608C000006	43		
0000CA000012	127		
08002B000020	28		
0000CA00001B	38		
08002B000022	25		

Address	Name	Vendor ID	Frames	Bytes	%	Ave	Errors
02608C00000C	Sleepy	3Com—00000C	127	66122	7.5	520	1

Stns	Frames	K bytes	Bdcast	FPS	Peak	CRC	Align	Coll	MU	Elapsed
21	1693	945	3	189	677	1	0	0		00:00:03

F2-Stn ID F3-Sort ID F4-Sort Cnt F5-Cnt/Kb/%/Av/Er F6-Sky F7-Stat F8-Clr

Exhibit 9-2-4. Monitoring Network Activity by Source Address with EtherVision

with a mechanism to view frame errors. EtherVision provides both an individual error log and a composite screen to view performance and frame errors.

Exhibit 9-2-6 illustrates the EtherVision composite screen display. This screen provides a horizontal bar showing network use as well as summary information on peak and current use, frame size distribution information, and a summary of frame errors. If one workstation has a large number of frame errors, this could indicate a failing or a failed adapter card or improper cabling to the workstation and can be used as a guide for examining the network infrastructure.

Server Monitoring

When abnormal performance occurs on a LAN , the problem can result from the use of the media or the server. Thus, a second type of LAN-based network management product, a server monitor, is required to obtain detailed knowledge of the status of a LAN. Server monitors, such as Frye Computer's NetWare Monitor, operate on a server and track server use,

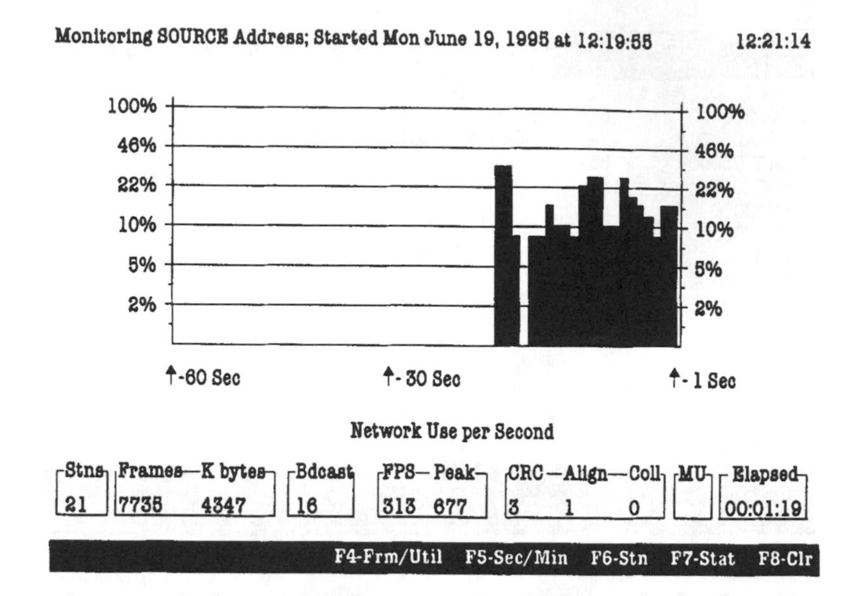

Exhibit 9-2-5. Using the EtherVision Skyline Display to View the Composite Network Use

server memory use, print job buffer use, and other server-related parameters that directly affect the server and indirectly affect network use.

To determine the cause of a LAN performance problem requires the use of both a media monitor and a server monitor. For example, EtherVision might indicate a LAN use level of 27%. However, if the data center manager receives many reports of poor network performance, he or she might, with a server network management product, note that several programs running in the server used all available memory, causing an excessive amount of program swapping to disk, which results in extended user response times. The use of media and server monitoring programs is required to obtain a full insight into the cause and potential alleviation of LAN-related problems.

When selecting LAN-based network management systems, several features should be considered, including:

- Autodiscovery.
- Autotopology.
- Trend analysis.
- User customization.

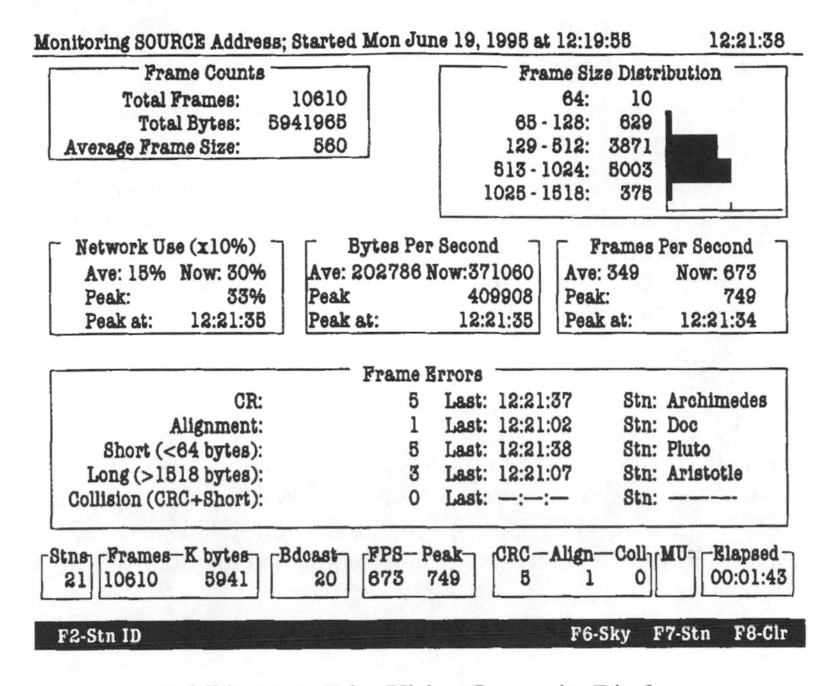

Exhibit 9-2-6. EtherVision Composite Display

Autodiscovery

Autodiscovery references the ability of the network management system to learn addresses automatically by examining the address in each frame. Some products require the user to select either the source or destination address; other products can use both addresses at the same time.

Autotopology

Autotopology is the process of creating a network map based on addresses obtained from the autodiscovery process. This enables the management system to provide a hierarchical view of the network.

Trend Analysis

Trend analysis is the system's ability to save and retrieve historical data, helping the manager determine whether a use problem repeats during certain periods. This would happen if users set up a log-in script to download or upload a file or group of files, resulting in network saturation each time they turned on their workstation.

User Customization

Although historical data can be important, sometimes it can be overwhelming. A few products permit users to accumulate requested data and transfer it to a file in a format supported by Excel, Lotus 1-2-3, or another electronic spreadsheet. This feature is invaluable in sorting through weeks or months of data when attempting to determine a trend.

SUMMARY

Although there are a few enterprisewide network management systems, the lack of interoperability between LAN and WAN vendor products requires network managers to operate separate systems. This also means that both LAN media and LAN servers must be monitored to obtain a true insight into LAN use.

9-3
Introduction to LAN Efficiency

William A. Yarberry, Jr.

CORPORATE DOWNSIZING HAS GROWN in proportion to network upsizing. As mainframe functions continue their migration to client/server, organizational investment in networks has grown dramatically. The network has become the computer.

During the 1980s, LANs were usually isolated. A few rings or segments were linked to a server and no communications capabilities to other LANs. As client/server technology began to proliferate, LANs needed to be linked to provide enterprisewide E-mail, electronic scheduling, data base access, and multimedia applications.

The IS manager should concentrate on areas in which costs are rising dramatically and service to users is vital. The LAN backbone (an electronic pathway connecting multiple LANs) meets this criteria.

ROLE OF BACKBONE MANAGEMENT

The technology to implement a backbone varies from simple Ethernet thinwire spanning three floors to a worldwide network of hubs, routers, fiber, and sophisticated network managers. In spite of these differences in implementation, many of the control and efficiency concepts are broadly applicable.

In a medium-to-large organization, individual LANs are connected by a corporate or enterprise backbone. An operations staff separate from daily LAN operations management is charged with maintaining viability of the backbone. The communications network manager should review the following functions to ensure they are in the recognized scope of the organizations backbone staff:

- *Proactively maintaining uptime within established service levels.*

- *Monitoring traffic on the backbone.* This includes IP packet volumes, exception conditions, identity of protocols, and invalid addresses.
- *Identifying and shutting down connection ports of LANs that are sending signals that can bring down the backbone.* These can be router or hub ports.
- *Providing analysis of protocols, traffic volumes, and device-specific information (e.g., server disk capacity, memory use).* Unfortunately high-level analysis is sometimes neglected for other duties (e.g., daily changing of hub cards or ferreting out a beaconing network interface card).
- *Providing a clearinghouse for network critical standards.* For example, Internet protocol addresses should be closely coordinated. The backbone team should provide leadership in developing a comprehensive subnetting scheme. Another example is the naming of devices. If LAN managers name servers without clearing them with the central group, duplicates can occur. This can result in irritated users who cannot get to their servers.
- *Maintaining a continuous review of the enterprise network design.* As technologies and organizations change, the backbone staff must examine the design to determine if it is adequate for business needs. For example, if the organization changes from LANs as office automation tools to LANs as full-production mainframe replacement systems, then the copper backbone may be inadequate. If the ethernet backbone has been added onto over time, a review of subnetting is probably necessary. Changes in wiring standards, hubs, routers, and protocols may be necessary. For example, are unroutable, protocols (e.g., NetBIOS) to be allowed?
- *Capacity planning.* In addition to reviewing design, the backbone team should examine volumetrics. (e.g., Does the current infrastructure have the capacity to handle increased volumes?)
- *Maintaining standards.* Standards for naming, connectivity, drivers, protocols, hardware, and software should be maintained. For example, if a particular version of Cheyenne Software's ARCserve does not function properly with newer versions of Novell NetWare, standards should dictate avoidance of that software combination.
- *Protecting missing critical applications.* This function varies by organization. The backbone team should have high-profile applications identified and steps available for protection, rerouting, and recovery.
- *Maintaining an up-to-date inventory of devices, addresses, and related organizational information.* If backbone services are charged back to individual business units, accounting information should be stored as well. Exhibit 9-3-1 shows a sample spreadsheet with hub inventory information. If the backbone team intends to eventually management down to the workstation (typically using simple network management protocol

Item	1094-ATRJ32S
Description	32 Port Token Ring Module
Serial #	9876-87
Hub serial #	77-345654
Location	Fifth Floor Annex
Slot	4
Comments	Upgraded by vendor on 11-1-94
Maintenance	Start on 1/1/95
Warranty expiration	12/31/94

Exhibit 9-3-1. Example Inventory for an ODS Hub Card

[SNMP]), real user names and locations need to be in a data base, associated with specific network interface card IDs.

- *Maintaining change and problem-management functions.* As in other areas of information technology, changes should be coordinated, assessed, and approved or disapproved using standard written procedures. The backbone team should exercise leadership by inviting the participation of all parties involved to make the functions of change control and problem management accepted throughout the organization. The network manager should be aware that old mainframe procedures may be ineffective, at least in the specifics. Network changes are much more dynamic than the mainframe environment; however, the principles should still apply.
- *Maintaining adequate security measures.* Client/server computing provides ample opportunity for security breaches. For example, many components of the network can be accessed through dialup. Devices have been manufactured to provide dial-up reboot, if permitted. Avoiding hub and router dial-up is generally safer. An alternative is putting a simple communication program on a palmtop computer (e.g., Hewlett-Packard's HP100 LX) and plugging it directly into the serial port. This provides the ability to configure hubs on the spot. The alternative is making the physical changes, writing down the configuration information, then entering them at the management station. Another reasonably secure option is using a dial-back modem.
- *Developing an equipment upgrade "philosophy."* If an organization is consistent with its upgrade philosophy, resource planning is easier and more effective.

TOOLS AND TRAINING

Efficient operation of the enterprise backbone depends heavily on appropriate use of software and hardware monitoring tools. The network manager should determine the availability of the tools and the proficiency of the backbone staff in using them.

Sniffers. Devices such as Data General's Network Sniffer provide diagnostic graphs and reports by analyzing packet data carried by various protocols. Typically several machines are needed, one for the central monitoring area and at least one more portable sniffer for local LAN traffic analysis. However, to obtain extended, summary reporting, remote monitoring capabilities in devices need to be exploited. These are information data bases contained within hubs that report on activity in a structured manner. These data bases can then be consolidated over time to provide trending information. For such reporting to be effective, enterprise communications devices should be RMON-enabled.

Spare Parts and Backup Equipment. For maximum uptime, the organization should not scrimp on spare parts. In practice, when the backbone goes down because of a router failure, the technicians on duty are under severe time pressure to minimize downtime. They need a router that has all potential cards available so that they can plug and play rather than having to switch and test individual router cards for Token Ring and Ethernet. The network manager should also review the physical placement of backup equipment. For example, are cables, routers, and hubs in a position inside the chassis that allows for ease of movement and connectivity in the event of a downed device? Are switching devices used to minimize physical cable relocation? How much physical effort is required to reconnect cables and shift equipment? Such seemingly minor details can have a significant impact on recovery time.

High-End Platform for Network Management. There are numerous network management packages on the market, including HP OpenView, Cabletron's Spectrum, and IBM's Tivoli. Using SNMP, these packages report on the elements of the network. They also extend control and auditing functions from the network operating center up to and including the workstation. Control of any specific element depends on whether an SNMP agent has been loaded in the device. Overall function of these high-end platforms include the following:

- *Traffic volume reporting by protocol.*
- *Alarms.* Various audio or visual alarms can be developed and run continuously by using a scripting language.

- *Automated pagers.* Software is available to detect specified events and notify technicians with a numeric or alphanumeric page. Scrips should be tailored so that they send out pages to the correct group and for the correct reasons. For example, sending out an alert every time an Ethernet runt is detected can result in many spurious alerts. The challenge is in defining those parameters that warrant immediate attention (e.g., zero traffic on a port that normally has a large volume, router not responding).
- *Real-time maps.* By periodic pinging of the network, managers can identify links that are functioning and those that are down. By clicking on a node of the network, more detail can be shown.
- *Statistics and history.* Potential failure points can sometimes be proactively identified using statistical information. Filters and packet capture parameters allow the network operator to narrow down the source of network errors.

STANDARDS

Standards for the enterprise network affect efficiency, security, and auditability. Although the specifics vary by technical environment, the network manager should review the major areas common to all networks.

Direct Attachment to the Backbone. Two theories address direct attachment. One theory states that the corporate backbone should be accessed only through routers and should have no devices (e.g., file server, printer) directly attached. Another theory states that enterprise devices accessed by users on most or all LANs should be directly attached to simplify access and reduce router hops. The control-oriented manager generally favors the first option, because devices other than routers may be more prone to bring down the backbone. In addition, router technology is rapidly advancing and hops are not as deleterious as in the past.

SNMP Compliance. Non-SNMP devices are not fully controllable or reportable from a network management system, unless the network manager and device are from the same vendor. Currently, SNMP is memory- and processing-intensive and is used primarily for hubs, routers, and servers, not desktop equipment. However, as desktop devices and operating systems become more powerful and integrated into the LAN, SNMP (or its equivalent) may become more common in the environment.

Standards to Access the Backbone. If the enterprise network has evolved to the mission-critical stage, adherence to standards becomes essential. Routers periodically poll the network to determine what devices are responding, then they build electronic tables that determine the path of traffic. For example, Novell specifies a routing information protocol frequency of 30 seconds. This means that all routers in the enterprise rebuild their tables that identify the existence of devices once per 30 seconds. If a device does not respond, the tables do not reflect its existence and it is therefore not addressable by anyone, including end users. A device that broadcasts every 2 minutes can (depending on exact timing) cause some other devices on the network to lose addressability. Other examples of devices and software that would not meet standards include:

- Out-of-date release of operating systems.
- Inappropriate protocols for a particular organization. (e.g., IS unencapsulated NetBIOS allowed?)
- Inappropriate variations of common protocols.
- Use of hubs that do not control Token Ring beaconing.
- Use of splitters that allow multiple users to share a single port in a hub (thus eliminating the ability of the hub to effectively control traffic from that port).

Conformity to Specifications. Fiber, copper, and wireless mediums all have well-defined requirements for optimal performance. For example, Ethernet backbones not only have a maximum length, they also have a specific length that reduces errors (e.g., runts, jabbers). Other examples of out-of-spec conditions include the following:

- Certain types of cabling placed near fluorescent lights or other equipment with similar electrical properties.
- Fiber optic cable run beyond distance limits.
- An excessive number of connectors.

NETWORK DOCUMENTATION

Documentation is as vital to efficiency as it is to internal control. The network management should review all documentation to ensure that it is adequate and current.

Network Diagrams. In a complex network, diagrams help both the backbone staff and LAN managers design optimum routes. Network diagrams are essential to planning new paths and identifying potential problem areas (e.g., routing loops), particularly if the cabling plant has evolved in a haphazard manner. The diagrams should be accessible to all parties in the organization that need them. Read-only versions should be put on an enterprise server that uses software available to most technicians in the organization. A standard package should be selected for network drawings so that each unit of the enterprise can contribute its section without the need to convert from one format to another.

Asset Listings. All assets of the central network control group (backbone team) should be recorded in a spreadsheet or with asset-management software. Serial numbers, purchase price, location, maintenance status (i.e., when is maintenance due and is there a maintenance contract in effect?), Internet protocol addresses, and internal owner (if applicable). Examples of assets include hubs, routers, modems, sniffers, servers, diagnostic and management software, and spare equipment.

Change Control and Problem Management Logs. Logs are necessary for the efficient isolation of error sources and events that affect the network. These should be maintained in electronic form (preferably in a data base accessible to interested parties) for a reasonable period of time.

Published Standards and Periodic Reports. General documentation needs to be readily available through the network. To be effective it should be brief, with the contents well organized.

Design Specifications. The enterprise should have a network philosophy and overall design. For example, the approach to hooking enterprisewide servers to the backbone should be specified. (e.g., Are servers directly attached or do they go through routers? How are WANs supposed to link into the building backbone?) The design may be obvious to the designers, but it should be clearly outlined in writing.

LAN EFFICIENCY TOOLS

The efficient backbone team should have a varied assortment of practical, time-saving devices. Examples of useful tools are addressed in the following sections.

Technicians' Communication Tools. In a large building, considerable time can be wasted if members of the backbone team have to be paged when problems occur. Short-range walkie-talkies provide instant communications so that the nearest backbone team member can go to the hub, router, or communications closet in which problems are occurring.

Alphanumeric Pagers. Using Internet facilities (such as AT&T's or SkyTel's home page) or PC-based software, short text messages can be transmitted to an alphanumeric pager, thus eliminating the need for a return telephone call. For example, a technician at a central command center could type a short message. The message is complete and does not require a response. With the introduction of two-way paging in some cities, the technician can key in short responses to a page.

INDIVIDUAL NETWORK MANAGEMENT

Although the backbone staff is responsible for keeping the enterprise highway operating efficiently, the manager responsible for one or more specific LANs also needs management tools. A large number of these tools are commercially available. Examples include NETsys Enterprise/Solver, Frontier's NETscout Probes and NETscout Manager, Anacapa's NetScore Intelligent Agent, and McAfee's Saber Tools. The IS manager should examine certain tools to determine if they are appropriate and if they are being used.

Graphical Views of the Network. A well-designed network management package provides a graphical network diagram that shows (through color and alerts) problems with servers of workstations. For example, if a server is sending bad packets, it would change colors from green to a warning yellow.

Software Distribution. For software that needs to be distributed across multiple LANs or servers simultaneously (so that all users within the organization have the same version), a software distribution method or tool is necessary. Some organizations E-mail (through attachments) new versions to LAN managers. Others use sophisticated distribution packages that intelligently forward files (i.e., executable and data files) across the enterprise and ensure that consistent versions of software are used. The distribution director then examines the format of each file distributed and copies it to the correct subdirectory on each LAN. Microsoft NT includes utilities that allow automated local drive distribution on software.

Hardware Inventories. Hardware inventories are sometimes only available when the user first logs onto the network. Packages are available that scan all attached hardware devices and report to a central data base on the LAN. Modems, network interface cards, printers, central processing unit chips, and other devices are listed. This simplifies upgrades and equipment planning. In addition, when troubleshooting network problems, hardware versions of network interface cards can be essential.

Statistics. When checking the overall health of the server and network, it is important to have such quantitative statistics as average server hard disk response time, memory available, average number of logins, and soft errors. Various packages provide these statistics (in addition to those provided by the operating system itself). The auditor should determine if these statistics are being reviewed periodically to allow for proactive maintenance and upgrades.

Dynamic Data Exchange Links to Spreadsheets. Dynamic data exchange (DDE) allows flexible reporting. Any reporting package should have DDE-linking ability. This means that the package can communicate directly with another DDE graphics and reporting package without cut and paste operations.

Pager Interface. Monitoring packages should have the ability to identify the specific events (previously identified) and should use a script to dial up a modem, which then sends a page (preferably alphanumeric) to one or more technicians' pagers.

Graphical Displays of User-Defined Events. Given the workload of most LAN managers, graphic displays can assist in discriminating the important from the trivial. Packages can flash messages based on predefined events (e.g., traffic loads or high number of soft errors).

Provide Detailed Workstation Information. When upgrading LAN software or troubleshooting, a detailed inventory of both application level and infrastructure level software is helpful. Many management packages provide such detailed information (usually at boot-up time). One disadvantage is that sometimes NetBIOS is used. NetBIOS is a protocol that should be avoided in larger networks because it generates a large amount of protocol advertising traffic.

Ability to Use Boolean Logic to Identify Potential Problem Devices. For example, potential problem devices include those nodes that have a particular network interface card driver and have a particular basic input-output system date. The more advanced network control packages have the ability to query the hardware and software inventory with Boolean-type logic. This can help in troubleshooting complex problems.

Ability to Track Cache, Disk Use, Memory Use, Number of Connections, and Buffer Information. Packages that can provide historical, trending data (in graphical form) are useful in providing proactive maintenance. For example, one company noticed that users were coming in to work earlier than usual. A review of the statistics for logon connections showed that the server was reaching the Novell limit of 256 connections early during the day. Users were coming in early to get connected before the others showed up for work.

Remote Control and Monitoring of Workstation. To support a large body of users, the LAN manager must have a means of remotely examining the user's workstation. By allowing the LAN manager to see what the user sees (e.g., both monitors show the same spreadsheet), the error or problem can be identified and resolved without traveling to the user's workstation.

Ability to Initiate a File Transfer from the LAN Manager's Console and a Specified Workstation. Again, the key to efficiency is to eliminate the need for a LAN manager to physically walk to a workstation to install a file. It can be performed remotely.

Network Printer Selection and Print Job Management. The LAN manager should have the option to up the priority of a print job in the queue or delete a long job if necessary. In addition, if a certain printer malfunctioning, the LAN management package should be able to eliminate it from the printer list. Some printer manufacturers offer sophisticated print management packages that offer detailed control of print queues.

Fast Ethernet and Switching. The manager should review the intraLAN design. Optimal solutions depend on traffic placed on the ring or segment. For example, if significant PC videoconferencing occurs, Fast Ethernet may not be appropriate because it is a shared-media technology (i.e., all users on that segment share the bandwidth). Switching, in which a limited number of users are on their own virtual segment, may be a better solution.

Effect of the Internet on LAN Efficiency. While access to Internet information can provide users with a wide array of business tools, the multimedia traffic generated can be a headache for the LAN manager. The following design elements should be carefully reviewed:

- Have Web tools that store commonly accessed sites been considered?
- Are there controls to ensure that IP addresses are centrally distributed and managed?
- If illegal addresses are used, are controls in place to ensure they are not inadvertantly transmitted to the outside?
- Has firewall capacity been thoroghly reviewed so that the firewall does not become an IP traffic bottleneck?

Efficient and Secure Remote Access. Many organizations have not standardized the manner in which remote users link to their LAN. As a result, some may dial in using a "PC Anywhere" technology while others may go in through the Internet using tunneling and perhaps access card encryption. To reduce costs and maintain adequate levels of security, the organization should adopt one standard means to access LAN servers remotely. Without such standards, valuable resources are wasted in trying to secure and service multiple entry points into the network.

CABLING PLAN

The impact of cabling on LAN efficiency and error rates is greater than is generally recognized. Horizontal (i.e., within a floor, not a backbone) cabling errors account for a large percentage of intermittent line noise. The network manager should review vertical (backbone) cabling and horizontal (intraLAN) cabling.

Vertical Cabling

The wiring structure of the backbone generally varies according to the volume of traffic, building topology, and budget. The auditor should determine if the design is appropriate for the existing and future network configuration.

Simple Loop. Exhibit 9-3-2 shows the simplest backbone design. The traffic hops from floor to floor, allowing insertion at any point (not desirable). Routers (preferable to bridges for larger networks) direct nonlocal traffic to the backbone. All enterprise traffic that is not local to the user's local LAN traverses the backbone. Advantages include simplicity of design, lower cost, and less cable length. However, the sum of all nonlocal, enterprise traffic can exceed the capacity of the backbone, particularly if frequently accessed servers are placed directly on the backbone. In addition, if a department has users and LANs scattered on several floors it may have several hops to get to common servers (two hops per router: one for sending the data, one for receipt). These hops can slow user response (roughly 10% to 30% depending on routers or bridges).

Collapsed Backbone. In a collapsed backbone configuration, all cabling is configured in a star structure terminating in a high power router (as shown in Exhibit 9-3-3). The backbone becomes the high-speed backplane of the router (typically in the 600M-bps range). Because all enterprise (i.e., nonlocal) traffic is supported by the high-speed back-plane of the router, consid-

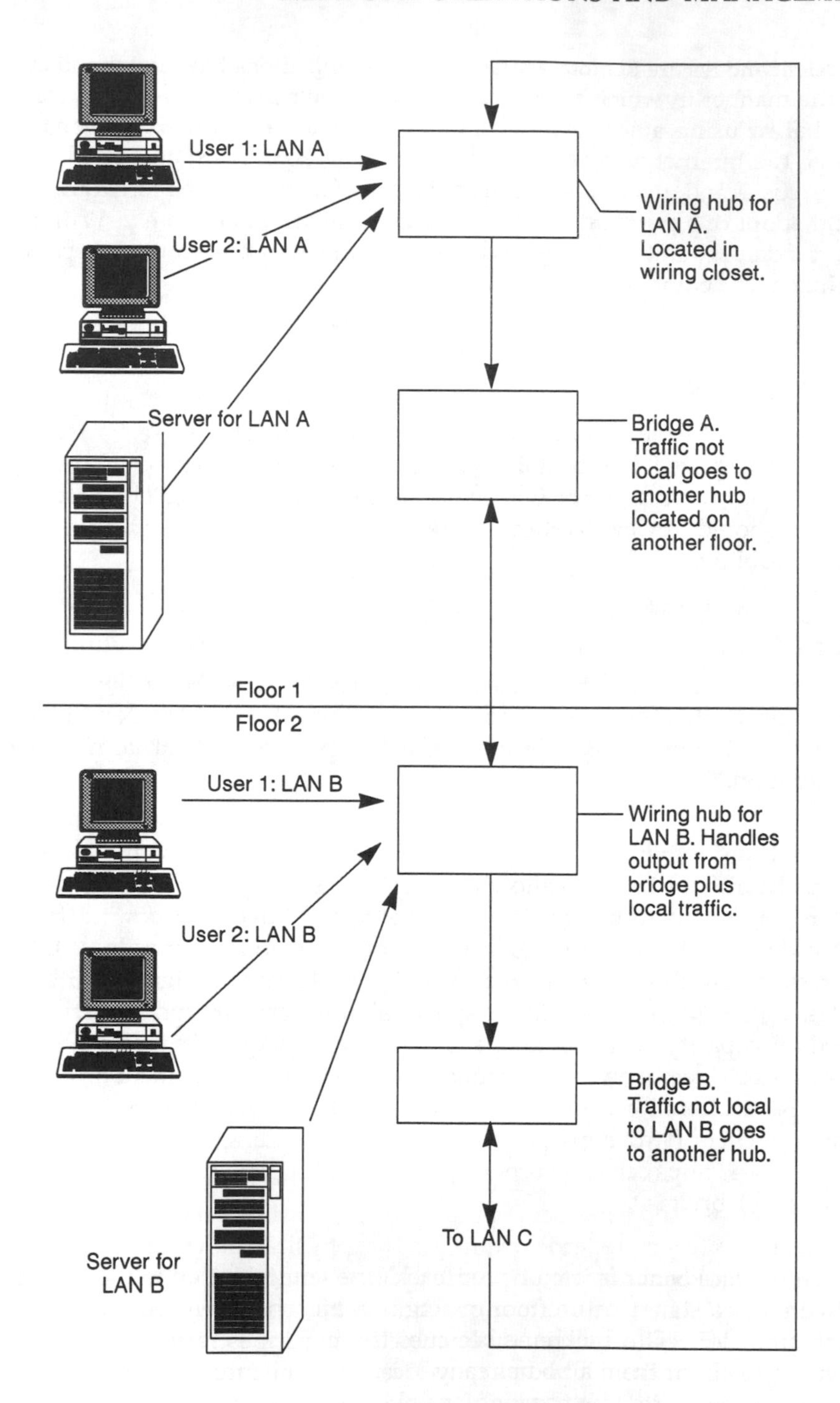

Exhibit 9-3-2. Simple Backbone (Bridged)

erably more enterprise volume can be accommodated than is the case of the simple loop. However, in larger organizations there may be more LANs than can be fitted into a single router.

Collapsed Backbone with Multiple Routers Linked by ATM or FDDI. This design (as shown in Exhibit 9-3-4) has all the advantages of the previously listed designs. In addition, it has the ability to support a larger number of LANs because of the linking of multiple routers. However, users may have a short hop across a minimal FDDI or ATM ring to another backbone.

Horizontal Cabling

Horizontal (i.e., intraLAN) wiring has a major impact on error rates as well as the cost of moving users. The IS manager should review standards and testing when considering this type of cabling.

Standards for Wiring. Exhibit 9-3-5 shows a typical wiring standard for a horizontal (nonbackbone) network with Token Ring, Ethernet, and ATM. Topics to be covered should include technical specifications for cabling, use of conduit, types of connectors and jacks, and topology (e.g., should wiring be deployed as a star from the wiring closet, or should intermediate pods be used so that permanent wiring goes from the wiring closet to the pods, then to the desktop?).

Testing During Installation. All cable should be tested with the appropriate equipment after installation. Technicians should be certified on such equipment as Token Ring and category 5 (CAT5) testers. The IS manager should pay particular attention to installation standards. For example, CAT5 has great potential for carrying error-free traffic if it is installed properly (e.g., standard connectors, housing), but it may be degraded if not installed according to specifications.

MOVES, ADDS, AND CHANGES

Physical moves of employees and contractors can be a significant source of errors, downtime, and expense. During a move, coordination of several players is required (e.g., properties staff for the physical move of desks and computers, voice communications for telephone service, backbone staff, and LAN personnel). If an entire floor or work area is reconstructed (e.g., conversion from offices to bullpen-style cubicles) cabling is rerun from either the wiring closet or from a pod already located in the area. The IS manager should carefully review all phases of a physical move.

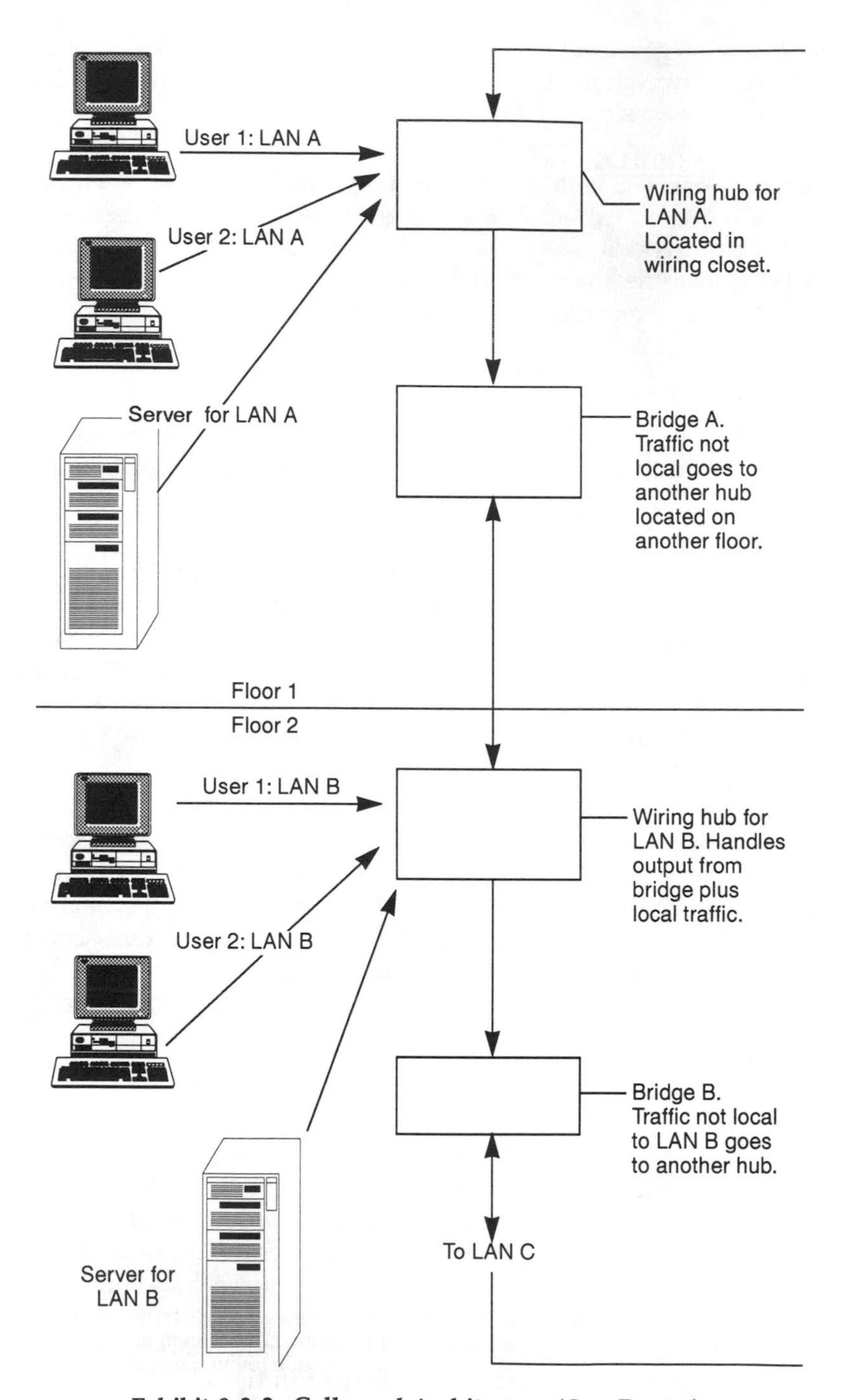

Exhibit 9-3-3. Collapsed Architecture (One Router)

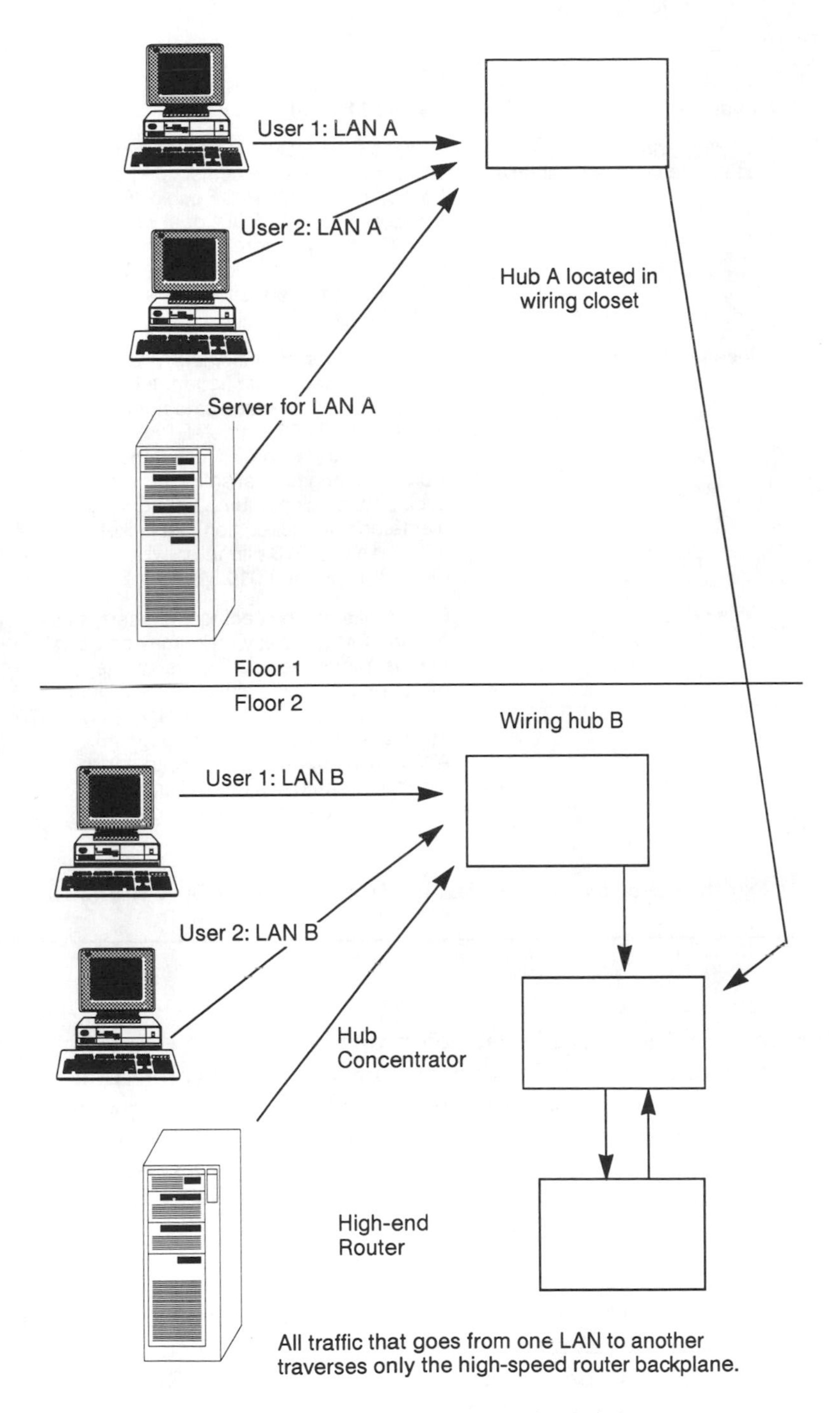

Exhibit 9-3-4. Collapsed Backbone with Multiple Routers

(Partial Listing Only)

Item	Detailed Specifications
Cabling (three components in one cable)	Trimese Cable. Dual Ultra Category 5/Duplex FDDI fiber, comscope part #0410, trimese CMP-OF cable. Fiber is duplex 62.5/125 FDDI dual wavelength. Ultra Category 5's are 4 pair unshielded twisted pairs compatible to EIA/TIA 568 extended frequency LANs up to 350 Mhz.
Token Ring Cabling	Comscope part #4901A, 22AWG solid data grademedia with fluorinated ethylenepropylene dielectric nominal diameter of 0.093 diameter, 38AWG 8 end tin/copper with 65% coverage individual longitudinal shielding foil for each pair, black plenum .355 nominal diameter over jacket, nominal jacket thickness of 0.013 diameter, with minimum spot of 0.010.
Relay Rack	For installs not exceeding 232 users, four existing 84 x 19 relay racks may be used. For installs exceeding 232 users it is necessary to install either one additional 84 x 19 rack or four 96 x 19 rack. All racks are to be grounded through #6 AWG solid copper wire to clean earth ground. Clean earth ground lug to be supplied by building property company.

Exhibit 9-3-5. Example Detail Horizontal Cabling Specifications

Labeling. Cables should be labeled properly to ensure connections go to the correct workstation. It is particularly important that connection points at the desktop are labeled correctly. For example, new or inadequately trained move personnel can plug a modem line into a Token Ring connection, resulting in LAN downtime. All connection points should be labeled with icon size pictures or words. Many organizations use temporary personnel to handle the irregular workload generated by moves.

Telephone Switch Changes. Modem lines frequently go through the organization's voice switch. Procedures should exist to closely coordinate switch software changes with LAN changes. For example, in the Rolm 9750 switch, modem lines can be activated or inactivated through terminal input. Users who need modem access may have their lines changed too early or too late.

Move Change Control. Physical moves usually compel significant changes in internal and external addresses for user equipment and other LAN devices (e.g., printers). Notification should occur beforehand, and all parties (e.g., backbone, properties company, voice communications, LAN managers) should coordinate to ensure that changes are implemented in the correct sequence. For example, if one department brings up a server with the name XYZ002 and that server name already exists, some users may not be able to access one of the servers. Either the back-bone team or a designated change control group should monitor, coordinate, and approve all moves and changes.

PHYSICAL ENVIRONMENT OF SERVERS

Servers are often located near the work group they serve. Unfortunately, these locations rarely optimize efficiency. In addition, dispersement of servers and other devices often means inefficient use of manpower. For medium to large enterprises, centralization has the following advantages:

- *Improved (and cheaper) physical security.*
- *Easier installation and changes for cabling because of raised flooring.*
- *Concentration of personnel.*
- *More effective planning for power use.* Servers should be on an uninterruptible power supply (to allow shutdown and backups). Building electricians and property managers can more easily plan for increases in electrical power consumption if they know that they are housed in a central location.
- *Synergy in use of various technical staffing.* A central network control center can monitor servers, perform wide area network monitoring, backbone monitoring, and other functions. For example, after hours the backbone staff could help reboot a downed server.
- *Easier to ensure adequate air conditioning.*
- *Disaster recovery, although expensive, is easier to implement.* This is true if at least the servers and network control center are physically centralized in a building.
- *Logistics of backup are simplified.*

For centralization to work, the enterprise must have an adequate infrastructure. The backbone needs to be reliable and routing should be relatively efficient.

SUMMARY

Efficient and reliable operation of a complex LAN network requires the right tools, training, and attention to detail. By using the auditing guidelines contained in this chapter, the network manager can provide important guidance to the organization. An objective and comprehensive review of network operations with an emphasis on management's goals (e.g., cost control and reliability) should be the network manager's goal.

9-4
Protecting Against Hacker Attacks

Ed Norris

THE PROBLEM OF PEOPLE BREAKING into computer and communications systems is not new. Such activities have occurred since the first introduction of these technologies. As new technologies become available, it can be expected that new methods of obtaining unauthorized access to these technologies will be developed.

To protect their systems against unauthorized intrusion, security practitioners need to understand hackers: who they are, what motivates them to break into systems, and how they operate. This chapter examines these issues, describes specific hacking techniques, and recommends actions that should be taken to reduce exposure to hacker attacks.

WHAT IS A HACKER?

The term *hacker* means different things to different people. The author of the book *Prevention and Prosecution of Computer and High Technology Crime* defines hackers as computer criminals and trespassers who view illegal computer access as an intellectual challenge and a demonstration of technical prowess. *The New Hacker's Dictionary* offers a more benign definition: "A person who enjoys learning the details of computer systems and how to stretch their capabilities—as opposed to most users of computers who prefer to learn only the minimum amount necessary." Indeed, many people consider themselves hackers, yet would never attempt to gain unauthorized access to a computer or telephone system.

Other terms are also used to refer to hackers. For example, phreaks are hackers that target telephone systems. The terms *computer intruder* and *hacker* are also commonly used for computer hackers. This chapter modifies the definition from *The New Hacker's Dictionary*, adding that some hackers may attempt to gain unauthorized access to computer and communications systems to achieve their goals.

A Hacker Profile

In the early 1980s, the hacker profile was of a highly intelligent, introverted teenager or young adult male who viewed hacking as a game; most were thought to be from middle- and upper-class families. Like most stereotypes, this profile has proved to be wrong. In reality, hackers can be very smart or of average intelligence, male or female, young or old, and rich or poor. And recent hacker arrests and convictions have taught hackers that, whatever they may have thought in the past, hacking is not a game.

To succeed, hackers need three things: motive, opportunity, and means. The motive may be increased knowledge, a joy ride, or profit. Many IS practitioners have the opportunity and means to hack systems but lack the motive.

The opportunity to hack systems has increased greatly over the years. Today, computer systems can be found everywhere. Hackers do not need state-of-the-art equipment to hack systems—used equipment is inexpensive and adequate to the task. Most companies allow some type of remote access by means of either dial-up lines or connections to external networks. For a relatively small monthly fee, anyone can have access to the Internet. Unfortunately, many corporations provide opportunities to access their systems by failing to provide adequate security controls. And many hackers believe that the potential for success outweighs the possible penalties of being caught.

The means of attack is limited only by the imagination and determination of the hacker. A basic law of hacking can be summarized as "delete nothing, move nothing, change nothing, learn everything."

Some hackers target such entities as corporations, security and law enforcement personnel, and other hackers. Kevin Mitnick allegedly electronically harassed a probation officer and FBI agents who got in his way. Some hackers target organizations for political reasons. For example, the Chaos Computer Club supports Germany's Green Party. And others target any machine that runs an operating system capable of executing a particular virus, worm, or Trojan horse.

The estimates of the number of people involved in hacking vary greatly. Estimates range from about one hundred serious hackers to hundreds of thousands. No one really knows the numbers of people involved. There are enough hackers to warrant taking precautions to prevent unauthorized access.

Hackers often use aliases, such as Shadow Hawk 1, Phiber Optik, Knight Lightning, Silent Switchman, Dark Avenger, and Rock Steady. These aliases allow them to remain anonymous, but retain a recognizable identify. They can change that identify at any time simply by choosing another handle. For example, Shadow Hawk 1 is known to have also used the handles Feyd Rautha, Captain Beyond, and Mental Cancer. Changing handles is intended to confuse security personnel as to the identity of the hacker, which makes it more difficult to monitor hacker activity. A hacker may also want the targeted organization to think that several people are attacking the target. Security practitioners need to be aware of these methods of operation in order to understand the identity and the true number of hackers involved in an attack.

Hacker Clubs

Some hackers and phreaks belong to such hacker clubs as Legion of Doom, Chaos Computer Club, NuKE, The Posse, and Outlaw Telecommandos. These clubs give a sense of companionship, although most members never physically meet. More importantly, they help members work as a team toward common goals. By bringing together unique technical skills of individual hackers (e.g., specialties in UNIX or TCP/IP), these teams can achieve goals that might be out of reach for an individual hacker. Some hackers may also view membership in hacker clubs as demonstrating to the hacker and security communities that they are skilled members of an elite.

Hacker clubs come and go. The more willing the members are to contribute to club activities, the longer such clubs remain active. Hack-Tic and the Chaos Computer Club have been in existence for a relatively long time, whereas, such groups as MAGIK (Master Anarchists Giving Illicit Knowledge) and RRG (Rebels' Riting Guild) lasted only a very short time.

Hacker Publications

Some hacker and phreak clubs produce publications. For example, the Legion of Doom produces *Phrack*, Phalcon/Skism produces *40Hex*, and the Chaos Computer Club produces *Chaos Digest*. These publications provide hackers with technical information as well as providing a social function. Some hacker publications can be received by means of electronic mail over the Internet; others are sent through the postal system. Some book stores and magazine stands sell *2600 The Hacker Quarterly*, which has been published for ten years. This publication periodically publishes a list of addresses of other hacker publications.

In order to keep informed of hacker interests and activities, security practitioners with access to the Internet should subscribe to the non-hacker publication *Computer underground Digest*. This electronic digest covers general issues related to information systems, and also covers hackers- and security-related topics. It often provides pointers to other sources of hacker information. Searching these sources can help the security administrator learn about the ways hackers obtain knowledge. *Computer underground Digest* can be subscribed to by sending an electronic mail message to listserv@vmd.cso.uiuc.edu; the message should be sub cudigest your-name.

Hacker Conventions

Some hacker groups sponsor hacker conventions. For example, the Chaos Computer Club sponsors the Chaos Congress, hack-Tic sponsors Hacking at the End of The Universe, and Phrack and the Cult of the Dead Cow sponsor HoHoCon. These conventions are held in the US and Europe. Hackers and phreaks, as well as security and law enforcement personnel, are featured speakers. The conventions are open to all interested parties.

Most of these conventions serve primarily as venues for hackers to brag, swap stories, and exchange information. They tend not to be highly organized; most substantive information is exchanged in hotel rooms and lobbies. There have been a few raids and arrests at some conventions.

Bulletin Boards and Newsgroups

Hackers and hacker clubs primarily communicate by means of bulletin board systems. It is estimated that there are about 1,300 underground bulletin boards in the US. The information found on bulletin board systems is usually current and state-of-the-art. Timeliness of this information is important; hacker techniques described in print publications are usually already well known by the time they appear in print, and these published methods may no longer work.

Even old information can be valuable, however. The LOD (Legion of Doom) Communications is selling old bulletin-board system archives. Even though these message bases are from the mid-1980s, many organizations are still being successfully attacked using methods described in those files.

It can be difficult to gain access to an underground bulletin board. A hacker has to have been active for some time and shared valid information in the hacker community. Some bulletin boards even require background checks and references. As new members become more trusted among their peers, their level of access to sensitive information increases. Above-ground hacker bulletin boards usually grant access to anyone, and may even invite security professionals to join the communications exchange. The telephone numbers for this type of bulletin board are sometimes published in the Internet newsgroup alt.bbs. Internet newsgroups are similar to bulletin board systems in that they allow people a vast forum for communication. Currently there are three active hacker newsgroups: alt.2600, de.org.ee, and zer.t-netz.blueboxing. The *Computer underground Digest* is also posted in the newsgroup comp.society.cu digest. The alt.2600 newsgroup is very active. Two other newsgroups that sometimes have information relevant in this arena are alt.wired and the Electronic Freedom Foundation's comp.org.eff.talk.

METHODS OF ATTACK

As technology has changed, so too have the challenges facing security professionals. It is important to keep informed of the many methods of attack used by hackers. The following sections describe some of the most popular techniques.

Social Engineering

Social engineering is used to describe techniques for getting someone to do something for an unauthorized individual. Hackers and phreaks are usually very adept at the art of social engineering. For example, a hacker may simply ask someone for help. He or she may not get the intended information, but the person asked will often provide at least some useful information. In one case, for example, an intruder told a guard that he needed access to an office because he had a report due on Monday; he went on to complain about having to work on the weekend. The guard accepted these facts and let him in without asking for his company identification or any other job-related information. The more sincere and knowledgeable the hacker sounds to the person being targeted, the more likely the hacker will get what he or she wants.

Knowledge can also be gained by many other methods: for example, by reading newspapers, magazines, and annual reports. An annual report might disclose the projects a corporation is pursuing and the people who are working on those projects. Reading Internet newsgroups may provide the hacker with a good idea of certain corporate projects; it may also let the hacker know which computer system is being used on which projects. Postings in newsgroups usually contain much useful information; telephone books or information can provide the hacker with a telephone number for the company. By assembling these various pieces of information, the hacker can appear as if he or she were an employee of the targeted company.

One hacker publication recommended that hackers write out a script before calling a target. The script should include the initial explanation, plus answers to questions that they might be asked.

Exhibit 9-4-1 illustrates how Internet newsgroups can reveal information useful to hackers. The newsgroup article in Exhibit 9-4-1 tells the reader that John Smith is a geologist in the MIS department of Eastern Mining Corporation and that the company is located in New York. The phone number to reach John is 212-555-1234. John is using the computer system emcmis and his user ID is jds. Eastern Mining has access to the Internet and has at least outgoing file transfer protocol (FTP) service as well as news and incoming mail access. Last, John's current project involves geological surveys. There is enough information in this one message for someone to start a social engineering attack.

Newsgroups: comp.programming
From: jds@emcmis.emc.com (John Smith (Geologist - MIS))
Subject: USGS Code ????
Reply-To: jds@emcmis.emc.com
Organization: Eastern Mining Corporation
Date: Fri, 6 May 1997 01:23:41 GMT

Does anyone out there know if it is possible to gain access to
the USGS (US Geological Survey) source code that I have seen
mentioned in journal, using the net.
If it is available on the net, can anyone provide me with a
site (or even better a list of sites) that I can access using ftp.

Thanks in advance ....

John

John D. Smith
jds@emcmis.emc.com

Eastern Mining Corporation
Phone: 212-555-1234
New York, NY

Exhibit 9-4-1. Gathering Information from a Newsgroup Message

To help limit disclosure of at least some of this information, the security manager might implement an application-level firewall to standardize the mail addresses to the corporation and other key access information.

Employees should also be made aware of social engineering and the ways to combat it. Because anyone can be the subject of such attacks, awareness programs should be directed to general employees as well as systems operators and help desk personnel.

Employees should also be informed as to whom to notify if they believe someone is asking for information or requesting them to perform a task that is suspect. External requests for corporate information should be passed to a trained public relations department staffperson. Requests to perform a task should not be carried out until that person can verify who the caller is and if the caller is authorized to request that the task be performed. If the verification cannot be completed successfully, someone in the organization should be notified of the attempted intrusion. Often, what appears to be an isolated attempt at unauthorized access is part of a larger social engineering attack. If such an attack is detected, advisories should be sent to employees warning them to be on guard.

Dumpster Diving

Dumpster diving is a term used to described the searching of garbage for information. Hackers frequently search dumpsters located outside of buildings for such information sources as operator logs, technical manuals, policies, standards, company phone books, credit card numbers, and dial-in telephone numbers, all of which might aid the hacker in gaining access to the organization.

The security administrator should make employees aware that confidential material should be disposed of in accordance with the organization's security standards. After material is destroyed, it may be recycled as part of the regular recycling program.

Hardware and Software Tools

Hackers also rely on hardware and software tools for carrying out attacks. For example, phreaks use hardware devices to generate tones that allow them to navigate the various telephone switches and gain free phone access. These devices, referred to as boxes, are known by their color.

For example, the red box is used to generate the coin tones used by pay phones. The newest type of red box makes use of Hallmark greeting cards that allow users to record a message; the phreaks use them to record coin tones. This is not the first time an innocent device has been used for illicit purposes; a whistle given away in Captain Crunch cereal boxes was used to generate the 2600 Hz telephone signaling tone.

War Dialers. Many software tools are available on the Internet or on bulletin board systems. The most infamous is the war dialer. War dialers scan a telephone exchange looking for modem tones. When one is found, the modem phone number is saved. Software tools do not need to be complicated to be of use to hackers.

Although it is not possible to stop attempts to connect to a modem, unauthorized successful connections are very easy to stop. Modem access can be protected by means of strong passwords, tokens, or other mechanisms. The protection mechanism should have the ability to log access attempts. If the organization is experiencing many unsuccessful access attempts, it may be the target of a hacker or hacker group.

A war dialer expects a quick answer to its call. Some war dialers can be thwarted by increasing the number of rings before the modem answers the call.

One problem that has become widespread is use of unauthorized modems. Internal modems can be purchased for under $30, well within the price range of an average employee. They are very easy to install in a PC workstation and can be used with the office phone line. Using a war dialer, the security manager should conduct a periodic check of telephone numbers that belong to the corporation. A check against the list of authorized modems will detect use of any unauthorized modems.

Password Crackers. Password crackers are another popular tool. With this approach, a hacker downloads a targeted password file (e.g., UNIX's /etc/passwd or OpenVMS's SYS$SYSTEM:SYSUAF.DAT) to his or her computer and then attempts to crack the passwords locally. The hacker can do this without triggering any alarms or having to run through the log-in sequence. Six-character UNIX-based passwords have been cracked in less than an hour. If only lower-case letters are used in the password, the password can be cracked in less than one minute. It should be recognized that what is considered a strong password with today's technology may not be adequate a year from now.

Many intrusions succeed because of weak passwords. The security administrator can run a password cracker program against system authorization data bases in order to find vulnerable passwords. The security manager should schedule and conduct such checks periodically. Anyone whose password is cracked should be instructed on how to select effective passwords. These persons should also be reminded that their failure to do so may jeopardize corporate assets.

Network sniffer software has also been used on the Internet to capture user IDs and passwords; TCP/IP packets were scanned as the packets passed through a node that a hacker already had under his control. (Some security professionals argue that use of Kerberos on the network cures the network sniffer problem, but this only protects the password when it travels between the Kerberos daemon and slave.)

The underlying problem presented by this hacker attack has to do with how the hacker was able to get control of one or more of the network nodes. Many corporations that have connected to the Internet fail to implement any security measures to counter the additional risk of public access. Any organization that plans to connect to the Internet should first install a firewall.

Firewalls. Firewalls are a collection of components placed between two networks. They have the following properties:

- All traffic in both directions must pass through the firewall.
- Only authorized traffic, as defined by the local security policy, is allowed to pass.
- The firewall itself is immune to penetration.

A firewall allows the organization to block or pass access to the internal and external networks based on application, circuit, or packet filtering.

In summary, the security manager should be familiar with the types of hardware and software tools used to attack computer and communications systems. By searching the Internet, he or she should be able to find the same tools that hackers are using. These tools can be used to verify that the organization is adequately protected against them.

Reverse Intent

Reverse intent refers to a phenomenon in which an object that is intended to perform an action is used to perform the opposite action. For example, a deliberate reverse intent message might state: "This product is not to be used to increase the octane in gasoline." The message is intended to warn us that use of the product is prohibited for the purpose of increasing octane levels, but it also discloses that the product is capable of boosting the octane rating.

Hackers and freaks can use this to their advantage. Computer Emergency Response Team (CERT) advisories and Computer Incident Advisory Capability (CIAC) information bulletins are intended to notify people of security problems. They contain information about a given product, the damage that can occur from use or misuse of the product, the solution, and additional information. As illustrated by Exhibit 9-4-2, this information can be useful to hackers. In this exhibit, it is reported that Sun Solaris V2.x. and SunOS V5.x have a security problem that gives local users the ability to gain root (full privilege) access. The local user can execute the expreserve utility that gives access to system files. If the computer system does not have expreserve disabled or the system administrator has not installed the patch solution provided by the vendor, the security of the system may be compromised. If a hacker has access to a Sun workstation, the hacker can find out how the exploit this security exposure, either on his or her own or with the help of others.

Hackers are quick to notify each other of these types of announcements. For example, a CERT advisory dated August 14, 1990 appeared in *Network Information Access* the following day.

It is important that the appropriate department within the organization receive security problem notification from software vendors. (Such notification should not be made to the purchasing department simply because it signed the check for the software.) CERT and CIAC information can be received by means of electronic mail over the Internet. Some vendors have their own advisory mailings over the Internet (e.g., *Hewlett Packard Security Bulletin*).

The information security manager should develop an action plan for installing security patches. The security manager should also conduct a postmortem after

The Computer Incident Advisory Capability
INFORMATION BULLETIN
Solaris 2.x expreserve patches available

July 1, 1996 0900 PDT
Number D-18
PROBLEM: The expreserve utility allows unauthorized access to system files.
PLATFORM: Sun workstations running Solaris 2.0, 2.1, and 2.2
(SunOS 5.0, 5.1, and 5.2).

DAMAGE: Local users can gain root access.
SOLUTION: Disable expreserve immediately, then install patch from Sun.

Exhibit 9-4-2. Example of Reverse Intent

the installation of security patches, noting which actions completed without problems and which actions did not. The action plan can then be adjusted to fix deficiencies and to reflect changes to the business environment.

In addition to obtaining information from advisories, hackers also seek out such sources of information as the system security manuals (or sections of other manuals) provided by software vendors with their products. These books are intended to instruct the system administrator on how to secure the product. Supplemental computer manuals found in almost every book store are another source for the hacker. A security manual might state: "Do not disable high-water marking on disk volumes." This statement tells a hacker that if one or more disks have disabled high-water marking, there is a potential problem to be exploited. In this case, the hacker may discover the art of disk scavenging and access the information contained in unallocated blocks.

Some professional organizations advocate sharing published security standards among their members. But it can be difficult to control the distribution of these standards, and they can also be used with reverse intent. The standards tell how a corporation secures its business. Many standards contain such sensitive information as group names, employee names and titles, phone numbers, electronic mailing address, and escalation procedures. A standard in the hands of a hacker becomes a powerful tool for social engineering.

It is impossible to stop reverse intent. Security practitioners must be aware of information that is available to hackers and ensure that they act appropriately according to the intent of the information. When an advisory or other piece of information reaches the security administrator, he or she should try to gauge how a hacker might use this information and modify his or her actions accordingly.

SECURITY MONITORING AND REVIEW

In order to stop a hacker, the security administrator must know when the hacker is knocking at the door or has already entered the system. Waiting until something has gone wrong may be too late. Auditing a system is more than turning on every auditable event, however. The security administrator must monitor enough events to be able to detect an attack, but not so many that the audit information becomes unmanageable. Too much data tends not to be analyzed properly and exceptions to normal behavior become more difficult to detect.

One of the first things a hacker attempts to do is delete the audit trail. Novice hackers may stop the audit processes and delete the entire audit data base; experienced hackers remove only their records from the data bases. If warranted, audit information should be printed directly to a hardcopy device or to a write-once storage device. The data should be analyzed on a regular basis with follow-up done on any suspect activity. Most hacker intrusions produce a few knocks on the door before a successful penetration takes place. It is easier to keep a hacker out than to recover after a successful intrusion.

SUMMARY

It is important to understand how hackers navigate throughout the electronic world and how they attack systems. IS management also needs to understand the threats they pose to the organization and implement appropriate security controls to counter those risks. A security program should also be monitored to ensure that it continues to be able to counter the risks created by hackers.

9-5 Documenting a Communications Recovery Plan

Leo A. Wrobel

TECHNICAL SERVICES PERSONNEL often respond tersely when asked by management to produce a disaster recovery plan for a communications system. They view such a request as a no-confidence vote and the plan as a test of their ability to perform a recovery in the event of a disaster.

Competency, however, is almost never the issue. In most large organizations, the technical services and communications staffs are capable of recovering from a disaster under virtually any circumstances. Often, the very people controlling the recovery process are those who actually designed and built the communications system in the first place. They know where every wire in the organization runs and have memorized the telephone number of every major equipment vendor and service supplier. In the event of a disaster, these employees would undoubtedly pull together in an almost superhuman effort.

Problems with disaster recovery occur when an organization's key personnel, such as the local area network (LAN) administrator, are unavailable for any reason during a disaster. Because of the growing complexity of LAN technology, one of the most effective tests of a disaster recovery plan is to assemble the disaster response team and remove the LAN administrator from the exercise. Loss of a LAN administrator who knows everything about a system can be devastating.

The need for a communications recovery plan should therefore be presented to technical services staff from the perspective that someone unfamiliar with the system may have to execute the plan. The goal of every organization should be to document its communications recovery plan in a systematic format that can be followed by any reasonably trained technical services employee, whether they be from a vendor, a rental company, a major supplier, or a carrier company.

DEVELOPING AN INVENTORY OF COMMUNICATIONS HARDWARE

Disaster recovery personnel need to know several details about the organization's hardware, including:

- Type and model number.
- Software packages residing on the hardware.
- Software revision numbers.
- Date of purchase and cost.
- Criticality to operations.
- Power requirements.

Other helpful items include:

- The name, address, and telephone number of the manufacturer.
- The local distributor or depot for the equipment.
- The location of secondary-market hardware suppliers, who can be instrumental in providing equipment during a disaster.

Exhibit 9-5-1 provides an example of a hardware inventory form that can aid recovery personnel in making quick command decisions. For example, if a four-year-old piece of equipment is depreciated over a five-year period, recovery personnel may decide to replace rather than to repair it. Such information is extremely useful when many decisions must be made rapidly and under tense circumstances.

A separate inventory should also be kept of all hardware maintained by an organization at a disaster recovery center. The inventory should contain some fairly minute detail, such as whether the hardware is on a movable rack or requires dollies, and the type of power plug the equipment requires.

MAINTAINING ACCURATE INVENTORIES

The best method for keeping track of hardware necessary for restoring communications is through a process called data importing.

Automated Data Importing

Importing is essentially a means of finding data bases and repositories of information within the organization that are reasonably up to date.

For example, when a piece of hardware is purchased, a document or file for the equipment is archived somewhere within the organization. Sometimes, the contract and the documentation for the equipment goes to the accounting department for amortization purposes. In other cases, the files become part of

Component:

Purpose:

Manufacturer:

Serial Number:

Associated Software:

Criticality Rating (1,2,3,4):

Date of Purchase:

Vendor Name and Telephone Number:

Remarks:

Exhibit 9-5-1. Hardware Inventory Form for Emergency Replacement and Restoration

the personal file of the communications manager or analyst. The optimal situation is for them to be stored and accessible on a LAN. In any case, recovery planners should locate and identify these repositories of inventory data so they can be automatically imported into the recovery plan.

There are good reasons for taking this approach. The price of some of today's hardware and the necessity for any mission-critical equipment to be protected by a disaster recovery plan means that the savvy technical services manager often includes disaster recovery in the selection criteria for major hardware purchases. Such equipment includes automated call distribution units; PBXs; major bridges, routers, and gateways; LAN networks; and mainframe computers. It is much more cost-effective to negotiate roll-in replacement guarantees when a vendor is vying for business than to try to add these services later. Yet, given today's levels of staff turnover, failure to import information on such guarantees and on maintenance contracts into the recovery plan could result in a future LAN manager needlessly paying for a disaster recovery plan for equipment already protected.

Organizations that make heavy use of internetworked LANs have an advantage because they can automatically transfer files between interconnected departments in several ways without human intervention. For example, when object linking Microsoft Word files, a technical services manager can key in on a specific file name in the accounting department to ensure that updates to a hardware repository file are transferred to the appropriate file in the recovery plan.

Manual Updating

Importing can also be accomplished through a sneaker net—assigning a key person or division to regularly go to the department containing files on equipment, make a floppy disk copy of the appropriate file, and update the recovery plan. But under the pressures of work, busy technologists can easily overlook this task, causing the recovery plan to be dangerously outdated. Updating this way therefore requires that staff realize the importance of the recovery planning process and that the process be enforced, to the point of withholding raises when the task is not performed.

ADDITIONAL COMPONENTS OF AN UP-TO-DATE PLAN

Importing Information on Personnel and Vendors

Up-to-date telephone numbers for personnel and critical equipment vendors are essential to the successful implementation of a disaster recovery plan. Once again, this means importing data from reliable sources regularly and, preferably, without human intervention.

Consider employees' home telephone numbers, which are found in several places, such as human resources and the company telephone directory. Care must be taken to ensure that the data contained in these sources is up-to-date. Here again, importing is best done by object linking files together, but it can also be accomplished through use of a sneaker net and floppy disk.

Telephone numbers for key hardware vendors and suppliers can often be found in the network control center, help desk, or other operational environment whose personnel have day-to-day contact with vendors and are often the first to know about changed telephone numbers. Because operations staff are usually regularly involved in escalation procedures, their departments generally document information on second- and third-level management within the vendor community.

Software Package:

Purpose:

Supplier Name and Serial Number:

License Number:

Version Number:

Date Purchased:

Criticality Rating (1,2,3,4):

Location Where Software Can Be Purchased or Replaced:

Remarks:

Exhibit 9-5-2. Software Inventory Form for Emergency Replacement or Restoration

Developing an Inventory of Software

Communications recovery planners must also develop an inventory of all software required for operation of mission-critical communications equipment. This inventory should include:

- Acquisition date.
- Original cost.
- License number.
- Version number.

Exhibit 9-5-2 provides an example of a software inventory form.

Emergency Phone Lists of Management and Recovery Teams
Vendor Callout and Escalation Lists
Inventory and Report Forms
Carrier Callout and Escalation Lists
Maintenance Forms
Hardware Lists and Serial Numbers
Software Lists and License Numbers
Team Member Responsibilities
Network Schematic Diagrams
Equipment Room Diagrams
Contract and Maintenance Agreements
Special Operating Instructions for Sensitive Equipment
Cellular Telephone Inventory and Agreements

Exhibit 9-5-3. Useful Appendices to a Disaster Recovery Plan

Equipment Room Diagrams

Equipment room diagrams should show all installed communications hardware and delineate any special environmental specifications such as air flow, temperature, and power needs. The diagrams should also outline equipment footprints, clearances, and any other information useful to a network installer. They form part of the appendices that should accompany a thorough recovery plan. See Exhibit 9-5-3 for a list of the information that should be contained in these appendices.

Importing Components of the Corporatewide Plan

Technical recovery planners should consider importing components of the corporatewide recovery plan dealing with global policy issues. It makes little sense for a technical recovery planner to write procedures for such companywide concerns as loss of a building, physical security, fire and bomb-threat procedures, purchasing, and media affairs. In these cases, the technical recovery planner should direct the reader of the communications recovery plan to the relevant section of the corporate plan. In most cases, the LAN or network recovery plan itself will probably end up being imported into a corporatewide recovery plan for execution by an emergency management team. The process is reciprocal.

Assigning Technical Teams

It is advisable to split a technical recovery plan into numerous subsections managed by technical teams. Such sections include LAN management, voice

1. Recognition of Need for Planning
   - Protect Human Life.
   - Recover Critical Operations.
   - Protect Competitive Position.
   - Preserve Customer Confidence and Goodwill.
   - Protect against Litigation.
2. Response
   - Reacting to Initial Report of a Disaster.
   - Notifying Police, Fire, and Medical Personnel.
   - Notifying Management
   - Establishing the Executive Management Team (EMT).
   - Filing Initial Damage Assessment Reports to the EMT.
3. Recovery
   - Assisting the EMT in Preparation of Statements.
   - Opening a Critical Events Log for Auditing Purposes.
   - Using Modified Signing Authority for Equipment Purchases.
   - Obtaining Necessary Cash.
   - Maintaining Physical Security.
   - Arranging Security at the Damaged Site, Recovery Center, and Emergency Funds Disbursement Centers.
4. Restoration
   - Coordinating Restoration of the Original Site.
   - Restoring Hardware Systems.
   - Restoring Software Systems.
   - Restoring the Power, Uninterruptible Power Supply, and Common Building Systems.
   - Replacing Fire Supression Systems.
   - Securing the Building.
   - Rewiring the Building.
   - Restoring the LAN.
   - Restoring the WAN.
5. Return to Normal Operations
   - Testing New Hardware.
   - Training Operations Personnel.
   - Training Employees.
   - Scheduling Migration Back to the Original Site.
   - Coordinating Return to the Original Site.

Exhibit 9-5-4. The Seven R's of a Successful Disaster Recovery Plan

6. Rest and Relaxation
   - Scheduling Compensatory Time off.
7. Reevaluate and Reassess
   - Reviewing the Critical Events Log.
   - Evaluating Vendor Performance.
   - Recognizing Extraordinary Achievements.
   - Preparing Final Review and Activity Report.
   - Aiding in Liability Assessments.

Exhibit 9-5-4. *(continued)*

communications, data communications, and the emergency network control center. Each of these functions has assigned day-to-day responsibilities, and the team assigned to each will have specific responsibilities during a disaster. A network control center or help desk, for example, could take on a very different function in a disaster by helping to maintain command and control.

THE REAL REASONS FOR DISASTER PLANNING

All companies use automated systems to conduct business and all suffer when these systems fail. Whether they are LANs or other communications systems, automated systems should not increase the risk to a company merely because they are convenient. A communications recovery plan should therefore accurately focus on restoring a business' core operations, or the items most crucial to its profitability, in the event of a disaster. Exhibit 9-5-4 presents the seven R's of a successful disaster recovery plan.

SUMMARY

A disaster recovery plan is a complex road map of how to rebuild an organization after a disaster. It should be written for execution by a reasonably well-educated technical person in the event key personnel are unavailable.

A thorough recovery plan includes input from all major vendors, suppliers, and departments and must import data from accurate sources. It should delineate recovery tasks systematically and clearly and strike a balance between cumbersome detail that discourages reading and a cursory explanation understandable only to people familiar with a system.

It can take two years or more to complete a successful recovery plan, but the effort is well worth the protection the plan affords an organization.

9-6
Points-of-Failure Planning

John P. Murray

THE MOVE TO CLIENT/SERVER PROCESSING brings new opportunities, challenges, and risks to the management of communications and information systems. Although client/server hardware and operating software differ from that used in mainframes, many methods and procedures that applied in the mainframe environment still apply to distributed systems.

Some organizations have successfully mastered the transition to client/server computing and removed all their mainframe hardware. Most organizations, however, are only currently considering how to make the transition. In these cases, the function of the mainframe is shifting from the primary processing tool to that of a large file server. As reliance on the mainframe lessens, a series of new management concerns arises for IS managers.

THE CLIENT/SERVER ENVIRONMENT'S EXPANDED SCOPE

A primary concern in client/server computing is managing a processing environment that consists of many more components distributed over a much larger geographical area. In addition, the work being processed on those components covers a greater span of business functions than was the case in the past.

Central site personnel may be unaware of the specific work being done at the remote sites. The first time the IS staff hears about difficulty with a particular processing problem may be the first time it has heard of the application. This situation highlights the scope of the work done in the client/server environment and the fact that much of this work is carried out beyond the direct control of any central site.

Support Issues. Although the scope of the activity in the client/server environment expands, the responsibility of the IS staff is not going to change. Although client/server processing gives more autonomy to business users outside the IS department, IS personnel remain responsible for support functions, which may include:

- The availability of the processing environment.
- Response time across the network.
- Easy access to files and training.
- The protection of the data on the various networks.

Hardware/ Software Reliability. Client/server hardware and software have become more reliable over the past few years, but they have not yet reached the levels found in mainframe processing. Whereas the IS staff has a great deal of experience handling the nuances of the mainframe processing environment—including being able to anticipate problems and solve them before they disrupt the environment—potential problems in client/server environments are less apparent. Not only are there more problems in the distributed environment, but they may take longer to resolve.

SERVICE ORIENTATION NEEDED

The client/server model presents new challenges and opportunities for IS operations. The challenges include managing a more diverse environment with greater levels of customer service and satisfaction. Delivering those services must be accomplished while moving from the traditional command-and-control environment to a service-oriented environment.

Because the effective use of increased IS processing power plays a key role in helping the enterprise grow and prosper, the development and management of a solid client/server infrastructure throughout the organization is a major opportunity area. IS personnel who cannot, for whatever reason, adjust to this environment face the prospect of outsourcing. Business users of information technology are increasingly sophisticated with little tolerance for anything less than high-quality data center service. Today, and increasingly in the future, IS customers are going to enjoy many processing options, and they will be able to do more for themselves—including choosing some source other than the in-house data processing function. The IS department must be positioned to respond to this new reality effectively and positively.

Management and Control Issues

Many business functions, especially those at higher levels in the organization, want shorter delivery time for information. Disruptions will assume greater levels of urgency than they did in the past. For example, given the work and travel patterns of executives, they expect the systems they want to use to be available, accurate, and fast, whenever and wherever they want to use them. It does not take too many calls from high-level people complaining about poor levels of network availability to convince the IS department that it is in a new ball game now.

More organizations are moving to processing environments where the IS resources must be available 24 hours a day, seven days a week for anyone who may want to use those resources. This is a customer service issue, because the ability of customers to contact an organization any time of the day or night is a competitive tool. Serious customer service and public relations issues must be overcome when the availability of these systems is impaired. Failure to provide the array of products and services expected by customers, because of technology-based constraints, opens opportunities for the organization's competitors.

RECOGNIZING THE POINTS OF FAILURE

A key issue for the IS department is how to move to a position that can ensure the continued operation of the organization's critical applications at all times. The challenge is to get to 100% availability and remain at that level. Nothing less than total availability is going to be acceptable. The reason is clear: the organization's competitors are also figuring out how to move to processing environments that do not fail. Once an organization attains that goal, it quickly becomes the standard for everyone in that particular industry.

IS management and staff can address the control and management of the client/server processing environment by thinking about the issue as a series of potential points of failure. The idea is that any entity within the client/server processing environment that has the potential to disrupt that environment—a point of failure—must be identified and managed. The next step is to develop a set of plans to help the organization anticipate the potential for damage from those point-of-failure entities. The final step in the process is to lay the groundwork to mitigate any point-of-failure damage should it occur.

Most IS personnel think first about the vulnerability of the IS hardware. Although the continued availability of the hardware is critical to the smooth functioning of the production processing environment, there are other items that have to be included in any consideration of the client/server points of failure.

The identification of the potential points of failure has to be broadened from that of the hardware components to ensure that those other items that carry the potential to disrupt the processing environment have been recognized. Recognition of the points of failure is not enough, however; IS has to be ready to move aggressively when any of the identified points of failure create difficulty. The potential damage, depending on the circumstances, may range from inconvenience to disaster.

The items that need to be considered as client/server processing points of failure are:

- *Hardware.* The mainframe and associated traditional components plus network hardware (e.g., servers, disk storage components, routers, and bridges) must be considered potential points of failure. Communications components such as telephone switches and lines must also be considered as points of failure under the rubric of hardware.
- *Software.* Both operating and applications (either products developed inhouse or purchased application software) are potential points of failure.
- *Change control.* This area is easily overlooked when identifying client/server processing points of failure. However, the ineffective management of the introduction of change into the client/server processing environment can create great difficulty.

No one point of failure category should be seen as more important than the other two. IS must position itself to be able to manage any disruption that might arise in any of the three point-of-failure areas. That management must be based on an understanding of each of the point-of-failure areas and the development of appropriate management controls for each area. In addition, the three categories cannot be thought of as separate; there are linkages that must be taken into consideration.

For example, the introduction of a new piece of hardware on the network raises on obvious point-of-failure concern. The primary issue is that of the reliability of the particular hardware component. However, concern with reliability cannot be the only concern. There is also the issue of the ability to integrate this piece of hardware with the rest of the network components, including the operating software.

In addition, change control problems have to be acknowledged. The timing of the placement of the hardware should be thought out so as to reduce the likelihood of the possibility of difficulty. Bringing in the new hardware during a period of heavy business activity may not be a sound approach. Another change control task is to make certain that those who may be affected by the changes are informed of those changes well before they occur.

Containing Risk

An important aspect of management and control of the points of failure is to limit the variety of hardware and software components installed throughout the organization within the limits of practicality. Risk can be reduced by reducing the overall variances within the processing environment.

Often a case will be made to move to some specialized piece of hardware or software to solve some limited problem. A careful analysis of the actual benefit of that component should be made against the risk associated with the increased processing complexity that may occur as a result of the installation. If the introduction of the piece of equipment or software increases the vulnerability of the processing environment to disruption, then the benefit associated with that installation must be sufficiently strong to help offset that risk.

IS installations have to come to grips with the result of the uncontrolled addition of different network components. That problem becomes apparent when network difficulties arise that require a considerable amount of time and effort to resolve. Usually the network environment has become encumbered with so many different types of hardware, software, and special tools, tracking the cause of problems is itself a problem.

MAPPING THE POINTS OF FAILURE

The place to begin the process of managing the hardware environment is with the development of a series of hardware maps. These maps should show, in some detail, all the IS hardware components within the organization. Hardware maps for one corporate data center are shown in Exhibits 9-6-1 and 9-6-2.

Because this effort should be a mapping process in every sense, not only must the location of each entity be shown, but the relationships of those entities to each other need to be identified. Identifying those relationships is important because the effect of the failure of one component on other areas of the processing environment may not be readily apparent.

Besides the graphic representation of those components, additional detail should be provided about the size and age of each component. The vendor of each piece of hardware should be identified. Should a critical component (i.e., a point of failure) crash, bringing the function back online as soon as possible is going to be crucial. Having as much detailed information as possible will help make that happen.

Hardware maps can be constructed easily with the use of PC-based drawing tools that provide a series of stencils and templates for laying out the environment. Well-designed drawing tool packages provide a full array of stencils for laying out all types of IS processing environments; all that is required is selecting an item from the stencil and dragging and dropping it into the map. If a particular item is not available in the stencil, other drawing tools are usually available in the package that can be used to represent a particular entity. The use of the drawing tools is also invaluable for keeping the maps current. It is possible to make changes to the environment quickly, as needed, which is especially useful in large organizations where there may be many changes to the components in the various client/server installations.

Beyond mapping the hardware components, the maps should also contain information abut other components of the processing environment that per-

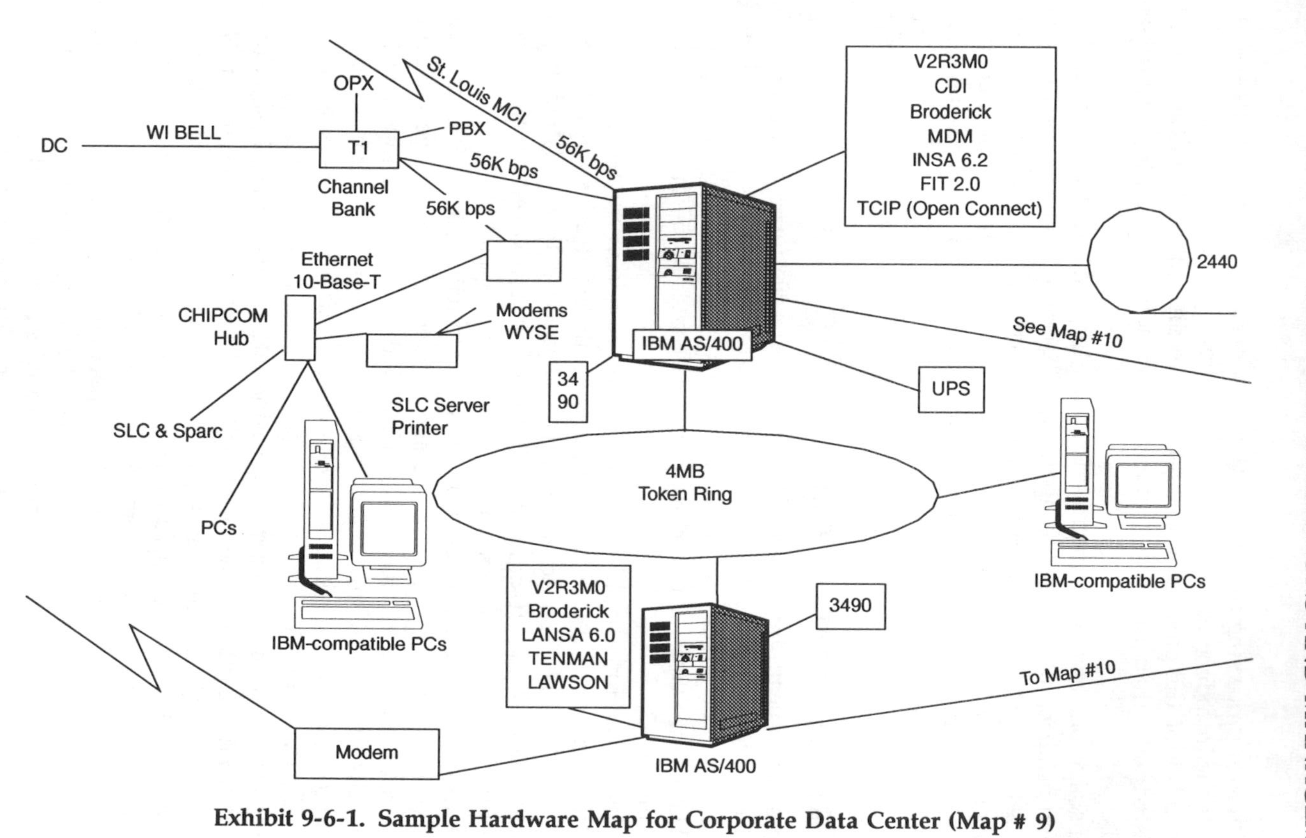

Exhibit 9-6-1. Sample Hardware Map for Corporate Data Center (Map # 9)

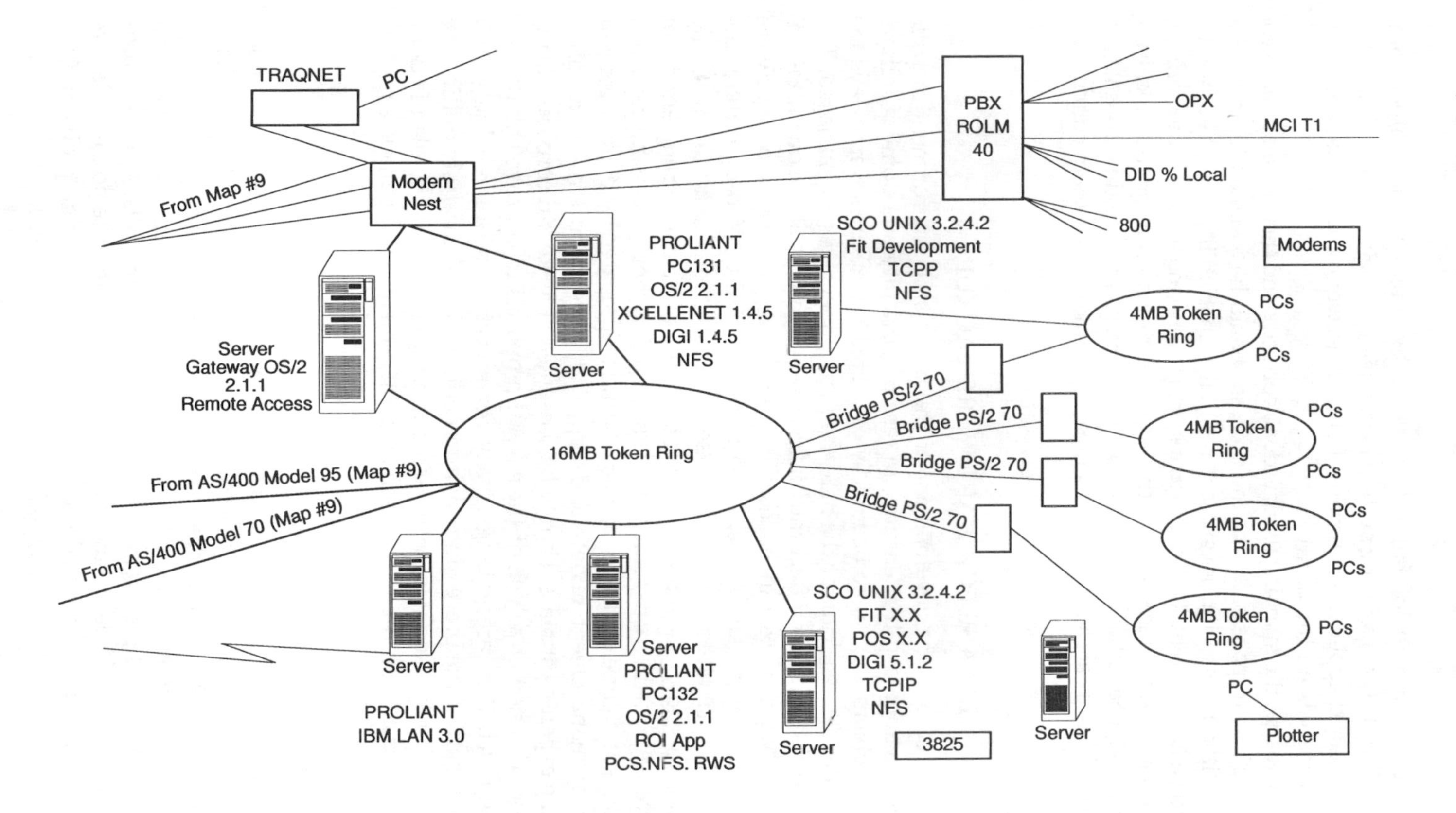

Exhibit 9-6-2. Sample Hardware Map for Corporate Data Center (Map # 10)

tain to managing point-of-failure situations. Examples of those other components include the operating system being used on a given network and the use of packaged applications products on the network and special hardware or software tools on the network.

To ensure that the mapping work is done correctly and that the maps are kept up-to-date, the responsibility for the process should be assigned to an IS employee. Without a specific assignment for that responsibility, the maps are not going to be kept current. Unless the maps are current and accurate, they are not going to be of much assistance in the event of a disruption of the processing environment. The initial effort to construct the maps may be time consuming, but once the maps are in place, it requires less work to keep them current.

BUILDING THE CASE FOR HARDWARE BACKUP

The next step in the development of the processing environment maps is to devise an approach that ensures, in case of difficulty at a point of failure, that the problem is going to be rapidly corrected. There are several ways the issue of adequate backup for the hardware components can be managed. The approach selected must reflect the business requirements of the organization.

The hardware maps can be used in developing the case for hardware backup. In many organizations, at best only a handful of people fully understand the magnitude of the client/server processing environment. Using the maps to show the scope and vulnerability of that environment is an effective way for IS management to acquire redundant processing capacity.

A point-of-failure contingency plan should be prepared by IS management for senior management review and approval. Several options with corresponding expense estimates should be presented. The key in corporate management's decision on which of the several proposals to approve will depend on how much they are willing to pay for a given level of reliability in a particular part of the processing environment. IS should carefully outline the expense and risk associated with each option presented. If the plan is done well, the decision on the part of senior management will be made easier.

Emphasizing Business Support

For example, with respect to a particular hardware component, one option would be to use some level of hardware redundancy within the processing environment. One way to deal with redundancy would be to provide, in the case of file servers and disk storage on the network, mirror processes that would immediately take over if the primary component fails. Another hardware approach would be to have a limited stock of duplicate point-of-failure hardware components deemed critical in case of a hardware problem. One more approach would be to have an agreement with a vendor to provide immediate replacement hardware.

Individual hardware components are relatively inexpensive in the client/server processing environments. Lower costs notwithstanding, it will likely take some convincing by IS management to sell senior management on the idea of spending money on what may turn out to be redundant hardware. If senior management is approached correctly, the IS department stands an excellent chance of success.

The way to begin is to develop a case for additional funds to support the idea of managing the points of failure within the client/server processing environment. That case can be built as a business rather than a technology support issue. One way to develop a business case is to calculate the payroll expense tied to a particular local area network operation.

For example, assume the LAN supports 60 professional employees and that the fully loaded personnel cost for those employees comes to $2,400 an hour. Using that figure, if the LAN is unavailable for four hours, the organization will suffer a loss of $9,600 of productive time. The lost productive time is itself a serious issue, but the negative effect on customer service might be a more serious one to the business.

Two other issues should be mentioned, however. First is that having a critical piece of LAN hardware out of service for four hours is not unheard of. The second is that, given the low cost of microcomputers, $9,600 can buy a great deal of security and peace of mind.

CONTROLLING OPERATING AND APPLICATIONS SOFTWARE

Procedures for the change of operating or applications software in the mainframe environment apply as well in the client/server environment. IS personnel should be kept fully apprised of changes made to the software environment. In the mainframe world, the staff could exert some influence on the testing and introduction of new or changed software. In the client/server environment, such a level of control is not always available, given the many different physical locations of the hardware.

Client/server products for control of software are not yet fully robust. But as appropriate tools are introduced, the sooner they are selected, installed, and worked into control procedures, the better.

The management of operating and applications software in client/server processing is within the purview of IS management and must be addressed. It is a mistake to wait until client/server tools have matured to the point of the available mainframe control tools. The rate of change and growth of client/server processing will not wait on the maturity of the required tools. Any delay in doing whatever is possible to gain control of the software environment is going to make the task of control that much more difficult in the future.

The choice of improper operating or applications software may represent as much a point-of-failure issue as hardware issues or change control. It is important that IS management or staff be advised about the software (operating or applications) being introduced into the client/server processing environment and that they be given the chance to object and recommend an alternative, where appropriate. If an application is introduced that will overtax the processing components, and as a result, the desired work cannot be accomplished, another aspect of the point-of-failure issue will have surfaced.

CHANGE CONTROL

One of the lessons learned in the mainframe environment has been to limit the number of changes introduced to the processing environment at one time. That number should be as few as possible, ideally only one. In addition to limiting the number of changes introduced, it helps if there is a period between changes during which there is an opportunity to determine what negative effect the changes bring in the production environment.

As a practical matter, controlling the number of changes, or the length of time between changes, may not be possible, given the demands of client/server processing. Although it is probably going to be difficult, it is prudent to manage the client/server change process as aggressively as possible.

Moving many changes at a rapid pace into the client/server processing environment almost guarantees difficulty, which only increases the point-of-failure risk. The central idea of a well-managed client/server processing environment should be to limit and manage the points of failure.

SUMMARY

Increasingly, organizations have the goal of what is sometimes described as bullet-proof IS systems. This goal of uninterrupted (i.e., average system downtime of less than one hour per month) IS processing availability is actually being met by a few well-run installations. Those organizations that intend to compete in the future are going to have to develop the discipline to ensure that nothing disrupts the IS processing environment.

There is nothing esoteric about moving to a client/server IS processing environment that virtually never fails. The issues involved have more to do with management than with technology. The ability to anticipate problems before they occur and to be able to move aggressively to correct problems if they arise is critical. The key is an understanding of the client/server points of failure and the establishment of a process to manage those areas.

Section 10
Directions in Communications Systems

IT SEEMS THAT THE COMMUNICATIONS SYSTEMS MANAGER is always standing at the threshold gazing across unknown terrain toward an uncertain future. Change in the telecommunications industry occurs very rapidly. Just about the time managers feel that they have a grasp of the current situation, it mutates into something else that puts them back at the beginning of the learning curve. It would be wonderful if the future of telecommunications could be predicted with something better than a coin-flip 50% chance of being correct. The best communications managers can do is to make educated guesses about factors that can influence the direction of the technology and its emerging requirements and use these to project expectations for the future.

The single, overriding factor that will likely influence the telecommunications landscape is the worldwide movement toward deregulation of the industry. Freeing telecommunication service providers from the bonds of tight government controls is now the norm in Europe, East Asia, Canada, and the US.

Europe is now deregulating telecommunications under a European Union directive that opened all leased-line services to competition in 1996. The "liberalization" of all services is scheduled for January 1998. There are still questions about opening the local access lines to competition and there are a half-dozen countries that have requested extensions to as late as 2001. Despite the issues of delay and local access, a number of alternative service providers have entered the leased- and long-distance service businesses. Many are utility companies, especially electric power companies, who already have the necessary right-of-way for transmission lines. Others are competitive access providers such as WorldComm, Cable and Wireless, and Racal.

In Japan, the Nippon Telephone and Telegraph monopoly is being slowly disbanded. Australia and New Zealand have been open to competitive services for some time. Canada has had deregulated long-distance service since 1992 and is considering opening the local service market in 1998.

The most ambitious experiment in deregulation is the result of the Telecommunications Act approved in the US in 1996. This will have a dramatic effect on communications systems managers. Any and all telecommunications services can now be provided by any service provider. Local exchange carriers, interexchange carriers, competitive access providers, Internet service providers, cable companies, and anyone else that decides to enter the business will be lined up to offer communications managers all sorts of wonderful bundled deals.

There are, however, a few boulders strewn on the road to telecommunications nirvana. The first is the Federal Communications Commission (FCC), which is charged with converting Congressional verbiage into implementable rules and orders. The second is the Courts, which will have the last say as the parties whose ox gets gored turn to the Courts for relief. All of this could take a very long time. The FCC is already delaying their rulemaking on the tricky issue of universal service.

Meanwhile, the service providers are busy positioning themselves for the upcoming wars. Mergers and alliances are occurring continuously. The regional local service companies are forming mergers. Bell Atlantic and NYNEX, SBC and Pacific Telesis are two examples. WorldComm acquired MFS Communications and Internet service provider UUNet. Long-distance provider MCI is merging with British Telecom. A dizzying number of global and regional alliances are being formed. Unisource, WorldPartners, Uniworld, Concert, and Phoenix are only a few of these. These groups are also making equity investments in the facilities of developing economies that promise much future growth.

Other trends that may have a more immediate effect on communications managers include the rapid acceptance of corporate intranets, the increasing requirement to support a more mobile and telecommuting workforce, the continued remarkable success of Ethernet and IP in the local and campus arenas, and the deployment of ATM in the wide area core networks. On the flip side, many managers will likely retreat from switched multimegabit data service, FDDI will fade as it loses out to Fast Ethernet and ATM, and Token Ring will be demoted to a niche rather than a mainstream solution.

A visible trend in communications is the increased requirement to support remote computing. The term "remote computing" is applied to those situations where a workstation, although geographically separated, appears to be locally connected. Chapter 10-1, "Remote Computing: Technologies and Trends," defines remote computing requirements and discusses related issues such as scalability, manageability, and performance.

Another trend is the rapid distribution of information technology beyond the secure confines of the corporate data center. This situation presents a formidable security problem that is the subject of Chapter 10-2, "Information Security and New Technology." This chapter examines a number of new information technologies ranging from smart cards to imaging systems. The unique security vulnerabilities of each are exposed followed by an explanation of appropriate defense mechanisms.

Many managers believe that the real explosion in deployment of personal information technology into the general population will occur when the users can talk to their devices. At this point, the devices will become true digital assistants. "Voice Recognition Interfaces for Multimedia Applications," Chapter 10-3, provides a primer on this exciting technology.

Chapter 10-4, "The Superhighway: Information Infrastructure Initiatives," addresses the possibilities and problems revolving around the concept of the "information highway." The chapter explores trends and issues associated with the high-speed transport of multimedia information. The convergence of video, data, and voice services is explained. Finally, the chapter details the various national and international initiatives being undertaken to move the information superhighway from concept to reality.

10-1

Remote Computing: Technologies and Trends

Randall Kennedy

REMOTE COMPUTING TECHNOLOGY (RCT) is rapidly becoming a necessary component of information systems strategies, its growth spurred on by the globalization of the market coupled with the move toward a more distributed, mobile work force. As networking and telecommunications vendors offer new RCT products, the competing standards and implementations provide customers with more choices, but also a level of confusion to the purchase decision.

Traditionally, remote computing products have been based on either remote control or remote node technology. Remote control technology provides a direct relationship between the remote terminal and its host (i.e., local) PC. The remote system acts as a form of dumb terminal, parroting the visual display of the host. Remote node technology, in contrast, is more akin to a virtual network connection—the remote PC is integrated seamlessly with the local network and appears as a local node to other connected PCs and servers.

The purpose of this chapter is to analyze the current state of the art in RCT, with an eye toward providing decision makers with an objective critique of the available solutions.

THE ENTERPRISE-LEVEL CHALLENGE

Across the spectrum of organizational entities, decision makers are recognizing the need for a new way of functioning. Loosely organized work groups are replacing the once rigid office/management paradigm. Decision-making power is being distributed, with local, focused teams gaining autonomy in order to react more efficiently to changes in their specific markets. This evolution places a strain on the existing organizational and communications infrastructure, with the needs of mobile users and smaller, more autonomous units given a high priority.

A natural part of this reengineering is the adoption of remote computing technology solutions at the enterprise level. Traditional RCT solutions—remote control and remote node—are appropriate for limited deployment within smaller work groups, but prove inadequate when applied to the larger problem of enterprise-level communications.

The issues are partly architectural. Many remote control and remote node solutions are inherently PC-centered and lack the degree of scalability and manageability required for enterprise deployment. Other problems can be linked to the applications being targeted for these solutions. Most traditional applications are poorly suited to the limited bandwidth and disconnected nature of an RCT environment. Exhibit 10–1–1 shows the remote computing landscape as it exists today. Network managers need a new approach that leverages the best aspects of existing technologies and offers new capabilities that address the concerns of the enterprise. Key computing trends and their effects are summarized in Exhibit 10–1–2.

REMOTE CONTROL

Remote control technology has been used in various forms since the early 1980s. Based on a direct connection between two dedicated PCs, remote control is a popular solution for users who have both portable and desktop systems and need access to their desktop while on the road.

The remote control PC connects to the host and mirrors its interface and input devices. In the case of a Windows PC, the remote user sees a replica of the host's GUI display. Mouse clicks, keystrokes, and other input are transmitted over the RCT connection from the remote terminal to the host, which responds as if the user was local. Similarly, screen updates and, in some cases, sound events are relayed to the remote terminal, allowing users to work as if they are at the local PC's console.

This remote/host communication is typically accomplished over traditional analog telephone lines using high-speed modems. When deployed on a small scale (i.e., less than 10 PCs in a small work group), a remote control solution is cost-effective. The components are inexpensive, and this solution has the added benefit of providing access to the personal desktop, complete with familiar icons and menu schemes.

Remote control is not as efficient, however, when deployed on a larger scale. Each remote control session requires a dedicated, local PC, which is prohibitively expensive. If the user requires network access, the cost of network interface cards, cables, and additional concentrators must be considered, as well as the need for multiple high-speed modems and their accom-

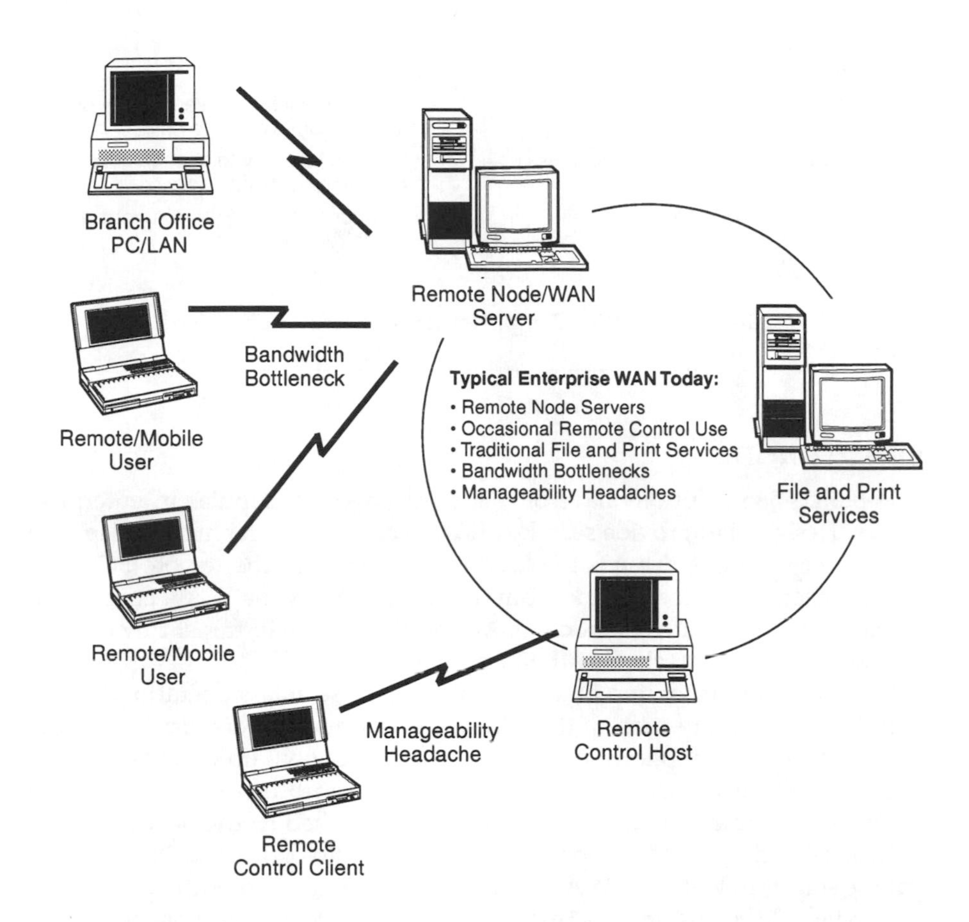

Exhibit 10-1-1. Today's Remote Computing Landscape

panying high-speed I/O solutions. Most PCs can barely handle the bandwidth of a V.34 connection without some sort of buffered serial port.

Expense issues aside, remote control solutions are also difficult to manage. Multiple, independent PCs with the prerequisite dial-up connections scattered throughout the organization, coupled with the possibility of breaches in security, create a potentially disaster-prone environment.

The network managers' aversion to remote control solutions is justifiable. Although the technology makes sense tactically, it is usually inappropriate when applied to the broader enterprise infrastructure. Although remote control offers reliable performance and a familiar desktop environment, it is difficult to manage and offers no scalability.

Trend	Effect
• Shift to a distributed decision making model	• Need for efficient dissemination of information to satellite offices/users
• Emergence of autonomous business units	• Require flexibility to choose best of breed products and services
• More mobile work force	• Greater strain on RCT solutions and infrastructure bandwidth

Exhibit 10-1-2. Key Computing Trends and Their Effects

REMOTE NODE

Remote node solutions have only recently become popular in enterprise circles. Designed to provide seamless LAN access to geographically disparate clients, remote node extends the local area network to the remote PC. Network packets are passed to and from the remote PC over a WAN connection (typically using a dial-up modem link), and the remote terminal functions as if it were directly attached to the local LAN.

Remote node's most important advantage is its seamless operation. To the remote PC, the connection to the wide area network appears just like a local LAN connection—applications interact with the remote link just as they do with a traditional LAN link, and remote users can browse and connect to network resources as if they were physically attached to the network.

Remote node's seamless operation translates into a flexible, highly compatible solution. With the WAN/LAN link transparent to both applications and the user, the customer is free to choose products because of their functional capabilities instead of for their ease of use with the WAN.

This seamlessness is also the source of remote node's single biggest drawback: poor application performance. Most traditional applications, such as data bases, are monolithic—they exist as one or more large files (i.e., executable, data, or both) and must be completely downloaded to the remote PC before executing. This requirement overwhelms the limited bandwidth of the remote node connection, which using even state-of-the-art modem technology is still 100 to 700 times slower than a typical LAN connection.

Even client/server applications tend to generate too much network traffic to perform well over a remote node link. One costly alternative is to load standalone versions of the necessary executable programs onto each remote PC, but this solution introduces potential management and version control problems.

The remote node, like remote control, is only a partial solution. Although the seamless nature of its operation and its broad application compatibility are attractive from a usability standpoint, remote node's poor performance, especially when dealing with legacy applications and the costly duplication executables, make it inadequate as an enterprise solution.

COMPROMISE SOLUTIONS

Vendors have been attempting to improve their remote computing solutions for years. The resulting hybrid products solve some older problems, but in some cases, introduce new ones.

Clustered CPUs

Clustered CPU technology is a good example. These solutions address the problem of remote control scalability by letting users concentrate two or more CPU cards—essentially complete PCs on an add-in board—into a single chassis. Each card is a self-contained communications engine capable of running an entire remote control session by itself. Users then dial into these remote control servers and are assigned a dedicated CPU card on which they execute their remote session.

The problem with the clustered CPU approach is that as user needs outgrow the capacity of the solution, and as PC technology evolves beyond the capabilities of the CPU cards, users often end up with a lot of useless proprietary hardware.

Other hybrid approaches include the bundling of third-party products to fill the gaps in an existing RCT solution. An example is the combination of a LAN modem solution (for remote node access) and an established remote control package. This solution could be problematic, however, because users may need to reconcile differing interfaces, configuration profiles, and security schemes.

NEW ENVIRONMENTS DEMAND A NEW APPROACH

The requirements of today's enterprise environment cannot be fully satisfied by remote control, remote node, or a hybrid approach. Enterprise-level solutions require a fresh approach that combines the best aspects of existing RCT solutions and addresses:

- *Scalability.* The product needs to grow with the customer, using off-the-shelf components running on established platforms.
- *Manageability.* The product must be easy to administer from a central site.
- *Seamless performance.* The product must provide both remote node and

remote control capabilities and be capable of mixing the two in a single session.

- *Flexibility.* The product must integrate well with a variety of existing network environments.

Scalability

To provide long-term value to a modern enterprise environment, an RCT solution must be scalable. In practical terms, a scalable solution must do more than simply provide for many incoming connections. The very foundation upon which the solution is built (i.e., the underlying OS or NOS) must be scalable.

SMP Support. Support for symmetrical multiprocessing (SMP) is one way to address this requirement. Many modern 32-byte operating system and network operating system solutions include support for SMP, and this inherent scalability allows these products to grow to meet the customer's ongoing requirements. For example, as the number of inbound connections grows, new CPUs can be introduced to compensate for the increased processing load.

SMP support also lets these platforms double as application servers. For example, an SMP-based RCT server can act as both a communications hub and a data base server, giving dial-up users a direct route to their data while reducing local network traffic. Such combined solutions are attractive to customers who want to centralize server resources. With only one server to administer, it becomes easier to monitor performance and adapt the underlying hardware to the actual workload.

Manageability

A solution is only as effective as it is manageable. GUI configuration tools are only part of the equation; a modern RCT solution must also provide the depth and breadth of management features that enterprise environments demand. (See Exhibit 10–1–3.)

For example, any potential RCT solution must include robust security and auditing features to safeguard critical data. These features should be account-based, allowing management to restrict access to resources based on user account privileges and access control lists (ACLs).

ACL-based security should extend to the file level with basic security measures, such as user-definable callbacks and case-sensitive passwords. To assist in tracking user activity, extensive auditing features allow the network administrator to keep a complete, virtual paper trail of the remote user's movements within the system.

Feature	Benefit
• User-accounts	• Provide a manageable security object
• Access control lists	• Let system administrator restrict resource access
• Auditing	• Virtual paper trail of user activity

Exhibit 10-1-3. Management Features and Benefits

Seamless Performance

Of the traditional RCT solutions, remote node comes closest to providing a seamless connection. However, the performance of most remote node solutions has proved inadequate for many application types (most notably monolithic and other client-processing intensive programs). Remote control provides a better alternative to remote node, but at the cost of seamlessness and manageability.

Solutions that combine both technologies, possibly packaging them in new and unique ways, would potentially give remote users the choice of how they wish to run their applications—either through local executables that access a centralized data source (remote node) or through on-screen emulation of an application/console running on a host system (remote control).

At the back end, the solution should be based around a centralized server technology. For example, the remote control component is a perfect candidate for deployment on a multisession application server. This gives the solution scalability and avoids the management quagmire of multiple host PCs servicing multiple incoming connections.

Ideally the product's operation should be invisible. Both remote node and remote control applications should integrate with the remote PC's native desktop as if they were running off the user's local hard disk. (See Exhibit 10–1–4.)

This technology convergence requires rethinking on the part of RCT vendors. New transport protocols and client-side presentation technologies need to be developed first. In addition, the back-end server technologies must be unified so that a single connection point provides access to the spectrum of remote node and remote control capabilities.

Flexibility

An RCT solution must be flexible enough to integrate with and add value to an existing networking infrastructure. Most IS shops are reluctant to discard the tried-and-true technologies they have invested in.

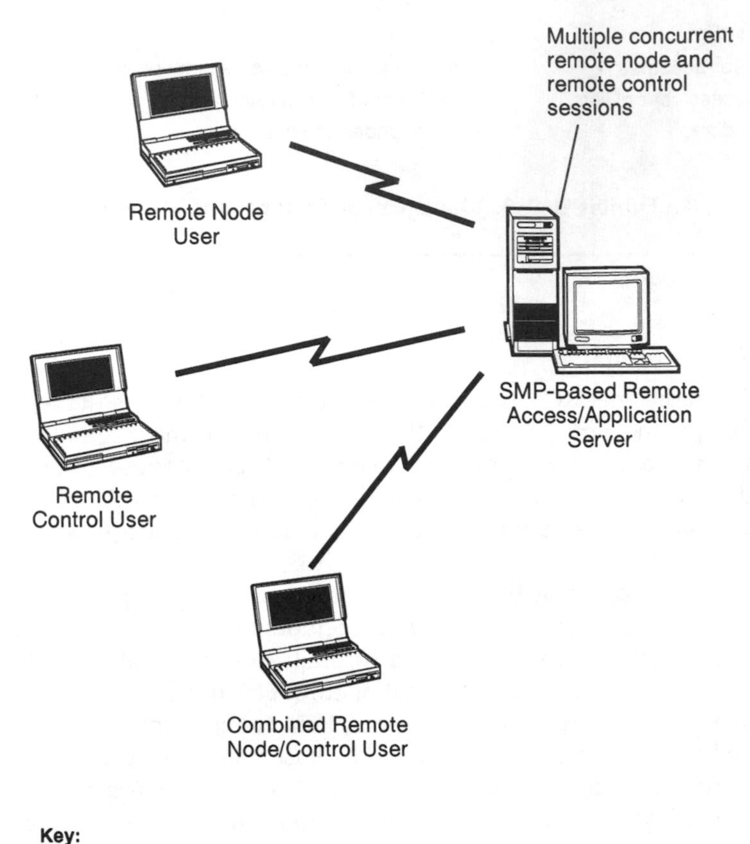

Exhibit 10-1-4. The Ideal RCT Solution

RCT solutions must easily integrate into the heterogeneous LAN environment. In the PC realm, this means tight integration with NetWare and Windows NT. Security, auditing, and systems management functions should support established protocols and standards. The overall solution must have broad hardware support so that customers can tailor it to their own needs.

SUMMARY

The perfect RCT solution is not really available. The technology exists to create it, however. Some vendors are heading in the right direction by hosting combined remote node/remote control solutions on scalable application servers.

In the interim, network managers should carefully scrutinize new product offerings to see whether they pass the four-part litmus test—scalability, manageability, seamless performance, and flexibility—before any widespread deployment is undertaken.

10-2 Information Security and New Technology

Louis Fried

THE JOB OF THE IS SECURITY SPECIALIST has gone from protecting information within the organization to protecting information in the extended enterprise. Controlled offices and plants have given way to a porous, multiconnected, global environment. The pace at which new information technology capabilities are being introduced in the corporate setting also creates a situation in which the potential of new security risks is not well thought out. Corporate managers and their information security specialists need to be aware of these threats before adopting new technologies so that they can take adequate countermeasures.

Information security is concerned with protecting:

- The availability of information and information processing resources.
- The integrity and confidentiality of information.

Unless adequate protection is in place when new business applications are developed, one or both of these characteristics of information security may be threatened. Availability alone is a major issue. Among US companies, the cost of systems downtime has been placed by some estimates at $4 billion a year, with a loss of 37 million hours in worker productivity.

The application of information security methods has long been viewed as insurance against potential losses. Senior management has applied the principle that it should not spend more for insurance than the potential loss could cost.

In many cases, management is balancing information security costs against the potential for a single loss incident, rather than multiple occurrences. This fallacious reasoning can lead to a failure to protect information assets continuously or to upgrade that protection as technology changes and exposes the company to new opportunities for losses.

Those who would intentionally damage or steal information also follow some basic economic principles. Amateur hackers may not place a specific value on their time and thus may be willing to put substantial effort into penetrating information systems. A professional will clearly place an implicit value on time by seeking the easiest way to penetrate a system or by balancing potential profit against the time and effort necessary to carry out a crime. New technologies that create new (and possibly easier) ways to penetrate a system invite such professionals and fail to deter the amateurs.

This chapter describes some of the potential threats to information security that organizations may be facing in the next few years. The chapter concludes by pointing out the opportunities for employing new countermeasures.

NEW THREATS TO INFORMATION SECURITY

Document Imaging Systems

The capabilities of document imaging systems include:

- Reading and storing images of paper documents.
- Character recognition of text for abstracting or indexing.
- Retrieval of stored documents by index entry.
- Manipulation of stored images.
- Appending notes to stored images (either text or voice).
- Workflow management tools to program the distribution of documents as action steps are needed.

Workflow management is critical to taking full advantage of image processing for business process applications in which successive or parallel steps are required to process the document. Successful applications include loan processing, insurance application or claims processing, and many others that depend on the movement of documents through review and approval steps.

Image processing usually requires a mainframe or minicomputer for processing any serious volume of information, although desktop and workstation versions also exist for limited use. In addition, a full image processing system requires document readers (scanners), a local area network (LAN), workstations or personal computers, and laser printers as output devices. It is possible to operate image processing over a wide area network (WAN); however, because of the bandwidth required for reasonable response times, this is not usually done. As a result, most configurations are located within a single building or building complex.

Two years ago, an insurance company installed an imaging application for processing claims. The system was installed on a LAN linked to a minicomputer in the claims processing area. A manager who had received a layoff notice accessed the parameter-driven workflow management system and randomly realigned the processing steps into new sequences, reassigning the process steps in an equally random fashion to the hundred or so claims processing clerks using the system. He then took the backup tapes, which were rotated weekly, and backed up the revised system files on all the tapes, replacing them in the tape cabinet. The individual did not steal any information or delete any information from the system. The next morning, he called the personnel department and requested that his final paycheck be sent to his home.

The cost to the insurance company? Tens of thousands of dollars in clerical time wasted and professional and managerial time lost in finding and correcting the problem. Even worse, there were weeks of delays in processing claims and handling the resultant complaint letters. No one at the company can estimate the loss of goodwill in the customer base.

The very techniques of workflow management that make image processing systems so effective are also their Achilles' heel. Potential threats to image processing systems may come from disruption of the workflow by unauthorized changes to sequence or approval levels in workflow management systems or from the disruption of the workflow by component failure or damage. Information contained on documents may be stolen by the unauthorized copying (downloading of the image to the workstation) and release of document images by users of workstations.

These potential threats raise issues that must be considered in the use of image processing technology. The legal status of stored images may be questioned in court because of the potential for undetectable change. In addition, there are the threats to the business from loss of confidentiality of documents, loss of availability of the system during working hours, damage to the integrity of the images and notes appended to them, and questions about authenticity of stored documents.

Minisupercomputers

Massively parallel minisupercomputers are capable of providing relatively inexpensive, large computational capacity for such applications as signal processing, image recognition processing, or neural network processing.

Massively parallel processors are generally designed to work as attached processors or in conjunction with workstations. Currently available minisupercomputers can provide 4,096 processors for $85,000 or 8,192 processors for $150,000. They can interface to such devices as workstations, file servers, and LANs.

These machines can be an inexpensive computational resource for cracking encryption codes or computer-access codes; consequently, organizations that own them are well advised to limit access control for resource use to authorized users. This is especially true if the processor is attached to a mainframe with WAN connectivity. Such connectivity may allow unauthorized users to obtain access to the attached processor through the host machine.

Even without using a minisupercomputer but by simply stealing unauthorized time on conventional computers, a European hacker group bragged that it had figured out the access codes to all the major North American telephone switches. This allows them to make unlimited international telephone calls at no cost (or, if they are so inclined, to destroy the programming in the switches and deny service to millions of telephone users).

Neural Network Systems

Neural network systems are software (or hardware/software combinations) capable of heuristic learning within limited domains. These systems are an outgrowth of artificial intelligence research and are currently available at different levels of capacity on systems ranging from personal computers to mainframes.

With their heuristic learning capabilities, neural networks can learn how to penetrate a network or computer system. Small systems are already in the hands of hobbyists and hackers. The capability of neural network programs will increase as greater amounts of main memory and processing power become easily affordable for desktop machines.

Wireless Local Area Networks

Wireless LANs support connectivity of devices by using radio frequency (RF) or infrared (IR) transmission between devices located in an office or an office building. Wireless LANs consist of a LAN controller and signal generators or receivers that are either attached to devices or embedded in them.

Wireless LANs have the advantage of allowing easy movement of connected devices so that office space can be reallocated or modified without the constraints of hard wiring. They can connect all sizes of computers and some peripherals. As portable computers become more intensively used, they can be easily connected to PCS or workstations in the office for transmission of files in either direction.

Wireless LANs may be subject to signal interruption or message capture by unauthorized parties. Radio frequency LANs operate throughout a transmitting area and are therefore more vulnerable than infrared transmission, which is line-of-sight only.

Among the major issues of concern in using this technology are retaining confidentiality and privacy of transmissions and avoiding business interruption in the event of a failure. The potential also exists, however, for other kinds of damage to wireless LAN users. For example, supermarkets are now experimenting with wireless terminals affixed to supermarket shopping carts that broadcast the price specials on that aisle to the shopper. As this technology is extended to the inventory control function and eventually to other functions in the store, it will not be long before some clever persons find a way to reduce their shopping costs and share the method over the underground networks.

WAN Radio Communications

WAN radio communications enable handheld or portable devices to access remote computers and exchange messages (including fax messages). Wireless WANs may use satellite transmission through roof-mounted antennas or regional radiotelephone technology. Access to wireless WANs is supported by internal radio modems in notebook and handheld computers or wireless modems/pagers on PCMCIA cards for optional use.

Many users think that telephone land lines offer some protection from intrusion because wiretaps can often be detected and tapping into a fiberoptic line is impossible without temporarily interrupting the service. Experience shows that most intrusions result from logical—not physical—attacks on networks. Hackers usually break in through remote maintenance ports on PBXs, voice-mail systems, or remote-access features that permit travelers to place outgoing calls.

The threat to information security from the use of wireless WANs is that direct connectivity is no longer needed to connect to networks. Intruders may be able to fake legitimate calls once they have been able to determine access codes. Users need to consider such protective means as encryption for certain messages, limitations on the use o wireless WAN transmission for confidential material, and enforcement of encrypted password and user authentication controls.

Videoconferencing

Travel costs for nonsales activities is of growing concern to many companies. Companies are less concerned about the costs of travel and subsistence than they are about the costs to the company of having key personnel away from their jobs. Crossing the US or traveling to foreign countries for a one-day meeting often requires a key employee to be away from the job for three days. Videoconferencing is used to reduce travel to only those trips that are essential for hands-on work.

The capabilities of videoconferencing include slow-scan video for sharing documents or interactive video for conferencing. Videoconferencing equipment is now selling for as little as $30,000 per installation. At that price, saving a few trips a year can quickly pay off. However, videoconferencing is potentially vulnerable to penetration of phone switches to tap open lines and receive both ends of the conferencing transmissions.

Protection against tapping lines requires additional equipment at both ends to scramble communications during transmission. It further requires defining when to scramble communications, making users aware of the risks, and enforcing rules.

Embedded Systems

Embedding computers into mechanical devices was pioneered by the military for applications ranging from autopilots on aircraft to smart bombs and missiles. In the civilian sector, process controls, robots, and automated machine tools were early applications. Manufacturers now embed intelligence and communications capabilities in products ranging from automobiles to microwave ovens. Computers from single-chip size to minicomputers are being integrated into the equipment that they direct. In factory automation systems, embedded systems are linked through LANs to area computers and to corporate hosts.

One security concern is that penetration of host computers can lead to penetration of automated factory units, which could interrupt productive capacity and create potential hazards for workers. In the past, the need for information security controls rarely reached the factory floor or the products that were produced because there was no connection to computers that resided on WANs. Now, however, organizations must use techniques that enforce access controls and segment LANs on the factory floor to minimize the potential for unauthorized access through the company's host computers.

Furthermore, as computers and communications devices are used more in products, program bugs or device failure could endanger the customers who buy these products. With computer-controlled medical equipment or automobiles, for example, potential liability from malfunction may be enormous. Information security techniques must extend to the environment in which embedded systems software is developed to protect this software from corruption and the company from potential liability resulting from product failures.

PCMCIA Cards

PCMCIA cards are essentially small computer boards on which chips are mounted to provide memory and processing capacity. (PCMCIA stands for the Personal Computer Memory Card International Association, which is the governing body standardizing the specifications for these cards worldwide.) They can be inserted (docked) into slots on portable computers to add memory capacity, processing capacity, data base capacity, or communications functions such as pagers, electronic mail, or facsimile transmission. PCMCIA cards can be expected to provide up to 20M bytes of storage. Removable disk drives, currently providing up to 260M bytes of storage in a 1.8-inch drive, can be inserted into portable devices with double PCMCIA card slots.

The small format of PCMCIA cards and their use in portable devices such as notebook or handheld computers makes them especially vulnerable to theft or loss. Such theft or loss can cause business interruption or breach of confidentiality through loss of the information contained on the card. In addition, poor work habits, such as failing to back up the data on another device, can result in the loss of data if the card fails or if the host device fails in a manner that damages the card. Data recovery methods are notoriously nonexistent for small portable computers.

Smart Cards

Smart cards, consisting of a computer chip mounted on a plastic card similar to a credit card, have limited intelligence and storage compared to PCMCIA cards. Smart cards are increasingly used for health records, debit cards, and stored value cards. When inserted into an access device (reader), they may be used in pay telephones, transit systems, retail stores, health care providers, and ATMs, as well as being used to supplement memory in handheld computers.

The risks in using this technology are the same as those for PCMCIA cards but may be exacerbated by the fact that smart cards can be easily carried in wallets along with credit cards. Because smart cards are used in stored value card systems, loss or damage to the card can deprive the owner of the value recorded. Both PCMCIA cards and smart cards must contain means for authenticating the user in order to protect against loss of confidentiality, privacy, or monetary value.

Notebook and Palmtop Computers

Notebook and palmtop computers are small portable personal computers, often supporting wireless connections to LANs and WANs or modems and providing communications capability for docking to desktop computers for uploading or downloading of files (either data or programs).

These devices have flat panel displays and may include microdisks with 260M-byte capacity. Some models support handwriting input. Smart cards, PCMCIA cards, or flashcards may be used to add functionality or memory. By the end of the decade, speech recognition capability should be available as a result of more powerful processors and greater memory capacity.

As with the cards that may be inserted into these machines, portable computers are vulnerable to loss or theft—both of the machine and of the information contained in its memory. In addition, their use in public places (such as on airplanes) may breach confidentiality or privacy.

It is vital that companies establish information security guidelines for use of these machines as they become ubiquitous. Guidelines should include means for authentication of the user to the device before it can be used, etching or otherwise imprinting the owner's name indelibly onto the machine, and rules for protected storage of the machine when it is not in the user's possession (as in travel or at hotel stays). One problem is that most hotel sales do not have deposit boxes large enough to hold notebook computers.

Portable computers combined with communications capability may create the single largest area of information security exposure in the future. Portable computers can go wherever the user goes. Scenarios of business use are stressing advantages but not security issues. Portable computers are used in many business functions including marketing, distribution field service, public safety, health care, transportation, financial services, publishing, wholesale and retail sales, insurance sales, and others.

As the use of portable computers spreads, the opportunities for information loss or damage increase. Exhibit 10-2-1 lists some of the potential uses of portable computers; these scenarios show that almost every industry can make use of the technology and become vulnerable to the implicit threats.

Portable computers, combined with communications that permit access to company data bases, require companies to adopt protective techniques to protect information bases from external access and prevent intelligence from being collected by repeated access. In addition, techniques are needed for avoiding loss of confidentiality and privacy by device theft and business interruption through device failure.

New uses create new business vulnerabilities. New hospitals, for example, are being designed with patient-centered systems in which the services are brought to the patient (to the extent possible) rather than having the patient moved from one laboratory to another. This approach requires the installation of LANs throughout the hospital so that specialized terminals or diagnostic devices can be connected to the computers processing the data collected. Handheld computers may be moved with the patient or carried by attendants and plugged into the LAN to access patient records or doctors' orders. It is easy to anticipate abuses that range from illegal access to patient information to illegal dispensing of drugs to unauthorized persons.

NEW OPPORTUNITIES FOR DEFENSE

New technology should not, however, be seen solely as a security threat. New technology also holds opportunities for better means of protection and detection. Many of the capabilities provided by the IT department can support defensive techniques for information or information processing facilities.

Expert Systems, Neural Networks, and Minisupercomputers. Used individually or in combination, these technologies may enable intrusion detection of information systems. These technologies can be used to recognize unusual

Industry or Function	Application	Benefits
Marketing	Track status of promotions	Better information on sales activities
	Identify purchase influencers and imminence of decision	Reports get done more quickly
	Prepare reports on site	
Distribution	Bill of lading data and calculations	More timely information on field operations
	Delivery and field sales data collection	Better customer service
	Enter and track parcel data	
Field service	Remote access to parts catalog and availability	Better service to customer
	Troubleshooting support	More efficient scheduling
	Repair handbooks	
	Scheduling and dispatching	
	Service records	
	Payment and receipt records	
Public safety	Dispatch instructions	Faster emergency response
	Police license and warrant checks	Identification and apprehension of criminals
	Building layout information	Improved safety of emergency personnel
	Paramedic diagnosis and treatment support	Better treatment to save lives
Transportation	Airline and train schedules	Convenience to customers
	Reservations	Replaces paper forms and records
	Rental car check-in and receipt generation	More timely information
	Report graffiti and damage	
	Monitor on-time performance	
Financial services	Stock exchange floor trader support	More accurate records trader support
		Reduces risk of fraud
Publishing	Electronic books and references	Flexible retrieval
		Compact size
Travel and entertainment	Language translators, travel guides, and dictionaries	Personal convenience to travelers
	Hotel and restaurant reservations	

Exhibit 10-2-1. Scenarios of Business Use for Portable Computers

Industry or Function	Application	Benefits
Wholesale sales	Record sales results	More accurate and timely information, both in the field and at corporate headquarters
	Send results to corporate host	
	Eliminate unnecessary phone contacts	
	Receive updates on product prices and availability	Cuts paperwork
		More productive use of staff
Retail sales	Capture sales and demographic data	Assess promotional results
	Update inventory data	Tighter control over field operations
Insurance	Access corporate data for quotes	Quicker quotations to customers
	Perform complex rate calculations	

Exhibit 10-2-1. *(continued)*

behavior patterns on the part of the intruder, configure the human interface to suit individual users and their permitted accesses, detect physical intrusion or emergencies by signal analysis of sensor input and pattern recognition, and reconfigure networks and systems to maintain availability and circumvent failed components. In the future, these techniques may be combined with closed-circuit video to authenticate authorized personnel by comparing digitally stored images of persons wishing to enter facilities.

Smart Cards or PCMCIA Cards. Used with card readers and carrying their own software and data, cards may enable authentication of a card owner through various means, including recognition of pressure, speed, and patterns of signatures; questions about personal history (the answers to which are stored on the card); use of a digitized picture of the owner; or cryptographic codes, access keys, and algorithms. Within five years, signature recognition capabilities may be used to limit access to penbased handheld computers to authorized users only, by recognizing a signature on log-in.

Personal Computer Networks (PCNs). PCNs, enabled by nationwide wireless data communications networks, will permit a personal phone number to be assigned so that calls may reach individuals wherever they (and the instrument) are located in the US. PCNs will permit additional authentication methods and allow call-back techniques to work in a portable device environment.

Voice Recognition. When implemented along with continuous speech understanding, voice recognition may be used to authenticate users of voice input systems—for example, for inquiry systems in banking and brokerages. By the end of this decade voice recognition may be used to limit access to handheld computers to authorized users only by recognizing the owner's voice on log-in.

Wireless Tokens. Wireless tokens used as company identity badges can pinpoint the location of employees on plant sites and monitor restricted plant areas and work check-in and check-out. They may also support paging capability for messages or hazard warnings.

REDUCING PASSWORD RISKS

The Obvious Password Utility System (OPUS) project at Purdue University has created a file compression technique that makes it possible to quickly check a proposed password against a list of prohibited passwords. With this technique, the check takes the same amount of time no matter how long the list. OPUS may allow prohibited password lists to be placed on small servers and improve password control so that systems are harder to crack.

Third-Party Authentication Methods. Systems like Kerberos and Sesame provide a third-party authentication mechanism that operates in an open network environment but does not permit access unless the user and the application are authenticated to each other by a separate, independence computer. (Third-party refers to a separate computer, not a legal entity.) Such systems may be a defense for the threats caused by portable systems and open networks. Users of portable computers may call the third-party machine and request access to a specific application on the remote host. The Kerberos or Sesame machine authenticates the user to the application and the application to the user before permitting access.

SUMMARY

The best way to stay ahead of the threats is to maintain a knowledge of technology advances, anticipate the potential threats and vulnerabilities, and develop the protective measures in advance. In well-run IS functions, information security specialists are consulted during the systems specification and design phases to ensure that adequate provisions are made for the security of information in applications. Information security specialists must be aware of the potential threats implicit in the adoption of new technologies and the defensive measures available in order to critique the design of new applications and to inform their senior management of hazards.

The combination of advanced computer capabilities and communications is making information available to corporate executives and managers on an unprecedented scale. In the medical profession, malpractice suits have been won on the grounds that treatment information was available to a doctor and the doctor did not make use of the information. In a sense, this means that the availability of information mandates its use by decision makers. Corporate officers could find that they are no longer just liable for prudent protection of the company's information assets but that they are liable for prudent use of the information available to the company in order to protect its customers and employees. Such conditions may alter the way systems are designed and information is used and the way the company chooses to protect its information assets.

10-3
Voice Recognition Interfaces for Multimedia Applications

Louis Fried

MOST OF THE TECHNIQUES ASSOCIATED with multimedia applications—video, text, graphics, object-oriented interfaces, handwritten input, touchscreens, scanning, and mouse point-and-click techniques—focus on a visually based human interface. Not enough has been said about the most natural of all human interfaces—speech. During the last few years, automatic speech recognition (ASR) technology has made substantial strides that warrant active consideration of speech as an effective interface for many emerging multimedia applications.

Although speech or voice recognition technology is not new, applications have often been confined by the limitations of the technology. In the past, most speech recognition systems could only recognize words spoken in isolation. If users wanted the computer to understand their speech, they had to . . . separate . . . the . . . words . . . by . . . brief . . . silences. In addition, most systems required that the speaker spend hours training the system to understand his or her individual pronunciation. These constraints still apply to many of the speech recognition products currently available for operation on personal computers.

Systems that require training by the user are generally limited to supporting one to four users and generally have limited or specialized vocabularies. Systems that operate in this mode include products from Kurzweil, Applied Intelligence, Inc., Dragon Systems, and IBM Corp. Although they are extremely useful for providing computer access to impaired people or to professionals who must keep their hands free for other work while accessing a computer, these systems do not meet the needs for general or public access to computer systems in what could be called a natural manner.

PRACTICAL SYSTEMS FOR WIDESPREAD USE

If a speech recognition system is to be of widespread, practical use, it must operate without individual training (i.e., be speaker independent), and it must understand continuous speech, without pauses, which is how people converse. Two approaches have been taken to solve the problem of speaker-independent recognition: the synthetic modeling approach and the sampling approach.

Speaker-Independent Recognition

Synthetic Modeling. Synthetic modeling builds words out of syllables that represent speech sounds. Using a data base of these subwords, words can be created in the recognition tables by phoneme transcription—much like constructing words using the pronunciation codes that are listed following a word in a dictionary.

Sampling. The sampling approach, which has proved to be more accurate for interpreting regional accents, involves gathering large numbers of spoken samples of utterances (referred to as tokens) and constructing words from them that match the anticipated user population. The broader a population of users, the more samples are needed. For example, in constructing a speaker-independent recognition Corona TM system, which is intended to be useful across the entire US, SRI International's researchers gathered more than 400,000 utterances from all regions of the country.

A system intended for widespread use must be capable of recognizing a large number of words. Given the current state of the art, a system that can recognize anyone saying yes or no is trivial. Early vocabulary-building applications could support up to about 2,000 words and phrases. More advanced systems can support vocabularies of 60,000 or more words. Vocabularies for widespread application must be capable of recognizing tens of thousands of words in essentially real-time as far as the speaker is concerned. Only in this way can a human's conversation with a computer appear to be natural.

The Complexities of Speech. New systems appearing commercially today do a creditable job of recognizing speaker-independent, continuous speech with an accuracy of better than 95% . However, there is a great deal of complexity that must be overcome. A sampling of the complex problems that need to be solved follows.

Homonyms. Most languages, especially English, have many homonyms, words that sound alike but may have different spellings and different meanings (e.g., two, to, and too). The ability to distinguish among homonyms depends on either limiting the vocabulary in some manner to eliminate homonyms or being able to understand the context in which the word is spoken.

Numbers. Numbers are difficult to recognize. Not only do some numbers sound like other words, but people say numbers in different ways. If asked for a telephone number, most people will say each number separately, but if asked for a dollar amount, most people will combine the numbers in some fashion, such as five thousand four hundred and thirty-six dollars and forty-seven cents. Long strings of numbers, or numbers and letters combined (such as in an automobile vehicle serial number) can be very difficult to understand. For this reason, many applications require the speaker to break up a longer number into smaller groups; a social security number might be broken into groups of three, two, and four, and a credit card number into groups of four digits.

Background Noise. Understanding speech is further complicated when a telephone is the instrument of voice capture. Over-the-phone speech is subject to the background hiss of phone lines. Good systems will include algorithms for cleaning the speech of background noise. Similar techniques must be applied if the speaker is talking in a noisy environment.

Spoken Language Understanding. Automatic speech recognition is not the same thing as spoken language, or natural language, understanding. An ASR system, for example, could pick out certain words in a stream of speech and, by spotting these words in the context of a limited dialog, respond as if it had understood a complete sentence. For example, if a customer calls a bank's speech recognition system and requests a "current account balance," a word-spotting system may simply pick out the word "balance" and provide the correct answer. Applications based on word-spotting techniques require that the dialog between the caller and the system be carefully constructed to limit the range of responses that the caller may make.

Spoken language understanding systems, by contrast, would operate without the limited context of word spotting and truly analyze the entire sentence before responding. Furthermore, such systems may have the capability to handle references to earlier sentences in the conversation or topic changes. For example, a conversation may include an inquiry about whether a particular check had cleared. After receiving a negative response, the speaker may say, "Put a stop payment on it." Here, a system could recognize that "it" refers to the check previously discussed.

Portability. Finally, for the technology to achieve widespread use, a system must be portable among computers. It must not rely on any special equipment for processing other than the chips for digitizing the analog signals from voice input, typically embedded in any computer to which a microphone may be attached. Some systems rely on the use of proprietary digital signal processing boards to both digitize the analog input and analyze the speech and convert it to digital components. This approach ties the application to a specific set of hardware.

CURRENT APPLICATIONS

The progress made in speech recognition technology means that developers of multimedia applications should seriously consider speech recognition in their human interface design. Two major classifications of applications exist:

- *Microphone applications.* The speaker talks into a microphone attached to a workstation and obtains spoken or visual responses (or both).
- *Telephony applications.* The speaker uses a telephone handset or equivalent to speak and hear responses.

Microphone-Input Applications for the PC

Microphone applications have the potential to provide impaired persons access to information facilities and to make people in many professions more productive. Because these applications are usually based on the older, speaker-dependent technology, many more of them have been implemented than have applications using speaker-independent recognition technology. Some examples of workstation or PC-based applications follow.

Kurzweil Applied Intelligence introduced its VoiceMED line of products for patient reporting almost 10 years ago. In these applications, the speaker (typically a physician who had trained the system) spoke into a microphone connected to a personal computer. The system translated the voice to written words and incorporated the specialized vocabulary of the physician. Extensions of the VoiceMED system were products such as VoicePATH (for pathology), VoiceEM (for emergency medicine), VoiceDIALYSIS (for kidney dialysis reporting), and VoiceCATH (for invasive cardiology). These systems grew out of Kurzweil's earlier work in voice-controlled typing systems.

West Publishing and Kolvox Communications have introduced the LawTalk large vocabulary speech recognition front-end to the WestLaw online legal research system. The system includes Dragon Systems' DragonDictate technology, which uses a microphone connection to the PC. The PC-based application then translates the speaker's query and interfaces to the online data base. Queries may be stated in either a formula-like Boolean expression or in natural language. The speech interface is further combined with a

WordPerfect speech interface that allows users to transfer downloaded information to documents using oral commands.

Syracuse Language Systems, Inc.'s TriplePlay Plus! software uses speech recognition to teach foreign languages by listening to a student's voice, evaluating the pronunciation, and replying in the foreign language used. The system can be used to learn French, German, English, or Spanish.

In the brokerage field, R.W. Pressprich & Co., Inc., has implemented voice recognition to replace keyboard entry for bond traders.

Under the banner of its VoiceType Dictation systems, IBM offers dictating systems not only on its PCs but also with a PCMCIA digital signal adapter for laptop and notebook computers. VoiceType dictation is available in English, German, Italian, French, and Spanish. Other vendors have entered the market with similar products. One of the more powerful features of speech recognition technology is that because words are made up of phonetic tokens or subwords, the technology does not have to be changed to provide support to new languages; only the vocabularies need to be changed.

Telephony Applications

Telephony applications provide information access to large numbers of people and may potentially replace many of the telephone service personnel who now provide information, take orders by phone, or otherwise serve as a human interface to callers. Telephony applications are often successful because they provide an acceptable interface for callers who do not have, or prefer not to use, touchtone input to a voice response system. In fact, many new applications use existing interactive voice response (IVR) units to answer the call and provide the voice response once the caller's speech is understood. Examples of such applications follow.

AT&T and most of the regional telephone companies in the US are in the process of providing speaker-independent recognition interfaces for callers. The most profitable target for these applications is the hundreds of directory assistance operators who provide telephone numbers to 411 callers. However, the application of speech recognition technology is also being used to create other applications such as a third-party billing system and processing of collect calls.

Even speaker-dependent systems have a place in telephone carrier applications. Sprint is testing a FONCARD application in which the caller verbally enters the private access code and the system verifies both the number and the caller's voice as valid for that code. Ameritech is testing a system that allows individual speakers at the same phone number to create speaker-dependent personal dialing directories.

Other industries may even be ahead of the telephone companies. One example is the customer service directory system implemented by Union

Electric in St. Louis MO. This system permits customers with rotary phones or those who prefer not to use touchtone input to connect to such service groups as installation and billing and speak to a live operator or leave a voicemail message. The script between the IVR and the caller carefully directs the responses and allows either touchtone or voice response.

EMERGING APPLICATIONS

Speech recognition applications have direct benefits to the organizations that adopt them in terms of productivity, safety, and reduced staffing. Efforts in development, and in some cases new products, are being announced to combine ASR with facsimile machines, imaging, and industrial controls. Assuming that the long-term trends in hardware continue to make intelligent devices cheaper, smaller, and faster, with more data storage capability, automatic speech recognition and especially speaker-independent recognition will play an important role in several areas; for example:

- For devices that are too small for other interfaces. Product development is now taking place on pocket-sized devices that incorporate the functions of a personal reminder and address book with a mobile telephone.
- For applications where keyboards are impractical. In industrial situations such as machine shops and assembly lines, keyboards rapidly become dirty and nonfunctional. ASR technology can also eliminate the need for manual entry in situations in which keyboards are subject to weather or other hazardous conditions. Examples of such use include public or customer access to automated teller machines, vending kiosks, and information kiosks.
- As an easy-to-learn means for untrained users and the general public to concurrently use a single information resource.
- As embedded technology in other devices for which other interfaces would detract from the design or efficiency. The technology is already being used in automobiles, and its use in home appliances is anticipated.
- For applications for the visually impaired as well as for communications under conditions in which the hands are in use or vision should not be distracted. One current application supports data entry by dock workers who wear small wireless microphones to dictate data about shipments and receipts for entry into the information system.
- As an alternative to systems that use touchtone input in conjunction with IVR. Between 20% and 40% of callers to such systems either do not have a touchtone instrument or refuse to use touchtone. In many cases, the aversion is not to the touchtone instrument but to the extensive menus that are recited before the caller can identify the sequence of

buttons to push. ASR techniques can substantially condense or eliminate menus.

SELECTING AN ASR PRODUCT

It is easy to overbuy and acquire technology that is too costly and sophisticated for real business needs. By the same token, if IS developers and users define current needs too narrowly and do not anticipate the future, they underbuy. Some buyers have opted for wordspotting systems only to find that their vendor did not support spoken language understanding (i.e., natural language). When full natural language capability is needed, a complete system replacement can be very expensive.

Once an application specification has been written, selection of the system depends first on who will be using it. When one or only a few specific users will use the system, the lowest-priced approach that meets their needs should be chosen. However, users must have the patience to train the system and be willing to take the isolated-word approach to dictation.

When an application involves so-called promiscuous speech (i.e., lots of people talking to the system), selection demands more stringent criteria. Obviously, the system must support speaker-independent, continuous speech recognition and the ability to translate speech into ASCII text. Additional criteria are usually called for. The weight placed on the criteria is, of course, a function of the intended application.

Functions and Features to Look For

When choosing a system for many users, IT buyers should look for certain features and functions:

- The system must be able to translate and respond with minimal perceptible lag time.
- If the system is intended for use in telephony applications, it must work with regular, unmodified telephone instruments, speakerphones, and cellular phones. It must also support barge-in (i.e., allowing speakers to talk while the system is still responding) and the IVR system in use or to be in use (i.e., it must be easily ported to that system). Finally, it must be able to alternate with touchtone input.
- If the system is intended to support broad-ranging inquiry, it must support a vocabulary of more than 60,000 words. *The Wall Street Journal* test is a good measure of performance. In contrast to most newspapers, which use a vocabulary of less than 15,000 words, the *Journal* uses a 40,000-word vocabulary. If the system performs with

reasonable accuracy in translating *Journal* articles, it should work for this type of application.

- Depending on the application (e.g., electronic engineering, programming, or medicine), the system should support or be able to add specialized vocabularies.
- If the system is intended to support multiple specialists, it should be able to switch among vocabularies.
- The system should support word spotting.
- The system should support spoken language understanding. Even if there is no plan to use this feature initially, it may be a desirable feature to have in a product to allow for future needs.
- If the system is intended for nationwide use, it should support translation of multiple dialects or regional accents for a single language (or the vendor should be able to adapt the system to accommodate regional accents).
- If intended for international use, the system must support multiple languages (or the vendor should be able to adapt the system to accommodate foreign languages).
- The system should detect silences and recognize when the speaker has stopped talking.
- The system should listen at all times (rather than only when the speaker is given a signal).
- The system should operate on a server in a client/server environment.
- The system should support performance analysis (e.g., recognition confidence factors, recording of exact word sequence, percentage of successful transactions, average transaction processing time).
- The system should maintain a time-stamped activity log.
- Individual sessions should be able to be recorded for later playback.

Performance Characteristics

An evaluation of performance characteristics should ask the following:

- How many concurrent users can access the system without degradation of response time?
- Can the system work in a noisy environment, filtering out external noise?
- Is the system portable among a variety of standard computers?
- Is the system composed of software only (i.e., it does not require special devices other than those for digitizing voice input)?

Vendor Capability

When selecting a vendor, buyers should be sure to ask the following questions:

- Is the provider of the ASR system the original developer?
- Can the provider create custom applications for the organization?
- Does the provider offer around-the-clock emergency support for critical applications?

BUILDING VOICE RECOGNITION APPLICATIONS

Most of the packages used in specialized applications such as law or medicine are for single users or a limited number of users. However, for promiscuous-speech applications that allow the general public or a company's customers to access the system, it is usually necessary to design and build the application using the ASR system and to implement the application using its accompanying tools.

Until automatic speech recognition becomes such a common component of the human-machine interface that it can be applied generally as a keyboard is used today, the technology will probably be reserved for high-payoff applications. Such applications, especially those that are telephony-based, provide a means for access that presents the company's image to the public or its customers. Whether the application takes orders for products or simply provides information, the owner must consider any customer interface to be mission-critical. For this reason, the approach to applications development must be a least-risk, conservative method.

The Development Team

The team assembled for building an ASR application needs specialized skills that members of the IS group may not possess. The team therefore must include several people not typically present on an applications development team.

Users or a Proxy for the Users. If the application is intended for internal use in the organization, a sample of users should participate in the development of a vocabulary, the design of an appropriate dialog (for an interactive application), and the testing of the application. If the application is intended for use by external parties, a proxy for those parties should participate. The proxy must be familiar with the business application and the usual dialogs conducted with external parties (e.g., customers).

A Corporate Public Relations Representative. For an application to be used by external parties, a representative of corporate public relations should participate to review the dialog and ensure that it conveys an appropriate image of the organization.

Technical Architects. Technicians familiar with the operation and architecture of the hardware elements of the system should design the interfaces among the components. Components may include:

- An ASR server.
- A local area network.
- An IVR system.
- The telephone interface to the IVR system.
- A host-resident data base for supplying information.
- Telephone line interfaces.

Technicians are needed to design the architecture, supervise installation and testing, and assist in expansion from pilot systems to full rollout.

Application Designers. These systems analysts design the application from a software perspective and continue with the iterative design cycle until the application is implemented. One of the application designers may be the project leader.

Linguistic Analyst. An interactive ASR application depends on an unambiguous dialog between the speaker and the system. Not only must the message to the speaker be understood, but it must not be misunderstood.

The linguist works with the users to design the dialog. In addition, the linguist helps to design any specialized vocabulary, specify words or phrases that may need to be spotted, and adapt the system for any regional accents or language usage.

Programmers. Programmers are needed to create the data base interfaces to the host data base and, depending on the amount of customization required, to modify or enhance the programs that directly interface to the ASR system.

The ASR system provider may provide the personnel needed to develop initial applications and concurrently train IS staff members to support the application and develop later applications. Some companies may find it best to buy a turnkey system and continuing maintenance from the ASR provider.

Six-Step Development Process

The iterative development process described here assumes that the application is a telephony-based system for providing information from a data base to customers. ASR applications are particularly suited to development through incremental prototyping—a technique that provides incremental feedback to the application requester and supports joint application development (JAD) approaches.

At each stage of prototype development, the system requester, simulating a user, is asked to evaluate the dialog provided by the system. This evaluation usually results in changes that grow fewer as the application converges on satisfactory results.

The following description of the development process assumes that the scope of the project is known and a preliminary specification has been developed. This specification would include:

- A description of the proposed application.
- A description of the operating environment, including existing hardware and software systems that will interface with the ASR application or any preference for hardware and software that imposes a constraint on the ASR solution.
- A preliminary description of the anticipated dialog between users and the system.
- A list of the specialized vocabulary words or phrases needed by the application.

Because speech recognition applications often interface with or support mission-critical systems, a cautious six-step approach that includes extensive testing before implementation is required. Exhibit 10-3-1 summarizes the tasks involved in the approach.

Task 1: Requirements Definition. The development team defines the application requirements in sufficient detail to create a more specific project plan and allocate assignments and target schedules. This documented definition includes:

- A definition of the dialog to be conducted between the user of the speech recognition system and the ASR system. This dialog anticipates such system behavior as points of dropout to a live operator, alternate use of touchtone input, and transfers to other voice routines.
- A definition of the application program interfaces (APIs) to cooperating voice response units or other equipment or software.
- A definition of message formats and message-passing protocols to sup-

Task	Description
1. Requirements Definition	Define dialog, APIs, text plan, load characteristics, measurement methods, and preliminary rollout plan.
2. Standalone Prototype Development	Iteratively develop and test standalone prototype and API, install vocabulary, tune for regional accents.
3. Integration	Integrate with IVR and test (or test with test data bases), simulate load testing.
4. Pilot Test	Test functional operation with a limited user set, modify as necessary.
5. Detailed Rollout Plan	Develop detailed rollout plan, install equipment for rollout.
6. Rollout	Full system test followed by implementation.

Exhibit 10-3-1. Development Task Sequence for ASR Applications

port any required interfaces, accesses, or updates to production data bases. To protect the integrity of the company's data bases, the ASR system should not directly access them in any manner except through a defined procedure/protocol provided by the technical staff responsible for the data base. The interface protocol permits sending and receiving messages between the voice recognition system and the data base access programs.

- Test plans for API testing, controlled environment testing, pilot testing, and acceptance testing.
- Identification and description of a pilot test group.
- Identification and description of anticipated load characteristics.
- A definition of performance measurement parameters for evaluating the performance of the system during pilot test and under full operation.
- A preliminary plan for rollout of the application to full use.

During task 1, the company should also acquire and install a workstation-based development environment for the creation and testing of the prototype and for use in future development and maintenance. The ASR application development toolkit is installed on the workstation.

Upon completion of task 1, the requester's representatives review and approve the resulting definitions and plans.

Task 2: Prototype Development. The team employs a technique of incremental prototyping to develop and iteratively test a prototype application in a standalone environment. Technical personnel assigned to the project team are trained in the use of the toolkit by the ASR system provider during this task. This basic training in use of the toolkit ensures that the staff becomes familiar with the tools and methods used in voice recognition applications development.

The team installs the vocabulary and any variations that are anticipated to handle regional accents, creates the dialog, implements word spotting if required, implements accuracy performance measurement routines, and, in general, establishes a complete prototype of the application in a standalone environment. Iterative development ensures that the prototype has continually been tested as it is developed.

At the conclusion of this task, the operating prototype should satisfy the functional requirements of the voice recognition portion of the application. The operating prototype is essentially the runtime version of the voice recognition portion of the application, ready for integration into the overall application environment.

Task 3: IVR Integration. In this task, the voice recognition application is integrated with the interactive voice response unit. Any software elements necessary to accommodate the interfaces, handle dropouts to live operators and touchtone signals, and provide message passing are developed and tested independently of the final implementation environment. This task results in tested interfaces to the IVR but not necessarily to host data bases.

Depending on the configuration of the environment, the IVR may access host data bases on the basis of touchtone input signals. In such a case, the ASR system returns appropriate messages to the IVR to initiate the data base access.

In the event that the access to host data bases is initiated directly from the voice recognition system, the following task is substituted for task 3.

Task 3a: Integration into Operating Environment. Here, the team integrates the voice recognition portion of the application into a controlled version of the operating environment. For this task, the team establishes test data bases (i.e., potentially copies of live files) so that the tests do not access live files.

The team may also install a communications connection to the ASR provider for remote maintenance of the application if required.

The team develops the message-passing routines between the voice recognition portion of the application and data base access, voice response, or other routines. Any software elements necessary to accommodate the interfaces, handle dropouts to live operators and touchtone signals, and provide message passing are developed and tested independently of the full environment. The team also develops program routines to evaluate performance of the system both in the pilot test and full rollout environments.

In addition, the team develops a test-bed to simulate peak loads on the system and conducts such simulation tests to determine concurrent voice recognition processing capacity for this application.

Functional simulations and testing of end-user interaction are conducted and adjustments or corrections made to the application until the application is considered ready for pilot testing. Such readiness is determined according to the predefined test plans developed in task 1.

Task 4: Pilot Test. On the basis of load tests, the company must purchase and install sufficient computer capacity to accommodate the projected load for the pilot test. Subsequent to the installation of this equipment, the team conducts another round of integration tests to ensure that the application and system are ready.

Following the integration tests, the application is connected to live files and tested with a limited number of users to ensure correct functional operation. An application designed for use by customers may be tested by a group of employees first.

The pilot test is conducted according to the plan developed in task 1 only after satisfactory controlled testing is completed. Measurements of performance are evaluated along with quality reports from users. Any adjustments necessary to the system are made until performance meets the agreed-on parameters.

The acceptance test described in task 1 is conducted as a part of the pilot test. At the conclusion of this task, the requester representatives examine test results and certify acceptance of the system.

Task 5: Rollout Planning. Concurrently with task 4, the project team, working with the application requester, develops a rollout plan. This plan describes, in detail, the steps necessary to ensure a satisfactory implementation of the application. The plan covers any additional equipment acquisition, architectural changes necessary to the system environment to accommodate full implementation, needed integration activities, performance management procedures, staging of implementation to users, feedback mechanisms to ensure quality of performance, the sequence and schedule of events, and assignments of responsibility.

The company must then acquire and install any additional equipment or software required to support full implementation.

Task 6: Rollout and Full Implementation. The team first performs a controlled test using the additionally installed equipment to ensure proper integration of the system. If possible, the pilot test is then extended using the newly acquired equipment instead of the previously used pilot test equipment. Full implementation is started only when all components and their integration have been tested.

Implementation is conducted according to the plans set forth in task 4. Members of the project team observe the implementation and contact the ASR or IVR provider if any major problems occur. Problems that cannot be handled by the company's trained staff are handled by the providers in accordance with warranty or maintenance agreements. Weekly performance reports to the requester management should be written by the project team until 60 days after full implementation of the system or 60 days of trouble-free operation.

Implementing other types of applications, such as workstation-based, microphone-input ASR applications requires a different development process; however, a similar cautious approach using repetitive testing is recommended. Conservative project approaches are mandatory not only because the systems are usually mission-critical, but because the introduction of new technology should not be exposed to the possibility of disappointing the requester.

SUMMARY

Automatic speech or voice recognition applications have the potential to provide high payoff and competitive advantage and to improve productivity, customer service, and the quality of business processes. The technology must be classified as emergent because most potential users do not yet have the skills to support their own applications development. However, the technology and many available products are currently robust enough and reliable enough for users and developers to seriously consider speech as an interface for multimedia applications. ASR systems may, at this stage, be compared to data base management systems in their early days of use. New skills must be acquired (or contracted) to maximize the benefits and conduct successful implementation projects.

10-4

The Superhighway: Information Infrastructure Initiatives

Keith G. Knightson

MANY COUNTRIES AND ORGANIZATIONS have developed initiatives aimed at establishing an electronic highway such as the National Information Infrastructure (NII) in the US and the European Information Infrastructure (EII). To cover global aspects, a Global Information Infrastructure (GII) is being developed. The outcome of these initiatives depends on the changes taking place in the information and communications industries because of converging technologies, deregulation, and business restructuring or reorganization based on economic considerations. This chapter explores some of the possibilities and problems associated with information infrastructures.

WHAT IS AN INFORMATION INFRASTRUCTURE?

The term *information infrastructure,* which is used interchangeably with the term *information superhighway* in this chapter, describes a collection of technologies that relate to the storage and transfer of electronic information, including voice, data, and images. It is often illustrated as a technology cloud with user devices attached, including broadband networks, the Internet, and high-definition TV.

However, problems emerge when users attempt to fit technologies together. For example, in the case of videophone service and on-demand video service, it is not clear whether the same display screen technology can be used, or whether a videophone call can be recorded on a locally available VCR. This example illustrates the need for consistency between similar technologies and functions.

Relevance of the Information Infrastructure

The information infrastructure is important because it provides an opportunity to integrate technologies that have traditionally belonged to specific industry domains, such as telecommunications, computers, and entertainment. (Integration details are discussed later in this chapter.) The information infrastructure also presents an opportunity to greatly improve the sharing and transferring of information. New business opportunities abound related to the delivery of new and innovative services to users.

Goals and Objectives of Information Infrastructures

The goals of most information infrastructures are to achieve universal access and global interoperability. Without corporate initiatives, the information infrastructure could result in conflicting and localized services, inefficient use of technology, and greater costs for fewer services. Some of the elements necessary to achieve such goals, including standards and open technical specifications that ensure fair competition and safeguard user interests, have yet to be adequately addressed.

BACKGROUND: TECHNOLOGY TRENDS

Two factors are often cited as driving the technology boom: the increase in computer processing power and the increase in the amount of available memory. Advances in these areas make a greater number of electronic services available for lower costs. This trend is expected to continue.

Bandwidth Pricing Issues

Unfortunately, comparable gains of higher bandwidths and decreasing costs are not as evident in the communications arena. Whether this is because of the actual price of technology or because of pricing strategies is debatable. Many applications requiring relatively high bandwidths have yet to be tariffed.

On-demand video is an interesting test case for the pricing issue. To be attractive, this service would have to be priced to compete with the cost of renting a videotape. However, such a relatively low price for high bandwidth would make the price of traditional low-bandwidth phone services seem extremely expensive by comparison. ATM-based broadband ISDN is likely to emerge as the vehicle for high-speed, real-time applications where constant propagation delay is required.

The lack of higher bandwidths at inexpensive prices has inhibited the growth of certain applications that are in demand. The availability of inexpensive high bandwidth could revolutionize real-time, on-demand applications, not only in the video entertainment area but also in the electronic publishing area.

Decoupling Networks and Their Payloads

One factor that is influencing the shape of the superhighway is the move toward digitization of information, particularly audio and video. Digitization represents a total decoupling between networks and their payloads.

Traditionally, networks have been designed for specific payloads, such as voice, video, or data. Digital networks may become general-purpose carriers of bit streams. In theory, any type of digital network can carry any and all types of information in digital format, such as voice, video, or computer data, thus banishing the tradition of video being carried on special-purpose cable TV networks and telephone service being carried only over phone company networks. All forms of information are simply reduced to bit streams.

The Service-Oriented Architecture

The separation of information services from bit-delivery services leads to the concept of a new service-oriented architecture. The most striking aspect of this service-oriented architecture is that the control and management entity may be provided by a separate service organization or by a distributed set of cooperating entities from different service organizations. The architecture represents a move away from the current world of vertical integration toward one of horizontal integration.

Deregulation of communications also plays a part in this scenario. Deregulation often forces an unbundling of components and services, which creates a business environment ideally suited to a service-oriented architecture.

KEY ISSUES IN CREATING THE SUPERHIGHWAY

Achieving a singular, seamless information highway is going to be a challenge, and whether users can influence development remains to be seen. Unless all interested parties act in harmony on the technical specifications (i.e., standards), market sharing, and partnering issues, the end user may be the biggest loser.

For provision of a given service (e.g., voice or data), it should not matter whether a user's access is through the telephone company, the cable company, or the satellite company. Similarly, it should not matter whether the remote party with whom a user wants to communicate has the same access method or a different one.

Several common elements exist in any end-to-end service. For example, there is a need for agreed-on access mechanisms, network platforms, addressing schemes, resolution of inter-provider requirements, and definition of universal services. The development of a generic framework would help to ensure that service requirements are developed equitably and to introduce innovative new services.

The User's Role

Users are becoming more technology literate. The use of technology in the home in recent years has increased. Many users already benefit from what can be achieved through the convergence and integration of user-friendly technologies. User perspectives, rather than those of a single industry or company, should be thoroughly considered in the development of infrastructure initiatives.

Government's Role

The private sector takes most of the risks and reaps most of the rewards for development of the information superhighway. However, government should assert some influence over the development of universally beneficial user services. The role of the government mainly involves:

- Encouraging industry to collaborate and develop universally beneficial user services.
- Mediating between competing industry factions.
- Solving problems involving cultural content, cross-border and customs issues, protection of the individual, obscene or illegal material, and intellectual property and copyrights.

Industry's Role

Three dominant technology areas—telecommunications, computers and related communications, and the entertainment industry—are converging. Although there has already been some sharing of technology among industries, a single integrated system has not been created.

For example, many existing or planned implementations of videophone service invariably involve a special-purpose terminal with its own display screen and camera. For a home or office already equipped with screens and loudspeakers for use with multimedia-capable computers, the need for yet another imaging system with speakers is a waste of technology. Apart from the cost of duplication associated with the industry separations, there is the problem of the lack of flexibility. For example, if a VCR is connected to a regular TV, it should also be able to be used to record the videophone calls.

A plug-and-play solution may soon be possible in which the components are all part of an integrated system. In such a case, screens, speakers, recording devices, computers, and printers could be used in combination for a specific application. The components would be networked and addressed for the purposes of directing and exchanging information among them. Similar considerations apply to computing components and security systems. Using the videophone example, if the remote videophone user puts a document in front of the camera, the receiving party should be able to capture the image and print it on the laser printer.

Plug-and-play integration is not simple; yet if the convergence is not addressed, the result will be disastrous for end users, who will be faced with a plethora of similar but incompatible equipment that still fails to satisfy their needs.

The Dream Integration Scenario

In the ideal configuration, there would be only one pipe into the customer's premises, over which all services—voice, video, and data—are delivered. User appliances can be used interchangeably. In this scenario, videophone calls could be received on the home theater or personal computer and recorded on the VCR.

The Nightmare Scenario

In the nightmare scenario, customer premises would include many pipes. Some services would only be available on certain pipes and not others. The premises would have duplicate appliances for generating, displaying, and recording information. End-to-end services would be extremely difficult to achieve because all service providers would not choose to use the same local- or long-distance delivery services. In addition, all the local- and long-distance networks would not be fully interconnected.

Purveyors of technology and services may argue that this means they can all sell more of their particular offerings, which is good for business. Users, on the other hand, are more likely to feel cheated, because they are being forced to subscribe to different suppliers for slightly different services.

Corporate Networks

Large corporations create networks that are based on their preferred supplier of technology. They are usually extremely conservative in their technology choices because many of their business operations depend totally on the corporate network.

Two factors are causing this traditional, conservative approach to be questioned:

- *The cost of maintaining private networks.* In many cases, several private networks operate within a single corporation, such as one for voice, one for IBM's SNA network, and one for a private internet using TCP/IP or Novell's IPX. The change taking place is sometimes referred to as consolidation. Consolidation involves network sharing by operating the different systems protocols over the same physical network.

- *The need for global communications.* Corporations cannot afford to remain electronically isolated from their customers. As every business tackles cost cutting by increasing the use of information technology, the need for intercompany communication increases. Companies now need to communicate electronically with the banking industry, their suppliers, their customers, and the government to carry out their business. The GII is going to increase in importance for corporations, particularly in terms of availability and reliability.

THE INTERNET AND B-ISDN

Many users consider the Internet the only true information highway. In many ways, this is true—the Internet is the only highway, at least in the sense that it is the only worldwide, seamless, and consistent end-to-end digital networking facility available. In addition, it has become a place where certain standardized applications can be used. It has a globally unique, centrally administered address space. The Internet provides national and international switched data services on a scale that would usually be associated with the major telecommunications carriers.

Not surprisingly, not everyone agrees that the Internet is the only highway. Technically, the Internet is a connectionless packet network overlaid on a variety of network technologies, such as leased lines, frame relay, asynchronous transfer mode (ATM), and LANs . However, it is difficult to imagine that at some point in the future, all voice and video traffic would be carried over such a network rather than directly over a broadband integrated services digital network (B-ISDN).

Thus, there may be a battle between the Internet and the traditional telecommunications carriers for control of the primary switching of data. The carriers may try to establish broadband ISDN as the primary method of switching data end to end, using telephone company-oriented number/addressing plans such as E.164.

The Internet community is interested in the use of broadband ISDN, primarily as a replacement for leased lines between Internet switching nodes (i.e., routers) where the real switching occurs. The deployment of broadband ISDN within the Internet may result in the migration of routers to the edges of the Internet, eliminating the need for intermediate routers. In any event, the interaction between the traditional router-based Internet style of operation and the emerging broadband ISDN switched services will be closely watched by corporate users.

The anarchic nature of the Internet will also be put to the test by commercial users who will want better service guarantees and accountability for maintenance and recovery. Despite these known deficiencies, the Internet remains the predominant information highway and it is difficult to imagine that it will lose its dominance in the near future.

TELECOMMUNICATIONS AND CABLE TV

Deregulation in many countries now permits cable TV companies to offer services traditionally offered by the telephone companies. The cable companies are just beginning to form plans on how new two-way services should be offered. Access to the telephone company network would also provide access to other services, such as the Internet.

A major issue is the kind of interface to be provided on the cable network for associated telephone apparatus. It is not clear whether a traditional phone could simply be plugged into the cable system. Other issues, such as numbering and access to 800 service, need to be resolved. Whether traditional modem, telephony, or ISDN interfaces could be used or whether new cable-specific interfaces would be developed is also under consideration. Both solutions could coexist through provision of appropriate conversion units.

Cable systems usually consist of a head end with a one-way subtending tree and branch structure. Whether the head end would provide local switching within the residential area has not been determined. Other topologies, such as rings, may be more appropriate for new services.

Conversely, deregulation also permits the telephone companies to offer services previously offered by the cable companies. In such a case, a video server would be accessed by the telephone company network, probably using broadband ISDN and ATM technology.

COMPUTER-INTEGRATED TELEPHONY

Computing and telecommunications are coming together in several ways. Computers can now be attached to telecommunications lines to become sophisticated answering machines, autodialers, and fax machines.

The availability of calling and called-line identification permits data bases to be associated with telephone calls. For example, the calling line identification can be used to automatically extract the appropriate customer record from a data base so that when the call is answered the appropriate customer information becomes available on a screen.

Computer-integrated telephony allows a variety of telephone service features to be controlled by the customer's computers. Intelligent network architectures that facilitate the separation of management and control are ideally suited to external computer control.

Public switched data networks have not been very efficient because of the costs of building separate networks and because the scale and demand for data proved nothing like that for voice services. A single digital network such as narrowband ISDN (N-ISDN) or broadband ISDN changes the picture significantly when coupled with the new demand for digital services.

COMPUTING AND ENTERTAINMENT

Most personal computers on the market have audiovisual capabilities. Movies and audio clips can be combined with text for a variety of multimedia applications. Video or images can be edited as easily as text.

With the advent of high-definition TV and digital encoding of TV signals, it is easy to imagine a system in which the traditional TV screen and the PC monitor would be interchangeable. Computers are already being used to produce movies and as a playback medium, even providing the possibility of real-time interaction with the users.

Integrating all the appliances into a single architecture is the difficult part. Home theater systems provide simple forms of switching between components—for example, video to TV or VCR, or audio from TV to remote speakers. Soon, no doubt, the personal computer will be part of this system.

NATIONAL AND INTERNATIONAL INITIATIVES

Many countries have prepared recommendations for their respective national information infrastructures, including the US, Canada, Europe, Japan, Korea, and Australia, among others. The major differences in each country's initiatives seem to revolve around to what extent government will fund and regulate the information infrastructure.

The US

The Information Infrastructure Task Force (IITF) launched the National Information Infrastructure (NII) initiative in early 1993. The IITF is composed of an advisory council and committees on security, information policy, telecommunications policy, applications, and technology. Government funding is being made available for the development of NII applications.

The IITF's goal is that the information infrastructure become a seamless web of communications networks, computers, data bases, and consumer electronics. The NII initiative is also closely associated with the passage of a new communications act, which outlines principles for the involvement of the government in the communications industry. According to the communications act, the government should:

- Promote private sector investment.
- Extend the universal service concept to ensure that information resources are available at affordable prices.
- Promote technological innovation and new applications.
- Promote seamless, interactive, user-driven operation.
- Ensure information security and reliability.
- Improve management of the radio frequency spectrum.
- Protect intellectual property rights.
- Coordinate with other levels of government and with other nations.
- Provide access to government information and improve government procurement.

International Initiatives

The G7 countries (Britain, Canada, France, Germany, Italy, Japan, and the US) are considering developing an information infrastructure that would offer, among others, the following services:

- Global inventory.
- Global interoperability for broadband networks.
- Cross-cultural education and training.
- Electronic museums and galleries.
- Environment and natural resources management.
- Global emergency management.
- Global health care applications.
- Government services online.
- Maritime information systems.

STANDARDS AND STANDARDS ORGANIZATIONS

It is difficult to imagine how objectives such as universal access, universal service, and global interoperability can be achieved without an agreed-on set of standards. However, some sectors of industry prefer that fewer standards are established because this gives them the opportunity to capture a share of the market with proprietary solutions. Regardless, several national and international standards development organizations (SDOs) throughout the world are initiating activities related to the information infrastructure.

ISO and ITU

Both the International Standards Organization (ISO) and the International Telecommunications Union (ITU—formerly the CCITT) are embarking on information infrastructure standards initiatives. The ISO and ITU have planned a joint workshop to address standards issues.

American National Standards Institute Information Infrastructure Standards Panel (ANSI IISP)

The ANSI IISP goals are to identify the requirements for standardization of critical interfaces (i.e., connection points) and other attributes and compare them to national and international standards already in place. Where standards gaps exist, standards development organizations will be asked to develop new standards or update existing standards as required.

ANSI IISP is developing a data base to make standards information publicly available. In its deliberations, the ANSI IISP has been reluctant to identify specific networking architectures or interconnection arrangements and appears to be confining its efforts to a cataloguing process.

Telecommunications Standards Advisory Council of Canada (TSACC)

The TSACC is an umbrella organization for all the standards organizations in Canada. It is a forum where all parties can meet to discuss strategic issues. The objectives of TSACC, in respect to the Canadian Information Infrastructure and the GII, are similar to those of the ANSI IISP. However, TSACC considers the identification of specific networking architectures and associated specific access and interconnection points essential to achieving the goals of universal access, universal service, and interoperability.

European Telecommunications Standards Institute (ETSI)

The Sixth Review Committee (SRC6) of ETSI published a report on the European Information Infrastructure (EII) that emphasizes the standardization of the EII. Many of the recommendations in the report concern the development of reference models for defining the particular services and identifying important standards-based interface points. Broadband ISDN is recommended as the core technology for the EII.

The Digital Audio Visual Council (DAVIC)

DAVIC was established in Switzerland to promote emerging digital audiovisual applications and services for broadcast and interactive use. DAVIC, which has a very pro-consumer slant, believes that these services will only be affordable through sufficient standardization. The council has formed technical committees in the following five areas:

- Set-top units.
- Video servers.
- Networks.
- Systems and applications.
- General technology.

DAVIC may be the only forum in which home convergence issues can be solved.

ALTERNATIVE INITIATIVES

Following are two interesting US-based information infrastructure initiatives.

EIA/ TIA

The Electronic Industries Association (EIA) and its affiliate Telecommunications Industry Association (TIA) have just released version 2 of their white paper titled "Global Information Infrastructure: Principles and Promise." The basic principles conclude that:

- The private sector must play the lead role in development.
- Enlightened regulation is essential.
- The role of global standards is critical.
- Universal service and access must support competitive, market-driven solutions.
- Security and privacy are essential.
- Intellectual property rights must support new technologies.

The Computer Systems Policy Project (CSPP)

The CSPP is not a standards organization but an affiliation of the chief executive officers of several American computer companies. The CSPP has published a document titled "Perspectives on the National Information Infrastructure: Ensuring Interoperability." The CSPP document identifies the following four key points-of-presence as candidates for standardization:

- The interface between an information appliance and a network service provider.

- The applications programming interface between an information appliance and emerging NII applications.
- The protocols that one NII application, service, or system uses to communicate with another application, service, or system.
- The interfaces among and between network service providers.

SUMMARY

The technical challenges of creating a Global Information Infrastructure are not insurmountable. The main difficulties arise from industries competing for the same business rather than sharing an expanding business, and from the lack of agreement on necessary open standards to achieve universal access and global interoperability that would expand the total business.

Interoperability requires agreed-on network architectures and the associated standards that could, in some cases, stifle innovation. A balance must also be struck between government regulation and private sector control over GII development. However, if each camp can cooperate, it is possible that in the future the communications, information, and entertainment industries could merge technology to provide plug-and-play components integrated into a single, coherent system that offers exciting new services that exist now in only the wildest imaginations.

List of Acronyms Used in This Book

THE COMMUNICATIONS INDUSTRY serves up an alphabet soup of abbreviations and acronyms to denote complex technologies and concepts. Some terms are commonplace, others arcane. To help readers decode some of the industry shorthand, the following list includes key networking terms used in this book.

A

ABR	available bit rate
ACD	automatic call distribution
ACF	advanced communications function
ACL	access control list
ACR	allowed cell rate
ADPCM	adaptive differential pulse code modulation
AMI	alternate mark inversion (coding)
ANI	automatic number identification
ANSI	American National Standards Institute
API	application programming interface
APPN	Advanced Peer-to-Peer Networking
ARP	address resolution protocol
ATM	asynchronous transfer mode

B

BECN	backward explicit congestion notification
BONDING	Bandwidth on Demand Interoperability Group
BPS	bits per second

BRI	basic rate interface
BSA	basic service area
BSC	binary synchronous communications
BT	burst tolerance

C

CAD	computer-aided design
CASE	computer-aided software engineering
CBR	constant bit rate
CCITT	International Telephone and Telegraph Consultative Committee
CCR	current cell rate
CDMA	code division multiple access
CDPD	cellular digital packet data
CD-ROM	compact disk with read-only memory
CDV	cell delay variance
CERT	Computer Emergency Response Team
CHAP	challenge-handshake authentication protocol
CI	congestion indication
CIAC	Computer Incident Advisory Capacity
CIDR	classless interdomain
CIM	computer-integrated manufacturing
CIR	committed information rate
CIX	Commercial Internet Exchange
CLLM	consolidated link layer message
CLNP	connectionless network protocol
CLNS	connectionless network service
CMIP	common management information protocol
CONS	connection-oriented network service
COS	Corporation for Open Systems
CPE	customer premises equipment
CPU	central processing unit
CRC	cyclic redundancy check
CSDC	circuit switched digital capability
CSF	critical success factor
CSMA/CD	carrier sense multiple access with collision detection
CSU	channel service unit

D

DACS	digital access and cross-connect system
DASD	direct access storage device
DCE	data communications equipment or data circuit-transmitting equipment, distributed computing environment
DDS	Dataphone Digital Service
DE	discard eligibility (bit)
DES	Data Encryption Standard
DES	destination end station
DLC	data link connection
DLCI	data link connection identifier
DLSw	data link switching
DME	Distributed Management Environment
DNA	Digital Network Architecture
DNIC	data network identification code
DNS	domain naming system
DOS	disk operating system
DSE	data switching equipment
DSP	domain specific part, digital signal processing
DSU	data service unit
DTE	data terminal equipment
DTMF	dual tone multifrequency
DXI	data exchange interface

E

EDI	electronic data interchange
EFCI	explicit forward congestion notification
EFT	electronic funds transfer
EGP	Exterior Gateway Protocol
EIA	Electronic Industries Association
EISA	Extended Industry Standard Architecture
EMI	electromagnetic interference
ESMR	enhanced special mobile radio

F

FCC	Federal Communications Commission
FDDI	fiber distributed data interface

FECN forward explicit congestion notification
FEP front end processor
FIPS Federal Information Processing Standards
FM frequency modulation
FRAD frame relay access device
FRAD frame relay assembler/disassembler
FTAM file transfer, access, and management
FT1 fractional T1
FTP file transfer protocol

G

GCRA generic cell rate algorithm
GOSIP Government Open Systems Interconnection Profile
GUI graphical user interface

H

HDLC high-level data link control
HEO high elliptical orbiting (satellite)
HPR high performance routing
HSSI high-speed serial interface
HTML hypertext markup language
HTTP hypertext transfer protocol

I

IC information channel
ICMP Internet Control Message Protocol
IDP initial domain part
IEC interexchange carrier (also IXC)
IEEE Institute of Electrical and Electronic Engineers
IETF Internet Engineering Task Force
IP Internet protocol
IPX internet packet exchange
IR infrared
IS information systems
IS-IS intermediate system to intermediate system
ISA Industry Standard Architecture

ISDN	integrated services digital network
ISO	International Standards Organization
ISP	Internet Service Provider
ISP	International Standards Profile
IT	information technology
ITU	International Telecommunications Union (formerly CCITT)
ITU-TSS	International Telecommunications Union-Telecommunications Standards Sector
IVR	interactive voice response
IXC	interexchange carrier (also IEC)

L

LAN	local area network
LAP	link access procedure
LAPB	link access procedure, balanced
LAPD	link access procedure for the D channel
LAT	local area transport
LATA	local access transport area
LEC	local exchange carrier
LED	light emitting diodes
LEO	low-earth orbit (satellite)
LLC	logical link control
LMI	local management interface
LU-LU	logical unit to logical unit

M

MAC	media access control
MAN	metropolitan area network
MAU	medium attachment unit, multistation access unit
MCC	mobile communications controller
MCU	multipoint control unit
MHS	message-handling service
MHZ	megahertz
MIB	management information base
MS-DOS	Microsoft disk operating system

MTSO	mobile telephone switching office
MUX	multiplexer

N

NETBIOS	network basic I/O
NCP	network control program
NDIS	Network Driver Interface Specification
NIC	Network Information Center, network interface card
NID	network interconnection device
NIST	National Institute of Standards and Technology
NLPID	network layer protocol identification
NLSP	network link services protocol
NOS	network operating system
NRZ	non-return to zero
NSAP	network service access point
NTSC	National Television Standards Committee

O

OCn	optical carrier level n
ODI	Open Data link Interface
OOP	object-oriented programming
OSPF	open shortest path first (protocol)
OS	operating system
OSF	Open Software Foundation
OSI	Open Systems Interconnection

P

PAD	packet assembler-disassembler
PAL	phase alternating by line (European TV standard)
PAP	password authentication protocol
PBX	private branch exchange
PC	personal computer
PCM	pulse-code modulation
PCMCIA	Personal Computer Memory Card International Association
PCN	personal computer network
PCR	peak cell rate

PCS	personal communication system (service)
PDA	personal digital assistant
PICS	protocol implementation conformance statement
PIN	personal identification number
PIR	protocol independent routing
POSIX	portable operating system interface for UNIX
POP	point of presence
POTS	plain old telephone service
PPP	point-to-point protocol
PPS	packets per second
PRI	primary rate interface
PSTN	public switched telephone network
PTT	postal, telegraph, and telephone
PVC	permanent virtual circuit

R

RAID	redundant array of inexpensive disks
RAM	random access memory
RBOC	regional Bell operating company
RF	radio frequency
RFI	radio frequency interference
RFC	request for comment
RIP	routing information protocol
RM	resource management (cells)
RMON	remote monitoring
RTP	real time transport protocol

S

SAP	service advertising protocol
SCR	sustainable cell rate
SCSI	small computer systems interface
SDDN	sofware-defined digital network
SDLC	synchronous data link control
SDN	Software Defined Network
SES	source end system
SIP	simple Internet protocol

SLIP	serial-line IP
SMDS	switched multimegabit data service
SMP	symmetrical multiprocessing
SMTP	simple mail transfer protocol
SNA	Systems Network Architecture
SNACP	subnetwork access control protocol
SNAFS	SNA file services
SNADS	SNA distribution services
SNAP	subnetwork access protocol
SNICP	subnetwork independent convergence protocol
SNMP	simple network management protocol
SNPA	subnetwork point of attachment
SNR	signal to noise ratio
SONET	synchronous optical network
SPF	shortest path first
SQL	structured query language
SR/TLB	source-route translation bridging
SRB	source-route bridging
SRT	source-route transparent
SS7	signalling system # 7
STP	shielded twisted pair
SVC	switched virtual circuits

T

TA	terminal adapter
TCP/IP	Transmission Control Protocol/Internet Protocol
TDM	time division multiplexing
TDMA	time division multiple access
TIA	Telecommunications Industry Association
TIC	Token Ring interface coupler
TLI	transport layer interface
TMN	Telecommunications Management Network
TSAP-ID	transport service access point identifier
TTCN	tree and tabular combined notation
TUBA	TCP/IP using bigger addresses

U

UBR unspecified bit rate
UDP User Datagram Protocol
UPS uninterruptible power supply

V

VAN value-added network
VBR variable bit rate
VC virtual channel
VPN virtual private network
VSAT very small aperture terminal
VTAM virtual telecommunications access method

W

WAN wide area network
WATS wide area telephone service
WWW World Wide Web

X

XNS Xerox network services

About the Editor

JAMES W. CONARD is the founder and president of Conard Associates, Inc., a firm specializing in telecommunications training and management consulting services. He has been extensively involved in the development and implementation of many major communications systems for government, banking, airline, and manufacturing industries. He has more than 25 years of experience with various firms including Control Data Corp., Marshall Communications, Collins Radio Company, and McDonnell Douglas Corp.

Jim is past chairman of American National Standards Institute Committee X3S3.4 for data link protocols. He is currently a member of ANSI as well as the Institute for Electrical and Electronic Engineers (IEEE). He has been involved in high-speed networking through his consulting, training, writing, and standards committee activities. Jim has participated in the development of ANSI, ISO, ITU and IATA communications standards.

He is the editor of two Auerbach publications, *Handbook of Communications Systems Management* and *Broadband Communications Systems,* and the consulting editor for Auerbach's *Data Communications Management.* In addition, Jim is the author of more than 50 articles on modems, communications standards, and communications protocols. Jim has earned an international reputation as a lecturer, seminar leader, and speaker at industry conferences. His clients include many of the industry's leading organizations. He has conducted seminars for the Data-Tech Institute, IEEE, McGraw Hill, Omnicom, Cap Gemini, and Technology Transfer.

A radio engineering graduate of Philadelphia Technical Institute, Jim also completed studies at Orange Coast College and California State College.

Index

A

Access providers
  Internet 2-1
  network 5-4
  wireless access 7-3
Addressing
  application-level addressing 4-2
  collapsed layer addressing 4-2
  internetwork addressing 4-2
  IP addresses 3-7
  LAN addressing 4-2
  mapping names to addresses 4-2
  mobile address considerations 4-2
  naming and addressing basics 4-2
  open network addressing 4-2
  subnetwork addressing 4-2
  X.25 and open network addresses 4-2
Applications development
  competitiveness of the corporation, enhancing 1-1
Architecture
  distributed computing 9-1
  enterprise networks 6-1
  wireless networks 7-3
Asynchronous transfer mode, *see* ATM
ATM
  available bit rate service 3-4
  broadband networks 5-2
  conformance definition using GCRA 3-4
  constant bit rate service 3-4
  frame relay-to-ATM support 3-3
  monitoring techniques 3-4
  switch behavior 3-4
  traffic control 3-4
  unspecified bit rate service 3-4
  variable bit rate service 3-4
Auditing
  CDPD costs 5-5
  frame relay costs 5-5
  ISDN costs 5-5
  telecommunications costs for data and video 5-5
  video communications costs 5-5
Automatic speech recognition products (ASR)
  building 10-3
  features 10-3
  performance 10-3
  vendor capability 10-3

B

Backbone networks
  managing 6-5
  multiprotocol backbones 6-5
Backup strategies
  frame relay backup 3-2
  general network 8-5
  hardware backup 9-6
  mobile computing 7-4
Bandwidth
  bandwidth on demand 3-2, 5-4
  bottlenecks in WANs 6-2
  frame relay 3-3
  imuxers (bandwidth controllers) 4-3
  information superhighway 10-4
  Intranets and bandwidth management 3-7
  pricing issues 10-4
Banyan WANs
  merging 8-3
Bastion hosts 6-5
Broadband networks
  ATM 5-2

Broadband networks (*cont*)
broadband ISDN 5-2
SMDS 5-2
SONET 5-2
T3 service 5-2
Bulletin boards
hackers 9-4
Business issues
communications systems managers, lessons for 2-3
continuity plans, testing 2-2
cost allocation for networks 2-5
downsizing 2-6
enterprise networking 6-1
hiring valued employees 2-4
information technology 1-1
Internet business trends 2-1
Internet security 6-4
profits, increasing through IT 1-2
recovery strategies 2-2
remote LAN/WAN connections 8-6
resumption planning 2-2
telecommuting, benefits of 7-5
vendor support policies, evaluating 8-2
videoconferencing, business benefits of 3-5
wireless networking costs 7-3

C

Cabling
fiber optics 8-6
horizontal cabling 9-3
vertical cabling 9-3
Career issues
communications systems managers, lessons for 2-3
downsizing and managing morale 2-6
employee empowerment 2-6
hiring and retaining valued employees 2-4
technical skills, improving 2-6
telecommuting 7-5
work teams 2-6
Case studies
remote LAN/WAN connections 8-6
WAN network integration 8-3
CDPD
cellular transmissions 3-6
costs 5-5
wireless option 7-1
Cellular digital packet data, *see* CDPD
Cellular transmission
advantages 7-2
basics 7-2
cellular digital packet data (CDPD) 3-6
challenges 7-2
costs 5-5
impairments to cellular data transmission 7-2
modem challenge, solutions to 7-2
standards 7-2
Change
controlling in a client/server environment 9-6
downsizing, managing morale 2-6
managing 2-3
Circuit-switched services
Fractional T1 5-2
ISDN 5-2
wireless 7-3
Client/server computing
business support 9-6
hardware backup 9-6
management 9-6
points-of-failure planning 9-6
risks 9-6
service orientation 9-6
software controls 9-6
Clustered CPUs 10-1
Communications industry
changing role of 2-3
consolidation of 1-1
downsizing 2-6
globalization of 1-1
hiring valued employees 2-4
managers, lessons for 2-3
Communications networks
costs 2-5
frame relay in an IBM environment 8-4
Communications systems managers
change, managing 2-3
downsizing and managing morale 2-6
hiring valued employees 2-4
lessons for 2-3
management styles 2-4
planning 2-3
risk management 2-3
training 2-3
telecommuters, managing 7-5

Compression
continuous mode operation 6-2
data in routed internetworks 6-2
dictionary encoding 6-2
header compression 6-2
link compression 6-2
payload compression 6-2
statistical encoding 6-2
Conferencing technologies
audio compression standards 3-5
bridging technology 3-5
Internet conferencing 3-5
management services 3-5
multipoint conferencing 3-5
network connectivity 3-5
standards 3-5
teleconferencing 3-5
transport issues 3-5
videoconferencing 3-5
Contracting
frame relay services 3-2
Corporate strategies
communications systems managers 2-3
competitive stance, enhancing with IT 1-1
cost allocation for networks 2-5
downsizing 2-6
enterprise networking 6-1
globalization of competition 1-1
hiring valued employees 2-4
Internet as a business tool 2-1
IT strategy, aligning with 1-1
profits, increasing through IT 1-2
research and development investments 1-1
security of corporate information 1-1
systems development 1-1
telecommuting 7-5
WAN integration 8-3
Costs
cellular digital packet data (CDPD) 5-5
chargeable resources 2-5
communications networks 2-5
data communications 5-5
distributed computing 9-1
expense categories 2-5
fiber optics 8-6
frame relay 3-2, 5-5, 8-4
frame relay in an IBM environment 8-4
infrared technology 8-6
international networks 3-3
Internet communications 5-5
Internet gateway access 2-5
inverse multiplexing 3-1
ISDN 5-5
low earth orbital satellite communications 5-5
network access 5-4
network resource categories 2-5
rate determination for communications services 2-5
remote LAN access 3-6
switched digital services 5-1
telecommuting 7-5
wireless communications 7-1, 7-3
vendor support policies, evaluating 8-2
video communications 5-5
WAN upgrades 8-3

D

Data bases
accessing 9-2
maintenance 9-2
report generation 9-2
Data compression
continuous mode operation 6-2
data in routed internetworks 6-2
dictionary encoding 6-2
header compression 6-2
link compression 6-2
payload compression 6-2
statistical encoding 6-2
WAN-bandwidth bottlenecks 6-2
Data storage 8-5
Datapath 5-1
DDS 5-2
Deregulation
Europe's open network provision 5-6
interconnection 5-6
Japan's interconnection rules 5-6
Network Reliability and Interoperability Committee (NRIC) 5-6
1996 U.S. Telecommunications Act 5-6

Deregulation (*cont*)
problems with 5-6
unbundled access 5-6
Dedicated T1 service 5-2
Digital data services, *see* DDS Digital services
circuit-switched digital capabilities 5-1
digitizing the network 5-1
switched digital services 5-1
Disaster recovery
automated data importing 9-5
electronic vaulting 2-2
equipment room diagrams 9-5
fault tolerance 1-5
frame relay 3-3
generators 1-5
insurance 1-5
inventorying hardware and software 9-5
LANs 1-5
links to remote sites 1-5
network security controls 8-5
off-site storage 1-5
planning 1-5, 9-5, 9-6
RAID technology 1-5
restoral capabilities 1-5
risk assessment 1-5
spare parts pooling 1-5
switched digital services 1-5
surge suppressors 1-5
uninterruptible power supplies 1-5
worst-case scenarios 1-5
Distributed computing
architecture 9-1
costs 9-1
data integration 9-1
managing 9-1
quality 9-1
risks 9-1
service delivery models 9-1
standards 9-1
Distributed computing environment (DCE)
managing 9-1
service delivery 9-1
Document imaging systems
security 10-2
Documentation
communications recovery plans 9-5
LAN management 9-3

E

Electronic mail
Internet applications 6-3
wireless communications 7-3
Embedded systems
security 10-2
Encryption 1-3, 8-5
Enterprise networks
applications 6-1
architecture 6-1
consolidating 6-1
frame relay in an IBM environment 8-4
global issues 6-1
integrating network technologies 6-1
Internet access 6-4
interoperability 6-1
LAN management 9-3
mobile enterprise applications 7-3
multivendor environments 8-1
protocols 6-1
remote 10-1
routers 6-1
security 6-4
vendor solutions 6-1
vendor support policies 8-2

F

Fiber optics
advantages 8-6
costs 8-6
disadvantages 8-6
cabling 8-6
data transmission 8-6
File transfer
Internet applications 6-3
Firewalls
Internet security 6-4
intranet security 3-7
monitoring 8-5
Fractional T1 and T3 5-2
Frame relay
ATM support 3-3
backup 3-2
bandwidth on demand 3-2
bandwidth optimization 3-3
benefits 8-4
congestion 6-6
contracting 3-2

costs 5-5, 8-4
disaster recovery 3-3
DLCI 8-4
economics 3-2
equipment providers 8-4
IBM environment, frame relay in 8-4
installation 3-2
international networks 3-3
Internet access over 3-2
ISDN support 3-3
jitter control 3-3
LAN transport 3-2
local access diversity 3-3
network design 3-2
packet-switched network management 3-2
redundancy 8-4
RFC 1490 6-6
router networks, compared with 6-6
SDLC, frame relay as a replacement for 6-6
SNA over 6-6
SNA/SDLC communications 3-2
switched 56K-bps dial backup 3-3
switch anatomy 3-3
system tuning 3-3
testing 8-4
throughput management 3-3
traffic prioritization 3-3
voice 3-3
wide area link sharing 3-2

Frame structure
channel assignment 3-1
information channel frame 3-1

H

Hacking
bulletin boards 9-4
conventions 9-4
dumpster diving 9-4
profile 9-4
reverse intent 9-4
security monitoring 9-4
social engineering 9-4
tools, hardware and software 9-4

Hardware
backup 9-6
hacking tools 9-4
hardware and software openness 3-6
inventorying for disaster recovery 9-5
hardware pricing 8-2

High-Speed Serial Interface, *see* HSSI

Hiring IS employees
communication styles 2-4
corporate culture 2-4
exit interviews 2-4
interviews, structuring 2-4
job definition 2-4
management styles 2-4
positions, creating 2-4
reference checking 2-4
work environment, defining 2-4

HSSI
bandwidth controllers 4-3
DSU/CSUs 4-3
EIA-422/449 compared with 4-3
implementation 4-3
multiplexers 4-3
pin assignments 4-3
signal definitions 4-3
V.35 compared with 4-3

I

IBM
frame relay in an IBM environment 8-4
IBM legacy devices 6-6

Information infrastructure
bandwidth pricing 10-4
decoupling networks 10-4
EIA/TIA 10-4
entertainment 10-4
goals 10-4
government's role 10-4
industry's role 10-4
information superhighway 10-4
integration issues 10-4
Internet 10-4
service-oriented architecture 10-4
standards 10-4
technology trends 10-4
telecommunications 10-4
telephony 10-4
US initiatives 10-4
users 10-4

Information superhighway
bandwidth pricing 10-4

Information superhighway (*cont*)
decoupling networks 10-4
goals 10-4
government's role 10-4
industry's role 10-4
integration 10-4
Internet 10-4
telecommunications 10-4
telephony 10-4
US initiatives 10-4
users 10-4
Information technology
business issues 1-1
economics 1-2
growth, enhancing 1-2
paced and pacing activities 1-2
people versus technology 1-2
profits, increasing 1-2
suppliers 1-1
trends 1-1
users 1-1
Information technology economics
paced and pacing activities 1-2
growth, enhancing 1-2
people versus computers 1-2
profit, enhancing 1-2
Infrared technology
advantages 8-6
costs 8-6
disadvantages 8-6
licensing requirements 8-6
Infrastructure
National Information Infrastructure 10-4
Integrated services digital network, *see* ISDN
Integration strategies
data integration 9-1
information superhighway 10-4
WANs 8-3
International networks
configuring 3-3
planning 3-3
service costs 3-3
Internet
access providers 2-1, 6-3
applications 6-3
Archie 6-3
business trends 2-1
conferencing over 3-5
connection methods 6-3
cost of access 2-5
electronic mail 6-3
file transfer 6-3
firewalls 6-4
frame relay, access over 3-1
Gopher 6-3
host computer access 6-3
information superhighway 10-4
IPv6 6-7
LAN interconnection 5-2, 6-3
marketing 2-1
publishing 2-1
remote login 6-3
security 2-1, 6-4
software developers 2-1
World Wide Web 6-3
Internet protocol, *see* IP
Internet security
bastion host 6-4
enterprise networks 6-4
firewalls 6-4
public access host 6-4
risks of connecting to the Internet 6-4
router-based packet filtering 6-4
tools 6-4
Internetworking
data in routed internetworks 6-2
information superhighway 10-4
internetwork addressing 4-2
SNA-LAN internetworks 6-5
switched digital services and 5-1
Intranets
ActiveX 3-7
applets 3-7
application servers 3-7
bandwidth management 3-7
circuit-level gateways 3-7
client applications 3-7
fat versus thin clients 3-7
firewall security 3-7
IntranetWare 3-7
IP administration 3-7
Java-enabled browsers 3-7
management 3-7
network performance improvement 3-7
operating system 3-7
packet filtering 3-7
scripting languages 3-7
switching 3-7
Inverse multiplexing
applications 3-1

channel assignment 3-1
costs 3-1
delay equalization 3-1
frame structure 3-1
fundamentals 3-1
operational modes 3-1
requirements 3-1
IP
addresses 3-7
administration 3-7
encapsulated routing 6-5
IPv6 6-7
IPv6
destination options header 6-7
fragment header 6-7
hop-by-hop options header 6-7
packet 6-7
ISDN
connections 3-6, 5-2
costs 5-5
disaster recovery 1-5
frame relay-to-ISDN support 3-3

J

Java
ActiveX 3-7
applets 3-7
browsers, Java-enabled 3-7
scripting languages 3-7

L

LAN backbone
cabling 9-3
management 9-3
network documentation 9-3
servers, physical placement of 9-3
standards 9-3
LANs
addressing 4-2
backbone management 9-3
data compression 6-2
disaster recovery 1-5
frame relay over 3-2
interconnecting 5-2
Internet connections 6-3
network management 9-2
parallel networks 6-5
remote access 3-6
remote connections 8-6
routed internetworks 6-2
security 10-2
SNA and LAN internetworks 6-5
wireless 7-1, 10-2
Legacy systems
frame relay over 3-2
IBM legacy devices 6-6
Licensing
infrared technology 8-6
microwave technology 8-6
Local area networks, *see* LANs

M

Mainframe systems
recovery strategies 2-2
Management
communications systems managers, lessons for 2-3
downsizing, managing morale when 2-6
hiring valued employees 2-4
telecommuters 7-5
Messaging systems
agent-based 7-3
Microwave transmission
advantages 8-6
disadvantages 8-6
licensing requirements 8-6
Middleware
wireless networking 7-3
Modems
cellular data transmission 7-2
gateway services 7-2
V.series modems 7-2
Mobile computing
addressing considerations 4-2
backup 7-4
commercial mobile data communications services 7-1
confidentiality 7-4
enterprise applications 7-3
notebook connectivity 7-4
satellite data 7-1
security 7-4
specialized mobile radio 7-1
Mobile computing security
backup 7-4
confidentiality 7-4

Mobile computing security (*cont*)
employee security 7-4
notebook connectivity 7-4
physical considerations 7-4
software considerations 7-4
Multimedia
applications 10-3
automatic speech recognition 10-3
microphone-input applications for the PC 10-3
speaker-independent voice recognition 10-3
telephony applications 10-3
voice recognition applications 10-3
voice recognition systems 10-3
Multiplexers
HSSI 4-3
inverse multiplexing 3-1
Multiprotocol backbones
IP-encapsulated routing 6-5
source-route bridging 6-5
synchronous passthrough 6-5
Multivendor environments
configurations, fitting 8-1
contractual criteria 8-1
customer engineering 8-1
quality of service 8-1
software vendors 8-1
vendor financials 8-1

N

Narrowband services
analog dialup lines 5-2
leased lines 5-2
Network access
alternate access 5-4
AT&T access value plan 5-4
bandwidth on demand 5-4
consolidated access 5-4
dark fiber 5-4
dial-up connections 5-4
discount pricing 5-4
MCI's access pricing plan 5-4
T3 service 5-4
Networks
access 5-4
applications 6-1
architecture 6-1
ATM 3-4
availability 1-5
consolidation 6-1
controls 8-5
cost allocation 2-5
decoupling 10-4
digitizing 5-1
disaster recovery planning for 1-5
enterprise networking 6-1
frame relay 3-2, 8-4
information superhighway 10-4
installation issues 8-5
integrated services digital network (ISDN) 1-5
international 3-3
inverse multiplexed 3-1
management 9-2
neural network systems 10-2
open network addressing 4-2
packet-switched 3-2
protecting 1-5
protocols 6-1
recovery strategies 2-2
reliability 1-5
remote control 3-6
routed internetworks 6-2
routers 6-1
security 10-2
SNA and LAN internetworks 6-5
SNA frame relay networks 6-6
value-added 1-3
virtual private 5-2, 5-3
WAN integration 8-3
X.25 networks 3-6, 5-2
Network management
configuration management 9-2
data base management 9-2
downtime, reducing 9-2
failures, isolating 9-2
filtering 6-1
LANs 9-2
intranets, improving network performance with 3-7
multivendor environments 8-1
NetView Management Services 6-6
open network addressing 4-2
operational goals 9-2
operations controls 8-5
packet-switched networks 3-2
password access 9-2
performance monitoring 9-2
remote access 3-6

RMON 4-4
security controls 8-5
server monitoring 9-2
SNA and LAN internetworks 6-5
SNA frame relay networks 6-6
status reporting 9-2
WAN-based systems 9-2
Network operations control
firewall maintenance 8-5
monitoring network performance 8-5
network backup and recovery 8-5
network logging 8-5
network testing 8-5
Neural network systems
security 10-2
Notebook computing
security 7-4, 10-2

O

Operating systems
intranet operating systems 3-7

P

Packet data networks
headers 6-7
IPv6 packets 6-7
packet cellular technology 7-3
packet radio technology 7-3
wireless applications 7-3
Packet-switching
frame relay 3-1, 3-2, 5-2
network management 3-2
services 5-2
SMDS 5-2
X.25 5-2
Passwords 8-5, 9-2, 9-4, 10-2
PCMCIA cards
security 10-2
Planning
business resumption planning 2-2
communications systems managers and 2-3
disaster recovery 1-5, 9-5
points of failure 9-6
Points of failure
business support 9-6
change control 9-6
client/server environment 9-6
hardware backup 9-6
managing 9-6
mapping 9-6
recognizing 9-6
risk management 9-6
software 9-6
Protocols
enterprise networking 6-1
IPv6 6-7
multiprotocol backbone 6-5
SNA-LAN interconnection 6-5

Q

Quality monitoring 5-5
Quality of service
vendors 8-1
Quotation schedules 8-2

R

Recovery strategies
automated data importing 9-5
business requirements 2-2
documenting 9-5
electronic vaulting 2-2
equipment room diagrams 9-5
frame relay 3-3
inventorying hardware 9-5
inventorying software 9-5
LANs 2-2
mainframes 2-2
midrange systems 2-2
personnel information 9-5
points-of-failure planning 9-6
vendor information 9-5
WANs 2-2
work group systems 2-2
Redundant arrays of inexpensive disks (RAID) 1-5
Remote access
cellular transmission 3-6, 7-2
connection devices 3-6
connectivity options 8-6
hardware and software openness 3-6
ISDN connections 3-6
LAN connections 8-6
mainframe remote terminal 3-6
modems 3-6
remote control networks 3-6

Remote access (*cont*)
remote login 6-3
remote node technology 3-6
security 3-6
strategies 3-6
telephone lines 3-6
WAN connections 8-6
X.25 networks 3-6
Remote computing
cellular transmission 3-6, 7-2
challenges 10-1
clustered CPUs 10-1
connection devices 3-6
connectivity options 8-6
flexibility 10-1
LAN connections 8-6
mainframe remote terminal 3-6
managing 10-1
modems 3-6
performance 10-1
remote control networks 3-6, 10-1
remote login 6-3
remote node technology 3-6, 10-1
scalability 10-1
security 3-6
strategies 3-6
WAN connections 8-6
X.25 networks 3-6
Remote monitoring, *see* RMON
Risk
assessment 1-5
distributed computing 9-1
Internet connection 6-4
management 2-3
RMON
alarms group 4-4
benefits 4-4
events group 4-4
filters group 4-4
history group 4-4
host table group 4-4
host Top N group 4-4
in-band and out-of-band activities 4-4
packet capture group 4-4
statistics group 4-4
traffic matrix group 4-4
Router-based packet filtering 6-4
Routers
enterprise networks 6-1
IP-encapsulated routing 6-5
protocol independent routing 6-5
source-route bridging 6-5
synchronous passthrough 6-5

S

Satellites
low earth orbital satellite communications 5-5
satellite data 7-1
satellite voice services 7-1
Scalability
remote computing networks 10-1
Security
access controls 8-5
administrative controls 8-5
backup 8-5
corporate information 1-1
disaster recovery plans, documenting 9-5
document imaging systems 10-2
dumpster diving 9-4
embedded systems 10-2
encryption 8-5
environmental controls 8-5
firewalls 3-7, 6-4
hacker attacks 9-4
Internet 2-1, 6-4
intranets 3-7
minisupercomputers 10-2
mobile user security 7-4
monitoring 9-4
neural network systems 10-2
notebook computers 10-2
operations controls 8-5
passwords 8-5, 10-2
PCMCIA cards 10-2
physical security 8-5
points-of-failure planning 9-6
recovery 8-5
remote access 3-6
reverse intent 9-4
social engineering 9-4
software controls 8-5
threats 10-2
value-added networks 1-3
videoconferencing 10-2
WAN radio communications 10-2
wireless LANs 10-2
Servers
application servers 3-7

browser and server software 2-1
monitoring 9-2
physical placement 9-3
Simple network management protocol, *see* SNMP
Smart cards
security 10-2
SMDS 5-2
SNA
frame relay, SNA over 6-6
LANs, SNA over 6-5
parallel networks 6-5
protocols 6-5
remote polling 6-5
source-route bridging 6-5
synchronous passthrough 6-5
SNA frame relay networks
APPN 6-6
AS/400 6-6
IBM legacy devices 6-6
Intermediate Network Node 6-6
private frame relay network 6-6
SNA Network Interconnect 6-6
SNMP
management 6-6
RMON 4-4
Software
browser and server software 2-1
hacking 9-4
Internet software developers 2-1
inventories for disaster recovery 9-5
mobile user security and 7-4
security software 8-5
vendor support policies 8-2
vendors 8-1
SONET 5-2
Source-route bridging
broadcast storms 6-5
SNA-LAN interconnection 6-5
Specifications
HSSI 4-3
Speech recognition devices 10-3
Standards
American National Standards Institute (ANSI) 10-4
cellular data transmission 7-2
conferencing 3-5
conformance testing 4-1
EIA/TIA 10-4
harmonization efforts 4-1
information superhighway 10-4
International Standards Organization 4-1
LAN backbone management 9-3
testing programs 4-1
test tool development 4-1
Standards conformance testing
harmonization efforts 4-1
infrastructures for 4-1
International Standards Organization 4-1
National Institute of Standards and Technology 4-1
national testing programs 4-1
test tool development 4-1
Switched digital services
applications 5-1
circuit-switched digital capabilities 5-1
costs 5-1
digitizing the network 5-1
internetworking 5-1
operation basics 5-1
service availability 5-1
service offerings 5-1
Switched multimegabit data service, *see* SMDS
Switching technology
ATM switch behavior 3-4
circuit-switched services 5-2
frame relay switch, anatomy of 3-3
intranet switching 3-7
packet-switched networks 3-2
switched digital services 1-5, 5-1
switched 56K-bps dial backup 3-3
switched multimegabit data service 5-2
Synchronous passthrough 6-5
Systems development
users and developers, improving communication 1-4
Systems network architecture, *see* SNA

T

Telecommunications
costs for data and video 5-5
deregulation 5-6
information superhighway 10-4
Telecommunications Act of 1996 5-6

Telecommunications Act of 1996
 deregulation worldwide 5-6
 interconnection 5-6
 Network Reliability and Interoperability Committee (NRIC) 5-6
 unbundled access 5-6
Telecommunications deregulation
 Europe's open network provision 5-6
 Japan's interconnection rules 5-6
 Telecommunications Act of 1996 5-6
 United Kingdom 5-6
Telecommuting
 benefits 7-5
 costs 7-5
 employee preference for 7-5
 implementing 7-5
 job satisfaction 7-5
 jobs suitable for 7-5
 managing 7-5
 organizational benefits 7-5
 productivity 7-5
 technology requirements 7-5
 telecenters 7-5
Teleconferencing
 audio system 3-5
 bridging technologies 3-5
 environments 3-5
 transmission facilities 3-5
Telephone communications 7-4
Telephony
 computer-integrated 10-4
 multimedia applications 10-3
Testing
 business continuity plans 2-2
 frame relay in an IBM environment 8-4
 standards conformance 4-1
Traffic control
 ATM networks 3-4
Training
 enterprise backbone management 9-3
 disaster recovery 1-5
Transmission services
 analog dialup lines 5-2
 ATM 5-2
 broadband ISDN 5-2
 dedicated T1 5-2
 digital data service 5-2
 fiber optics 8-6
 fractional T1 5-2
 frame relay 5-2
 ISDN 5-2
 leased lines 5-2
 SMDS 5-2
 T3 service 5-2
 X.25 5-2

U

Upgrades
 routers 8-3
 WANs 8-3
Users
 assessing 1-4
 communication between system developers and 1-4
 cooperation, fostering 1-4
 customer needs 1-1
 goals 1-4
 long- and short-term issues 1-1
 mobile 7-4
 monitoring 1-4
 mutual partnering 1-4
 participation 1-4
 students 1-4
 systems development, involvement in 1-4
 telecommuters 7-5

V

Value-added networks (VANs)
 business operations 1-3
 control-related VAN services 1-3
 customer information security 1-3
 encryption 1-3
 fee-based services 1-3
 password protection 1-3
 security 1-3
 service auditor reports 1-3
 third-party assessments 1-3
 vulnerable features 1-3
Vendors
 automatic speech recognition products 10-3
 benchmarks 8-1
 contractual criteria 8-1
 enterprise network solutions 6-1
 financial condition of 8-1
 frame relay 8-4
 hardware pricing 8-2

multivendor environments 8-1
performance tracking 8-1
quotation schedules 8-2
selecting 8-1
software 8-1
software pricing 8-2
support policies, evaluating 8-2
wireless support 7-1
Videoconferencing
costs 5-5
desktop systems 3-5
management services 3-5
midrange systems 3-5
multipoint conferencing 3-5
network connectivity 3-5
room-based systems 3-5
security 10-2
standards 3-5
transport issues 3-5
videophones 3-5
Virtual private networks
benefits 5-3
data networks 5-3
frame relay 5-3
hybrid networks 5-3
international services 5-2
LAN interconnection 5-2
Software Defined Network 5-3
Voice
applications 10-3
cellular voice 7-1
digitization 3-3
frame relay, switching voice over 3-3
microphone-input applications 10-3
multimedia applications 10-3
satellite voice services 7-1
speaker-independent voice recognition devices 10-3
voice recognition applications, building 10-3
voice recognition interfaces 10-3

W

WANs
Banyan WANs 8-3
bandwidth bottlenecks 6-2
integrating 8-3
integration case study 8-3
network management 9-2
radio communications 10-2
recovery strategies 2-2
remote connections 8-6
security 10-2
upgrade costs 8-3
upgrade modeling 8-3
Wide area networks, *see* WANs
Wireless communications
agent-based messaging systems 7-3
applications 7-3
architecture 7-3
CDPD 7-1
cellular voice 7-1
circuit-switched networks 7-3
commercial mobile data communications services 7-1
costs 7-1, 7-3
E-mail based systems 7-3
infrared technology 8-6
LANs 7-1, 10-2
microwave transmission 8-6
packet data networks 7-3
PBX systems 7-1
products, selecting 7-3
remote access 7-3
satellite data 7-1
satellite voice services 7-1
service providers 7-3
specialized mobile radio 7-1
vendor support 7-1
wireless radio 7-1
Work teams
composing 2-6
disbanding 2-6
manager's role 2-6
strategies 2-6
World Wide Web
Internet applications 6-3
security 7-4

X

X.25
address configuration 3-6
networks 3-6